D0138751

Applications & Design with Analog Integrated Circuits

Applications & Design with Analog Integrated Circuits

J. Michael Jacob
Purdue University

RESTON PUBLISHING COMPANY, INC.
A Prentice - Hall Company
Reston, Virginia

Library of Congress Cataloging in Publication Data

Jacob, J. Michael.
Applications & design with analog integrated circuits.

Includes index.
1. Integrated circuits. 2. Operational amplifiers. I. Title. II. Title: Applications and design with analog integrated circuits.
TK7874.J3 621.381′735 81-15367
ISBN 0-8359-0245-5 AACR2

A Prentice-Hall Company
Reston, Virginia 22090

5 7 9 10 8 6

PRINTED IN THE UNITED STATES OF AMERICA

to my Wife and Children

Contents

Preface

The integrated circuit, particularly the operational amplifier, has become the true building block of analog electronics. Transistors, like vacuum tubes (remember those?) before them, are being used only in those applications where very high frequency or high power demand their unique abilities. The bulk of analog signal processing, however, is now being done with integrated circuits.

This book presents a detailed overview of the *use* of operational amplifiers, integrated circuit regulated power supplies, Norton amplifiers, waveform generator integrated circuits, active filters, log amplifiers, multipliers and other analog integrated circuits. Characteristics and limitations of each integrated circuit are discussed. Extensive *applications* and designs are outlined; operation of each application is derived or analyzed; and its performance limitation are pointed out. This book will enable you to select the appropriate device and circuit configuration for your need, calculate the component values required, build, and analyze the overall circuit's behavior. Throughout, the book is tutorial, not overly brief (as with many handbooks).

Several years experience teaching analog integrated circuits courses at both the associate degree and bachelor degree engineering technology levels precipitated the writing of this book. A reasonable ability in algebra is needed. Passages using calculus have been kept to a minimum, and can be skipped without loss in your ability to *apply* that information. Although originally intended as a sophomore or upper class electrical engineering technology text, extensive transistor circuit knowledge is not required. Combined with a basic diode and transistor characteristics supplement, the first four or five chapters work well for the first course in analog electronics in engineering technology, industrial electronics or for personal instruction.

Integrated circuits are treated as functional devices. Analysis of internal transistor circuitry is minimized; used only briefly when discussing device limitations. This is *not* a book on the design of analog integrated circuits. It is a text on the *application* of the integrated circuit devices. As such, circuits using these devices are rigorously analyzed and all performance or design equations are derived. This math is grouped to allow you to go directly to the use of the equations, if you choose.

To help organize your thoughts, give you specific goals, and help you evaluate what you have learned, each chapter begins with a *detailed* list of performance objectives. These list precisely what you will learn to do as you read the chapter. The problems at the end of the chapter are based on these objectives. Not only are you expected to repeat what was done in the chapter, but many of the problems require you to apply the techniques used to new circuits. This gives you the chance not only to drill, but also to think and stretch your ability.

As compared to other linear integrated circuit texts, this book has several unique features. The coverage is as broad as most handbooks, but provides the rigor to help you understand *why* circuits are built as they are, as well as *how* they are built. Chapter three gives a unique, practical, simple design and analysis approach to regulated power supplies, which consistently produces results easily verifiable in the lab. The use of operational and Norton amplifiers, biased from a *single* supply is given an entire chapter. Active filters are presented rigorously. All transfer functions and tuning parameters are completely derived. However, application of these results is given equal emphasis with many design and analysis examples. First through sixth order filters with varying damping for low pass, high pass, band pass, notch and state variable circuits are illustrated. The characteristics and applications of nonlinear circuits are given extended treatment in a full chapter.

I must express my appreciation to those students whose ideas and suggestions lead to this book. The *critical* evaluation of those students who suffered through the original manuscript and modifications was very helpful. My deepest gratitude and apologies must go to my wife and children who were more than understanding of all of the stolen hours.

J. M. J.

Applications & Design with Analog Integrated Circuits

CHAPTER 1

Introduction to The Operational Amplifier Integrated Circuit

Integrated circuit technology has allowed the construction of inexpensive, reliable digital systems, such as the pocket calculator and digital watch. The same basic techniques have allowed the construction of inexpensive, complex analog amplifiers with very stable characteristics. These integrated circuits (ICs) are amazingly easy to use. Because of this simplicity, and because of a high performance cost ratio, analog integrated circuits have replaced most discrete transistor amplifier circuits. This chapter introduces the operational amplifier, the most widely used analog IC. It is treated as a functional block. Its characteristics are described and techniques and considerations for breadboarding are presented without a detailed analysis of the internal configuration.

OBJECTIVES

After studying this chapter, you should be able to do the following:

1. Briefly describe the two IC fabrication techniques.
2. State the differences between an analog and a digital circuit. Give examples of each.
3. List the characteristics of an ideal amplifier.
4. Compare actual operational amplifier characteristics to those of an ideal amplifier.
5. State the requirements and precautions associated with operational amplifier power supplies.
6. Identify IC packages and lead convention.
7. Describe good breadboarding techniques.

1-1 ANALOG INTEGRATED CIRCUIT

What is an analog integrated circuit? To begin the study of analog integrated circuits, you should first understand what makes them unique. What is the difference between an integrated circuit and a circuit built with discrete transistors? How do analog circuits differ from digital circuits?

An integrated circuit (IC) is a group of transistors, diodes, resistors, and sometimes capacitors wired together on a very small substrate (or wafer). From five to tens of thousands of components can be contained in a single integrated circuit. Functions as simple as single-stage amplifiers to those as complex as complete computers have been built in a single integrated circuit. This drastic decrease in size has also yielded a similar reduction in weight, power consumption, and production cost, while giving a proportional increase in reliability. Integrated circuits have swept every field of electronics and are strongly altering the automotive, medical, entertainment, and business industries.

There are two main IC fabrication techniques monolithic and hybrid. Monolithic integrated circuits are built in a wafer of silicon. Transistors and resistors (rarely capacitors) are produced by diffusion processes, just as discrete transistors are. However, instead of the individual transistors being cut apart, they are interconnected by very thin metal runs. Isolation between transistors relies on reverse-biased junctions between the transistors and the silicon substrate (wafer).

Design of monolithic integrated circuits requires the tedious preparation of a large number of photographic masks. These are used to control the areas of diffusion for isolation, collectors, bases, emitters, (or channels and gates for field effect transistors), resistors, surface insulation, and connection metalization. These masks are reduced in size and used, one at a time, to produce the desired parts of

Figure 1-1 Monolithic IC fabrication mask (*Courtesy of Bell Laboratories*)

the monolithic IC. Of course, many separate integrated circuits may be made in each wafer, with many wafers processed simultaneously. The wafers are tested, the good ICs (called chips at this point) are cut apart and placed in one of many different style packages.

For monolithic integrated circuits, the initial design of the masks needed to produce the IC is very time-consuming and expensive. However, hundreds to thousands of monolithic ICs can be produced simultaneously. Consequently, large-volume production keeps the price per chip quite low.

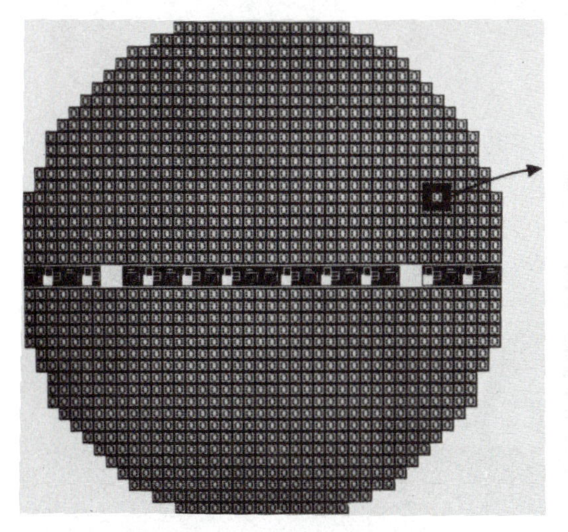

Figure 1-2 Monolithic IC wafers
(*Courtesy of Bell Laboratories*)

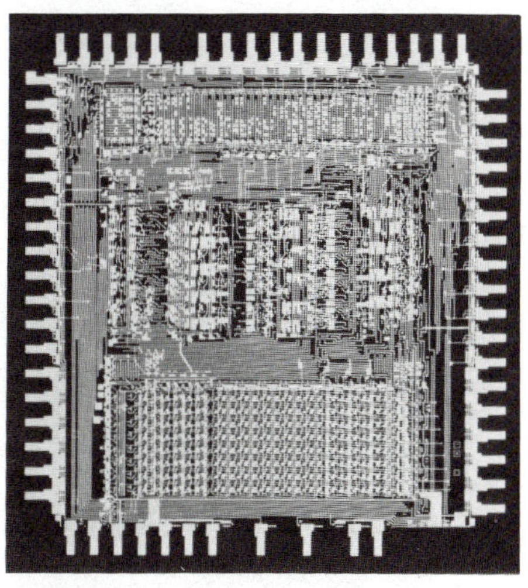

Figure 1-3 Monolithic IC
(*Courtesy of Bell Laboratories*)

The major advantage of monolithic integration, then, is low cost per chip if many of the ICs are to be produced.

However, there are several disadvantages. Capacitors are very difficult to produce in monolithic ICs. Tuning the IC by adjusting (trimming) internal resistor values to meet custom or very demanding specifications is also difficult. Isolation between transistors often is not adequate. It is hard to integrate both digital and analog functions, both bipolar transistors and field effect transistors on the same wafer. All of these disadvantages place some limits on the use of monolithic ICs.

In applications where these limitations can not be tolerated, hybrid integrated circuits can be used. In hybrid ICs, discrete resistors, capacitors, diodes, transistors, and even monolithic ICs are placed on a ceramic wafer, and then interconnected. The wafer actually serves as a miniature chassis with discrete components attached and interconnected. This technique provides excellent iso-

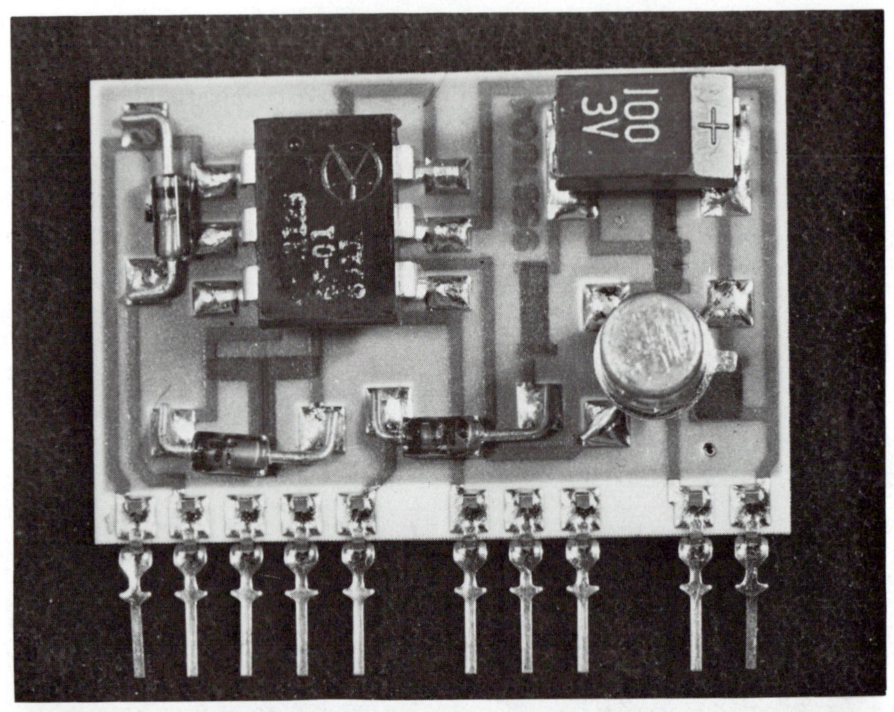

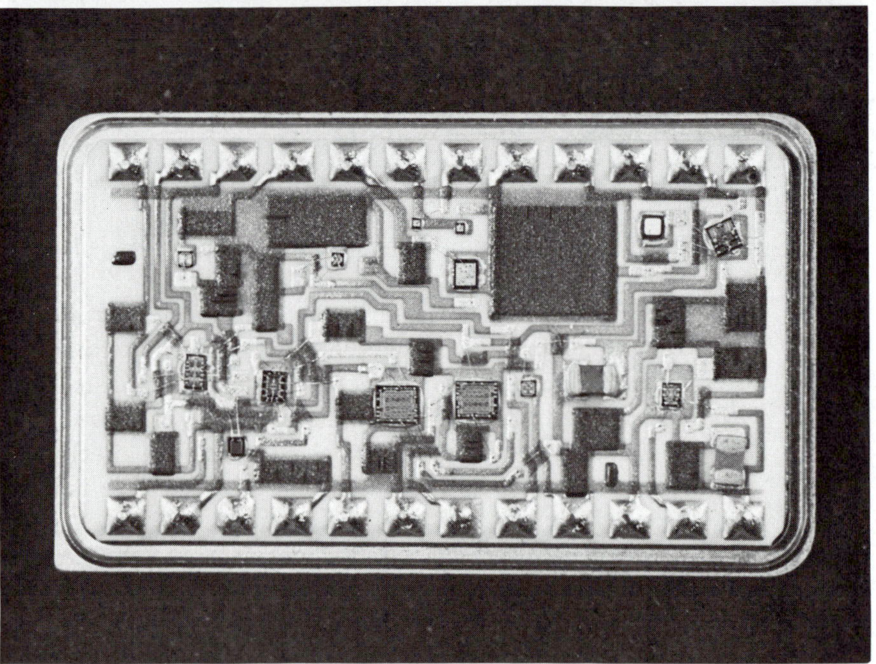

Figure 1-4 Hybrid IC modules (*Courtesy of CTS Microelectronics*)

lation between components, allows the use of practically any device, and can be easily custom tuned before encasing.

Although the hybrid IC can eliminate most of the monolithic IC's disadvantages, the hybrid IC is considerably more expensive. This is because much of the assembly of hybrid ICs must be done by hand.

Most of the devices you will see in this book will be monolithic ICs, and fairly inexpensive. However, a few hybrid ICs will be discussed with applications in Chapter 9, "Nonlinear Circuits."

Integrated circuits are divided into two main groups according to the way they are built (monolithic or hybrid). Two divisions may also be made according to the function of the integrated circuit (analog or digital). The output of an analog IC may be any value whatever, and may change continuously from one value to another. The output of a digital IC will be recognized only as either a logical low level, or a logical high level. Only two output values are accepted. Furthermore, in many digital systems, changes from one level to the other can occur only at specific times. An analog IC output and a digital IC output are shown in Fig. 1-5.

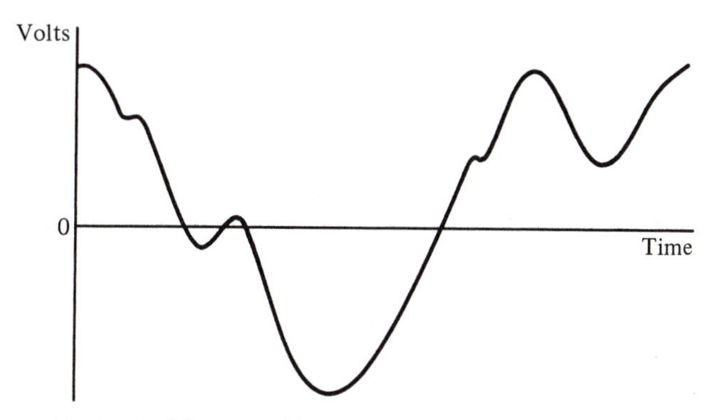

(a) Analog IC output. There is continuous variation between values.

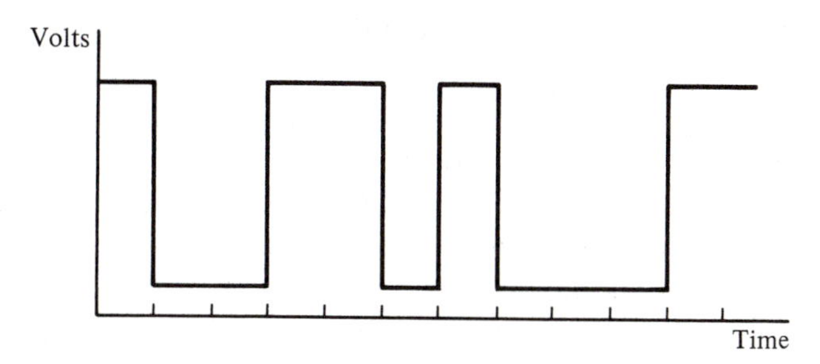

(b) Digital IC output. Only two output levels are allowed.

Figure 1-5 Comparison of analog and digital IC outputs

Computers, calculators, digital watches, communication switching, digital instruments, and some industrial control systems use digital ICs. Analog ICs are used in radio, television, stereo, amplifiers, voltage regulators, signal generators, filters, test instruments, and extensively in industrial measurement and control.

1-2 THE IDEAL OPERATIONAL AMPLIFIER

The operational amplifier, or op amp for short, is a high gain, wide band, DC amplifier with high noise rejection ability. The schematic symbol for an op amp is shown in Fig. 1-6. It has one output and two inputs. The output voltage, with respect to circuit common (ground), depends on the difference in potential between the two inputs. This is illustrated in Fig. 1-7.

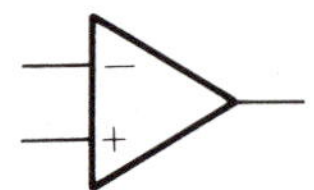

Figure 1-6 Schematic symbol of an operational amplifier

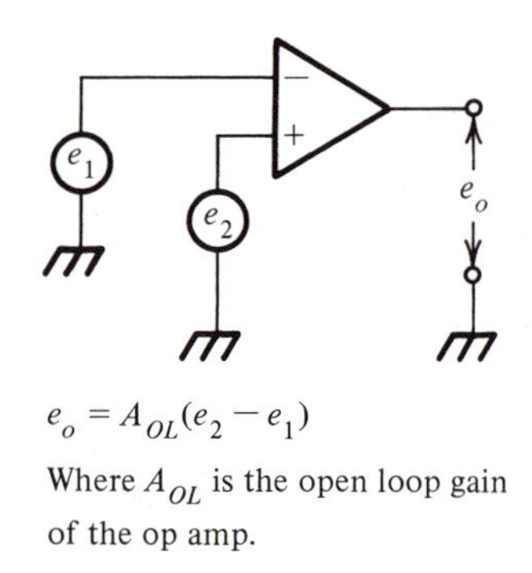

$e_o = A_{OL}(e_2 - e_1)$

Where A_{OL} is the open loop gain of the op amp.

Figure 1-7 Output/input voltage relationship of an operational amplifier

Operational amplifiers were first used in analog computers to do addition, subtraction, integration, and scaling. They were made from vacuum tubes and required a full chassis for each amplifier. Later, the tubes were replaced by transistors. A single op amp then required only one printed circuit card full of components. This reduction in size lowered the price of the op amp and broadened its application. Op amps were used in many signal conditioning areas, test and measuring equipment, and industrial control.

With the advent of integrated circuit techniques, it became possible to place the entire op amp in a single eight-lead mini-DIP package. Surprisingly, the cost of these op amp ICs is comparable to a single discrete transistor. Most small signal analog circuits designed today use the operational amplifier as the basic active device, and fall back on discrete transistors only when the op amp will not solve the problem.

If you were going to design an ideal operational amplifier, what characteristics would you want? Since the purpose of the circuit is to amplify a signal, you would want an arbitrarily large gain. That is, no matter how small the input signal ($e_2 - e_1$ in Fig. 1-7), A_{OL} would be large enough to provide an output signal (e_o) of adequate size.

$$A_{OL} = \infty \tag{1-1}$$

This means that the difference between the inputs ($e_2 - e_1$) can be negligibly small.

Second, it is important that all of the output signal be applied to the load. However, the output impedance of the op amp (Z_0) forms a voltage divider with the load. This is shown in Fig. 1-8. For all of the op amp's output voltage to be applied to the load (none dropped across Z_0), the output impedance must be zero.

$$Z_0 = 0 \tag{1-2}$$

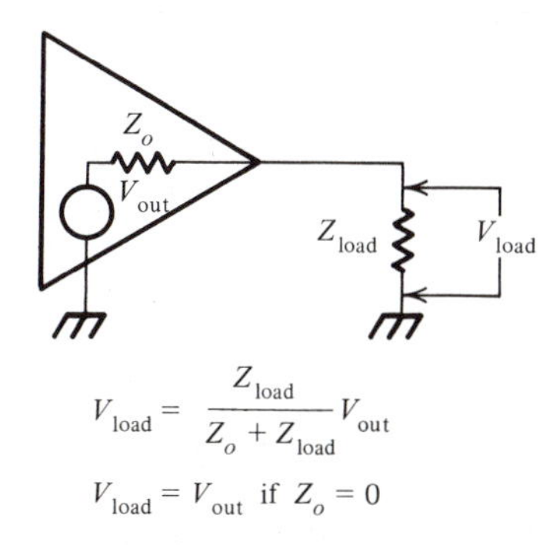

Figure 1-8 Effect of output impedence on load voltage of an operational amplifier

On the input side of an ideal operational amplifier, it is important to assure that the amplifier does not load down the source. The input impedance of the op amp forms a voltage divider with the output impedance of the source. This is illustrated in Fig. 1-9. If all of the source voltage is to be applied to the inputs of the op amp, and none lost across the source output impedance (R_s), then the op amp's input impedance must be arbitrarily large compared to R_s.

$$Z_{in} = \infty \tag{1-3}$$

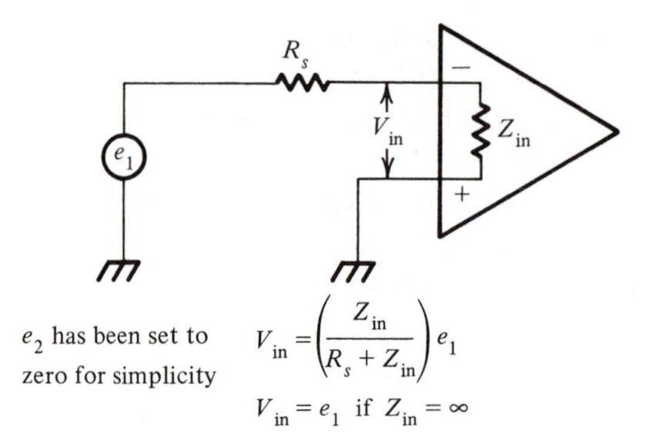

Figure 1-9 Effect of input impedence on input voltage and source loading for an operational amplifier

This means that no current will flow into (or out of) the inputs of an ideal op amp.

Finally, it is necessary for an ideal amplifier to work equally well at all frequencies. At zero hertz (DC) you would want an input difference in potential of zero volts to produce an output of zero volts. Any deviation from this is called *offset*.

$$V_{in} = e_2 - e_1 = 0$$

leads to

$$V_{out} = 0$$

or

$$\text{Offset} = 0 \tag{1-4}$$

Ideal AC amplifiers would provide the same gain at all frequencies. You would not want the gain to decrease at low frequency (low frequency cutoff), or decrease at high frequency (high frequency cutoff). Since the difference between the low and high frequency cutoffs is the bandwidth, an ideal operational amplifier would have unlimited bandwidth.

$$\text{Bandwidth} = \infty \tag{1-5}$$

1-3 BASIC SPECIFICATIONS AND REQUIREMENTS

You realize, of course, that the ideal op amp just described can not actually be built. But, for a remarkably low cost, monolithic op amps can be manufactured which, when properly used, closely approximate the ideal. In this section, key nonideal characteristics will be discussed, along with simple techniques and precautions for minimizing their effects.

1-3.1 Specifications of Initial Importance

Table 1-1 lists the ideal op amp characteristics in the first column and the typical characteristics in the second column for the 741C, a popular general-purpose monolithic op amp IC.

TABLE 1-1. Ideal and typical characteristics of a monolithic operational amplifier integrated circuit

		Ideal	*741C (typical)*
Voltage gain	A_{OL}	∞	2×10^5
Output impedance	Z_0	0	75 ohms
Input impedance	Z_{in}	∞	2 megohms
Offset current	I_{os}	0	20 nanoamps
Offset voltage	V_{os}	0	2 millivolts
Bandwidth	BW	∞	1 megahertz

Although the open loop gain of the op amp definitely is not infinite, generally less than ten percent error will be introduced by assuming that it is infinite. This will be true if for any circuit using op amps you

limit the overall (closed loop) gain of the circuit to *100* or less.

The typical input and output impedances, though not ideal, need to be considered only for a few demanding industrial measurement or high-speed pulse applications.

The output amplitude for most op amps is not limited by the output impedance, typically 75 ohms for the 741C. Instead, the amount of current the op amp can source or sink to the load limits the output voltage. In an attempt to make these integrated circuits as foolproof as possible, a short circuit current limit circuit has been added to many op amps. This means that the load impedance can be shorted without overheating the IC. Instead, the output current will be limited to a safe level (25 milliamps for the 741C). The output voltage drops appropriately to keep the current below the safe limit as the load impedance drops. This is very handy. However, be careful! Not all op amps are short-circuit protected. You must read the manufacturer's literature. Shorting the output of an unprotected op amp will probably damage it.

Use an op amp which can withstand a short circuit indefinitely (such as the 741C).

Interestingly, it is the last three specifications in Table 1-1 along with several others to be described which limit the op amp's applications.

Some current must flow into the input of a general-pupose op amp to bias its input transistors. This is illustrated in Fig. 1-10. If

$$I_1 = I_2$$

then

$$e_2 = e_1$$

and

$$V_{out} = A_{OL}(e_2 - e_1) = 0$$

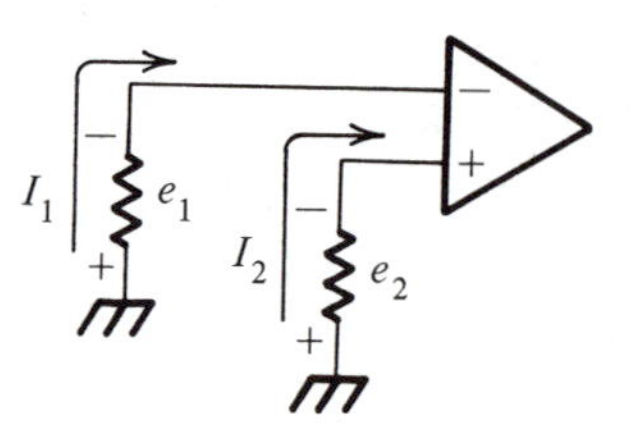

Figure 1-10 Bias and offset currents of an operational amplifier

The effect on the output voltage because of these bias currents is zero. Normally, I_1 does not equal I_2. There will be some small difference between the two bias currents.

$$I_{offset} = I_2 - I_1$$

For the 741C, you can see in Table 1-1 that the offset current is typically 20 nanoamps.

To determine the effect of this offset current, let the resistors in Fig. 1-10 equal one megohm. The offset current will then cause a difference in potential between the two inputs $(e_2 - e_1)$ of

$$1\ \text{M}\Omega \times 20\ \text{nA} = 20\ \text{mV}$$

The voltage at one input will be 20 millivolts larger than the voltage at the other input. That may be significant. To minimize the effect of offset currents,

Reduce the size of the resistance seen from each input terminal (R) as much as practical.

Usually, keeping R in the kilohm range is quite adequate.

The offset voltage results from the impossibility of perfectly matching the input transistors. The result is that even if both inputs are shorted together (to cause $e_2 = e_1$), the output will not exactly equal zero. For the 741C, typically the output will be 2 millivolts. Usually, a 2-millivolt error in the output voltage can be tolerated. However, if it can not, there are zeroing techniques to eliminate the offsets. These will be discussed in detail in Chapter 4.

The bandwidth specification is actually a measurement of the maximum possible gain at a given frequency. As the frequency increases, the gain falls off proportionally. For the 741C, at no point can the product of the voltage gain and the operating frequency exceed one megahertz. This is shown in Table 1-2. From the table you can see that the open loop gain becomes unusably small above about 10 kilohertz. For that reason,

Use the 741C only at DC to upper-audio frequencies. Use a wide band op amp above 10 KHz.

TABLE 1-2. Possible gains at various frequencies for the 741C operational amplifier

Frequency (Hz)	*Maximum Possible Gain*
10	100k
100	10k
1k	1k
10k	100
100k	10

Bandwidth = 1 megahertz = $A \times f$

The LM 318 has a bandwidth of 15 megahertz, giving useful gain up to at least 150 kilohertz.

Certain operational amplifiers have exceptionally high gain at high frequencies (for example, 709). These are usually uncompensated. At first glance, you may think that this is an advantage. However, high-frequency noise is amplified, coupled back into the input through interwiring capacitance, amplified, coupled back to the input, amplified, coupled back, ... Oscillations result. External compensation components must be added to make these op amps usable. This will be discussed in detail in Chapter 4. Generally, to simplify circuits,

Use an internally compensated op amp (741C, LM318). Avoid uncompensated op amps (709).

1-3.2 Power Supply Requirements

In order for an op amp to operate, power supply voltages must be provided. Normally, bipolar voltages are required. That is, two voltages are needed, one positive and the other equal in magnitude, but negative. Often, these supply voltages are omitted from the schematics. However, Fig. 1-11 illustrates the proper schematic and connections.

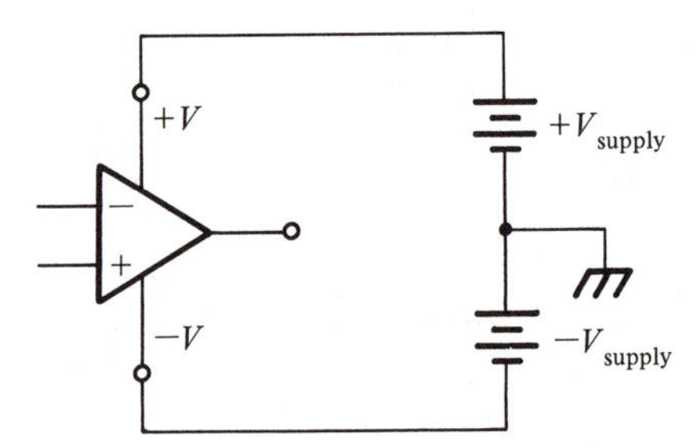

Figure 1-11 Bipolar power supply connections for an operational amplifier

Three op amp characteristics which are dependent on the power supply voltages are listed in Table 1-3. The manufacturer specifies that the supply voltages ($+V$ and $-V$) may not exceed ± 18 volts for the 741C. The output voltage magnitude is limited by these supply voltages. The maximum possible output will be about two volts less than the supply voltages. Consequently, when selecting the power supply voltages, you normally want to keep them as low as possible (to minimize power dissipated by the op amp and, therefore, burnup possibilities), while ensuring enough "head room" for the output signal. For example, if the output signal must be ± 6 volts, you might choose to use two 9 volt batteries ($+V = +9$ V, $-V = -9$ V).

Keep the supply voltages small, but they must be at least ± 2 volts larger than the maximum output desired.

It is specified that the difference in potential between the two inputs (differential input voltage) may be no greater than 30 volts for the 741C. Also, neither

TABLE 1-3. Power supply dependent specifications for the 741C operational amplifier

Absolute Maximum	
power supply voltage	± 18 volts
differential input voltages	± 30 volts
common mode input voltages	± 15 volts

input may be ± 15 volts away from circuit common (common mode input voltage). In addition, though not specifically stated, should either input exceed the power supply voltage, damage may occur. If the op amp has a ± 9-volt supply, neither input should exceed 9 volts, even though the specifications indicate ± 15 volts.

Limit the input amplitude to less than the power supply voltages.

One certain way to damage an op amp is to reverse the supply voltages (i.e., $+9$ volts connected to the $-V$ terminal, -9 volts connected to the $+V$ terminal). It is very discouraging to spend several hours wiring a complex circuit using many op amps and then to accidently cross the power supply wires, wiping out hours of work and possibly several dollars' worth of ICs.

Use extreme care not to reverse power supply connections.

1-3.3 Packaging

Analog integrated circuits are available in many different package styles. These are illustrated in Fig. 1-12.

The dual inline pack (DIP) of Fig. 1-12(a) is popular for commercial applications. It is easy to handle, fits standard mounting hardware, and is inexpensive when molded in plastic. Ceramic DIPs are used for high-temperature, high-performance (usually military) equipment.

The mini-DIP [Fig. 1-12(b)] provides a low cost/space efficient package. For circuits where space is critical, the flatpack shown in Fig. 1-12(c) gives the most compact package. However, flatpacks are much more difficult to handle than DIPs and often do not dissipate power as well.

The metal can package [Fig. 1-12(d)] allows easy connection to a heatsink, and is often chosen when heat dissipation is a chief consideration.

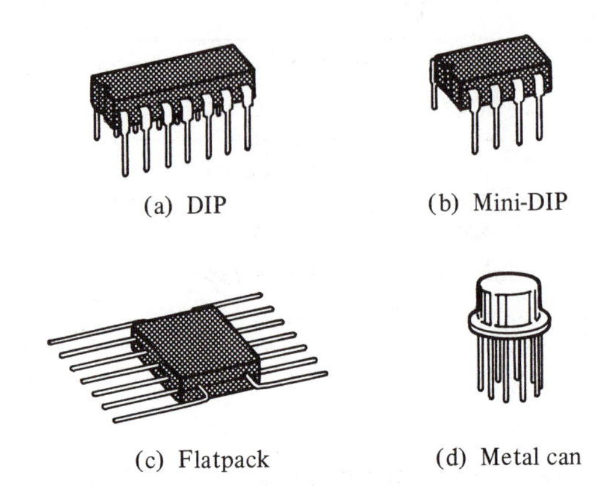

Figure 1-12 Analog IC package styles (*Courtesy of Motorola Semiconductors*)

Pins are identified on the pin diagrams included with the manufacturer's specifications. It is standard, however, to view the IC from the *top* and count pins counterclockwise from the key, mold mark, or tab. Each package style is identified with a different letter suffix. This is shown in Fig. 1-13.

The op amp terminals connect to different pins in each different package style. For example, the output of the 741C is connected to pin 10 of the DIP, pin 6 of the mini-DIP, pin 7 of the flatpack, and pin 6 of the eight-lead metal can. Since it is critical that the proper connections be made to the op amp, the op amp number, package style suffix, and pin numbers **should *always* be recorded on the schematic.** This is illustrated in Fig. 1-14.

1-4 GENERAL BREADBOARDING HINTS

Once a circuit has been designed, it must be tested. To do this quickly and reliably, a good breadboarding system is needed. It should allow for the easy interconnection and removal of the analog ICs, discrete components, power supplies, and test equipment. It is *absolutely critical* that connections between the breadboard, the components' power supplies, *and the **test equipment*** be mechanically and electrically sound. Most beginners spend more time running down poor or wrong breadboarding connections than they spend actually evaluating the circuit they have built. In this section you will find breadboarding hints which will help you minimize problems and errors in building your circuit for testing.

The universal breadboard socket illustrated in Fig. 1-15 provides a popular and convenient technique. It gives two to four busses for power supplies and ground running along the edges. The body provides an array of solderless connections properly spaced and sized for most analog and digital ICs, transistors,

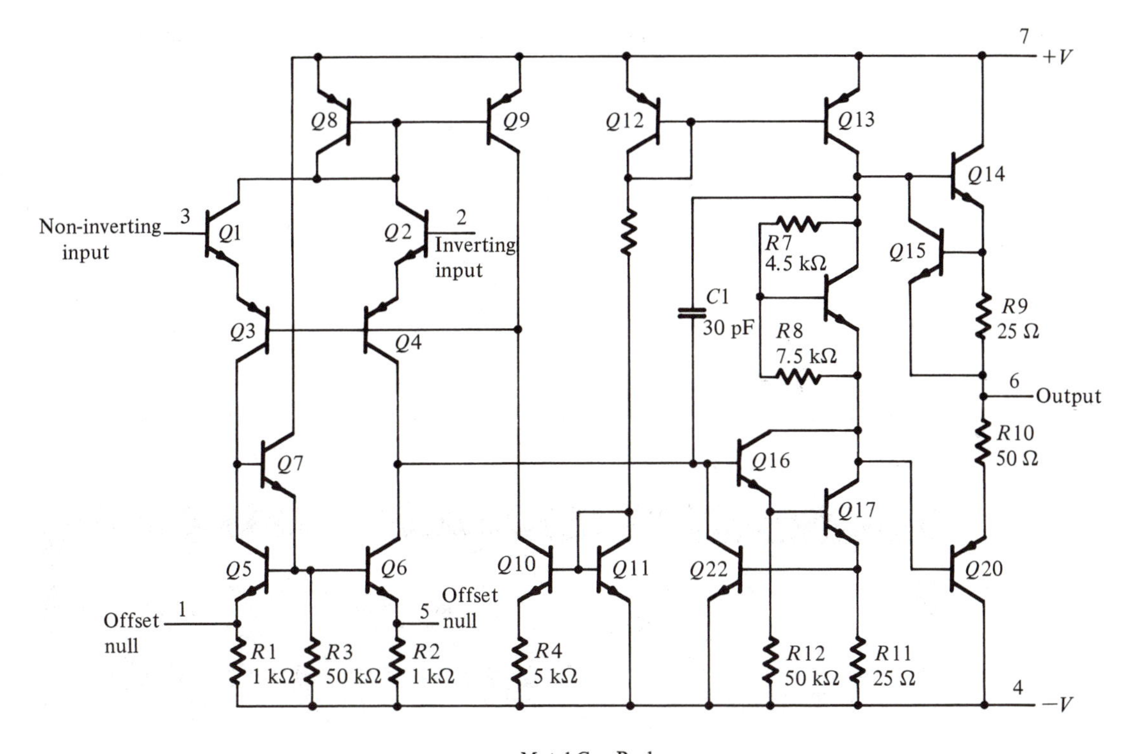

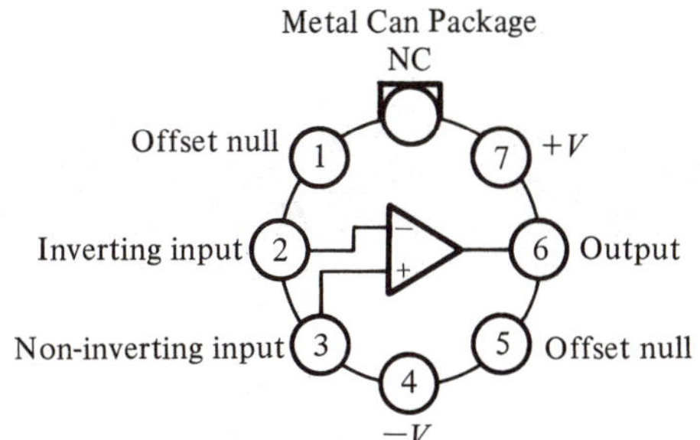

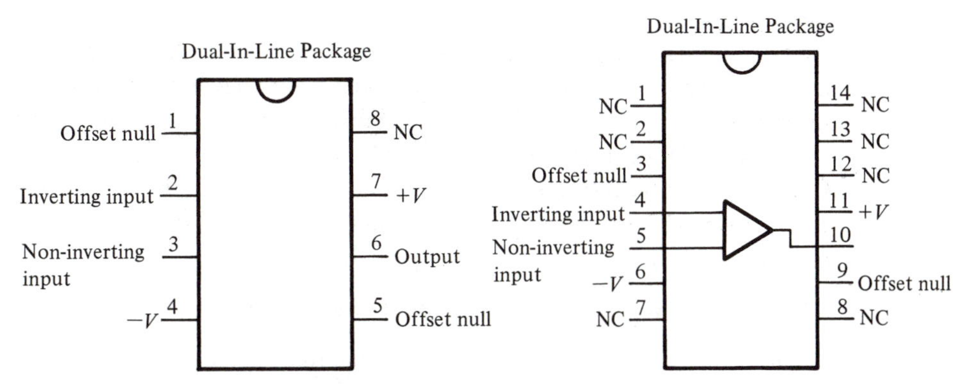

Figure 1-13 Pin diagram for the 741 operational amplifier (*Redrawn by courtesy of National Semiconductors*)

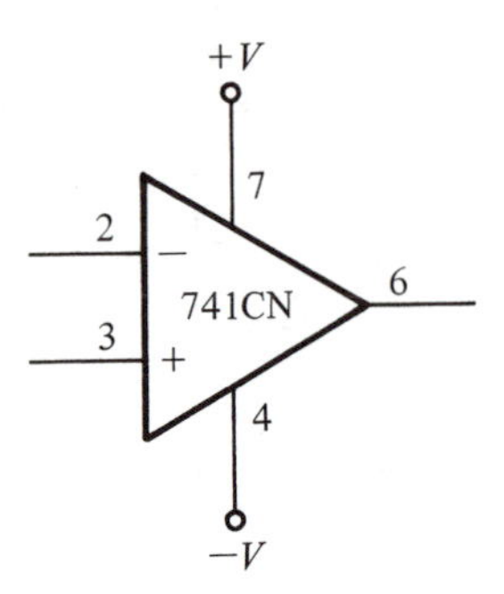

Figure 1-14 741C op amp schematic diagram for the mini DIP style package

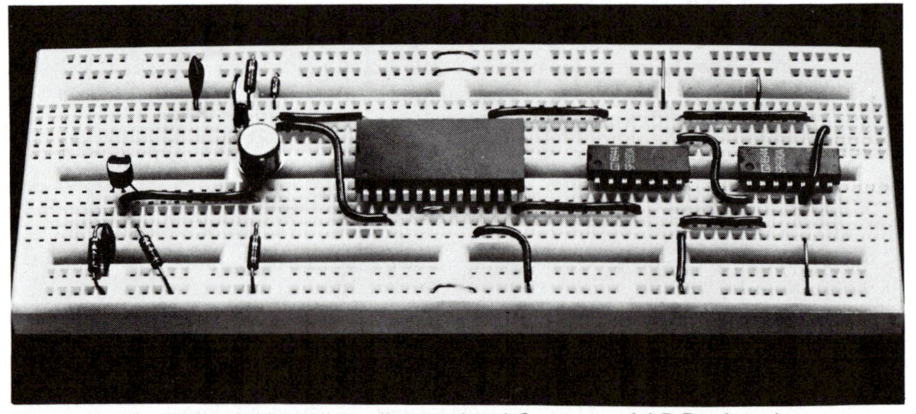

Figure 1-15 Universal breadboarding socket (*Courtesy of AP Products*)

diodes, small capacitors, $\frac{1}{4}$-watt resistors, and 22AWG solid wire. Using it, you can construct circuits quickly, compactly, and reliably. These breadboards are available in a variety of styles and qualities from most electronic component suppliers.

The universal breadboarding sockets provide a good interface between components of the circuit, but care must be taken when they are connected to power supplies and test equipment. First, the breadboards are small and quite lightweight. As cables are attached, the board will slide around, turn over, and too often short out power supply or test leads. To prevent this, all breadboards should be mounted on some larger, sturdier base. These are available commercially as shown in Fig. 1-16, or they can be easily built.

Mount the breadboarding socket securely to a base.

Just as a chain is only as good as its weakest link, test equipment can perform no better than the technique used to connect it to the circuit under test.

Figure 1-16 Commercial breadboards with physical supports (*Courtesy of AP Products*)

Excellent standard leads supplied with banana plugs, BNC connectors, or probes are common. Use them. Hours of careful design and breadboarding can literally go up in smoke because of a shorted or open wire to a power supply or from two alligator clips which accidentally touch, or jump off at just the wrong time. Alligator clips are a major source of trouble. They are often too large for use on a breadboard, short together, or fail to hold adequately. Instead of connecting test equipment to the breadboard with alligator clips, mount several five-way binding posts, BNC receptacles or screw terminals around the edge of the breadboard support. Connect from the test instrument to the breadboarding system, using standard cables with either BNC, banana plugs, or spade lugs. Then wire from the binding posts to the breadboarding socket with 22AWG wire, inserting the wire into the desired connector. This technique, will provide an electrically and mechanically sound and professional way to build circuits, eliminating the cause of most breadboarding headaches, bad connections.

Use only standard connectors to connect test equipment to the breadboard. *Never* alligator clips.

Probes must also be used carefully. It is far too easy, when you are trying to touch a pin on an IC, for the probe to slip between two pins, shorting them together. This could damage the IC or supporting equipment. Instead of probing

IC pins directly, there are two better choices. One is offered by the IC test clip, which easily slips over a DIP and provides well-spaced connections for the probe. This is shown in Fig. 1-17. A simpler solution is to connect a wire from the point you want to probe to a vacant part of the socket, where it can be secured and safely probed.

Never probe IC pins directly.

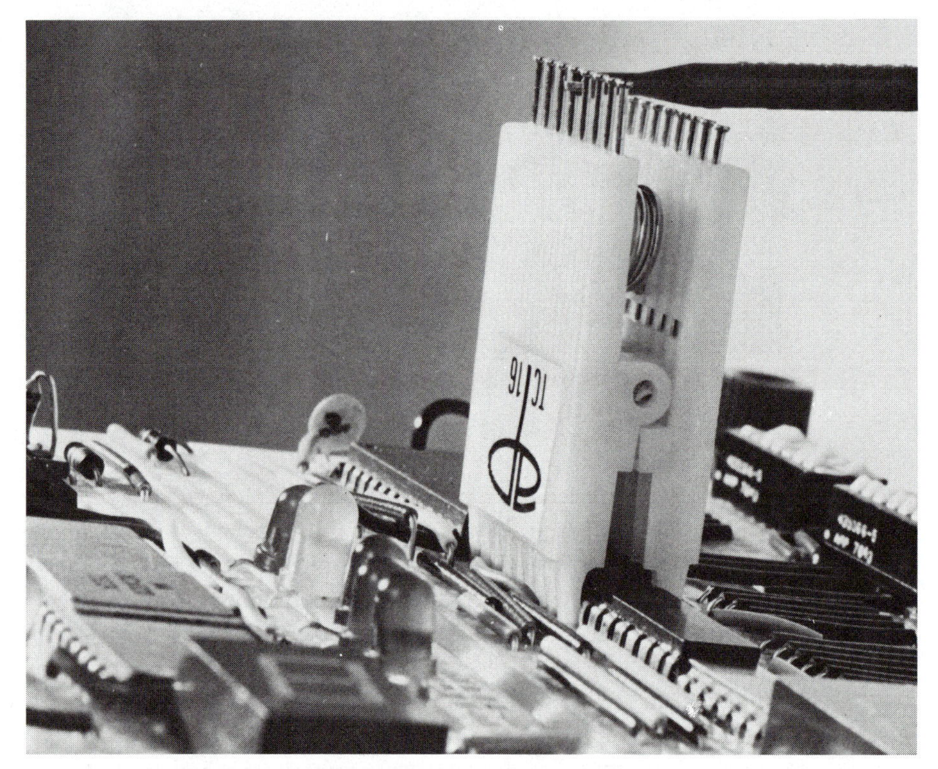

Figure 1-17 IC test clip (*Courtesy of AP Products*)

Power supply connections to the breadboard and to the individual ICs, though necessary, can cause some problems. As was stated earlier, one sure way to damage an analog IC is to reverse the power supply connections. This can be easily prevented when you are breadboarding by first labeling each bus in some highly visible way. This should prevent you from connecting the IC to the wrong supply bus. The busses can be protected by two diodes, as shown in Fig. 1-18. Should the power supply be incorrectly connected to the busses, the diodes will go off, protecting the components wired to the bus.

The power supply bus must present a good AC ground. But, because of the

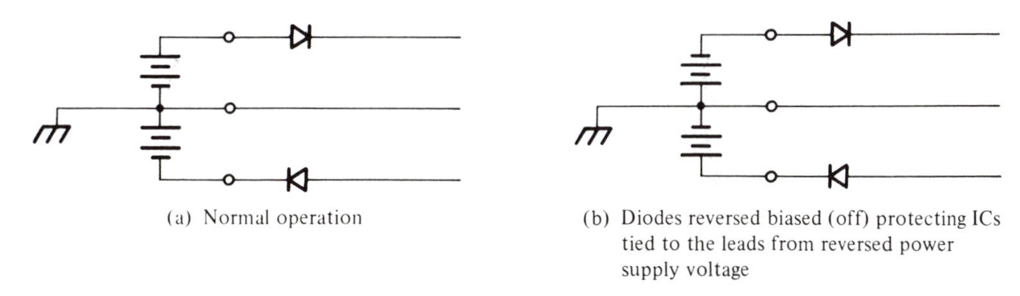

(a) Normal operation

(b) Diodes reversed biased (off) protecting ICs tied to the leads from reversed power supply voltage

Figure 1-18 Diode protected breadboard power supply busses

inductance presented by the cables running from the breadboard to the power supply, high-frequency (often noise) signals can exist on the bus. When coupled from one IC to another and amplified, these rf signals on the busses can cause the entire circuit to oscillate. This can be prevented by placing a 0.1 microfarad capacitor from each power supply pin of each IC to ground. The rf signals are passed to ground as they leave the IC, before they can contaminate the busses.

Use 0.1 microfarad decoupling capacitors at *each* power supply *pin* of *every* IC.

In addition to the precautions mentioned concerning connecting test equipment and power supplies to the breadboarding system, there are several things to be aware of when you are interconnecting components.

1. Component Selection
   Beware of potentiometers.
   a. Wire-wound pots can cause trouble at high frequencies.
   b. Many printed-circuit-size pots will not properly fit into breadboarding sockets.
   c. Multiturn pots, although good for making fine adjustments, are often confusing, because you may not know where the pot is set.
   d. Be sure you know where every pot is set before applying power.
2. Layout Plan
   a. Simplify the schematic and layout as much as possible for initial testing. Fine tuning, zeroing, and additional stages can easily be added after you have the basic circuit working.
   b. Be sure to include IC number, package type suffix, and **pin numbers** on each IC on the schematic (see Fig. 1-14).
   c. Make the layout look as much as possible like the schematic.
   d. Keep the inputs well separated from the outputs to prevent oscillations.
3. Circuit Construction
   a. Always clear the breadboard of any old circuits before beginning to build a new circuit.

b. Exercise care in inserting and removing ICs. Pins are easily bent and jabbed into your fingers.
c. Solder 22AWG solid wire to the leads of components with large leads.
d. Devise, post and carefully follow a color code scheme for $+V$, $-V$, ground, and signal wires.
e. Avoid jungles. Make all components lie flat. Trim and bend leads and wires to fit the layout. Neat, flat layouts work better and are far easier to troubleshoot than a jumble of components and wires.
f. Select one connector as the ground point. Tie the breadboarding socket's ground bus, power supply ground, and all test instruments' grounds to that single point.

4. Circuit Test
a. Analyze the circuit before applying power to ensure that you know what to expect.
b. Double check all connections, especially power supply connections, before applying power.
c. Apply power to the IC before applying the signals.
d. Remove the signal from the IC before removing the power.
e. Change components and connections with the power off.

SUMMARY

This chapter has introduced the operational amplifier integrated circuit. An integrated circuit is defined as a group of transistors, diodes, resistors, and occasionally capacitors built and interconnected on a very small substrate. All of the components in a monolithic IC are produced simultaneously by using diffusion techniques similar to those used in producing discrete transistors. Hybrid ICs consist of discrete components mounted on a ceramic wafer and wired together. The output of an analog IC may be any value and may change at any time. However, a digital IC output is recognized as either of two levels only.

The ideal operational amplifier has an inverting input (−), a noninverting input (+), and a single output referenced to circuit common (ground). The value of this output depends on the difference between the two inputs,

$$e_O = A_{OL}(e_2 - e_1)$$

The open loop voltage gain, the input impedance, and the bandwidth are extremely large, while the output impedance and offset are negligibly small.

The fact that no real device can meet these ideal characteristics can be, more or less, ignored if several precautions are observed. These are listed in boldface and are indented in the body of the chapter. Care must also be taken when supplying the op amp with power. These precautions are also identified by indented boldface.

For the beginner, most of the problems encountered in getting a circuit to work are caused by the way it is built, not by the design. Good breadboarding practices can solve many of these problems. Use a universal breadboarding socket, mounted to a base. Connect to test equipment *only* through commercial BNC, banana plugs or spade lugs, *never* alligator clips. Exercise care when probing the circuit, using a remote test point rather than the IC pin itself. Ensure that *power supplies are not reversed*, and provide each IC with two decoupling capacitors. Select reliable components, plan the breadboard layout, wire the circuit neatly, and pay close attention to what you are doing (and the order of steps) while testing.

This chapter has introduced the op amp, its characteristics, and basic precautions. In Chapter 2, "Amplifiers and Feedback," you will see how to use the op amp to build comparators, LED drivers, inverting and noninverting amplifiers and summers, and differential amplifiers.

PROBLEMS

1-1 Describe the major physical differences between a monolithic IC and a hybrid IC.

1-2 What factors would cause you to select a hybrid IC rather than a monolithic IC?

1-3 Generally, monolithic ICs cost less than hybrid ICs. Explain why.

1-4 List three specific analog circuit application examples not given in this chapter. Why are these considered analog?

1-5 List three specific digital circuit application examples not given in this chapter. Why are these considered digital?

1-6 If an op amp has $e_1 = 0$ (grounded), what must the voltage at e_2 be to cause a 5 volt output if

a. $A_{OL} = 50{,}000$

b. A_{OL} is ideal

c. What do the results of (a) and (b) mean about the importance of a nonideal A_{OL}?

1-7 In Fig. 1-9, find V_{in} if

a. $R_s = 10$ kilohms, $Z_{in} = 1$ megohm, $e_i = 10$ microvolts

b. $R_s = 1$ megohm, $Z_{in} = 1$ megohm, $e_i = 10$ microvolts

c. Z_{in} is ideal, $e_i = 10$ microvolts

d. What do the results of (a), (b), and (c) suggest about the relationship of R_s and Z_{in} if you want the same effect that an ideal Z_{in} would give?

1-8 What must be the relationship of the two resistors in Fig. 1-10 to eliminate the effects of the bias currents on the output voltage? Assume that $I_1 = I_2$.

1-9 You need a gain of 150 at a frequency of 12 kilohertz. Will the 741C op amp work? Explain.

1-10 Explain what internal compensation means. Why is it needed?

1-11 A 741C op amp is connected to a ± 12 volt power supply.

- **a.** What is the maximum possible output voltage?
- **b.** What is the maximum allowable input voltage with respect to ground?
- **c.** The supplies are raised to ± 18 volts. What is the maximum allowable input voltage with respect to ground?

1-12 Under what conditions would you choose each of the following packages:

- **a.** flat pack
- **b.** metal can
- **c.** mini DIP

1-13 Design a breadboard system to meet (or exceed) the following specifications.

- **a.** Universal breadboarding socket with at least three power supply busses and a matrix of at least 100 terminals each with at least four tie points.
- **b.** Sturdy mounting base.
- **c.** Two BNC cable receptacles.
- **d.** Two pairs (red and black) 5-way binding posts (3/4 inch center-to-center spacing between the red and the black post of each pair).
- **e.** Two nine volt transistor batteries to provide $\pm V$ supply.

Refer to electronic parts catalogs to obtain part number, price and size information.

Draw a mechanical layout diagram (with dimensions) of the system, showing how you would mount all parts.

Draw a schematic showing how you would permanently wire the batteries and the connectors to the breadboarding socket.

What is the total cost of your breadboarding systems?

CHAPTER 2

Amplifiers and Feedback

Practically all circuits using operational amplifiers are based around one of a few fundamental configurations. In this chapter, you will learn about these building blocks. Their schematics, operation, analysis, design, and limitations are discussed. Examples of how these circuits are actually used in laboratory and industrial equipment are given. Every following chapter assumes that you have mastered the basic concepts of comparators and amplifiers, the building blocks of this chapter.

OBJECTIVES

After studying this chapter, you should be able to do the following:

1. Describe the open loop operation of an operational amplifier.
2. Discuss the effects of negative feedback on an amplifier.
3. For each of the circuits listed below, do the following:

   noninverting comparator
   inverting comparator
   nonzero reference comparator
   LED driver
   inverting amplifier
   inverting summer
   noninverting amplifier
   voltage follower
   noninverting summer
   difference amplifier

   a. Draw the schematic.
   b. Qualitatively describe the operation.
   c. Analyze a given circuit to determine quantitative operation.
   d. Design a circuit to meet given performance specifications.
   e. Discuss specific applications.
   f. State limitations.

2-1 OPEN LOOP OPERATION

The simplest way to use an op amp is shown in Fig. 2-1. The power supply terminals are properly connected. The inverting input (−) is tied to circuit common (ground), and the noninverting input (+) is connected to the source (e_2). As with any op amp, the output voltage is

$$e_0 = A_{OL}(e_2 - e_1) \tag{2-1}$$

But e_1 is the voltage applied to the inverting terminal, which for the circuit in Fig. 2-1 is ground, or 0 volts. So equation (2-1) becomes

$$e_0 = A_{OL}e_2$$

For an ideal op amp, A_{OL} is arbitrarily large. This means that the circuit in Fig. 2-1 can have only three possible outputs.

Case 1

$$e_2 = 0$$

$$e_0 = A_{OL} \times 0$$

$$e_0 = 0$$

If the input exactly equals zero, the output will also go to zero.

Case 2

$$e_2 > 0$$

$$e_0 = \infty \times e_2$$

$$e_0 \to \infty$$

If the input becomes even slightly greater than zero, the arbitrarily large voltage gain will force the output voltage to its maximum positive value, $V_{\text{saturation}}$. So if

$$e_2 > 0$$

then

$$e_0 = +V_{SAT}$$

Case 3

$$e_2 < 0$$

$$e_0 = \infty \times -e_2$$

$$e_0 \to -\infty$$

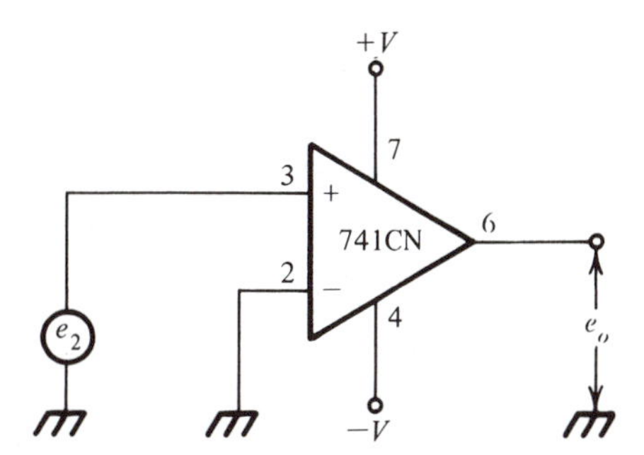

Figure 2-1 Simple op amp comparator

If the input becomes even slightly less than zero, the large voltage gain will drive the output to the maximum negative value, $-V_{SAT}$. So if

$$e_2 < 0$$

then

$$e_0 = -V_{SAT}$$

To summarize, if the noninverting input voltage is even slightly above ground, the output will go to positive saturation. When the noninverting input drops to just below ground, the output goes to negative saturation. It is practically impossible to hold e_2 to *exactly* zero. Consequently, though the output will pass through ground, it will rarely (if ever) rest at zero volts. Even the slightest noise at e_2 is enough to force the output to $\pm V_{SAT}$. No output levels other than $\pm V_{SAT}$ are possible.

2-1.1 Op Amp Comparators

The circuit in Fig. 2-1 is called a noninverting comparator. Although it definitely does not provide linear amplification, it can be of major use. Figure 2-2 is a plot of an input voltage applied to the noninverting comparator of Fig. 2-1 with the resulting output.

When e_2 is greater than zero volts, the comparator outputs $+V_{SAT}$. When e_2 drops below zero volts, the comparator's output switches to $-V_{SAT}$. Consequently, even a change of a millivolt or less on the input may cause the output to change as much as 30 volts. To determine whether e_2 is above or below ground, you simply have to look at the output of the op amp.

The noninverting comparator amplifies (to the extreme) any differences between its input, e_2, and ground, giving you a large voltage indication ($\pm V_{SAT}$). Also, the op amp serves as a buffer. Its input impedance is arbitrarily large

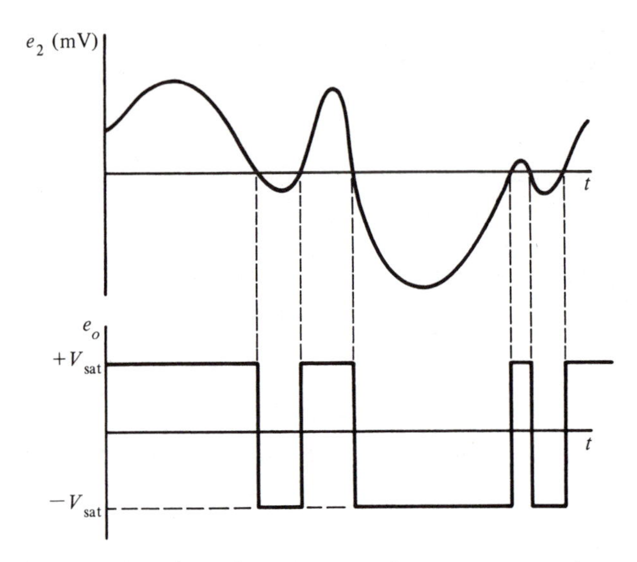

Figure 2-2 Example of noninverting comparator's response to an input signal

(ideally), so it does not load down e_2. However, it can provide 20 milliamps or more to drive whatever load or indicator you have attached to the output.

How would the circuit in Fig. 2-3 respond? The ground at the inverting terminal has been replaced by a reference voltage (V_{ref}). Applying the basic open loop op amp equation, you have

$$e_0 = A_{OL}(e_2 - e_1)$$

$$e_1 = V_{ref}$$

$$e_0 = A_{OL}(e_2 - V_{ref}) \tag{2-2}$$

Case 1

$$e_2 = V_{ref}$$

$$e_0 = A_{OL} \times 0$$

$$e_0 = 0$$

If e_2 *exactly* equals the reference voltage, the output will go to zero.

Case 2

$$e_2 > V_{ref}$$

$$e_0 = \infty \times \text{some positive number}$$

$$e_0 \to \infty$$

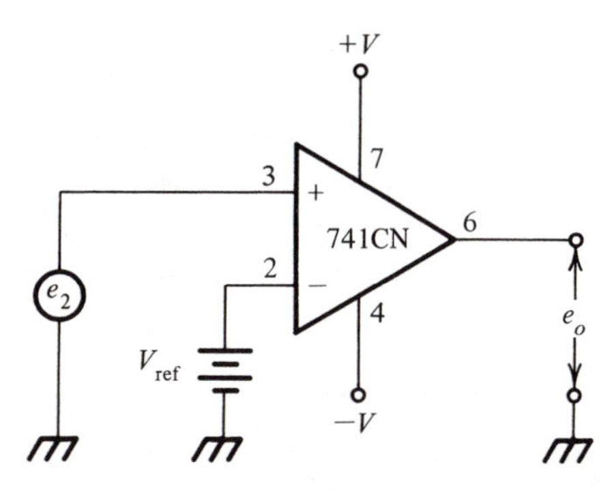

Figure 2-3 Noninverting op amp comparator with reference voltage

If the input becomes *slightly* greater than the reference, the open loop voltage gain will drive the output to the maximum positive.

$$e_2 > V_{ref}$$

$$e_0 = +V_{SAT}$$

Case 3

$$e_2 < V_{ref}$$

$$e_0 = \infty \times \text{some negative number}$$

$$e_0 \rightarrow -\infty$$

If the input becomes *slightly* less than the reference, the open loop voltage gain will drive the output to the maximum negative.

$$e_2 < V_{ref}$$

$$e_0 = -V_{SAT}$$

Figure 2-4 shows the input/output response of a noninverting comparator with a reference voltage. Figure 2-5 gives the schematic of several practical references.

Replacing the ground at the inverting terminal with a reference voltage changes the comparison level. For the simple comparator of Figs. 2-1 and 2-2 switching occurs when the input crosses *ground*. Switching, for the referenced comparator of Figs. 2-3 and 2-4, occurs when the input crosses the *reference voltage.*

The noninverting comparator outputs $+V_{SAT}$ when the input (+ input terminal) exceeds the reference (− input terminal) and outputs $-V_{SAT}$ when the input is less than the reference.

The schematic in Fig. 2-6 is an inverting comparator. The input signal is applied to the inverting input (−) while the noninverting input (+) is tied to

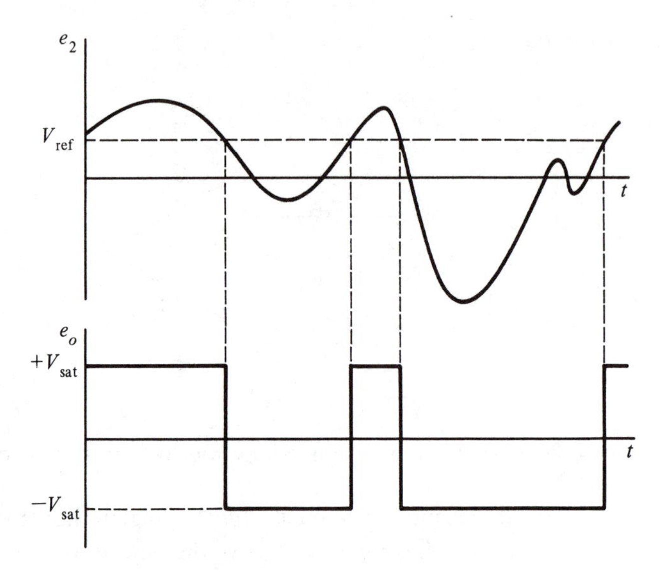

Figure 2-4 Input/output response of a noninverting comparator with a reference voltage

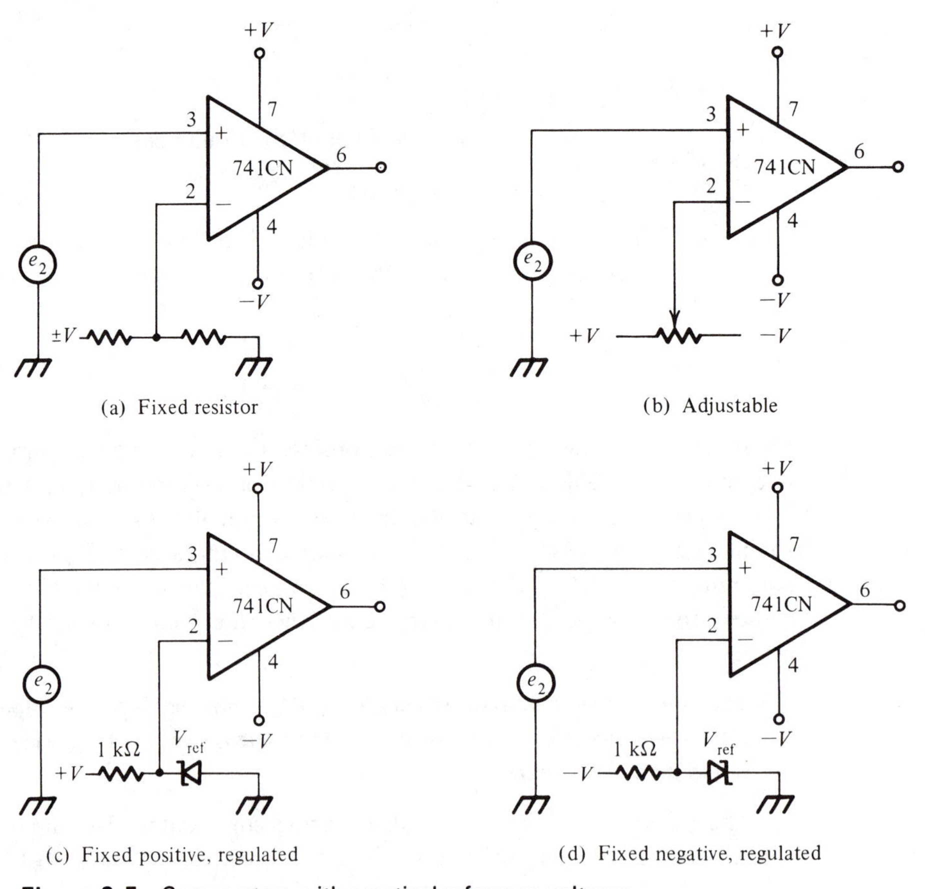

Figure 2-5 Comparators with practical reference voltages

circuit common. An input slightly above zero will force the output to $-V_{SAT}$, while an input just below zero will cause the output to go to $+V_{SAT}$. The analysis equations are given in Fig. 2-6. Figure 2-7 illustrates the input/output response of the inverting comparator.

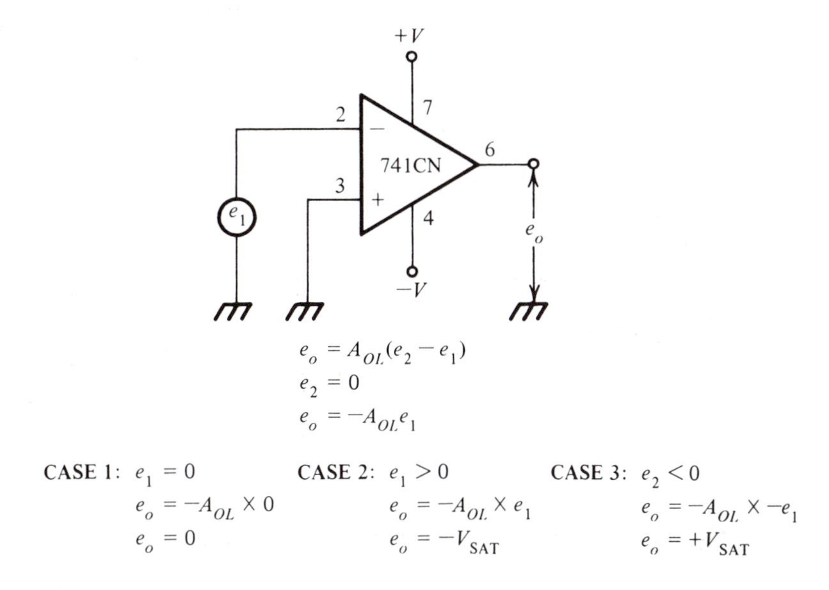

Figure 2-6 Inverting comparator

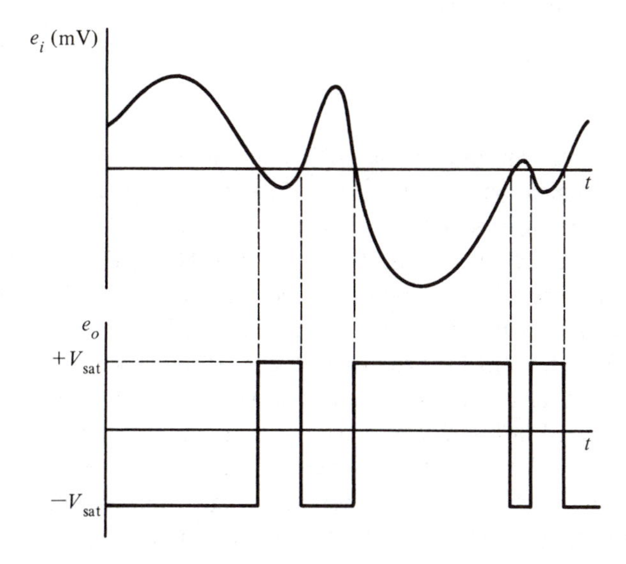

Figure 2-7 Example of inverting comparator response to an input signal

A reference voltage can be applied to the inverting comparator, as it was to the noninverting comparator. The schematic and input/output response of an inverting comparator with a nonzero reference voltage are shown in Fig. 2-8.

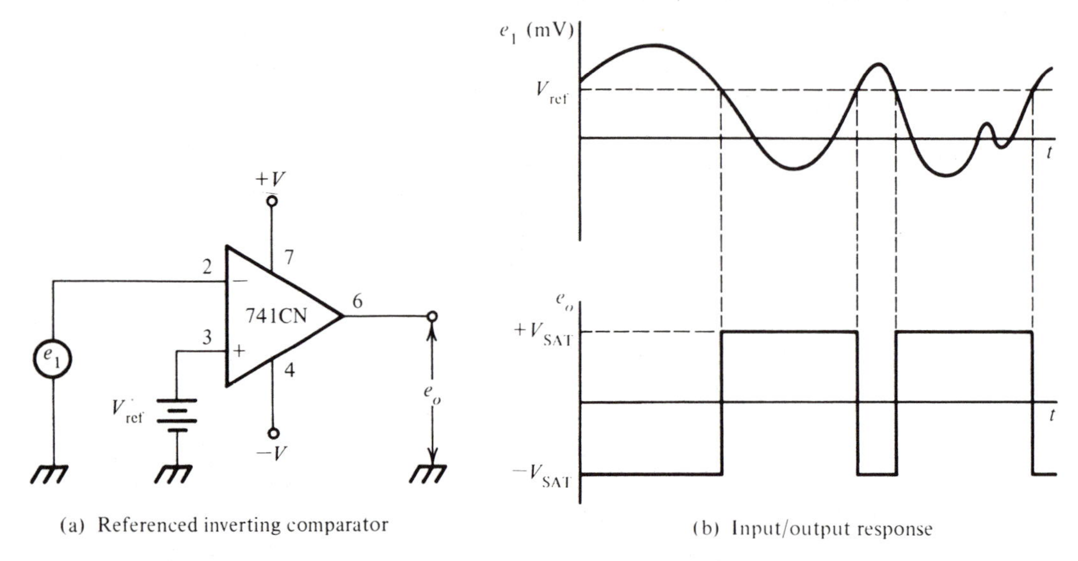

(a) Referenced inverting comparator

(b) Input/output response

Figure 2-8 Inverting comparator with a non-zero reference

The inverting comparator outputs $-V_{SAT}$ when the input (−input terminal) exceeds the reference (+ input terminal) and outputs $+V_{SAT}$ when the input is less than the reference.

2-1.2 LED Driver

The light emitting diode (LED) has found wide application as an indicator, communications interface, and isolation device. To properly drive the LED, its anode must be made 1.8 volts more positive than the cathode *and* about 10 to 25 milliamps of current must be provided. However, often the signal being sensed cannot provide this voltage at the needed current. A driver circuit is needed. A transistor could be used to drive the LED. However, bias resistors, reference offsets, and a LED current limiting resistor are needed.

Any of the comparators of section 2-1.1 can drive a LED directly. This is illustrated in Fig. 2-9. When e_2 rises above V_{ref}, the output of the op amp climbs toward $+V_{SAT}$. However, when the output reaches +1.8 V, CR1, the red LED begins to go on. It demands more and more current from the op amp as it illuminates. The op amp enters its short circuit current limiting mode, maintaining enough output voltage to keep the output current at about 20 mA. This is enough to turn the red LED on brightly, but should not damage it. The output voltage, instead of being at $+V_{SAT}$, is held at 1.8 volts by the diode and the short circuit current limit circuitry within the IC. Whenever the input voltage (e_2) exceeds the reference voltage, the red LED is turned on. The red LED could be a panel indicator warning of a dangerously high voltage (or other condition).

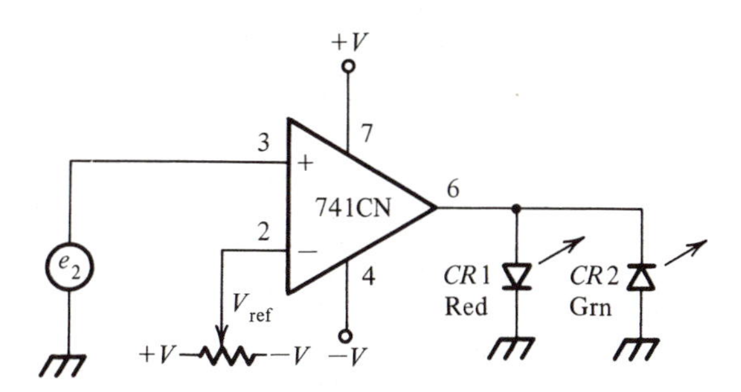

Figure 2-9 Comparator driving two light emitting diodes

When e_2 falls below V_{ref}, the output of the op amp drops toward $-V_{SAT}$. The red LED goes off. When the output reaches −1.8 volts, the green LED's cathode is now 1.8 volts below its anode (which is tied to circuit common) and CR2 begins to go on. It demands more and more current, which flows from ground, through CR2, and into the op amp (sink current). The current limit level is reached and the op amp maintains the output at −1.8 volts, 20 milliamps. The green LED glows. This could be taken to be a panel indication that the parameter represented by e_2 is below the dangerous level (set by V_{ref}).

A three-level comparator and indicator circuit can also be designed. There are three indicators: yellow for input too low, green for safe input, and red for input too high. They are turned on and off as indicated in Table 2-1. The schematic for the circuit is in Fig. 2-10. There are two comparators. Op amp U1 serves as the upper level comparator and has 6 volts on its inverting input. Op amp U2 is the lower level comparator with 3 volts on its inverting input. The input signal is applied to the noninverting inputs of both op amps. Resistors R4 and R5 set the cathode of CR1 at about +5 volts, while resistors R6 and R7 hold CR3 anode at −5 volts.

When the input is below 3 volts, the outputs of both comparators go to $-V_{SAT}$ (−7 V). This turns CR3 on (−5 V on its anode and −7 V on its cathode), while turning CR1 and CR2 off.

When the input is between 3 volts and 6 volts, the output of U2 goes positive and the output of U1 stays negative. This places a positive voltage on the cathode

TABLE 2-1. Three level comparator LED specifications

Input (volts)	*Yellow CR3*	*Green CR2*	*Red CR1*
Less than 3V	On	Off	Off
3V-6V	Off	On	Off
Greater than 6V	Off	Off	On

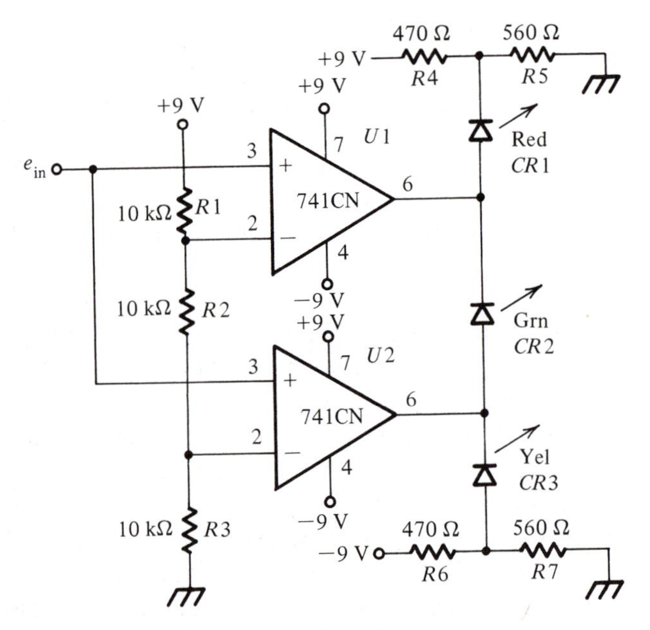

Figure 2-10 Three level comparator with LED indicators

of CR3, turning it off. That same positive voltage is on the anode of CR2. Since CR2 also has a negative on its cathode (from the output of U1), CR2 goes on. Diode CR1 is held off by the negative on the output of U1.

An input of greater than 6 volts will cause the outputs of both U1 and U2 to go to $+V_{SAT}$. This positive voltage out of U2 keeps CR3 off. The positive out of U1 turns CR2 off and CR1 on.

Of course, there is no reason that you must stop with three levels. Op amps can be stacked up, one on top of the other, to provide a multilevel comparator and display circuit. This would be useful not only in sensing out-of-tolerance conditions (as with the circuit in Fig. 2-10), but would also allow you to monitor the level. Audio VU monitoring could be done this way.

One word of caution: Be sure that the op amps you use to drive LEDs this way are internally short circuit current limited (as the 741C). Otherwise, the op amp may be damaged. If you are using an IC that contains several op amps, carefully read the manufacturer's specifications. You may not be allowed to short-circuit the output of all op amps in the IC at the same time.

2-2 EFFECTS OF NEGATIVE FEEDBACK

The comparator will not provide linear amplification. This is because of the extremely large open loop gain of the op amp. Also, this gain varies drastically. For the 741C it is typically 200,000, but may be as small as 20,000 with no maximum level specified. Although this factor of 10 or more variation in gain will not affect the comparator (since the output is being driven into saturation), variations of this magnitude could cause very serious problems for linear amplifiers.

The same negative feedback techniques used in transistor circuits can be used with op amps to reduce the overall circuit gain to a usable level, to allow the user to set this circuit gain precisely, and to make the circuit gain independent of changes in the op amp's open loop gain.

Figure 2-11 is a block diagram of a negative feedback amplifier. The output signal is e_O. It is fed back through an attenuator, which reduces it by a factor of beta.

$$e_f = \beta \times e_O \tag{2-3}$$

where

$$\beta < 1$$

This feedback signal, e_f (a reduced version of the output), is subtracted from the input signal, e_i by the summer (Σ).

$$e_d = e_i - e_f \tag{2-4}$$

The difference between the circuit input and the feedback signals is applied to the op amp. Here it is increased by a factor of A_{OL}.

$$e_O = A_{OL}\, e_d \tag{2-5}$$

The output is, in turn, sampled and fed back, beginning the process again.

Properly designed, this circuit can become virtually independent of the A_{OL} (if A_{OL} is large). For example, should A_{OL} increase, the output signal (e_O) would also increase. An increase in e_O will increase e_f, the negative feedback. Increasing

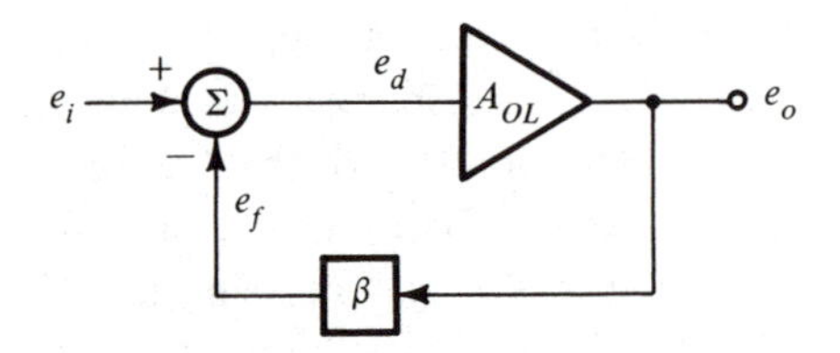

Figure 2-11 Block diagram of a negative feedback amplifier

e_f increases the amount *subtracted* from the input e_i. This causes a smaller difference signal, e_d. A decrease in e_d will cause e_0 to go down. Since the increased gain, A_{OL}, originally increased the output signal, this drop in output should bring e_0 back to approximately where it was before A_{OL} changed. This process is outlined in Fig. 2-12.

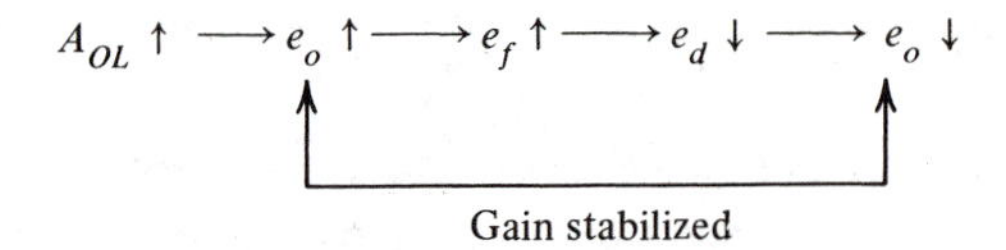

Figure 2-12 Gain stabilization caused by negative feedback

This same effect can be proven more rigorously by deriving the overall closed-loop gain (A_v). From the definitions

$$e_0 = A_{OL}\,e_d \tag{2-5}$$

$$e_f = \beta e_0 \tag{2-3}$$

$$e_d = e_i - e_f \tag{2-4}$$

The closed-loop gain is defined as

$$A_v = \frac{e_0}{e_i} \tag{2-6}$$

Substituting equation (2-5) into (2-6), you obtain

$$A_v = \frac{A_{OL}\,e_d}{e_i} \tag{2-7}$$

Equation (2-4) can be rearranged to give

$$e_i = e_d + e_f \tag{2-8}$$

Substituting equation (2-8) into equation (2-7), you obtain

$$A_v = \frac{A_{OL}\, e_d}{e_d + e_f} \tag{2-9}$$

Substituting equation (2-3) into equation (2-9) gives

$$A_v = \frac{A_{OL}\, e_d}{e_d + \beta e_O} \tag{2-10}$$

Substituting equation (2-5) into equation (2-10), you obtain

$$A_v = \frac{A_{OL}\, e_d}{e_d + \beta A_{OL}\, e_d} \tag{2-11}$$

Grouping and rearranging terms in the denominator, you have

$$A_v = \frac{A_{OL}\, e_d}{e_d(1 + \beta A_{OL})}$$

or

$$A_v = \frac{A_{OL}}{1 + \beta A_{OL}} \tag{2-12}$$

Since A_{OL} is very large for an op amp,

$$\beta A_{OL} \gg 1$$

so the 1 can be ignored in the denominator. Equation (2-12) becomes

$$A_v = \frac{A_{OL}}{\beta A_{OL}}$$

or

$$A_v = \frac{1}{\beta} \tag{2-13}$$

The parameter (open loop gain) of the op amp fell out, and the closed loop (circuit) gain is solely dependent on the feedback attenuation factor (β). Setting beta with stable, fixed resistors external to the op amp will allow you to establish a fixed, linear gain, independent of the op amp.

Use of negative feedback will give stable, linear circuit gains which are independent of the op amp. The gain is determined solely by feedback resistors. This assumes that adequate negative feedback ($\beta A_{OL} \gg 1$) is provided.

2-3 INVERTING AMPLIFIER

One of the most common *linear* op amp circuits (one which uses negative feedback) is the inverting amplifier. Before beginning the analysis of this circuit, however, you must recall two facts about the op amp itself. These are illustrated in Fig. 2-13.

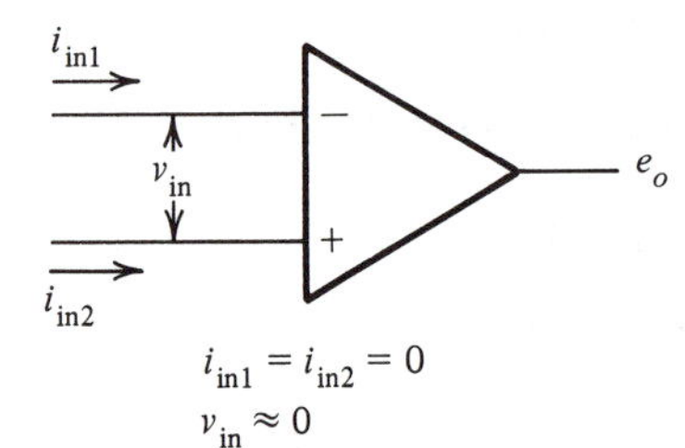

Figure 2-13 Fundamental op amp characteristics

First, since the input impedance is very high, the *signal* current that flows into either input terminal (− or +) is negligibly small.

$$i_{in1} \approx i_{in2} \approx 0 \tag{2-14}$$

Second, the open loop gain (e_0/v_{in}) is extremely large. This means that for any reasonably sized output (any output in the linear region, not $\pm V_{SAT}$), v_{in} is also negligibly small.

$$v_{in} \approx 0 \tag{2-15}$$

The two input leads are at virtually the same potential. You certainly will never be able to measure any significant potential difference between the two inputs during linear operation.

The schematic of an op amp inverting amplifier is given in Fig. 2-14. The op amp provides the gain. Resistors R_i and R_f attenuate the feedback signal (from e_0 to the − input), establishing both beta and the circuit's closed-loop gain. The output signal is 180 degrees out of phase with the input e_s, since the inverting terminal of the op amp is being driven. Consequently, when the feedback is combined with the input at the inverting terminal, subtraction takes place (as is required for negative feedback amplifiers). All the requirements for a negative feedback amplifier outlined in section 2-2 are met.

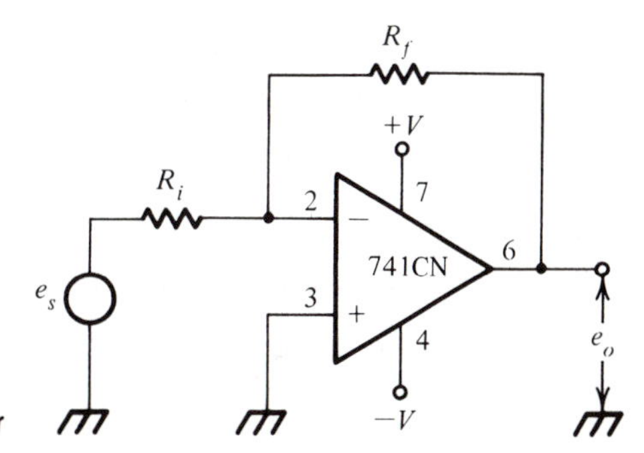

Figure 2-14 Op amp inverting amplifier

The output signal (e_0) will be 180 degrees out of phase with the input (e_s). The circuit's voltage gain is strictly determined by the feedback components R_i and R_f.

$$A_v = -\frac{\mathrm{R}_f}{\mathrm{R}_i} \tag{2-16}$$

A negative gain indicates the 180 degree phase shift. The input impedance is set by R_i.

$$Z_{\text{in}} = \mathrm{R}_i \tag{2-17}$$

You can derive these characteristics from the following analysis. Also, please refer to Fig. 2-15. You have seen that

$$v_{\text{in}} = 0$$

the voltage at the junction of R_i, R_f and the inverting input, must be zero volts. The inverting input is said to be at virtual ground.

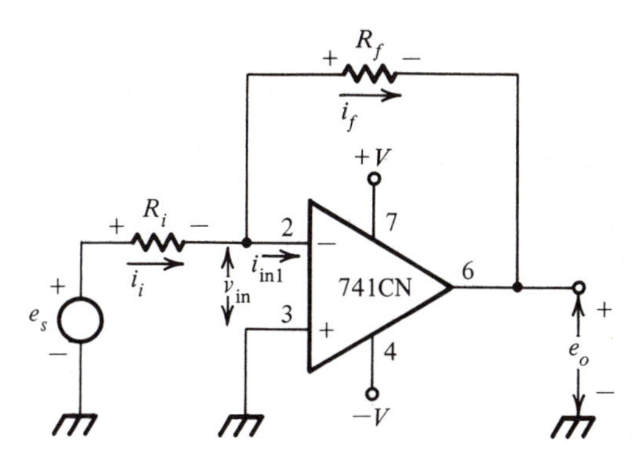

Figure 2-15 Analysis of an op amp inverting amplifier

With the inverting input at virtual ground, the input impedance is the resistance between the circuit input (e_s) and ground. This is just R_i.

$$Z_{in} = R_i$$

The current flowing through R_i is

$$i_i = \frac{e_s}{R_i}$$

or

$$e_s = i_i R_i \tag{2-18}$$

Since

$$i_{in1} = 0 \tag{2-19}$$

all of the current flowing through R_i (i.e., i_i) must flow through R_f. So

$$i_i = i_f = i \tag{2-19}$$

The current i_f flowing through R_f must develop a voltage drop equal to the output voltage.

$$e_0 = -i_f R_f \tag{2-20}$$

(The negative sign is caused by the direction of flow of i_f and the virtual ground at the left end of R_f.)

The circuit voltage gain is

$$A_v = \frac{e_0}{e_s} \tag{2-21}$$

Substituting equations (2-18) and (2-20) into equation (2-21), you have

$$A_v = -\frac{i_f R_f}{i_i R_i}$$

But

$$i_i = i_f = i \tag{2-19}$$

So

$$A_v = -\frac{i\,R_f}{i\,R_i}$$

or

$$A_v = -\frac{R_f}{R_i}$$

With the inverting amplifier you can set the input impedance by picking R_i and the voltage gain by adjusting R_f. However, there are several practical considerations.

1. Setting R_i too high will give problems with the bias current. It is usually restricted to less than 10 kΩ.
2. Remember, there is an upper limit to the bandwidth ($A_v \times f$). Do not try for a gain which is so high that at upper frequencies the bandwidth criterion is violated. Usually A_v is held below 100.
3. The peak output is limited by the power supplies. At about two volts less than the supply, the op amp goes into saturation.
4. The output current may or may not be short-circuit limited. If it is not, heavy loads (low load resistance) may damage the op amp. If there is short-circuit protection, a heavy load may drastically distort the output *voltage* wave shape.

2-4 INVERTING SUMMER

A simple addition to the inverting amplifier produces a circuit with much versatility. This inverting summer is shown in Fig. 2-16. An additional input (e_2) and input resistor (R2) have been added to the inverting amplifier of Fig. 2-14.

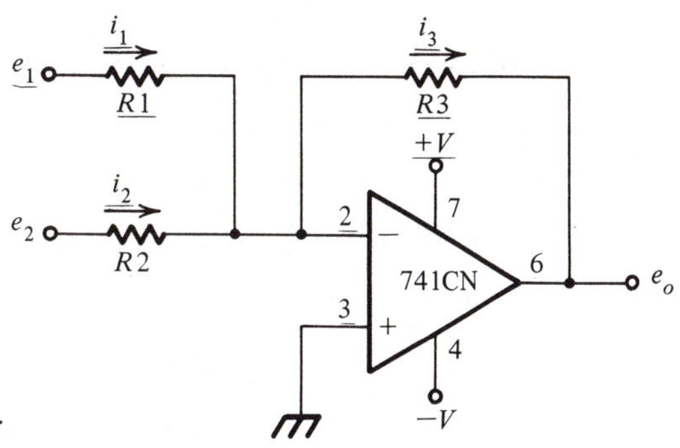

Figure 2-16 The inverting summer

The input impedances and the output voltage equation are derived very much as the inverting amplifier equations were. Since the circuit uses negative feedback to obtain linear operation, there is no difference in potential between the input pins of the op amp; the inverting input pin is at virtual ground. The input impedance seen from e_1 is the resistance from that point to ground. This is just R1. Similarly, the input impedance seen from e_2 is R2.

$$Z_{in1} = R1$$

$$Z_{in2} = R2$$

The current drawn from e_1 and flowing through R1 is

$$i_1 = \frac{e_1}{R1} \tag{2-21}$$

Similarly, for i_2

$$i_2 = \frac{e_2}{R2} \tag{2-22}$$

As you saw with the inverting amplifier, no significant signal current flows into the op amp itself. This means that i_1 and i_2 must combine at the junction of R1, R2, and R3 to produce i_3. Or, applying Kirchoff's current law to that node, you have

$$i_1 + i_2 = i_3 \tag{2-23}$$

The voltage drop produced by i_3 flowing through R3 must equal the output voltage.

$$e_0 = -i_3 \, R3 \tag{2-24}$$

Substituting equation (2-23) into equation (2-24) gives

$$e_0 = -(i_1 + i_2)R3 \tag{2-25}$$

Substituting equations (2-21) and (2-22) into equation (2-25), you obtain

$$e_0 = -\left(\frac{e_1}{R1} + \frac{e_2}{R2}\right)R3 \tag{2-26}$$

Equation (2-26) can be rearranged to look a bit more familiar.

$$e_0 = -\left(\frac{R3}{R1}e_1 + \frac{R3}{R2}e_2\right) \tag{2-27}$$

where $\frac{R3}{R1}$ is the gain for e_1.

$\frac{R3}{R2}$ is the gain for e_2.

Equation (2-27) tells you that the inverting summer outputs a voltage (e_0) which is the inverted sum of the gain associated with each input ($\frac{R3}{R1}$ or $\frac{R3}{R2}$) times that input. The output is the weighted or scaled sum of the inputs.

Since this is only a slight modification of the inverting amplifier, you should observe all of the precautions listed in section 2-3. Also, although three, four, or more inputs can be summed just as two were, the amplifier noise goes up with each input added. So you should limit the number of summing inputs to as few as possible.

The inverting summer will allow you to alter the zero and the full scale (span) levels of a signal. This is of particular importance in matching the output of industrial transducers, allowing them to be displayed in convenient engineering units.

Example 2-1

The LM3911 temperature transducer outputs ± 10 millivolts per degree Kelvin. Alter this output so that $+100$ millivolts/degrees Celsius can be displayed. Both the zero and the span must be altered. The inverting summer amplifier solution to this is shown in Fig. 2-17.

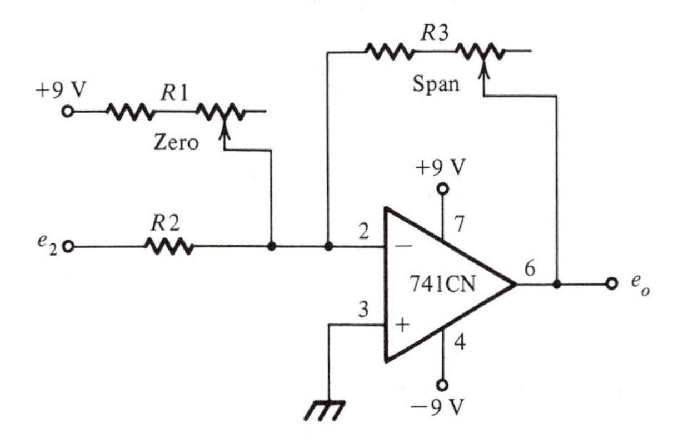

Figure 2-17 Zero and span adjustment for an industrial transducer

The output equation is

$$e_0 = -\left(\frac{R3}{R1}e_1 + \frac{R3}{R2}e_2\right)$$

The degree Kelvin and the degree Celsius are the same size, but to go from −10 millivolts/degree to 100 millivolts/degree a gain of −10 is needed for e_2.

Therefore,

$$\frac{R3}{R2} = 10$$

We choose R2 = 10 kΩ to keep from loading down the transducer

R3 = R2 × 10 = 100 kΩ

Resistance R3 should be chosen as a 82 kΩ resistor with a series 50 kΩ potentiometer to allow for fine tuning.

At freezing, you want the circuit to output 0 volts. However, the transducer is outputting

$$e_2 = -10 \text{ mV/°K} \times 273\text{°K}$$

$$e_2 = -2.73 \text{ V}$$

Resistor R1 can be adjusted to compensate for this.

$$e_0 = -\left(\frac{R3}{R1}e_1 + \frac{R3}{R2}e_2\right)$$

Substituting the given values at 0°C, you obtain

$$0 \text{ V} = \frac{-100 \text{ k}\Omega}{R1}(9 \text{ V}) - \frac{100 \text{ k}\Omega}{10 \text{ k}\Omega}(-2.73 \text{ V})$$

Combining terms gives

$$-27.3 \text{ V} = \frac{-900 \text{ k}\Omega\text{V}}{R1}$$

Multiplying by R1 and dividing by −27.3 V, we have

$$R1 = \frac{-900 \text{ k}\Omega\text{V}}{-27.3 \text{ V}} = 32.9\text{k}\Omega$$

Resistance R1 should be chosen as a 27 kΩ resistor with a series 10 kΩ potentiometer to allow fine tuning.

Does the circuit work? Check at mid-range. At 50°C, the transducer is outputting

$$e_2 = -10 \text{ mV/°K} \times 323\text{°K}$$

$$e_2 = -3.23 \text{ V}$$

You want the circuit to output

$$e_0 = 100 \text{ mV/°C} \times 50\text{°C}$$

$$e_0 = 5 \text{ V}$$

$$e_0 = -\left(\frac{\text{R3}}{\text{R1}}e_1 + \frac{\text{R3}}{\text{R2}}e_2\right)$$

Substituting component and voltage values, we obtain

$$e_0 = -\left[\frac{100 \text{ k}\Omega}{32.9 \text{ k}\Omega}(9 \text{ V}) + \frac{100 \text{ k}\Omega}{10 \text{ k}\Omega}(-3.23 \text{ V})\right]$$

$$e_0 = -(27.3 \text{ V} - 32.3 \text{ V})$$

$$e_0 = 5 \text{ V}$$

It works!

The output is the inverted sum of the input e_2 and the zero offset, weighted by the span.

Example 2-2

The function generator produces sine waves, square waves, and triangle waves. The amplitude can be altered by a front panel control, as can the DC offset (or DC level). This is also the job for an inverting summer. The schematic is shown in Fig. 2-18.

If the inputs are 1 volt peak to peak and you want to be able to adjust their amplitude with the amplitude control to as much as 10 volts peak to peak, what value of R2, R3, and R4 would you choose?

To increase a 1 volt signal from the function switch to 10 volts at the output, the R4/R3 gain must be 10.

As with Example 2-1, the feedback resistor can be chosen at 100 kΩ, with R3, the input resistor, 10 kΩ.

The DC offset requires a bit more thought. The voltage at the wiper of R1 can be varied from ±12 volts. However, you do not want it to be able to force

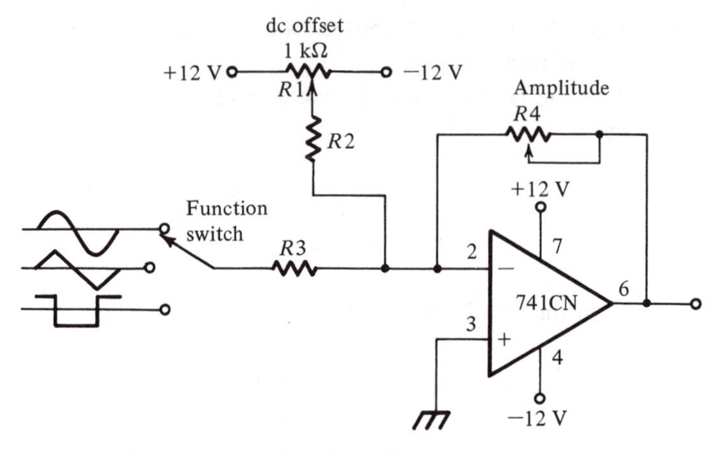

Figure 2-18 Function generator signal conditioner

the amplifier into saturation. (If it did, the output wave would be lost.) A reasonable compromise would be to allow ±12 volts at the wiper of R1 to cause ±8 volts DC out. This would provide a reasonable range of DC level adjustment while providing the signal some "headroom." So, with zero volts in from the function switch,

$$e_0 = -\left(\frac{\mathrm{R4}}{\mathrm{R2}}e_1 + \frac{\mathrm{R4}}{\mathrm{R3}}e_2\right)$$

$$e_2 = 0$$

$$-8\ \mathrm{V} = -\frac{100\ \mathrm{k\Omega}}{\mathrm{R2}}(+12\ \mathrm{V})$$

$$\mathrm{R2} = -100\ \mathrm{k\Omega}\left(\frac{12\ \mathrm{V}}{-8\ \mathrm{V}}\right)$$

$$\mathrm{R2} = 150\ \mathrm{k\Omega}$$

Example 2-3

Turning an analog signal on or off with a digital control voltage proves very useful in building any type of system. Although this could be done with an analog switch, these are relatively expensive. The circuit in Fig. 2-19 offers an inexpensive alternative for controlling AC signals.

When the control voltage is at zero volts,

$$e_0 = -\left(\frac{R3}{R1}e_1 + \frac{R3}{R2}e_2\right)$$

$$e_0 = -\left[\frac{100\ k\Omega}{10\ k\Omega}(0\ V) + \frac{100\ k\Omega}{100\ k\Omega}(e_s)\right]$$

$$e_0 = -e_s$$

The AC signal is passed. Even if e_1 is not *exactly* zero volts, the RC coupler at the output will filter out any DC offset introduced by the control voltage.

When the control voltage goes to 2.2 volts,

$$e_0 = -\left[\frac{100\ k\Omega}{10\ k\Omega}(2.2\ V\ DC) + \frac{100\ k\Omega}{100\ k\Omega}(1\ V\ p\text{-}p)\right]$$

$$e_0 = -[10\ (2.2\ V\ DC) + (1\ V\ p\text{-}p)]$$

$$e_0 = -22\ V\ DC - 1\ V\ p\text{-}p$$

$$e_0 = -V_{SAT}$$

The output is forced to negative saturation. The input e_s is not large enough to bring it out. Since the RC coupler at the op amp output does not pass DC, the *circuit* output will be zero.

When the control voltage is a logic low, the AC signal passes unaffected. When the control voltage switches to a logic high, the AC signal is turned off.

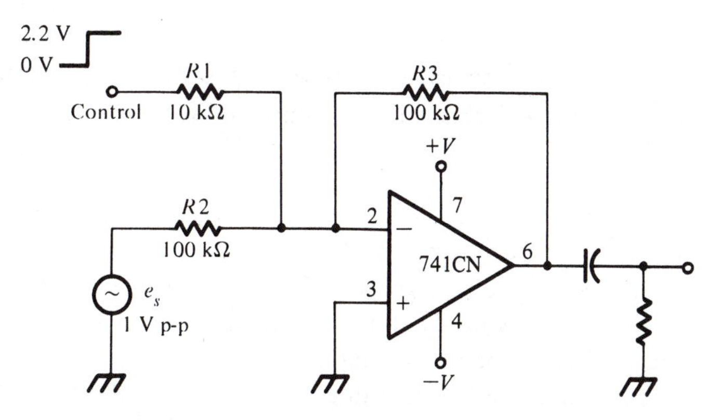

Figure 2-19 Digital control of an AC signal

2-5 NONINVERTING AMPLIFIER

Both the inverting amplifier and the inverting summer have fairly low input impedances. Also, in some applications, the phase inversion can create a problem. The noninverting amplifier solves both of these objections. Its schematic is given in Fig. 2-20. As with the inverting amplifier, the op amp provides the gain, while R_i and R_f attenuate the feedback signal (from e_0 to the − input), establishing both beta and the circuit's closed-loop gain. The output signal is *in phase* with the input e_s, since the noninverting terminal of the op amp is being driven.

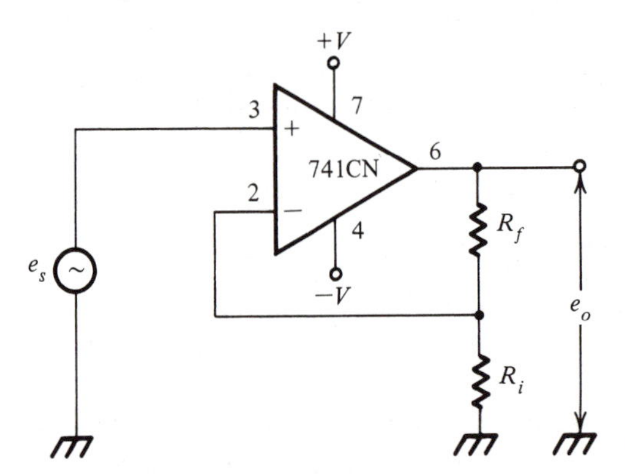

Figure 2-20 Noninverting amplifier

The circuit's voltage gain is strictly determined by the feedback components R_f and R_i.

$$A_v = 1 + \frac{R_f}{R_i} \tag{2-26}$$

The input impedance depends on the op amp's impedance and βA_{OL} of the circuit. Although this will vary considerably from one op amp to another, generally

$$Z_{\text{in circuit}} > Z_{\text{in op amp}} \tag{2-27}$$

To derive the voltage gain of the noninverting amplifier, you must recall that there is virtually no difference in potential between the two inputs of the op amp (during linear operation using negative feedback). This means that the inverting terminal and the noninverting terminal are both at e_s. But the voltage at the inverting terminal is the voltage across R_i. The op amp must, therefore, provide enough output voltage to cause the voltage across R_i to equal e_s.

Applying the voltage divider law to e_0, R_i, and R_f, you obtain

$$e_{R_i} = \frac{R_i}{R_i + R_f} e_0$$

But

$$e_{R_i} = e_s$$

$$e_s = \frac{R_i}{R_i + R_f} e_0 \tag{2-28}$$

The voltage gain of the circuit is

$$A_v = \frac{e_0}{e_s}$$

Equation (2-28) can be rearranged to give

$$A_v = \frac{e_0}{e_s} = \frac{R_i + R_f}{R_i}$$

or

$$A_v = 1 + \frac{R_f}{R_i}$$

Compare this to the voltage gain of the inverting amplifier. Using the same components, the noninverting amplifier will always give a voltage gain which is 1 more than the gain of the inverting amplifier.

By using a noninverting amplifier rather than an inverting amplifier, you will notice three significant effects:

1. There is no phase inversion with the noninverting amplifier.
2. The input impedance has increased into the megohm range, without having to raise R_i or R_f to impractically large values.
3. You no longer can precisely determine the input impedance as you could with the inverting amp. All you know for sure is that it is larger than $Z_{\text{in op amp}}$.

2-6 VOLTAGE FOLLOWER

A recurring problem in electronics is loading down high impedance voltage sources. This is illustrated in Fig. 2-21. If the source's internal impedance is 10 kΩ, and the load impedance is 10 kΩ, only 50 percent of the source voltage is applied to the load. If the load is raised to 100 kΩ, 91 percent of the source voltage is applied to the load. Either of these two cases would cause significant error in transferring the voltage from the source to the load. If the source impedance could be lowered drastically, or the load impedance raised, loading could be eliminated. Both of these are accomplished by the voltage follower of Fig. 2-22.

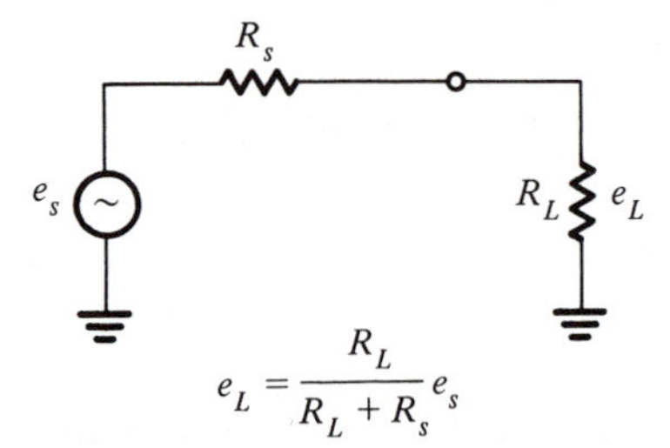

Figure 2-21 Source loading

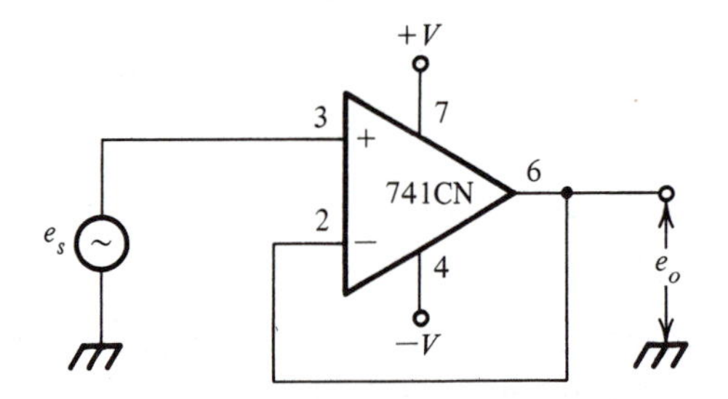

Figure 2-22 Voltage follower

The input is applied directly to the noninverting input. The output is fed back directly to the inverting input, so there is no attenuation of the negative feedback signal.

$$\beta = 1$$

$$A_v = \frac{1}{\beta}$$

$$A_v = 1$$

From negative feedback theory, a voltage gain of 1 is obtained. This means that the input and output voltages are identical.

$$e_0 = e_s$$

Compare the voltage follower in Fig. 2-22 with the noninverting amplifier in Fig. 2-20. Resistor R_i has been removed and R_f has been replaced with a short circuit.

$$A_v = 1 + \frac{R_f}{R_i}$$

$$A_v = 1 + \frac{0}{\infty}$$

$$A_v = 1$$

If you consider the voltage follower a special case of the noninverting amplifier, you still obtain a voltage gain of 1.

One final way to analyze the voltage follower is to recall that there is no difference in potential between the two inputs. This means that since the noninverting input is at the source potential, so is the inverting input. But the inverting input is tied *directly* to the output. The output voltage equals the voltage at the noninverting input, which is the input voltage. So output voltage equals input voltage, giving a gain of 1.

Although the voltage follower does not increase the *voltage* amplitude of the signal, it gives good current drive and low output impedance to the load while presenting a very high impedance to the source. The voltage follower makes an excellent impedance transformer or buffer.

2-7 NONINVERTING SUMMER

The voltage follower can be considered a special case of a noninverting amplifier. The noninverting summer adds additional inputs to the noninverting amplifier. A two input noninverting summer is shown in Fig. 2-23. Resistors R1 and R2 have been added to separate the two sources.

The output voltage is dependent, in a more complex way, on the circuit component values.

$$e_0 = \left(1 + \frac{R3}{R4}\right)\left(e_1 \frac{R2}{R1 + R2} + e_2 \frac{R1}{R1 + R2}\right) \tag{2-29}$$

$$e_0 = e_1 \frac{R2(R3 + R4)}{R4(R1 + R2)} + e_2 \frac{R1(R3 + R4)}{R4(R1 + R2)} \tag{2-29a}$$

As with the inverting amplifier, each input voltage is multiplied by some weight-

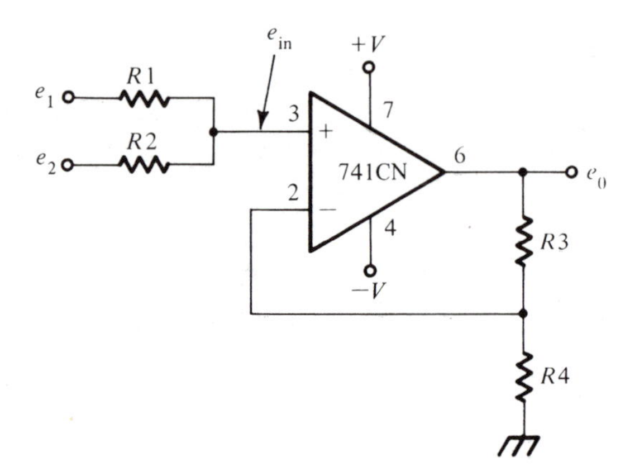

Figure 2-23 Noninverting summer

ing factor, and then the inputs are added together. However, these weighting factors for the noninverting summer include both the gain (1 + R3/R4) *and* a voltage divider ratio of the input resistances, R2/(R1 + R2) or R1/(R1 + R2).

Although these input resistors complicate the output expression, they are necessary. You cannot tie two different voltage sources directly in parallel. Each will try to drive the other to its own potential, probably damaging itself and the other source. Also, the input impedance seen from each source now consists of the parallel combination of

$$Z_{in} = Z_{noninverting} // (R1 + R2 + R_s)$$

where R_s is the output impedance of the *other* voltage source.

To obtain the output voltage equation [equation (2-29)], you must first realize that the voltage present at the noninverting input (e_{in}) will be amplified by the factor (1 + R3/R4).

$$e_0 = e_{in}\left(1 + \frac{R3}{R4}\right) \tag{2-30}$$

This input voltage is established by e_1, e_2, R1 and R2. Since no signal current flows into the op amp, the input circuit can be simplified as shown in Fig. 2-24. Applying Kirchhoff's voltage law to the input loop, we obtain

$$e_1 - iR1 - iR2 - e_2 = 0$$

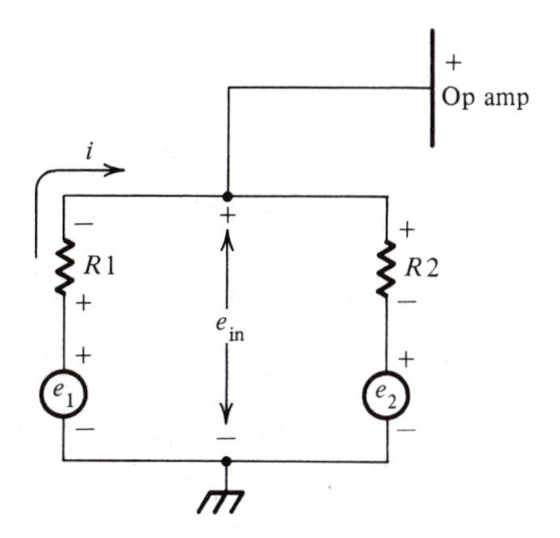

Figure 2-24 Noninverting summer input circuit

Rearranging and grouping gives

$$i(\mathrm{R1} + \mathrm{R2}) = e_1 - e_2 \tag{2-31}$$

Equation (2-31) can be solved for the current.

$$i = \frac{e_1 - e_2}{\mathrm{R1} + \mathrm{R2}} \tag{2-32}$$

Applying Kirchhoff's voltage law to the loop consisting of e_{in}, R2, and e_2, we have

$$e_{in} - i\mathrm{R2} - e_2 = 0$$

Solving for e_{in}, we obtain

$$e_{in} = i\mathrm{R2} + e_2 \tag{2-33}$$

Substituting equation (2-32) into (2-33), you obtain

$$e_{in} = e_2 + \frac{\mathrm{R2}(e_1 - e_2)}{\mathrm{R1} + \mathrm{R2}} \tag{2-34}$$

Equation (2-34) can now be substituted into the original gain equation, equation (2-30).

$$e_0 = \left(1 + \frac{R3}{R4}\right)\left[e_2 + \frac{R2(e_1 - e_2)}{R1 + R2}\right]$$

$$e_0 = \left(1 + \frac{R3}{R4}\right)\left[\frac{e_2(R1 + R2) + R2(e_1 - e_2)}{R1 + R2}\right]$$

$$e_0 = \left(1 + \frac{R3}{R4}\right)\left[\frac{e_2 R1 + e_2 R2 + e_1 R2 - e_2 R2}{R1 + R2}\right]$$

$$e_0 = \left(1 + \frac{R3}{R4}\right)\left(\frac{e_2 R1 + e_1 R2}{R1 + R2}\right)$$

or

$$e_0 = \left(1 + \frac{R3}{R4}\right)\left(e_1 \frac{R2}{R1 + R2} + e_2 \frac{R1}{R1 + R2}\right)$$

The first factor can be rewritten to give

$$e_0 = \left(\frac{R4 + R3}{R4}\right)\left(e_1 \frac{R2}{R1 + R2} + e_2 \frac{R1}{R1 + R2}\right)$$

Distributing the first factor over the terms of the second, you obtain

$$e_0 = e_1 \frac{R2(R3 + R4)}{R4(R1 + R2)} + e_2 \frac{R1(R3 + R4)}{R4(R1 + R2)}$$

Example 2-4

Although the output voltage equation is a bit more complex than the one for the inverting summer, you can use it for the same tasks. To show how, Example 2-1, a temperature transducer signal conditioner, will be reworked, using a noninverting summer.

The temperature transducer outputs ± 10 millivolts per degree Kelvin. Alter this output to give $+100$ millivolts/degrees Celsius. The noninverting summer of Fig. 2-23 is to be used. Input e_1 is the zero offset and may be set at anything handy. Pick

$$e_1 = -1 \text{ V}$$

Input e_2 is the transducer input. To use a noninverting summer, the positive transducer output should be chosen.

To alter the span from 10 millivolts/degree to 100 millivolts/degree, the gain of e_2 must be set to 10.

$$\frac{R1(R3 + R4)}{R4(R1 + R2)} = 10 \tag{2-34}$$

At freezing, the transducer outputs 2.73 volts (10 millivolts/°K × 273°K), but you want 0 volts (0°C) out of the signal conditioner.

$$0\ \text{V} = -1\ \text{V}\,\frac{R2(R3 + R4)}{R4(R1 + R2)} + 2.73\ \text{V}\ (10)$$

or

$$\frac{R2(R3 + R4)}{R4(R1 + R2)} = 27.3 \tag{2-35}$$

You now have two equations and four unknowns. Your only option is to pick two values and solve the simultaneous equations for the other two. This can be simplified by dividing equation (2-35) by equation (2-34).

$$\frac{R2}{R1} = 2.73 \tag{2-36}$$

To provide fairly high input impedance, pick

$$R1 = 100\ \text{k}\Omega$$

Equation (2-36) yields

$$R2 = 273\ \text{k}\Omega$$

This can be obtained with a 200 kΩ fixed resistor in series with a 100 kΩ potentiometer to provide fine tuning.

Now, substitute R1 and R2 into equation (2-34).

$$\frac{100\ \text{k}\Omega\ (R3 + R4)}{R4\ (100\ \text{k}\Omega + 273\ \text{k}\Omega)} = 10$$

$$100\ \text{k}\Omega\ (R3 + R4) = 3730\ \text{k}\Omega\ R4$$

$$R3 + R4 = 37.3\ R4$$

$$R3 = 36.3\ R4 \tag{2-37}$$

To keep bias and offset current effects small, pick R4 fairly low. Pick

$$R4 = 1\ \text{k}\Omega$$

Equation (2-37) then yields

$$R3 = 36.3\ \text{k}\Omega$$

Use a 33 kΩ resistor with a series 10 kΩ potentiometer to provide fine tuning.

As with the inverting summer, this solution must be checked.

At 50°C, the circuit should output 5 V (100 millivolts/°C × 50°C). The transducer outputs

$$e_2 = 10\ \text{millivolts/°K} \times 323\text{°K}$$

$$e_2 = 3.23\ \text{V}$$

$$e_1 = -1\ \text{V (You picked this.)}$$

$$R1 = 100\ \text{k}\Omega,\ R2 = 273\ \text{k}\Omega,\ R3 = 36.3\ \text{k}\Omega,\ R4 = 1\ \text{k}\Omega$$

$$e_0 = \left(1 + \frac{R3}{R4}\right)\left(e_1 \frac{R2}{R1 + R2} + e_2 \frac{R1}{R1 + R2}\right)$$

$$e_0 = \left(1 + \frac{36.3\ \text{k}\Omega}{1\ \text{k}\Omega}\right)\left(-1\ \text{V} \frac{273\ \text{k}\Omega}{100\ \text{k}\Omega + 273\ \text{k}\Omega} + \frac{3.23\ \text{V}\ 100\ \text{k}\Omega}{100\ \text{k}\Omega + 273\ \text{k}\Omega}\right)$$

$$e_0 = (1 + 36.3)(-1\ \text{V} \times 0.732 + 3.23\ \text{V} \times 0.268)$$

$$e_0 = (37.3)(-0.732\ \text{V} + 0.866\ \text{V})$$

$$e_0 = 5\ \text{V}$$

It works, too!

2-8 DIFFERENCE AMPLIFIER

To this point you have seen how to amplify or amplify and add signals. Often, it is also important to be able to obtain the difference between two signals. This could be used to tell you how far apart the two signals are, or to eliminate (subtract out) any portion of the two signals which are the same (or common), giving an output only when the two signals are different.

The schematic for a difference amplifier is shown in Fig. 2-25. Input signals are connected through identical resistors (R_i) to both the inverting and the noninverting op amp terminals. A sample of the output is fed back through the upper R_f to the inverting terminal to ensure negative feedback. The lower resistor R_f attached between the noninverting terminal and ground, balances operation, ensuring zero volts out when e_1 equals e_2, and significantly simplifies circuit analysis and design. This resistor has been shown as adjustable to emphasize the point

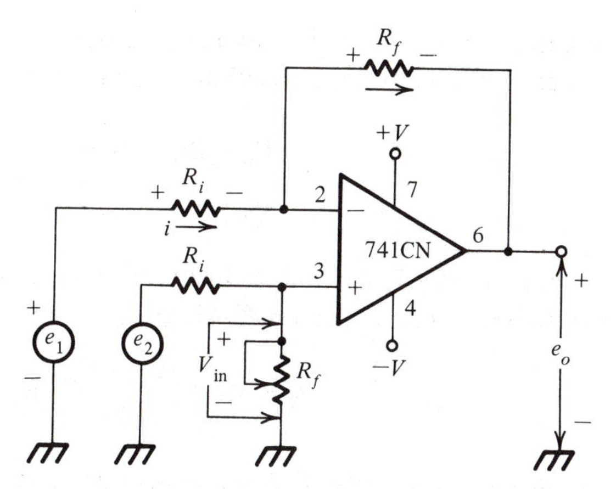

Figure 2-25 Difference amplifier

that it must *exactly* equal the feedback resistor, R_f, for proper operation. This is usually done by adjusting the potentiometer.

With the two input resistors (R_i) identical and the other two resistors (R_f) adjusted to be equal, the output voltage equation is

$$e_0 = \frac{R_f}{R_i}(e_2 - e_1) \tag{2-38}$$

The output is the difference between the two inputs times a gain set by R_f and R_i.

You can derive this relationship by first noting that the voltage at the noninverting op amp input is set by the R_i/R_f voltage divider.

$$V_{in} = e_2 \frac{R_f}{R_i + R_f} \tag{2-39}$$

From this point, the derivation is the same as with the inverting amplifier. However, the equations are just a bit messier.

Since for a linear, negative feedback amplifier, there is no significant difference in potential between the two op amp inputs, the inverting input is also at V_{in}.

Current i then is

$$i = \frac{e_1 - V_{in}}{R_i} \tag{2-40}$$

But this current flows through the feedback resistor R_f. The output voltage e_0 must go to a level to satisfy Kirchhoff's voltage law.

$$V_{in} - iR_f - e_0 = 0$$

$$e_0 = V_{in} - iR_f \qquad (2\text{-}41)$$

From here on you carefully substitute, gather terms, and simplify. Substituting equation (2-40) into equation (2-41) gives

$$e_0 = V_{in} - \left(\frac{e_1 - V_{in}}{R_i}\right) R_f \qquad (2\text{-}42)$$

Substituting equation (2-39) into equation (2-42), you obtain

$$e_0 = \frac{e_2 R_f}{R_i + R_f} - \frac{R_f}{R_i}\left[\frac{e_1(R_i + R_f) - e_2 R_f}{R_i + R_f}\right]$$

$$e_0 = \frac{e_2 R_f}{R_i + R_f} + \frac{-R_f e_1(R_i + R_f) + e_2 R_f^2}{R_i(R_i + R_f)}$$

$$e_0 = \frac{e_2 R_i R_f - e_1 R_f(R_i + R_f) + e_2 R_f^2}{R_i(R_i + R_f)}$$

$$e_0 = \frac{e_2 R_f(R_i + R_f) - e_1 R_f(R_i + R_f)}{R_i(R_i + R_f)}$$

$$e_0 = \frac{R_f(R_i + R_f)(e_2 - e_1)}{R_i(R_i + R_f)}$$

$$e_0 = \frac{R_f}{R_i}(e_2 - e_1)$$

Example 2-5

The temperature transducer signal conditioner of Examples 2-1 and 2-4 can be easily built with a difference amplifier. Recall, the problem is to convert a 10 millivolt per degree Kelvin signal to 100 millivolt per degree Celsius. Both zero and span (offset and gain) must be altered.

Since the degree Kelvin and the degree Celsius are the same size, changing from 10 millivolts/°K to 100 millivolts/°C means that the amplifier must have a gain of 10.

Set $\qquad \dfrac{R_f}{R_i} = 10$

Pick $\qquad R_f = 1\ M\Omega$

So $$R_i = \frac{R_f}{10} = 100\ k\Omega$$

When the temperature is at freezing, the signal conditioner should output zero volts, but the transducer signal is 2.73 V. So 2.73 volts must be subtracted from the transducer signal.

Figure 2-26 gives the schematic for the difference amplifier/transducer signal conditioner. Although this approach is a good deal simpler than the other two examples, two adjustments to the circuit must be made. First, points A and B must be shorted together (ensuring an input difference of zero), and R6 must be adjusted for a zero output. Second, R2 must be tuned to set the voltage at point A to 2.73 volts.

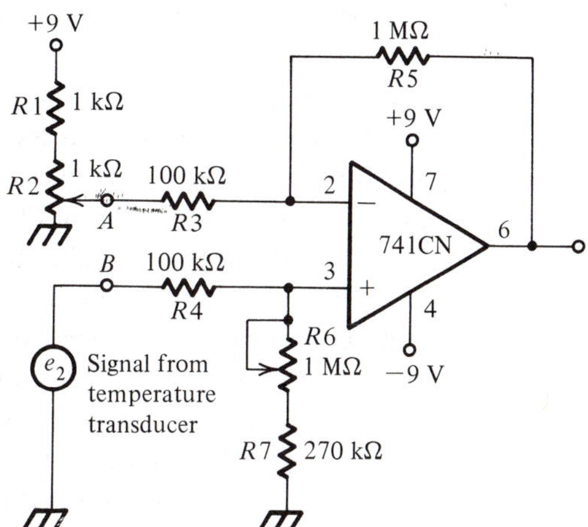

Figure 2-26 Transducer signal conditioner using a difference amplifier

To check this circuit, at 50°C, the transducer outputs 3.23 volts and the signal conditioner should output 5.0 volts.

$$e_2 = 3.23 \text{ volts}$$

$$e_1 = 2.73 \text{ volts}$$

$$e_O = \frac{R_f}{R_i}(e_2 - e_1)$$

$$e_O = \frac{1\ M\Omega}{100\ k\Omega}(3.23\ V - 2.73\ V)$$

$$e_O = 10(0.5\ V)$$

$$e_O = 5.0\ V$$

It also works.

SUMMARY

Operating open loop (no feedback), an op amp's output will normally either be $+V_{SAT}$ or $-V_{SAT}$. When the voltage at the noninverting input is larger than the voltage at the inverting input, the op amp's output voltage will be forced to positive saturation. On the other hand, when the voltage at the inverting input is larger, the output goes to negative saturation. No other outputs will normally occur.

This characteristic makes the open-loop op amp a good voltage comparator. Inverting and noninverting comparators can both be built which compare the input with either ground or with a reference voltage.

Short-circuit protected op amps usually limit the output current to about twenty milliamps. This is just right to drive a light emitting diode directly. By properly stacking op amps and LEDs, multilevel comparators and displays can be easily built.

Negative feedback is necessary to cause an op amp to operate linearly. This also, to a large degree, removes a circuit's dependence on the op amp's open-loop gain. The closed-loop gain of the circuit can be strictly determined by the characteristics of the negative feedback network (usually fixed, stable, linear resistors).

Inverting, noninverting and difference amplifiers as well as summers can be built with op amps and negative feedback networks. Analysis of each shows that the relationship between input and output voltages is linear. It is set solely by the resistors providing the negative feedback.

There is a wide variety of applications for these linear op amp circuits. They allow you to build circuits which directly implement mathematical equations. Multiplication by a constant is amplification. Addition gives you mixing, DC level shifting, or DC offset and analog switching. Subtraction allows for common mode rejection. All of these can be done for AC signals, DC signals, or combinations. Also, the circuits can present a high impedance to the source (to prevent loading it down), while giving a low output impedance to the load.

One of the most widely spread areas of op amp and analog IC application is in the control of power supply output voltage. Voltage regulators, using these ICs, automatically adjust themselves to provide a fixed output voltage over wide changes in output current demand and input voltage supplied. Much electronics today draw their energy from an IC regulated power supply. The principles, analysis, and design of these regulated power supplies are the topic of Chapter 3.

PROBLEMS

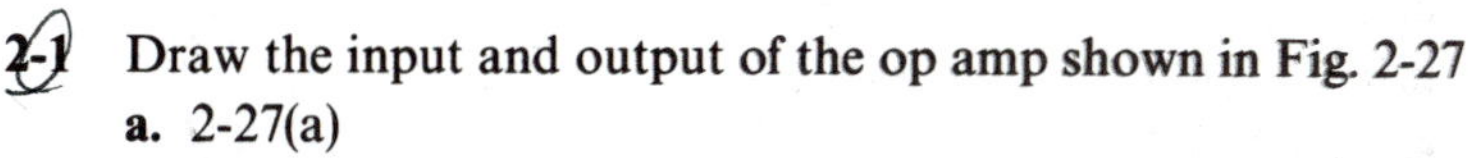

2-1 Draw the input and output of the op amp shown in Fig. 2-27

a. 2-27(a)
b. 2-27(b)
c. 2-27(c)
d. 2-27(d)

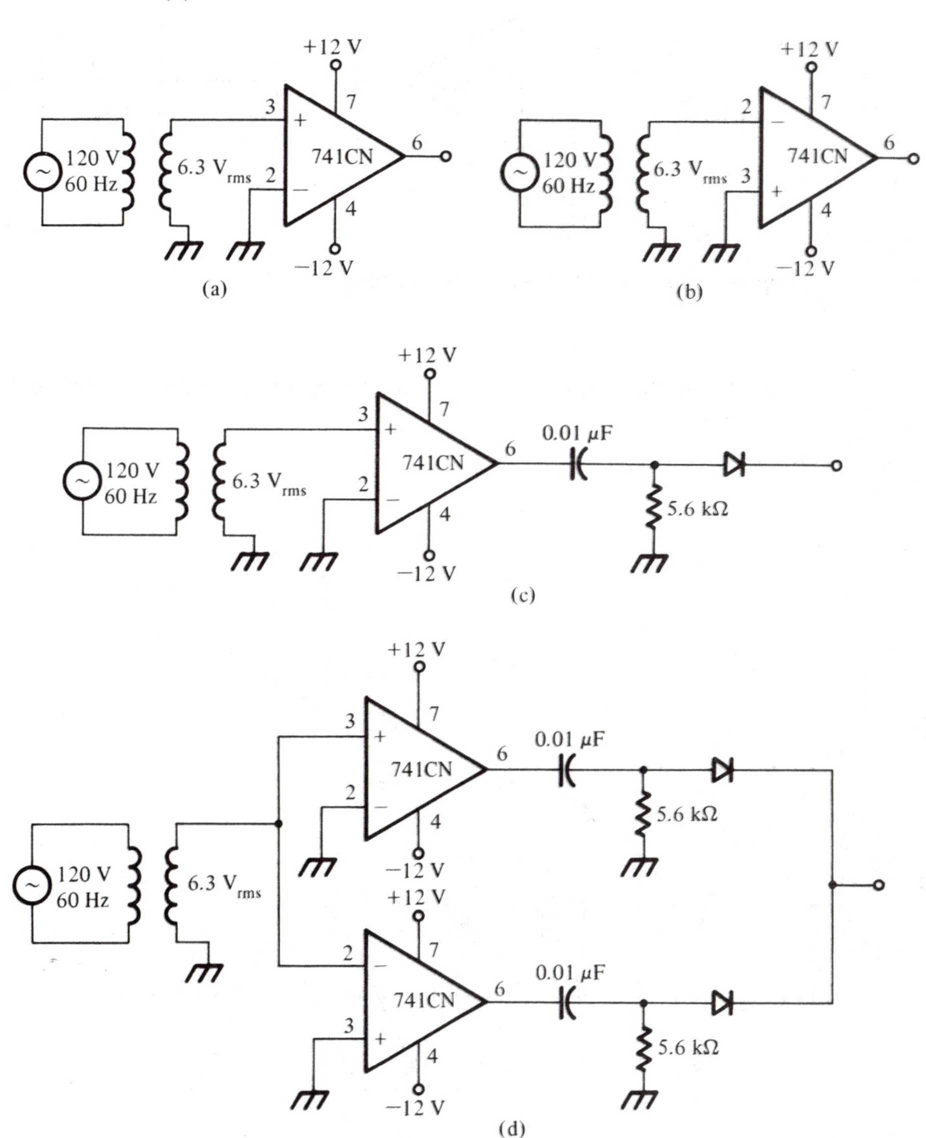

Figure 2-27 Schematics for problem 2-1

2-2 Draw the input and output of the op amp shown in Fig. 2-28 for

a. $e_1 = 0$ V
b. $e_1 = -4$ V
c. $e_1 = +3$ V
d. $e_1 = +6$ V

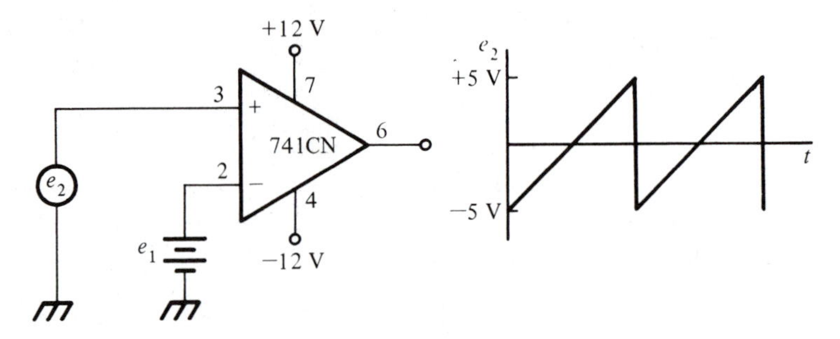

Figure 2-28 Schematic and input for problem 2-2

2-3 Draw the outputs for the three sets of inputs of the op amp shown in Fig. 2-29.

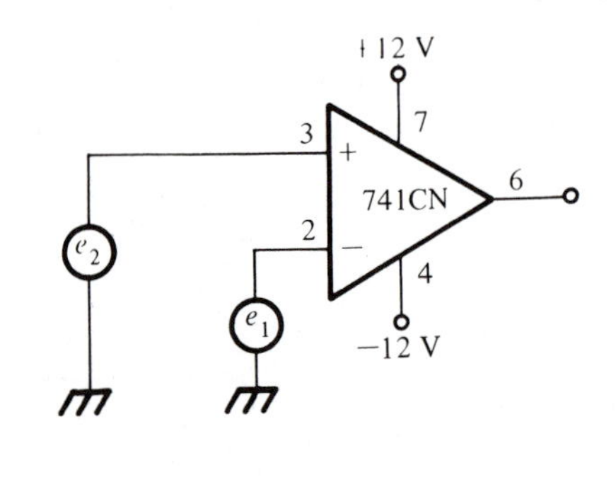

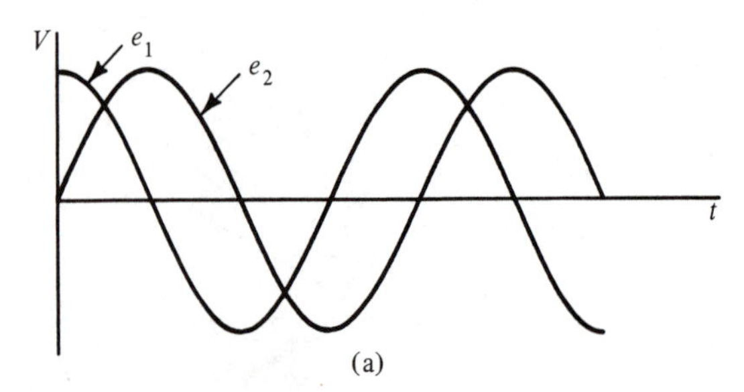

Figure 2-29 Schematic and inputs for problem 2-3

Figure 2-29 *continued*

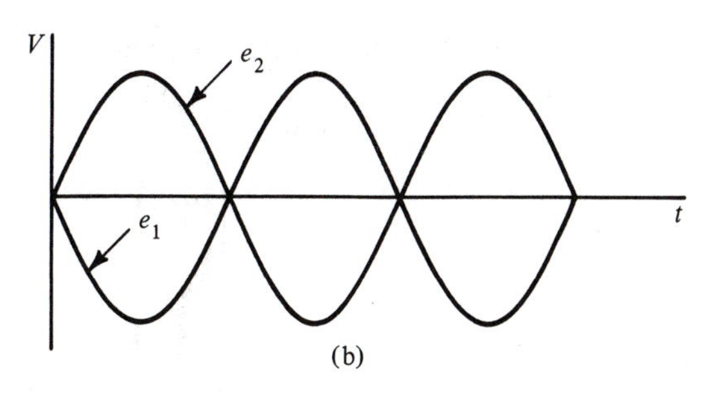

(b)

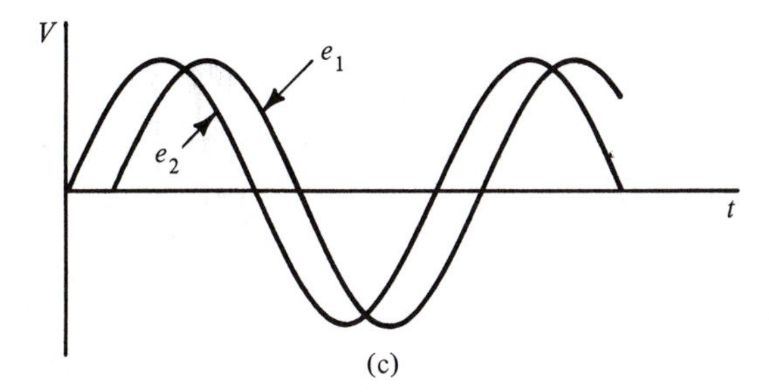

(c)

2-4 For the circuit in Fig. 2-30(a), what is the condition of each of the LEDs for

a. $V_{in} = 1$ V
b. $V_{in} = 2$ V
c. $V_{in} = 3$ V

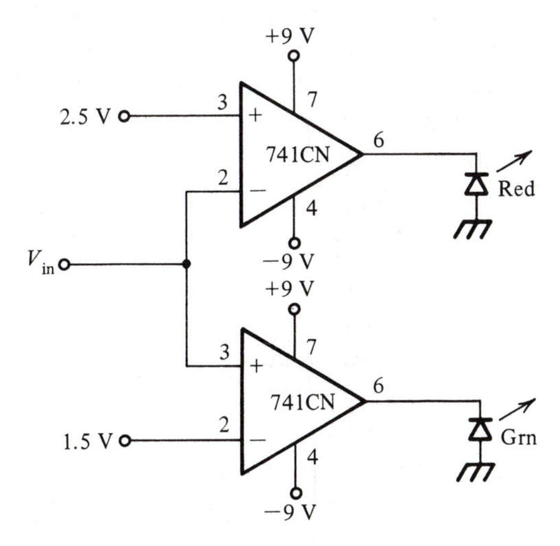

Figure 2-30 a Schematic for problem 2-4

2-5 For the circuit in Fig. 2-30(b), what is the condition of each of the LEDs for
a. $V_{in} = 3$ V
b. $V_{in} = 5$ V
c. $V_{in} = 7$ V

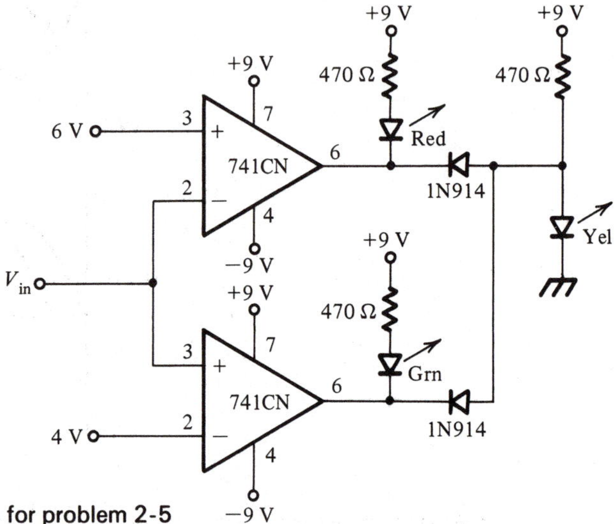

Figure 2-30 b Schematic for problem 2-5

2-6 Design an eight-level comparator (8 LEDs) similar to the three-level comparator discussed. However, set the comparison levels so that the circuit can be used as a VU meter on one channel of a stereo. (A VU meter will require a logarithmic relationship between comparison levels.)

2-7 Describe how negative feedback compensates for a decrease in the open-loop gain of an op amp. (See Figs. 2-11 and 2-12.)

2-8 Calculate the circuit voltage gain and input impedance for the amplifier in Fig. 2-31.

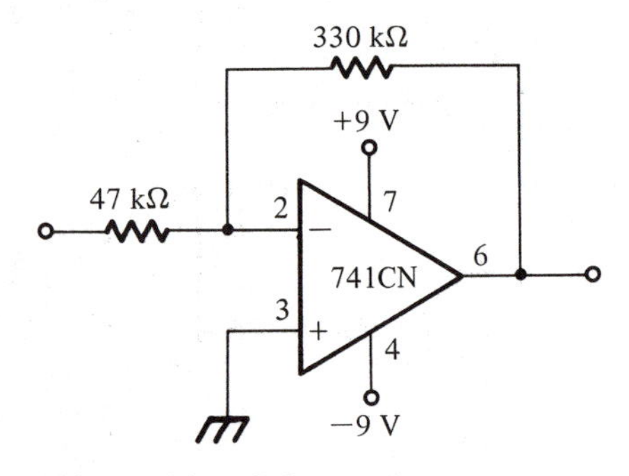

Figure 2-31 Schematic for problem 2-8

2-9 Design an inverting amplifier with an input impedance of precisely 10 kΩ, and a gain which can be adjusted between −10 and −20 (without affecting the input impedance).

2-10 Draw the output of the op amp in Fig. 2-32 for

a. $e_1 = 0.5\ V_{p\text{-}p}$ sine wave $\quad e_2 = +2$ V DC
b. $e_1 = 0.5\ V_{p\text{-}p}$ sine wave $\quad e_2 = -4$ V DC
c. $e_1 = 1\ V_{p\text{-}p}$ sine wave $\quad e_2 = +2$ V DC
d. $e_1 = 1\ V_{p\text{-}p}$ sine wave $\quad e_2 = -4$ V DC

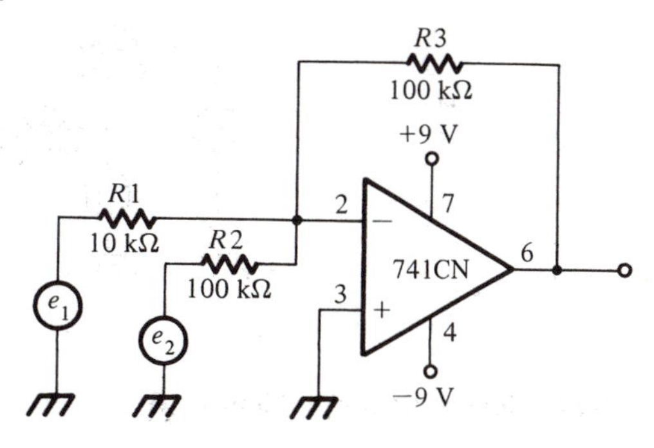

Figure 2-32 Schematic for problem 2-10

2-11 Draw the output of the op amp in Fig. 2-32 if R1 = 100 kΩ, R2 = 1 kΩ, R3 = 100 kΩ.

a. $e_1 = 0.5\ V_{p\text{-}p}$ sine wave $\quad e_2 = 0$ V DC
b. $e_1 = 0.5\ V_{p\text{-}p}$ sine wave $\quad e_2 = +3$ V DC

2-12 Design a signal conditioner using an inverting summer which will convert a 10 millivolt/degree Celsius signal (from a temperature transducer) to a 10 millivolt/degree Fahrenheit output voltage. Both zero and span must be considered.

2-13 What range of gains and what input impedance would you expect from the circuit in Fig. 2-33?

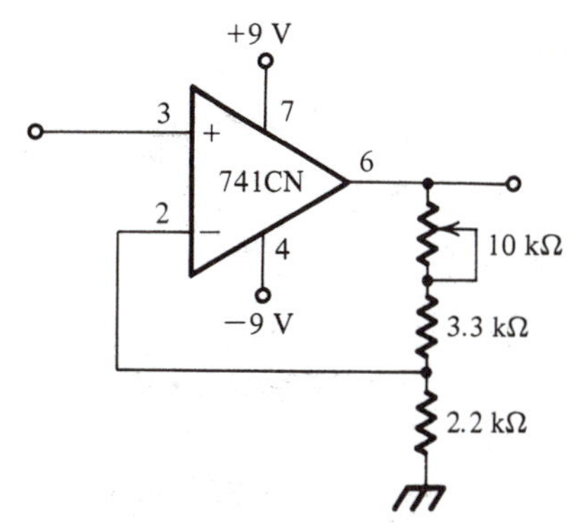

Figure 2-33 Schematic for problem 2-13

2-14 Design an amplifier which will give a gain which can be adjusted from 1.0 to 1.2.

2-15 Draw the output from the op amp in Fig. 2-34 for

a. $e_1 = 0.5\ V_{p\text{-}p}$ sine wave $e_2 = +2$ V DC
b. $e_1 = 0.5\ V_{p\text{-}p}$ sine wave $e_2 = -4$ V DC
c. $e_1 = 1\ V_{p\text{-}p}$ sine wave $e_2 = +2$ V DC
d. $e_1 = 1\ V_{p\text{-}p}$ sine wave $e_2 = -4$ V DC

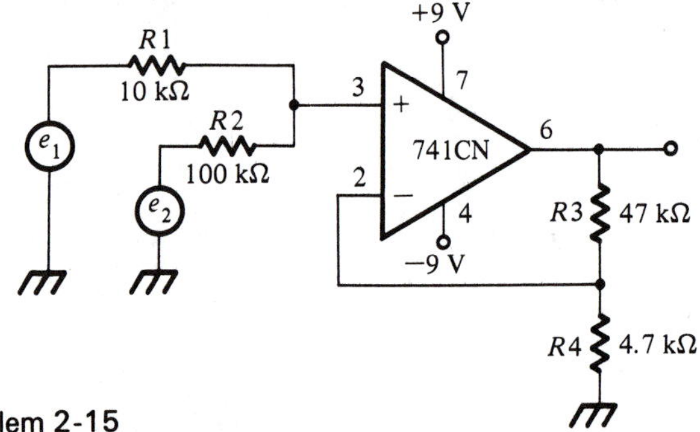

Figure 2-34 Schematic for problem 2-15

2-16 Draw the output from the op amp in Fig. 2-34 for R1 = 100k ohms, R2 = 10k ohms, R3 = 47k ohms, R4 = 4.7k ohms.

a. $e_1 = 1\ V_{p\text{-}p}$ sine wave $e_2 = 0.2$ V DC
b. $e_1 = 0.5\ V_{p\text{-}p}$ sine wave $e_2 = 2$ V DC
c. $e_1 = 1\ V_{p\text{-}p}$ sine wave $e_2 = -2$ V DC

2-17 Design a signal conditioner using a noninverting summer which will convert a 10 millivolt/degree Celsius signal (from a temperature transducer) to a 10 millivolt/degree Fahrenheit output voltage. Both zero and span must be considered.

2-18 What is the output voltage from the circuit in Fig. 2-35 if the input signal is a 7.5 milliamp current?

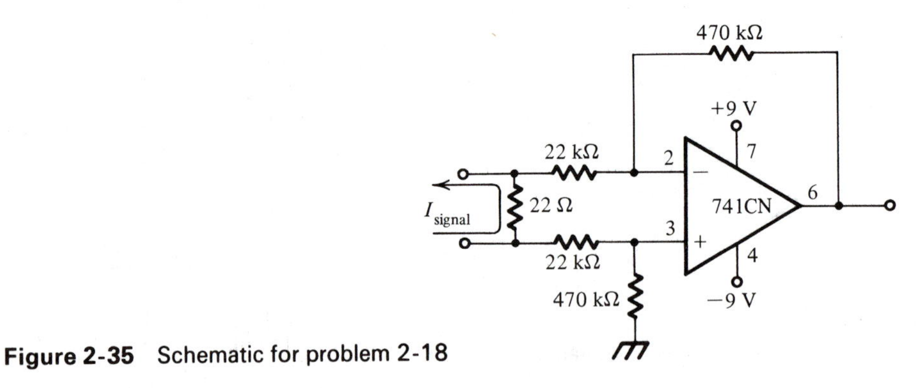

Figure 2-35 Schematic for problem 2-18

2-19 For the circuit shown in Fig. 2-36, the span potentiometer has been set to 41 kΩ. The zero potentiometer has been set to give a voltage at point B of 1.67 volts.

Calculate the voltage across the 22 Ω current sense resistor, the voltage at point A and the voltage at the output for

a. $I = 5$ milliamps
b. $I = 20$ milliamps

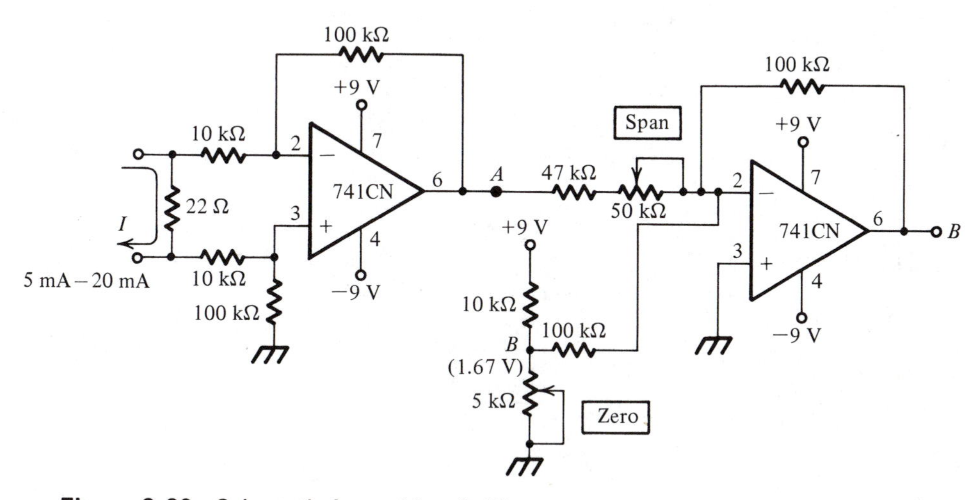

Figure 2-36 Schematic for problem 2-19

2-20 What is the voltage at point A and point B for the circuit in Fig. 2-37 if

a. $e_1 = 6$ V $e_2 = 6.2$ V
b. $e_1 = 5$ V $e_2 = 5.1$ V
c. $e_1 = 4$ V $e_2 = 4.25$ V
d. $e_1 = 3$ V $e_2 = 3.3$ V

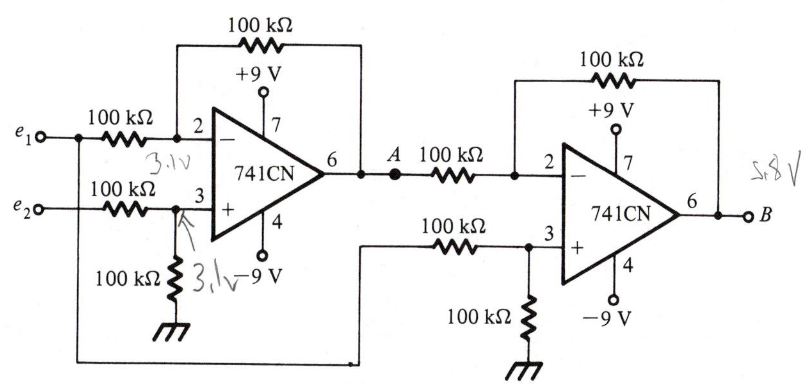

Figure 2-37 Schematic for problem 2-20

2-21 For the circuit in Fig. 2-38, calculate the voltages at A, B, upper limit and lower limit for

a. Input = 6 V Limit adjust = 940 Ω
b. Input = 6 V Limit adjust = 470 Ω
c. Input = 3 V Limit adjust = 470 Ω
d. Input = 5 V Limit adjust = 470 Ω
e. Describe the purpose of this circuit (i.e., the relationship among input, limit adjust, upper limit and lower limit).

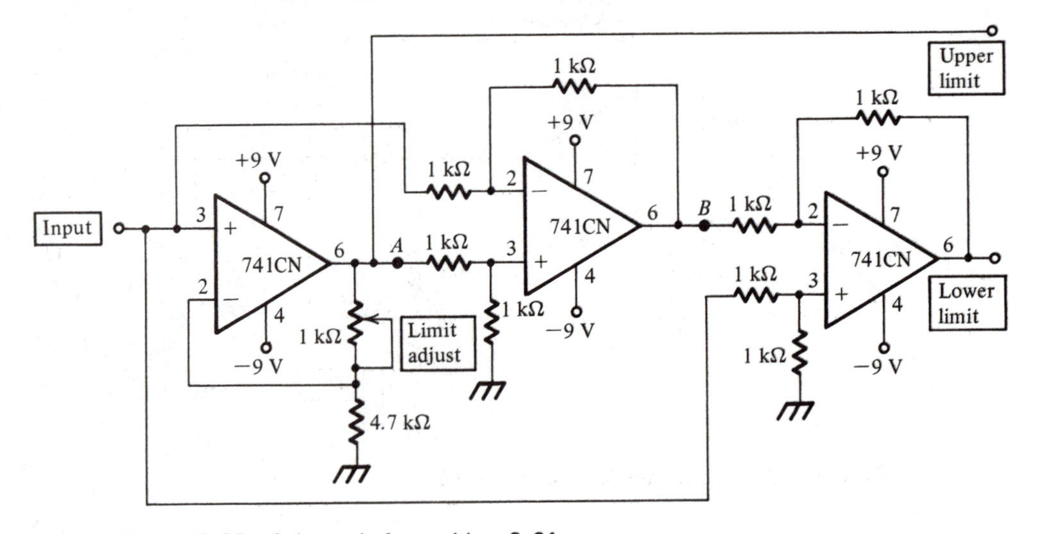

Figure 2-38 Schematic for problem 2-21

2-22 Derive the equation for the output voltage (e_0) for the circuit in Fig. 2-39. Careful! This is not trivial.

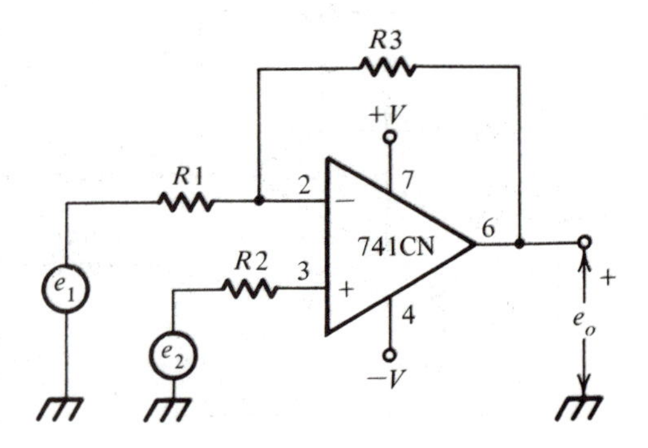

Figure 2-39 Schematic for problem 2-22

2-23 Use the equation derived in Problem 2-22 to simplify the circuit in Fig. 2-38 (Problem 2-21).

2-24 Design a circuit which will give the following output:

$$e_0 = A_3 e_3 - A_2 e_2 - A_1 e_1$$

where

e_0 is the output voltage.

e_1, e_2, and e_3 are input voltages.

A_1, A_2, and A_3 are gains.

CHAPTER 3

Power Supplies and Integrated Circuit Regulators

The single requirement common to all phases of electronics is the need for a supply of DC power. With increases in the specifications, performance, and level of complexity of the circuits in virtually every field of electronics has come more demand on the power supplies. It is unusual indeed to find a system today operating from an unregulated power supply. To meet the demands for stable, protected voltages over a wide range of load current demands, input voltages, and temperatures, analog integrated circuits have been widely incorporated into regulated power supplies. In this chapter you will find the characteristics, schematics, analysis, design, and limitations of practical power supplies with integrated circuit regulators.

OBJECTIVES

After studying this chapter, you should be able to do the following:

1. Draw the schematic and explain the operation of a power supply using an op amp-based regulator.
2. Draw the schematic of a power supply, using a three-terminal regulator.
3. Given the required performance specifications, calculate all required component values needed to build a three-terminal regulated power supply.
4. Draw the schematics and describe the operation and precautions of each of the following variations on the basic three-terminal regulator:
    a. Dual power supplies
    b. Adjustable output voltage
    c. High load current
5. Analyze and design a regulated power supply using a 723 analog IC in each of the following configurations:
    a. Low voltage positive regulator

b. High voltage positive regulator
c. Simple negative regulator
d. Current limited regulator
e. Current foldback regulator
f. Current boosted regulator

6. Explain the difference between a dual and a dual tracking regulated power supply.
7. Draw the schematic and explain the operation of a fixed and an adjustable dual tracking regulated power supply.
8. Draw the schematic and explain the operation of a switching regulator.
9. Discuss the advantages and limitations of switching regulators.

3-1 FULL WAVE RECTIFIED, CAPACITIVE FILTERED POWER SUPPLIES

The purpose of a DC power supply is to convert the widely available AC power to DC, normally for bias power. This is done in several stages, as shown in Fig. 3-1.

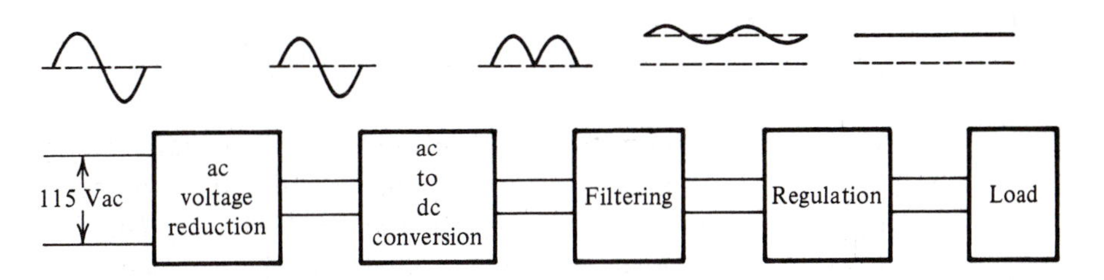

Figure 3-1 DC power supply block diagram

The first block, AC voltage reduction, is often handled with a step-down power transformer. Here the 117 volts rms AC input is reduced to potentials in the range of 1.5 volts to 28 volts. The AC to DC conversion is most efficiently handled by a full wave bridge rectifier. There is a wide variety of epoxy bridges available, providing all four diodes already interwired in a single, easy-to-use, inexpensive package. The output of the bridge is indeed DC. Current always flows in the same direction. However, the output voltage from the rectifier varies drastically in every cycle. To smooth out this variation a filter is needed. Although many complex filter configurations can be built, a simple, single, shunt capacitor works very nicely for most applications (especially when followed by an IC regulator). The characteristics of these first three blocks of a regulated power supply will be discussed in this section. You will see the relationships between circuit performance and load current under light and under heavy load.

3-1.1 Light Load Capacitive Filtering

The schematic for the first three blocks of a power supply is shown in Fig. 3-2. The typical waveforms out of each block are given in Fig. 3-3. The signal out of the filter will become the input to the regulator. Figure 3-4 illustrates the filter output in more detail.

As the incoming voltage from the rectifier rises, the capacitor charges to the peak value of that voltage.

$$V_{pk} = V_{sec_{rms}} \times 1.414 - 1.4V \tag{3-1}$$

The $-1.4V$ term accounts for losses across the diodes. When the rectified signal passes peak and begins to fall, the capacitor discharges slowly through the load. This continues until the incoming wave begins to recharge the capacitor. The minimum level to which the capacitor's voltage falls is called V_{min} and depends on the size of the capacitor, V_{pk}, and the load current. The distance between the peak and the minimum output voltages is the peak-to-peak ripple voltage:

$$V_{ripple}\,(\text{p-p}) = V_{pk} - V_{min} \tag{3-2}$$

The average value of the capacitor's output voltage lies somewhere between V_{pk} and V_{min}. This is the DC value of voltage passed to the regulator.

Light load conditions will cause the *ripple voltage* to be *less than ten percent of the output DC voltage.* Under this condition, the capacitor discharges only slightly, and almost linearly. This approximation greatly simplifies the analysis of the circuit. For *light loads,*

$$V_{min} = V_{pk} - \frac{I_{DC}}{fC} \tag{3-3}$$

where I_{DC} is the load current.

f is 120 Hz for a full wave rectifier.

C is the filter's capacitance.

The DC value of the output from the filter lies halfway between V_{pk} and V_{min} or

$$V_{DC} = V_{pk} - \tfrac{1}{2}V_{ripple}\,(\text{p-p}) \tag{3-4}$$

With this information, you can now analyze the operation of a capacitive filtered power supply under *light load.*

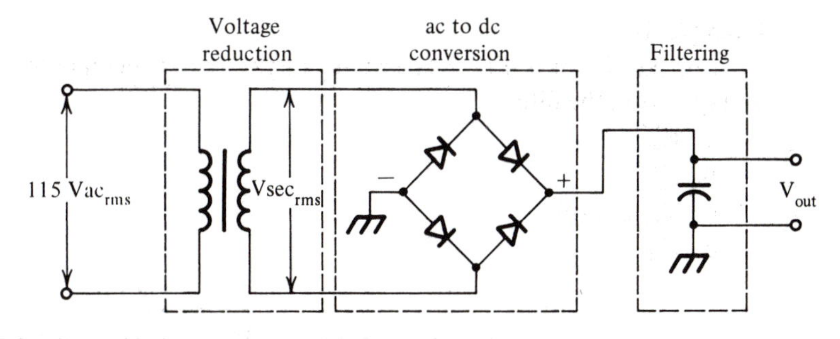

Figure 3-2 DC power supply schematic

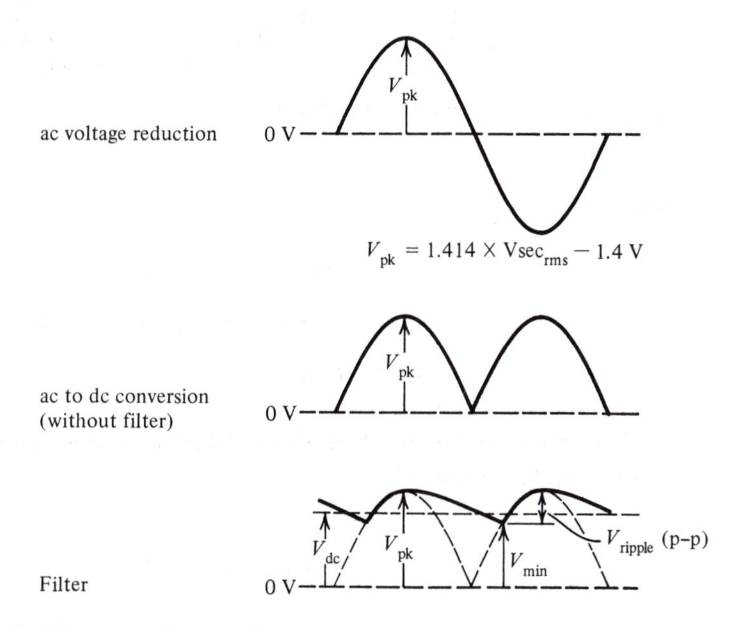

Figure 3-3 Power supply wave forms

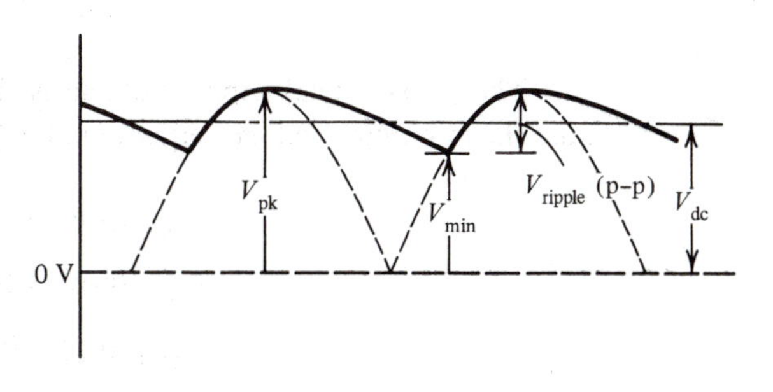

Figure 3-4 Capacitive filter output

Example 3-1

For the circuit in Fig. 3-5, assume a light load. Calculate the following output voltages from the filter:

a. V_{pk}
b. V_{min}
c. V_{ripple} (p-p)
d. V_{DC}

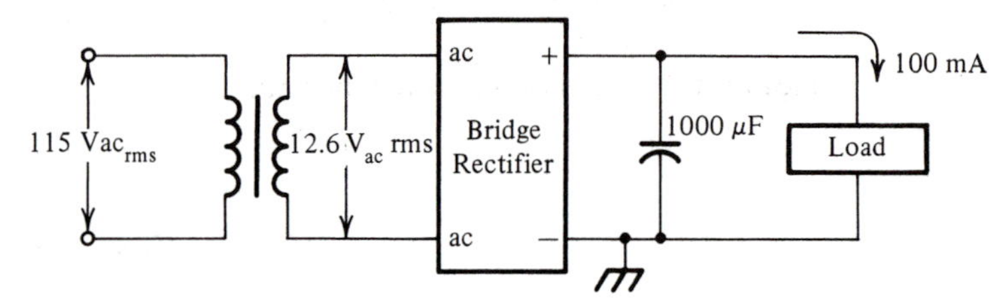

Figure 3-5 Schematic for Example 3-1

Solution

$$V_{pk} = 1.414 \times V_{sec_{rms}} - 1.4\text{V}$$

$$V_{pk} = 1.414 \times 12.6\text{V AC rms} - 1.4\text{V}$$

$$V_{pk} = 16.4\text{V}$$

The minimum voltage depends on the peak voltage, the load current, and the capacitance (equation 3-3).

$$V_{min} = V_{pk} - \frac{I_{DC}}{fC}$$

$$V_{min} = 16.4\text{V} - \frac{0.1\text{ A}}{(120\text{ Hz})\ (1000\ \mu\text{f})}$$

$$V_{min} = 15.6\text{V}$$

The peak-to-peak ripple voltage is the difference between the peak voltage and the minimum voltage [equation (3-2)].

$$V_{ripple}\ (\text{p-p}) = V_{pk} - V_{min}$$

$$V_{ripple}\ (\text{p-p}) = 16.4\text{V} - 15.6\text{V}$$

$$V_{ripple}\ (\text{p-p}) = 0.8\text{ V}$$

The average output voltage, the DC value of the output, lies halfway between the peak and the minimum [equation (3-4)].

$$V_{DC} = V_{pk} - \tfrac{1}{2}V_{ripple}\,(\text{p-p})$$

$$V_{DC} = 16.4\ \text{V} - \tfrac{1}{2}(0.8\ \text{V})$$

$$V_{DC} = 16.0\ \text{V}$$

The output waveform is drawn in Fig. 3-6.

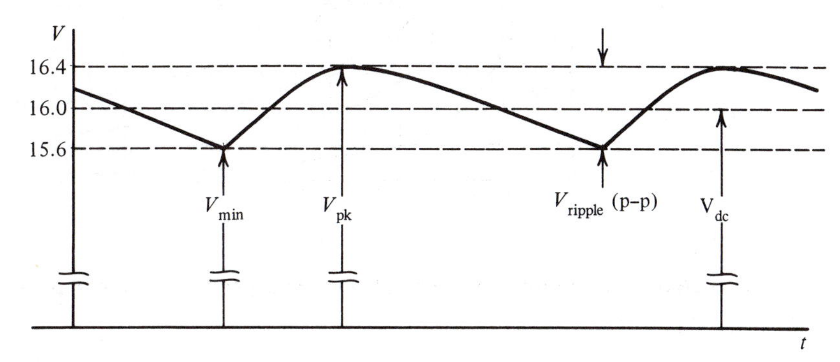

Figure 3-6 Output from the filter in Example 3-1

3-1.2 Heavy Load Capacitive Filtering

To continue operating in a light load condition as load currents increase, capacitance must be increased and the regulator IC (to be added) will have to dissipate more power. This results in power supplies that are bulkier and more expensive than necessary. However, if you allow the ripple voltage to increase beyond ten percent, the capacitor can be made smaller, and the required IC heat dissipation falls. Unfortunately, for such high ripple conditions, equations (3-3) (for V_{min}) and (3-4) (for V_{DC}) are no longer valid.

The analysis for the capacitive filter power supply under high ripple (heavy load) requires the solution of an equation with an exponential function on one side, and a sinusoidal function on the other. Instead of such complex transcendental equations or extensive computer approximations, Figs. 3-7 through 3-10 are plots of V_{min} and V_{DC} from values measured in the lab. Example 3-2 illustrates their use.

Example 3-2

For the circuit in Fig. 3-5, increase the load current to 1A; then determine:

a. V_{pk} **b.** V_{min} **c.** V_{ripple} (p-p) **d.** V_{DC}

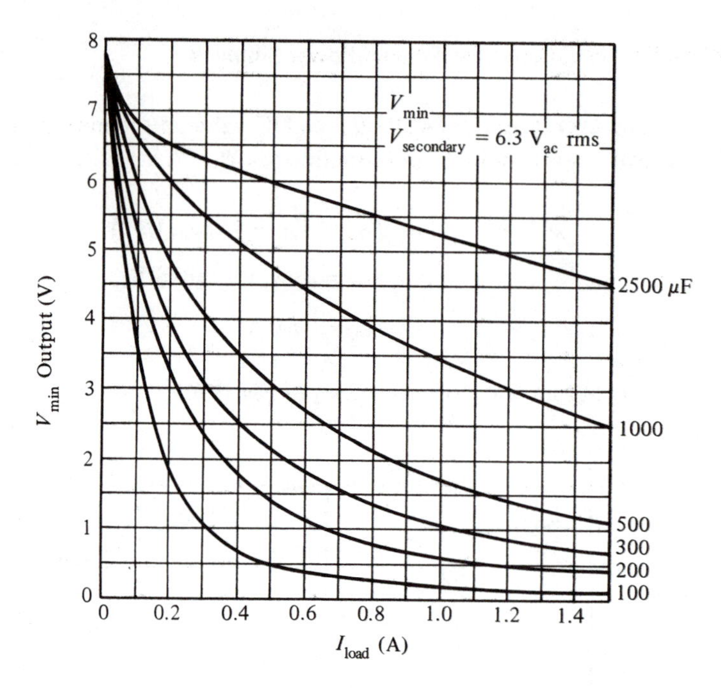

Figure 3-7 Minimum output of a 6.3V rms capacitive filtered supply

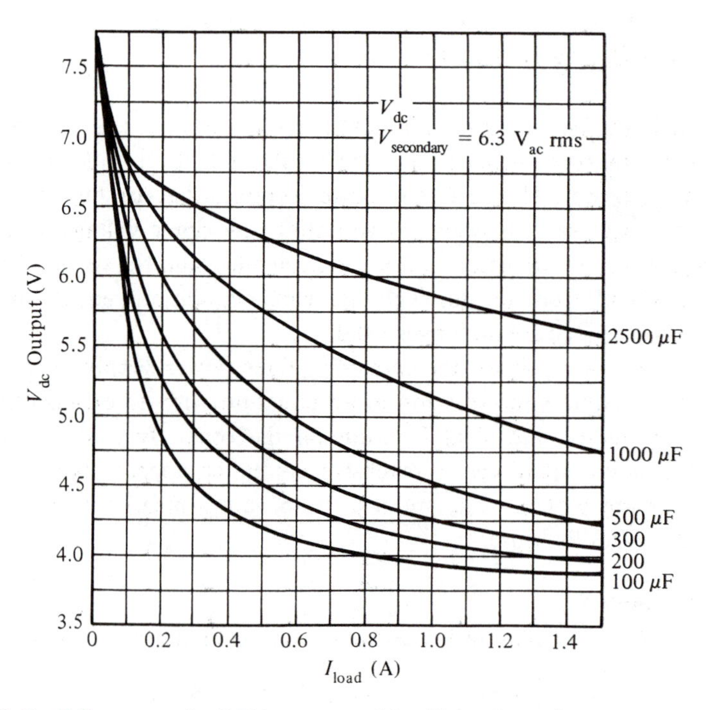

Figure 3-8 DC output of a 6.3V rms capacitive filtered supply

Figure 3-9 a Minimum output of a 12.6V rms capacitive filtered supply

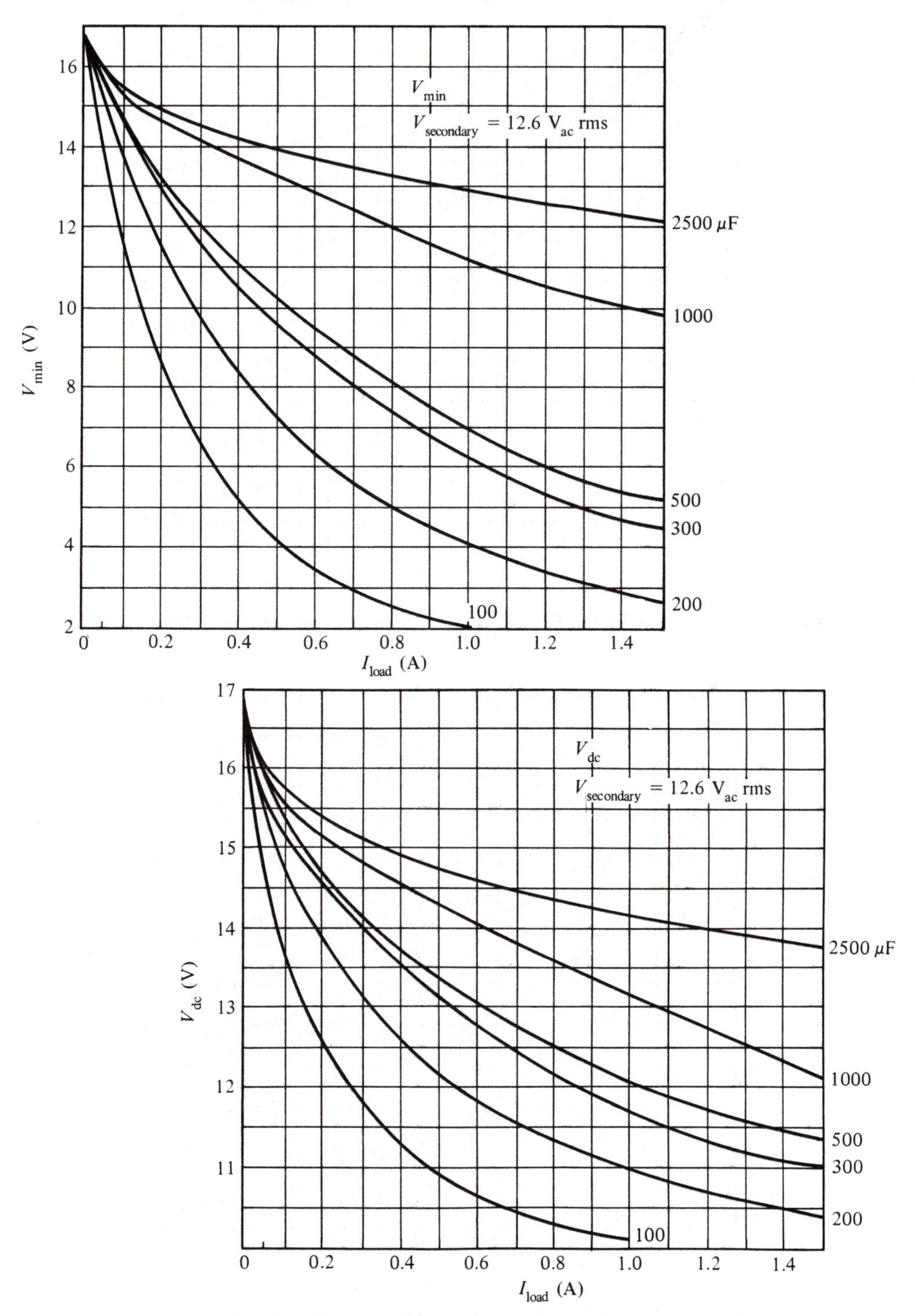

Figure 3-9 b DC output of a 12.6V rms capacitive filtered supply

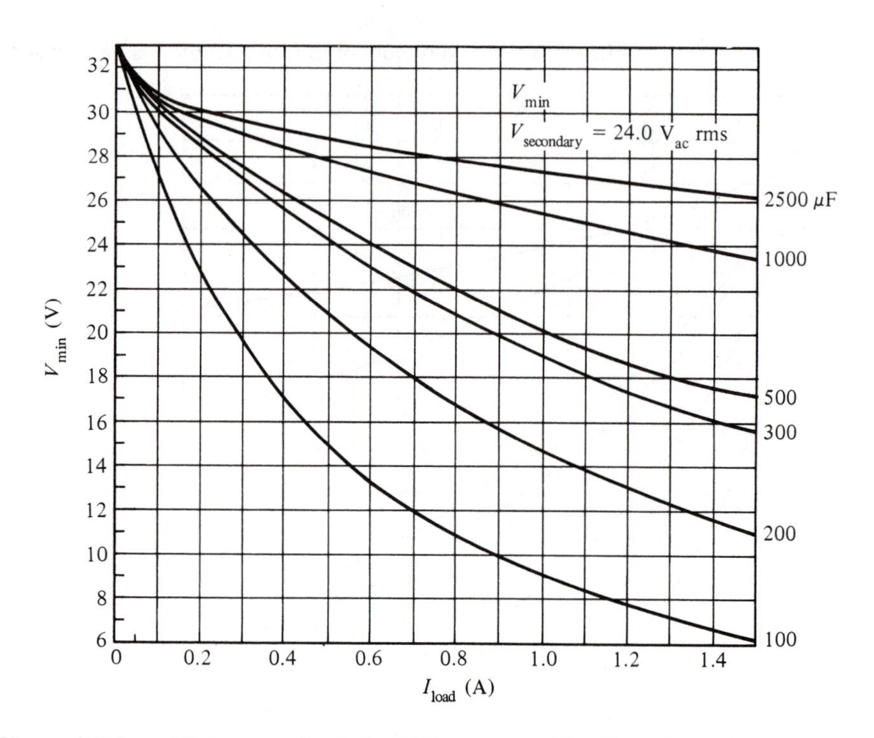

Figure 3-10 a Minimum output of a 24V rms capacitive filtered supply

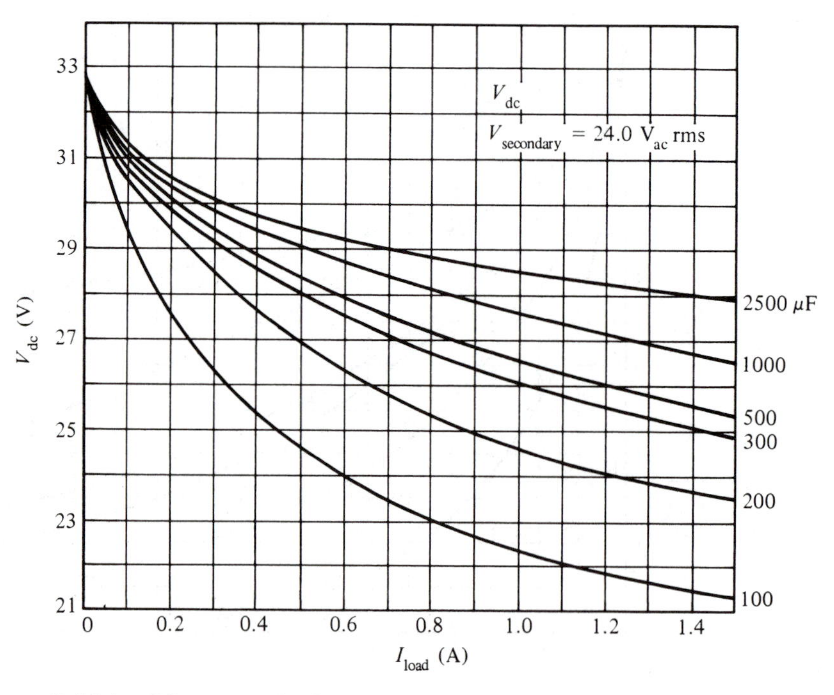

Figure 3-10 b DC output of a 24V rms capacitive filtered supply

Solution

The peak voltage depends on the transformer secondary. So, as with Example 3-1,

$$V_{pk} = 1.414 \times V_{sec_{rms}} - 1.4\ \text{V}$$

$$V_{pk} = 16.4\ \text{V}$$

The minimum voltage is found from Fig. 3-9a. Move vertically from the 1 A point on the horizontal axis until you intersect the 1000 μF line. Project this point horizontally to intersect the vertical axis:

$$V_{min} = 11.2\ \text{V}$$

$$V_{ripple}\ (\text{p-p}) = V_{pk} - V_{min}$$

$$V_{ripple}\ (\text{p-p}) = 16.4\ \text{V} - 11.2\ \text{V}$$

$$V_{ripple}\ (\text{p-p}) = 5.2\ \text{V}$$

The DC value of the output voltage can be read from Fig. 3-9b (just as V_{min} was read from fig. 3-9a):

$$V_{DC} = 13.3\ \text{V}$$

Had you used the light load equations by mistake, you would have calculated:

	Correct	*Light Load Equations*
V_{pk}	16.4 V	16.4 V
V_{min}	11.2 V	8.1 V
V_{ripple} (p-p)	5.2 V	8.3 V
V_{DC}	13.3 V	12.3 V

The surest and easiest way to know what a capacitive filtered power supply will do under heavy load is to measure the circuit's performance in the lab. Figures 3-7 through 3-10 were developed to cover the more popular size capacitors and transformers. However, if your application is not covered, carefully build your own curve. Current I_{load} was measured with a good analog ammeter, V_{min} with a DC coupled oscilloscope, and V_{DC} with a DVM on the DC scale.

3-2 SIMPLE OP AMP REGULATOR

You can see from Examples 3-1 and 3-2 that the output DC voltage from a simple capacitive filtered power supply varies quite a bit as you change the current going to the load. In most applications this could cause major problems. The power supply must maintain a constant output voltage, no matter how much load current is required. This calls for a regulator.

The simplest regulator consists of a series resistor and a zener diode in parallel with the load. This is shown in Fig. 3-11. When biased into the reverse breakdown region, the voltage across the zener diode (and therefore across the load) varies only a small amount even with relatively large changes in zener (and load) current. For simple applications this may be adequate, but often the change in zener voltage (and therefore load voltage) is too large. Better regulation is required.

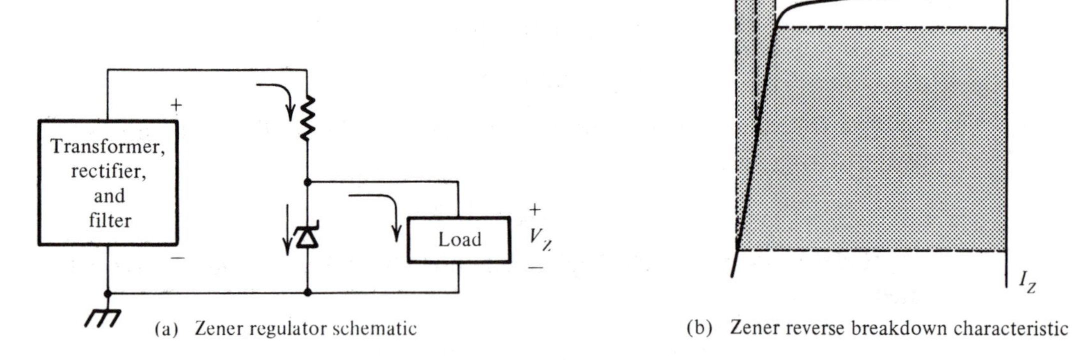

Figure 3-11 Zener regulator and characteristic

The circuit in Fig. 3-12 is the next step in better regulation. A voltage follower op amp circuit has been placed between the zener (reference) diode and the load. Now, the op amp provides the load current. The zener does not have to absorb wide variations in load current. This restricts I_z to a much smaller range, significantly lowering variations in V_z. Since the op amp is working as a voltage follower,

$$V_{out} = V_{in}$$

or

$$V_{load} = V_z$$

Two words of caution about the input voltage. First, its peak value must be kept below the maximum supply voltage for the op amp. (Op amps with rather high supply voltage ratings are available.) Second, the voltage from the filter must

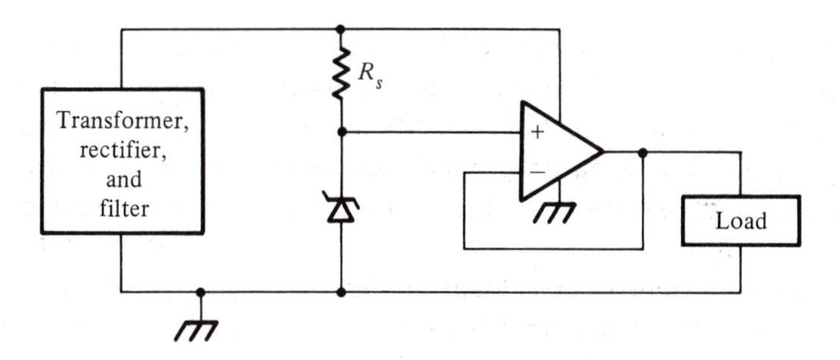

Figure 3-12 Voltage follower regulator

always be at least two volts above the output (load) voltage. This is necessary to keep the op amp out of saturation and to keep the zener diode in its breakdown region.

$$V_{\text{pk}} \leq \pm V_{\text{supply rating of op amp}} \tag{3-5}$$

$$V_{\text{min}} \geq V_{\text{load}} + 2\text{ V} \tag{3-6}$$

To obtain an adjustable, regulated output voltage, you only have to add two resistors, as shown in Fig. 3-13. The voltage follower of Fig. 3-12 has been replaced with a noninverting amplifier (with adjustable gain). The load or output voltage is

$$V_{\text{load}} = V_z\left(1 + \frac{R_f}{R_i}\right) \tag{3-7}$$

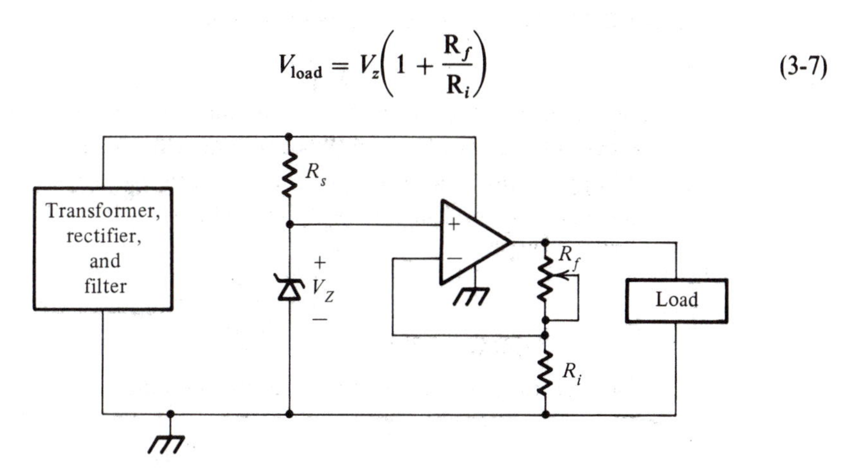

Figure 3-13 Adjustable output regulator

Equation (3-7) comes directly from the gain equation for a noninverting amplifier. The cautions expressed by equations (3-5) and (3-6) must be observed.

Also, notice that the minimum output voltage is V_z (when R_f is shorted). So be sure to pick a zener voltage smaller than the lowest output you want. Finally, be careful to make R_f the potentiometer. Making R_i a potentiometer and running the wiper all the way down (to ground) will remove all negative feedback. The output voltage will be driven to $+V_{SAT}$ (since the op amp would be operating as a comparator).

The output current (to the load) in Figs. 3-12 and 3-13 is limited by the short-circuit limit of the op amp. Unless you can find (and afford) a special power op amp, you will have to boost the load current with a pass transistor. This is shown in Fig. 3-14.

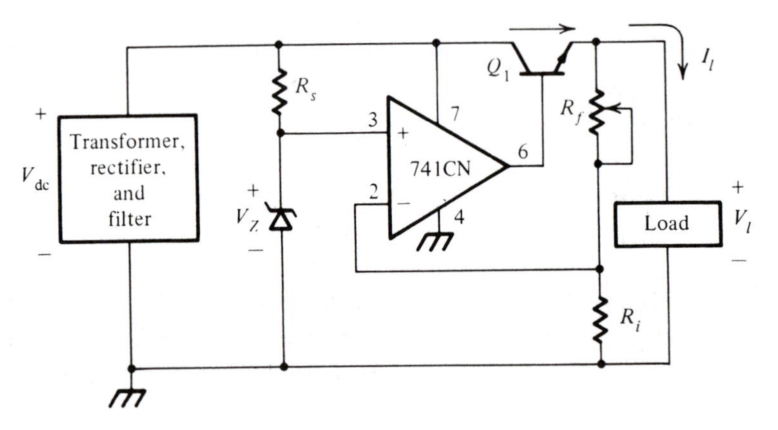

Figure 3-14 Pass transistor boosted regulator

The op amp now only has to provide the base current. This is then multiplied by the beta of the transistor to produce load current. Equations (3-5), (3-6), and (3-7) all apply. Also, the transistor must be able to provide adequate current to the load. This is determined, in part, by the beta:

$$\beta \geq \frac{I_{\text{load}}}{I_{\text{op amp max}}} \tag{3-8}$$

Also, the transistor must be able to dissipate the power produced:

$$P_Q = (V_{DC} - V_L)I_L \tag{3-9}$$

where V_{DC} is the DC output voltage from the filter.

V_L is the load DC voltage.

I_L is the load DC current.

The schematic in Fig. 3-15 will provide a bit better regulation than in Fig. 3-14. The only change is in driving the zener diode from the regulated (load) voltage rather than the unregulated (filter output) voltage. This further decreases the variation of I_z, allowing V_z and, therefore, V_{load} to vary less.

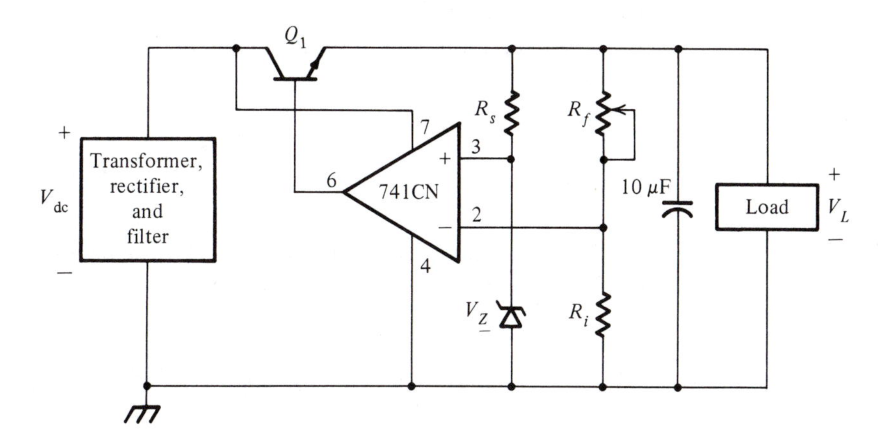

Figure 3-15 Improved current boosted regulator

Example 3-3

Design a regulated power supply to meet the following specifications. Use a 24 V rms transformer, a full wave bridge rectifier, and a 1000 μF filter capacitor.

$$V_{load} : \text{adjustable 5 V to 15 V}$$

$$I_{load} : \text{0 milliamps to 500 milliamps}$$

Solution

1. Filter output characteristics

$$V_{pk} = 1.414 \times 24 \text{ V rms} - 1.4 \text{ V} = 32.5 \text{ V}$$

$$V_{min} = 27.8 \text{ V} \quad \text{(Fig. 3-10a)}$$

$$V_{DC} = 29 \text{ V} \quad \text{(Fig. 3-10b)}$$

$$V_{ripple}(\text{p-p}) = V_{pk} - V_{min} = 4.7 \text{ V}$$

2. Regulator design
   a. Since the zener diode determines the minimum output voltage, select an 1N 748. It has a zener voltage of 3.9 volts at 20 milliamps. This will

allow adjustment of R_f to bring the output voltage below the minimum 5 volts specified.

The resistor R_s sets the current into the zener. It must be small enough to allow adequate current to keep the zener in its breakdown region, but large enough to keep the diode current below the rated maximum of 100 milliamps.

When the input from the filter is at V_{min}, minimum zener current will flow.

$$R_s \leq \frac{27.8 \text{ V} - 3.9 \text{ V}}{20 \text{ mA}} = 1.2 \text{ k}\Omega$$

When the input from the filter is at V_{pk}, maximum zener current will flow. This must be limited to less than 100 milliamps.

$$R_s \geq \frac{32.5 \text{ V} - 3.9 \text{ V}}{100 \text{ mA}} = 286 \ \Omega$$

Since $286 \ \Omega < R_s < 1.2 \text{ k}\Omega$, pick $R_s = 1 \text{ k}\Omega$

b. Op amp selection

The peak voltage from the filter is

$$V_{pk} = 32.5 \text{ V}$$

which must be less than the op amp's maximum supply voltage. For the 741C the maximum supply voltage is 36 volts, so the 741C will be used.

$$V_{min} = 27.8 \text{ V} > V_{load} + 2 \text{ V}$$

Equation (3-6) is satisfied.

c. Gain design

With an input of 3.9 volts from the zener diode, and a maximum output of 15 volts specified, the maximum gain is

$$\frac{15 \text{ volts}}{3.9 \text{ volts}} = 1 + \frac{R_f}{R_i}$$

Solving for R_f, you obtain

$$R_f = (3.85 - 1) \times R_i$$

Picking $R_i = 3.3\ k\Omega$ gives $R_f = 9.4\ k\Omega$,

so you can use a 10 kΩ potentiometer for R_f.

d. Transistor specifications

The 741C op amp will output at least ten milliamps before it begins to limit current. This is the transistor's base current. It must be able to output 500 milliamps:

$$\beta_{\text{minimum}} = \frac{500\ \text{mA}}{10\ \text{mA}} = 50$$

The worst case of power dissipation for the transistor is when the output voltage has been dropped to its minimum level (because this puts most of the voltage across the transistor) and maximum load current is being demanded:

$$P_Q = (V_{DC} - V_{\text{load}}) \times I_{\text{load}} = (29\ \text{V} - 3.9\ \text{V}) \times 0.5\ \text{A}$$

$$P_Q = 12.6\ \text{W}$$

Select a NPN silicon transistor that has a beta of 50 or more, and a power dissipating ability of 15 W. The complete power supply schematic is shown in Fig. 3-16.

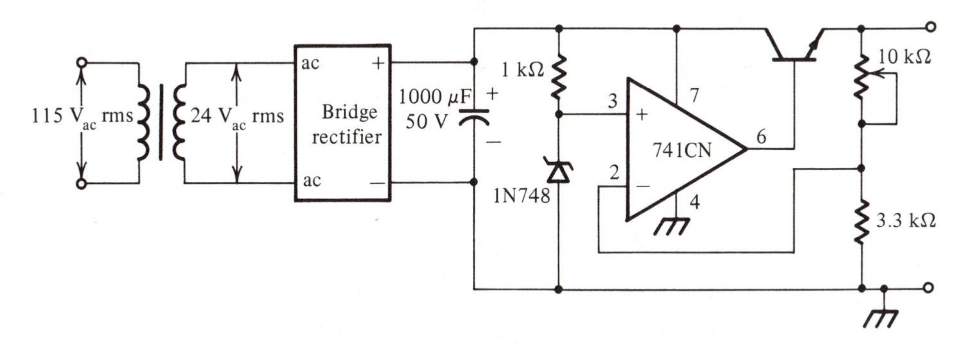

Figure 3-16 Schematic designed in Example 3-3

3-3 THREE-TERMINAL REGULATORS

There are many integrated circuit voltage regulators available. Because of its simplicity and easy application the three-terminal regulator is very popular. The schematic of the three-terminal voltage regulator is shown in Fig. 3-17.

Figure 3-17 A three terminal regulator IC schematic symbol

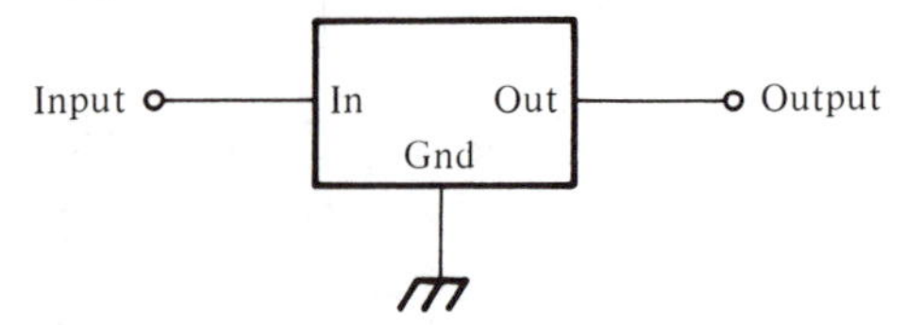

These units are designed to output a fixed voltage at currents normally at or below 5 A. The ICs find major application as on-card regulators. This means that an unregulated power supply voltage is provided to each printed circuit board in a system, and a three-terminal regulator on each printed circuit board regulates the voltage for the circuits on that card. This distribution of regulators throughout the equipment has several advantages over a single large centralized voltage regulator. Several three-terminal regulators each outputting 1 A are cheaper to build, or buy, than one regulator outputting several amps. The current from a centralized regulator must flow through a significant amount of series resistance and inductance to reach the board which will use it. This will seriously affect the value of the load voltage (as illustrated in Fig. 3-18).

$$V_{load} = 10\text{ V} - I_L \times 1\ \Omega$$

$$I_L = 1\text{ A} \qquad I_L = 5\text{ A}$$

$$V_{load} = 9\text{ V} \qquad V_{load} = 5\text{ V}$$

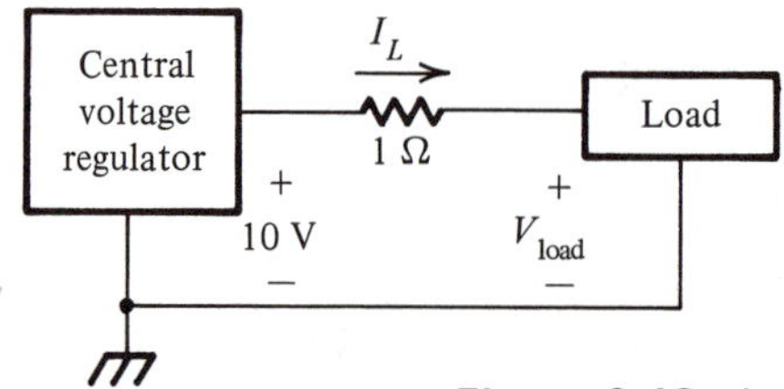

Figure 3-18 Loading of high current central voltage regulator

Also, any change in one load, connected to the centralized voltage regulator, will affect the input voltage to several other physically close loads. This is referred to as *coupling* and is highly undesirable. The on-card, three-terminal regulator eliminates all three of these problems.

3-3.1 Characteristics

There are four characteristics of the three-terminal regulators which must be outlined.

1. V_{out} Fixed

   The regulated output voltage for a given three-terminal regulator is fixed at the value specified by the manufacturer for the particular model you are using. Should a different regulated output voltage be required, a different model regulator must be used.

2. $|V_{in}| \geq |V_{out}| + 2\text{ V}$ (3-10)

   The unregulated input voltage must be at least two volts larger than the regulated output voltage. Examples:

   $V_{out} = 5\text{ V} \qquad V_{in} = 7\text{ V}$

   $V_{out} = -5\text{ V} \qquad V_{in} = -7\text{ V}$

3. $I_{out\ max}$

   The output or load current may vary anywhere from zero up to the rated maximum output. However, if adequate provisions are not made to ensure the removal of heat from the package, the unit may go into thermal shutdown. At what output current thermal shutdown occurs depends on the input voltage, output voltage, ambient temperature, and output current. **Unless adequate heat sinking is provided, the IC may not provide rated maximum output current.**

4. Thermal shutdown

   The IC has a temperature sensor built in. When the chip becomes too hot (usually 125°C to 150°C) the unit will turn off. The output current will drop and remain there until the IC has cooled significantly. Although this does not provide a true short-circuit limit capability, the three-terminal regulator can, if adequately heat sunk, withstand a sustained short circuit without damage.

Table 3-1 provides a summary of the important parameters of the major three-terminal models.

The first column is **Output Current**. This is the absolute maximum that the IC will output. As stated before, I_{out} may never be reached if proper heat sinking is not provided.

The second column lists the **device** number. Often there are three series listed for the same model. Note 1 explains that series 100 devices can withstand a wider ambient temperature range than series 200 devices. Series 300 devices, those normally found in commercial applications, are guaranteed to perform with ambient temperature ranging only from 0°C to 70°C.

TABLE 3-1a. Data sheet summary

Output Current	Device[1,2]	VOUT (V)	TA = 25°C (± %)	Max Regulation Line[3]	(%VOUT/V) Load[4]	Max VIN (V)	Ripple (dB)[6]	Typ Dropout Voltage (V)	Device	Pkg Style	Typ θJC (°C/W)	Typ θJA (°C/W)	Max PD (W)
5.0	LM138, LM238	1.2-32 (adj)	N/A	0.005	0.1	35	86	2	LM138K STEEL series	TO-3	2	35	30
	LM338	1.2-32 (adj)	N/A	0.005	0.1	35	86	2					
3.0	LM150, LM250	1.2-32 (adj)	N/A	0.005	0.1	35	86	2	LM150K STEEL (series)	TO-3	2	35	30
	LM350	1.2-32 (adj)	N/A	0.005	0.1	35	86	2					
	LM123K, LM223K	5	6	0.01	0.5	20	75	1.7-2	LM123K series	TO-3	2	35	30
	LM323K	5	4	0.01	0.5	20	75	1.7-2					
1.5	LM117, LM217	1.2-37 (adj)	N/A	0.01	0.1	40	80	2	LM117, LM317K STEEL	TO-3	2.3	35	20
	LM317	1.2-37 (adj)	N/A	0.01	0.1	40	80	2	LM317K STEEL	TO-3	2.3	35	20
	LM117HV, LM217HV	1.2-57 (adj)	N/A	0.01	0.1	60	80	2	LM117HV, LM217HVK STEEL	TO-3	2.3	35	20
	LM317HV	1.2-57 (adj)	N/A	0.01	0.1	60	80	2	LM317HVK STEEL	TO-3	2.3	35	20
									LM317T	TO-220	4	50	20
	LM109K, LM209K	5	6	0.004	1.0	35	80	1-2	LM109K series	TO-3	3	35	20
	LM309K	5	4	0.004	1.0	35	80	1-2	—	—	—	—	—
	LM140K	5, 6, 8, 10, 12, 15, 18, 24	4	0.02	0.5	35, 40 (24V)	66-80	1.6-2	LM140K	TO-3	4	35	20
	LM140AK	5, 6, 8, 10, 12, 15, 18 24	2	0.002	0.1	35, 40 (24V)	66-80	1.6-2	LM140AK	TO-3	4	35	20
	LM340	5, 6, 8, 10, 12, 15, 18, 24	4	0.02	0.5	35, 40 (24V)	66-80	1.6-2	LM340K, LM340AK	TO-3	4	35	20
	LM340A	5, 6, 8, 10, 12, 15, 18 24	2	0.002	0.1	35, 40 (24V)	66-80	1.6-2	LM340AK	TO-3	4	35	20
									LM340AT	TO-220	4	50	20
	LM78XXC	5, 6, 8, 10, 12, 15, 18,	4	0.03	0.5	35, 40 (24V)	66-80	1.6-2	LM340K, LM78XXKC	TO-3	4	35	20
								1.6-2	LM340CT, LM340T LM78XXCT	TO-220	4	50	18
0.5	LM117H, LM217H	1.2-37 (adj)	N/A	0.01	0.1	40	80	1.5	LM117H, LM217H	TO-39	15	150	2
	LM317H	1.2-37 (adj)	N/A	0.01	0.1	40	80	2.0	LM317H	TO-39	15	150	2
	LM117HVH, LM217HVH	1.2-37 (adj)	N/A	0.01	0.1	40	80	1.5	LM117HVH, LM217HVH	TO-39	15	150	2
	LM317HVH	1.2-37 (adj)	N/A	0.01	0.1	40	80	1.5	LM317HVH	TO-39	15	150	2
	LM317M	1.2-37 (adj)	N/A	0.01	0.1	40	80	2.0	LM317MP	TO-202	12	85	12
	LM341	5, 6, 8, 10, 12, 15, 18 24	4	0.02	0.5	35, 40 (24V)		1.2-1.7	LM341P	TO-202	12	80	12
	LM78MXX	5, 6, 8, 10, 12, 15, 18 24	4	0.03	0.5	35, 40 (24V)		1.2-1.7	LM78MXXCP	TO-202	12	80	12
0.25	LM342	5, 6, 8, 10, 12, 15, 18 24	4	0.03	0.5	35, 40 (24V)	53-64	1.5-2	LM342P	TO-202	12	80	10
0.20	LM109H, LM209H	5	6	0.004	0.4	35	80	1-2	LM109H, LM209H	TO-39	15	150	2
	LM309H	5	4	0.004	0.4	35	80	1-2	LM309H	TO-39	15	150	2
0.10	LM140L, LM240L	5, 6, 8, 10, 12, 15, 18, 24	2	0.02	0.25	35, 40 (24V)	48-62	1.5-2	LM140LAH, LM240LAH	TO-39	40	140	3
	LM340L	5, 6, 8, 10, 12, 15, 18 24	2	0.02	0.25	35, 40 (24V)	48-62	1.5-2	LM340LAH	TO-39	40	140	3
	LM78LXXA	5, 6, 8, 10, 12, 15, 18 24	4	0.03	0.25	35, 40 (24V)	45-60	1.5-2	LM78LXXACH LM78LXXACZ	TO-39 TO-92	40 40	140 180	3 1

1. Operating temp range:
   LM100 series −55°C to +125°C
   LM200 series −25°C to +85°C
   LM300 series 0°C to +70°C
2. Max T_J = 150°C except 125°C for LM309, 320, 323, 345
3. Typ at 50–100% of rated I_{OUT}, 25°C, max V_{IN} change
4. Near zero to max rated I_{OUT}, 25°C pulse test
5. Max mV per volt of out voltage rating
6. Subtract (20 log V_{OUT}) for ripple rejection factor
7. ±4% available for LM140A and LM340A
8. ±10% available as LM78L CH and LM78L CZ
9. DIP = 14-pin dual-in-line plastic pkg
   SGS = special DIP with heat sink
10. V_{IN} = 40V for LM120H15 & LM120K15 series

(Courtesy of National Semiconductor)

TABLE 3-1b. Data Sheet summary

Output Current	Device[1,2]	VOUT (V)	TA = 25°C (± %)	Max Regulation Line[3] (%VOUT/V)	Max Regulation Load[4] (%VOUT/V)	Max VIN (V)	Ripple (dB)[5]	Typ Dropout Voltage (V)	Device	Pkg Style	Typ θJC (°C/W)	Typ θJA (°C/W)	Max PD (W)
3.0	LM145K, LM245K	−5.0, −5.2	2	0.008	0.6	20	68	2	LM145K, LM245K	TO-3	2	35	25
	LM345K	−5.0, −5.2	4	0.008	0.6	20	68	2	LM345K	TO-3	2	35	25
1.5	LM137, LM237	−1.2− −37 (adj)	N/A	0.006	0.3	40	77	2	LM137, LM237K STEEL	TO-3	2	35	20
	LM337	−1.2− −37 (adj)	N/A	0.007	0.3	40	77	2	LM337K STEEL LM337T	TO-3 TO-220	2 3	35 50	20
	LM137HV, LM237HV	−1.2− −47 (adj)	N/A	0.006	0.3	50	77	2	LM137HV, LM137HVK STEEL	TO-3	2	35	20
	LM337HV	−1.2− −47 (adj)	N/A	0.007	0.3	50	77	2	LM337HVK STEEL	TO-3	2	35	20
	LM120K, LM220K	−5, −5.2, −6, −8, −9, −12, −15, −18, −24	2	0.02	0.3	25 35 (9V, 12V) 40 (15V, 18V) 42 (24V)	64 80 75 70	2 2 2 2	LM120K series	TO-3	3	35	20
	LM320K	−5, −5.2, −6, −8, −9, −12, −15, −18, −24	4	0.02	0.3	25 35 (9V, 12V) 40 (15V, 18V) 42 (24V)	64 80 75 70	2	LM120K series	TO-3	3	35	20
	LM320T	−5, −5.2, −6, −8, −9, −12, −15, −18, −24	4	0.02	0.3	25 35 (9V, 12V, 15V, 18V) 40 (24V)	64 75−80 70	2 4 4	LM320T	TO-220	3	50	20
	LM79XXC	−5, −5.2, −6, −8, −9, −12, −15, −18, −24	4	0.03	0.4	35, 40 (24V)	66−70	2−4	LM79XXCT	TO-220	3	50	20
0.5	LM137H, LM237H	−1.2− −37 (adj)	N/A	0.006	0.3	40	77	2	LM137H, LM237H	TO-39	15	150	2
	LM337H	−1.2− −37 (adj)	N/A	0.007	0.3	40	77	2	LM337H	TO-39	15	150	2
	LM137HVH, LM237HVH	−1.2− −47 (adj)	N/A	0.006	0.3	50	77	2	LM137HVH, LM237HVH	TO-39	15	150	2
	LM337HVH	−1.2− −47 (adj)	N/A	0.007	0.3	50	77	2	LM337HVH	TO-39	15	150	2
	LM337M	−1.2− −37 (adj)	N/A	0.007	0.3	40	77						
	LM120H, LM220H	−5.0, −5.2, −6, −8	2	0.02	0.6	25	64	2	LM120H, LM220H	TO-39	15	150	2
	LM320H	−5.0, −5.2, −6, −8	4	0.02	0.6	25	64	2	LM320H	TO-39	15	150	2
	LM320M	−5, −5.2, −6, −8, −9, −12, −15, −18, −24	4 4	0.02	0.6	25 35 (9V, 12V, 15V, 18V) 40 (24V)	60−64 70−80	2 2	LM320MP	TO-202	12	80	12
	LM79MXX	−5, −6, −8, −12, −15, −24	4	0.03	0.7	35, 40 (24V)	58−60	2	LM79MXXCP	TO-202	12	80	12
0.25	LM320ML	−5, −6, −8, −10, −12, −15, −18, −24	4	0.01	0.5	35, 40 (24V)	50−60	2	LM320MLP	TO-202	12	80	12
0.20	LM120H, LM220H	−9, −12,	2	0.02	0.1	35 (9V, 12V)	70−80	2	LM120H, LM220H	TO-39	15	150	2
	LM320H	−15, −18, −24	4	0.02	0.1	40 (15V, 18V) 42 (24V)		2	LM320H	TO-39	15	150	2
0.10	LM320L	−5, −6, −8, −9, −12, −15, −18, −24	4	0.01	0.5	35, 40 (24V)	60−65	2	LM320LZ	TO-92	40	180	1
	LM79LXXA	−5, −12, −15, −18, −24	4	0.02	0.6	35, 40 (24V)	50−55		LM79LXXACZ LM79LXXACH	TO-92 TO-39	40 40	180 140	1 2

(Courtesy of National Semiconductor)

The third column is V_{out}. This is the regulated output voltage of the three-terminal regulator. The LM340 appears to have eight output voltages. Actually, there are eight versions of the LM340 regulator: LM340-5 (output 5 V), LM340-6 (output 6 V), LM340-8 (output 8 V), etc.

The fourth column is ($\pm$%). All regulators of the same model number will not have exactly the same output voltage. From unit to unit V_{out} may vary by this percentage. However, each unit will regulate at its own V_{out}. For example, one LM309 may provide exactly $V_{out} = 5$ V, another LM309 may yield $V_{out} = 5$ V $+ 5\% = 5.25$ V, while a third may regulate at $V_{out} = 5$ V $- 5\% = 4.75$ V. Each will keep the output regulated. However, the precise regulated output voltage will vary from unit to unit.

The seventh column is **Max V_{in}**. Should the input voltage exceed this level, the regulator will be damaged.

The eighth column is **Ripple**. These IC voltage regulators not only keep the output voltage constant, but they will also drastically reduce the amount of ripple voltage.

$$V_{r\,out} = (V_{ripple\ in}) \times 10^{-[(dB - 20 \log V_{out})/20]} \tag{3-11}$$

See note 6 at the bottom of Table 3-1.

Example 3-4

$$\begin{aligned} V_{r\,out} &= (2\ V_{p\text{-}p}) \times 10^{-[(80 - 20 \log 5)/20]} \\ &= (2\ V_{p\text{-}p}) \times 10^{-(80-14)/20} \\ &= (2\ V_{p\text{-}p}) \times 10^{-(3.3)} \\ &= (2\ V_{p\text{-}p}) \times .501 \times 10^{-3} \\ V_{r\,out} &= 1.002\ mV_{p-p} \end{aligned}$$

The two volt ripple on the input has been reduced to 1.0 millivolts by the regulator. This can drastically reduce the filter requirements. However, since

$$|V_{in}| \geq |V_{out}| + 2\ V$$

some filtering will always be required to keep the input instantaneous voltage above the minimum required for operation.

The ninth column is the **TYP Dropout Voltage**. This is another statement of equation (3-10). The number listed is the difference between the input and output voltages.

The second half of the table lists those parameters which deal with the power rating and dissipation characteristics of the regulator. The eleventh column is **Pkg Style**. This specifies the type of enclosure used for that particular model. Notice that the letter following the model number (H, K, T, P, Z) in column 10 differentiates between packages. Package styles are illustrated in Fig. 3-19.

The remaining three columns will be dealt with in the heat sink calculation section of the chapter.

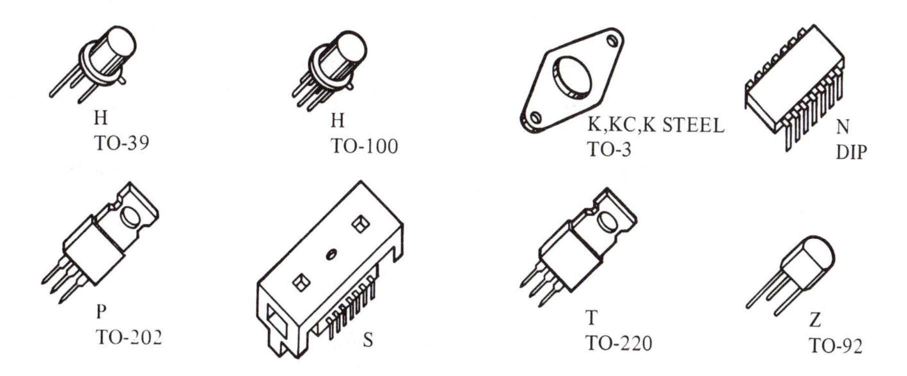

Figure 3-19 Available regulator package style (*Courtesy of National Semiconductor*)

3-3.2 Regulator Circuit Components

What, then, is necessary to use a three-terminal voltage regulator? First, you must ensure that the input voltage is at least two volts above the desired output voltage *at* the *rated* maximum load *current*; however, it may not exceed the rated maximum input voltage under no-load conditions:

$$V_{out} + 2\text{ V}|_{I_{L\,max}} < V_{in} < V_{in\,max}|_{I_L=0} \tag{3-12}$$

Second, input and output capacitors may be needed in addition to the normal power supply filter capacitors. If the regulator is more than 5 cm (2 in.) (as the electron flows) from the main filter capacitor, a capacitor must be placed, physically next to the regulator and connected between the regulator's input and ground. It should at least be a 0.22 μF ceramic disc, 2 μF solid tantalum, or 24 μF aluminum electrolytic. Also, a 0.1 μF ceramic disc capacitor should be located next to the regulator and connected between the output and ground. This will improve high frequency and transient (pulse) operation. The LM120 series and 7900 series of negative regulators require an additional 1 μF tantalum or 25 μF aluminum electrolytic output capacitor.

The final requirement to have a working three-terminal IC regulator is a heat sink. Although in many applications a heat sink may not be necessary, it is always wise to verify that with a calculation.

3-3.3 Heat Sinks

Any semiconductor device which carries current will dissipate power. That power is dissipated normally in the form of heat. The heat is generated at the wafer (junction) and must flow through the package to the surrounding air. The packaging material presents a certain opposition to this flow of heat (thermal resistance, θ_{JA}). The temperature of the junction (wafer) will rise. Should the junction temperature become too high, the semiconductor will be damaged (or the IC will undergo thermal shutdown). Equation (3-13) gives the key relationship:

$$T_J = T_A + \theta_{JA} P \tag{3-13}$$

where T_J is the temperature of the junction.

T_A is the surrounding, ambient temperature.

θ_{JA} is the thermal resistance from the junction to ambient.

P is the power dissipated by the device.

It is necessary to keep T_J below the maximum specified by the manufacturer of the device being used. This can be done by minimizing the power dissipated through careful circuit design, by decreasing the ambient temperature (T_A) with a fan, or by decreasing the thermal resistance (θ_{JA}) with a heat sink. We will discuss the last option (reducing θ_{JA}) in detail.

$$\theta_{JA}(\max) = \frac{T_J(\max) - T_A}{P} \tag{3-14}$$

When a heat sink is used, the junction to ambient thermal resistance consists of two major terms.

$$\theta_{JA} \approx \theta_{JC} + \theta_{SA} \tag{3-15}$$

where θ_{JC} is the thermal resistance, junction to case.

θ_{SA} is the thermal resistance, heat sink to ambient.

The use of the various charts and equations to determine which heat sink, if any, is needed is best illustrated with an example.

Example 3-5

Determine which three terminal regulators and which heat sink (if any) is needed to provide:

$$V_{out} = 5\ \text{V}$$

$$I_{out} \leq 0.7\ \text{A}$$

$$V_{in} = 15\ \text{V}$$

$$T_A = 60°\text{C}$$

Solution

1. First look in Table 3-1 to determine which regulators appear to meet the requirements.

   LM309K is rated at 1.5A output current.

   LM340-5 K and T are also rated at 1.5A output current.

   No others have adequate $I_{out\ max}$ ratings. The LM123K was skipped because 3A is certainly not needed.
2. Calculate the maximum allowable θ_{JA} under normal conditions.

$$\theta_{JA}(\text{max}) = \frac{T_J - T_A}{P}$$

$$P = (V_{in} - V_{out})I_{out} = (15\ \text{V} - 5\ \text{V}) \times 0.7\ \text{A} = 7\ \text{W}$$

Note 2 at the bottom of Table 3-1 indicated that:

$$\text{LM309}\ T_J = 125°\text{C}$$

$$\text{LM340}\ T_J = 150°\text{C}$$

So

$$\theta_{JA}\ (\text{max}) = \frac{125°\text{C} - 60°\text{C}}{7\ \text{W}} = 9.29°\text{C/W LM309}$$

$$\theta_{JA}\ (\text{max}) = \frac{150°\text{C} - 60°\text{C}}{7\ \text{W}} = 12.9°\text{C/W LM340}$$

3. Can either of these regulators be used at all?

$$\theta_{JC} < \theta_{JA}\ (\text{max})$$

The junction to case thermal resistance must be less than the maximum allowable junction to ambient thermal resistance, since the heat sink will also offer some resistance.

	θ_{JC}		θ_{JA} (max)
LM309K	3	<	9.29
LM340-5	4	<	12.9

Either regulator will work.

4. Can either of the regulators be used without a heat sink?

$$\theta_{JA} < \theta_{JA}\ (\text{max})$$

If the rated θ_{JA} for the *bare* regulator is below the maximum thermal resistance, a heat sink will not be needed.

TABLE 3-2. Thermal Resistance Requirements for Example 3-5

	θ_{JA}	θ_{JA}(max)
LM309K	35	9.29
LM340-5K	35	12.9
LM340-5T	50	12.9

All three models have bare case thermal resistances (junction to ambient) which exceed the allowable maximum. A heat sink will be needed.

5. At this point it is time to select a heat sink. There are three to choose from. The decision as to which regulator/heat sink to use will depend on regulator/heat sink combination price and size considerations.

The thermal resistance of the heat sink itself is called θ_{SA}. The *maximum* allowable heat sink thermal resistance is

$$\theta_{SA}\ (\text{max}) = \theta_{JA}\ (\text{max}) - \theta_{JC}$$

For the LM309K this becomes

$$\theta_{SA}\text{ (max)} = 9.29°\text{C/W} - 3°\text{C/W}$$

$$\theta_{SA}\text{ (max)} = 6.29°\text{C/W}$$

This is the *largest* thermal resistance the heat sink may have. Any larger resistance would cause the regulator to overheat.

From Table 3-3 select a heat sink which fits the regulator package and which has a thermal resistance *lower* than that calculated above.

TABLE 3-3. Heat Sinks for Example 3-5

	PKG	$\theta_{SA\ REG}$	*SINK*
LM309K	T03	6.29	Wakefield 390
LM340-5K	T03	8.9	Staver V3-5
LM340-5T	T220	8.9	Staver V3-5

The heat sink thermal resistances and possible models are given in Table 3-4.

Example 3-6

Design a power supply using a three-terminal regulator which will output 5 volts at currents up to 500 mA with a ripple of 15 millivolts peak-to-peak or less at 30°C. Select transformer filter capacitor, regulator, heat sink (if needed) and stabilization capacitors.

Solution

1. Initially select a 12.6 V rms secondary transformer and an LM 340-5T regulator IC. These are both inexpensive and commonly available.

$$V_{pk} = 1.414 \times 12.6\text{ V rms} - 1.4\text{ V} = 16.4\text{ V}$$

(Below the maximum input voltage for the LM 340-5T).

2. Ripple voltage

$$V_{ripple}\text{ (p-p)out} = V_{ripple}\text{ (p-p)in} \times 10^{-[(\text{dB} - 20 \log V_{out})/20]} \quad (3\text{-}11)$$

$$V_{ripple}\text{ (p-p)in} = V_{ripple}\text{(p-p)out} \times 10^{[(\text{dB} - 20 \log V_{out})/20]}$$

TABLE 3-4. Heat sink selection guide

No attempt has been made to provide a complete list of all heat sink manufacturers. This list is only representative.

For TO-202 Packages

θ_{SA} Approx[1] (°C/W)	Manufacturer & Type
12.5 - 14.2	Staver V4-3-192
13	Staver V5-1
15.1 - 17.2	Staver V4-3-128
19	Thermalloy 6106 Series
20	Staver V6-2
25	Thermalloy 6107 Series
37	IERC PA1-7CB with PVC-1B Clip
40 - 42	Staver F7-3
40 - 43	Staver F7-2
42	IERC PA2-7CB with PVC-1B Clip
42 - 44	Staver F7-1

For TO-220 Packages

θ_{SA} Approx[1] (°C/W)	Manufacturer & Type
4.2	IERC HP3 Series
5 - 6	IERC HP1 Series
6.4	Staver V3-7-225
6.5 - 7.5	IERC VP Series
8.1	Staver V3-5
8.8	Staver V3-7-96
9.5	Staver V3-3
10	Thermalloy 6032, 6034 Series
12.5 - 14.2	Staver V4-3-192
13	Staver V5-1
15	Thermalloy 6030 Series
15.1 - 17.2	Staver V4-3-128
16	Thermalloy 6106 Series
18	Thermalloy 6107 Series
19	IERC PB Series
20	Staver V6-2
25	IERC PA Series
26	Thermalloy 6025 Series

For TO-92 Packages

θ_{SA} Approx[1] (°C/W)	Manufacturer & Type
30	Staver F2-7
46	Staver F5-7A, F5-8-1
50	IERC RUR Series
57	Staver F5-7D
65	IERC RU Series
72	Staver F1-7
85	Thermalloy 2224 Series

For TO-5 Packages

θ_{SA} Approx[1] (°C/W)	Manufacturer & Type
12	Thermalloy 1101, 1103 Series
12 - 16	Wakefield 260-5 Series
15	Staver V3A-5
22	Thermalloy 1116, 1121, 1123 Series
22	Thermalloy 1130, 1131, 1132 Series
24	Staver F5-5C
26 - 30	IERC Thermal Links
27 - 83	Wakefield 200 Series
28	Staver F5-5B
30	Thermalloy 2227 Series
34	Thermalloy 2228 Series
35	IERC Clip Mount Thermal Link
39	Thermalloy 2215 Series
42	Staver F5-5A
45 - 65	Wakefield 296 Series
46	Staver F6-5, F6-5L
50	Thermalloy 2225 Series
50 - 55	IERC Fan Tops
51	Thermalloy 2205 Series
53	Thermalloy 2211 Series
55	Thermalloy 2210 Series
56	Thermalloy 1129 Series
58	Thermalloy 2230, 2235 Series
60	Thermalloy 2226 Series
68	Staver F1-5
72	Thermalloy 1115 Series

For TO-3 Packages

θ_{SA} Approx[1] (°C/W)	Manufacturer & Type
0.4 (9" length)	Thermalloy (Extruded) 6590 Series
0.4 - 0.5 (6" length)	Thermalloy (Extruded) 6660, 6560 Series
0.56 - 3.0	Wakefield 400 Series
0.6 (7.5" length)	Thermalloy (Extruded) 6470 Series
0.7 - 1.2 (5 - 5.5" length)	Thermalloy (Extruded) 6423, 6443, 6441, 6450 Series
1.0 - 5.4 (3" length)	Thermalloy (Extruded) 6427, 6500, 6123, 6401, 6403, 6421, 6463, 6176, 6129, 6141, 6169, 6135, 6442 Series
1.9	IERC E2 Series (Extruded)
2.1	IERC E1, E3 Series (Extruded)
2.3 - 4.7	Wakefield 600 Series
4.2	IERC HP3 Series
4.5	Staver V3-5-2
5 - 6	IERC HP3 Series
5.2 - 6.2	Thermalloy 6103 Series
5.6	Staver V3-3-2
5.8 - 7.9	Thermalloy 6001 Series
5.9 - 10	Wakefield 680 Series
6	Wakefield 390 Series
6.4	Staver V3-7-224
6.5 - 7.5	IERC UP Series
8	Staver V1-5
8.1	Staver V3-5
8.8	Staver V3-7-96
8.8 - 14.4	Thermalloy 6013 Series
9.5	Staver V3-3
9.5 - 10.5	IERC LA Series
9.8 - 13.9	Wakefield 630 Series
10	Staver V1-3
13	Thermalloy 6117

Staver Co, Inc: 41-51 N. Saxon Ave, Bay Shore, NY 11706
IERC: 135 W. Magnolia Blvd, Burbank, CA 91502
Thermalloy: PO Box 34829, 2021 W. Valley View Ln, Dallas TX
Wakefield Engin Ind: Wakefield MA 01880

1 All values are typical as given by mfgr. or as determined from characteristic curves supplied by mfgr.

(Courtesy of National Semiconductor)

$$V_{\text{ripple}}\ (\text{p-p})\text{in} = 15\ \text{mV} \times 10^{[(74-20\log 5)/20]}$$

$$V_{\text{ripple}}\ (\text{p-p})\text{in} = 15\ \text{volts}$$

So the capacitive filter must limit the ripple voltage to 15 volts peak-to-peak. The IC will reduce the ripple from 15 volts to 15 millivolts.

3. V_{min} from the filter:

$$V_{\text{min}} = V_{\text{load}} + 2\ \text{V} \tag{3-10}$$

$$V_{\text{min}} = 5\ \text{V} + 2\ \text{V} = 7\ \text{V}$$

4. Filter capacitance selection

You want to select the *smallest* capacitor which will meet the requirements in steps 1–3. The smaller the capacitor, the less it costs, the smaller physically it is, and the less power the IC must dissipate.

From Fig. 3-9a a 200 μF capacitor gives $V_{\text{min}} = 7.5$ volts, at 500 milliamps, while a 100 μF capacitor is too small.

Check the ripple:

$$V_{\text{ripple}}\ (\text{p-p}) = V_{\text{pk}} - V_{\text{min}}$$

$$V_{\text{ripple}}\ (\text{p-p}) = 16.4\ \text{V} - 7.5\ \text{V} = 8.9\ \text{V}$$

Step 2 indicates that the ripple from the filter must be 15 volts or less. So

$$C_{\text{filter}} = 200\ \mu\text{F}$$

5. From Fig. 3-9b, the average value of voltage from the filter will be

$$V_{\text{DC}} = 12.2\ \text{volts}$$

6. Heat sink calculation

a. Is a heat sink needed?

$$\theta_{JA}(\text{max}) = \frac{T_J - T_A}{(V_{\text{DC}} - V_{\text{out}})I_{\text{load}}} \tag{3-14}$$

$$\theta_{JA}(\text{max}) = \frac{150°\text{C} - 30°\text{C}}{(12.2\ \text{V} - 5\ \text{V})0.5\ \text{A}}$$

$$\theta_{JA}(\text{max}) = 33.3°\text{C/W}$$

This calculation tells you that at a junction to ambient thermal resistance of 33.3°C/W or more, the IC will overheat (150°C or more). Since the LM 340-5T has a $\theta_{JA} = 50°C/W$, a heat sink *is* needed.

b. Which heat sink?

$$\theta_{SA} \approx \theta_{JA}(\max) - \theta_{JA} \tag{3-15}$$

$$\theta_{SA} \approx 33.3°C/W - 4°C/W$$

$$\theta_{SA} = 29.3°C/W$$

A heat sink with a thermal resistance less than 29.3°C/W must be *properly* mounted to the LM 340-5T.
The Thermalloy 6106, 6107, IERC PB Series or Staver V6-2 all meet this requirement.
The full schematic is shown in Fig. 3-20.

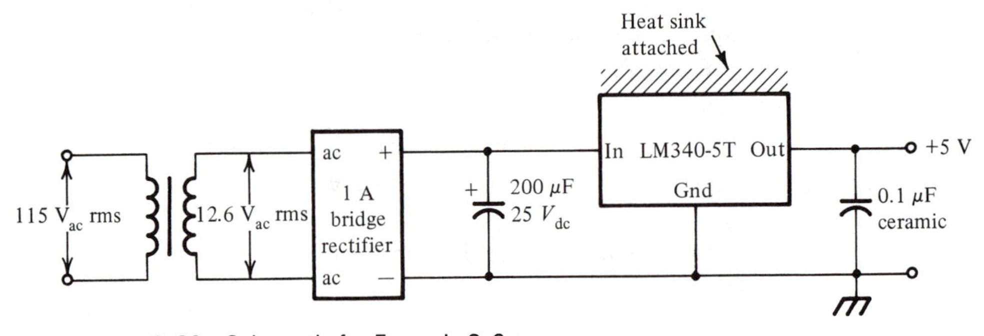

Figure 3-20 Schematic for Example 3-6

3-3.4 Dual Voltage Supplies

Many discrete and integrated circuits require bipolar (dual, or ±V) supplies. This can be easily done with two three-terminal regulators, as shown in Fig. 3-21.

Opposite-phase AC is provided by the transformer's secondary and a grounded center tap. The single full wave bridge turns these into positive and negative DC voltages (with respect to the grounded center tap). Filtering (with respect to ground) is provided by C1 and C2. Be careful to get the electrolytic capacitors' polarities correct!

The LM340 provides regulation of the positive voltage, while the LM320 regulates the negative voltage. Warning: **The LM320 has a different pin configuration than the LM340.** The case of the LM320 is *not* ground. Take care when you mount the negative regulator.

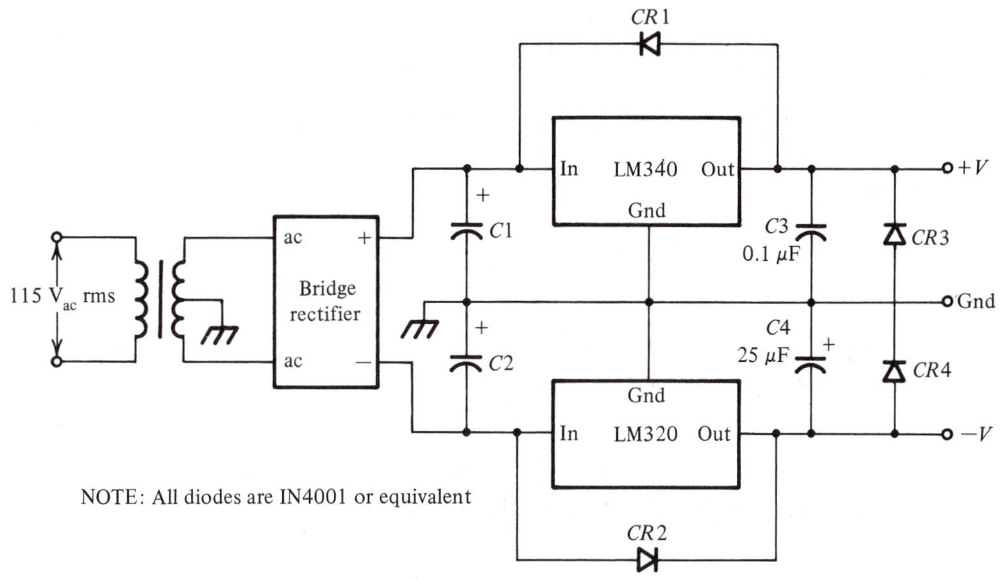

Figure 3-21 Simple dual power supply

The diodes provide protection, but be sure you do not accidentally reverse them. Diodes CR1 and CR2 ensure that transients on the regulator outputs do not drive the outputs to a potential above their inputs and damage the regulators. Also, the two regulators may not turn on simultaneously. If this occurs, the output of the slower regulator may be driven toward the potential of the faster. Diodes CR3 and CR4 prevent these reverse polarities on start up.

3-3.5 Adjustable Voltage Supplies

The precise output voltage you need may not be available in a standard three-terminal regulator, or you may want to vary the voltage. There are two simple ways to do this with three-terminal regulators.

The ground potential of one of the fixed, three-terminal regulators can be floated. This is shown in Fig. 3-22. The voltage across the 330 Ω resistor is fixed by the regulator. Changing the resistance of the potentiometer changes V_{pot}. Since

$$V_{out} = 5\text{ V} + V_{pot}$$

this changes the regulated output voltage. There are several precautions. The component values shown in Fig. 3-22 should not be changed a great deal, or circuit performance will be affected. The minimum output voltage is the value of

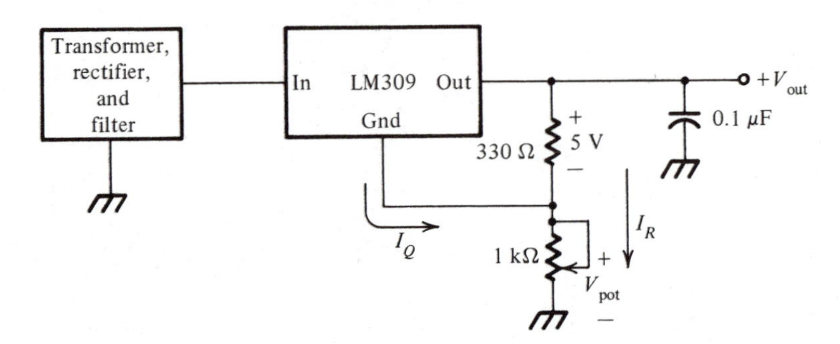

Figure 3-22 Adjustable output voltage from a three terminal regulator (*Courtesy of National Semiconductor*)

the regulator's fixed voltage (5 volts for the circuit in Fig. 3-22). The V_{min} from the filter must be at least two volts above the *maximum* desired output:

$$V_{min} \geq V_{out\ max} + 2\ \text{V} \tag{3-16}$$

Finally, the regulator is dissipating maximum power when the output voltage is adjusted to its *minimum* level. So in heat sink calculations use:

$$\theta_{JA\ max} = \frac{T_J - T_A}{(V_{DC} - V_{out\ min})I_{load\ max}} \tag{3-17}$$

The LM 117, 217, 317 series regulators are three-terminal, adjustable output voltage ICs. Their performance has been optimized for adjustable operation. The output voltage can be adjusted over a range of 1.2 volts to 40 volts, at load currents up to 1.5 amps (with proper heat sinking). The basic schematic is given in Fig. 3-23. The output voltage is

$$V_{out} = 1.2\ \text{V} \times \left(1 + \frac{\text{R2}}{\text{R1}}\right) + 50\ \mu\text{A} \times \text{R2}$$

If R2 is kept at or below 3 kΩ, the second term is negligible, reducing the equation to

$$V_{out} = 1.2\ \text{V} \times \left(1 + \frac{\text{R2}}{\text{R1}}\right) \tag{3-18}$$

Since there is no ground connection, all internal IC biasing currents must flow to the load. There must always be, then, at least a 10 milliamp load connec-

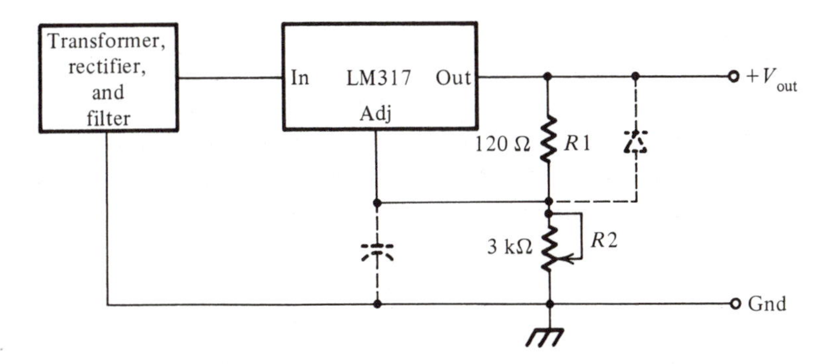

Figure 3-23 Adjustable voltage regulator (*Courtesy of National Semiconductor*)

ted between the output and ground. You can provide this by keeping R1 at or below 120 Ω.

The capacitor shown in Fig. 3-23 is optional, but does improve the ripple rejection of the IC. *If* the capacitor is used *and if* the output voltage is to be adjusted above 25 volts, the rectifier diode (as shown) must be added across R1 to protect the IC.

Input voltage and heat sink calculations are the same as for the fixed voltage three-terminal regulators. Equations (3-16) and (3-17) must be observed. One result, which is not first obvious, is that less load current is available at lower output voltages than at higher output voltages. With 40 volts on the input, and 1.2 volts at the output, a maximum of only 0.4 amps (not 1.5 amps) is available. This is a consequence of equation (3-17).

3-3.6 Higher Current Regulation

If your application demands more than one amp at a regulated voltage, there are several possible solutions.

The first, and best if feasible, is to partition your system. Supply unregulated power to each part (or PC board). Then use a one amp, three-terminal regulator on each portion. This eliminates the distribution problem described earlier and provides good isolation between parts (PC boards or subsystems).

Never try to run two regulators in parallel to improve current drive. Any two regulator ICs will have slightly different output voltages. When connected in parallel, the higher output regulator will drive current into the output of the lower regulator. At the same time, the lower regulator will try to regulate the voltage down to its level. The result could be failure of one or both ICs.

If you simply *must* supply more than one amp from a single regulator, you can use one of the 3 A or 5 A regulator ICs listed in Table 3-1. For the real current hogs, there is a series of 5 A, 10 A, 20 A, and 30 A *hybrid* regulator modules from Lambda Electronics. These high-current ICs come in fixed (±5 percent tunable) voltages. Specifications of the LAS7000 30 amp hybrid power regulator are given in Table 3-5. The connection diagram and positive voltage regulator circuit schematic are given in Fig. 3-24. Filter capacitor C1 should be selected, depending on the load current:

$$\text{C1} = I_0 \times 2000\ \mu\text{F/A}$$

The output capacitance also is determined by the load current:

$$\text{C2} = I_0 \times 100\ \mu\text{F/A}$$

The voltage adjustment resistor R1 should be selected as the next higher standard value above:

$$\text{R1} \geq 0.25\text{V}_{\text{out}} \times 100\ \frac{\Omega}{\text{V}}$$

When the load demand exceeds the maximum safe rating, the module will automatically drop output current (and therefore voltage) to about ten percent of maximum.

Finally, a lower current regulator, such as the LM340 or LM317, can be current-boosted with external components. This is shown in Fig. 3-25. The voltage dropped across the diode should equal the voltage dropped across the base emitter junction of the transistor. This effectively puts R1 and R2 in parallel. Current flowing into the junction of R1 and R2 from the filter will divide according to the ratio of the resistance:

$$\text{I}_1 = \left(\frac{\text{R2}}{\text{R1}}\right) I_{\text{reg}}$$

If R2 is made five times as large as R1, the current flowing through the pass transistor will be five times larger than the current flowing through the regulator. So, although the regulator controls the output voltage, the pass transistor supplies most of the load current. The total current to the load is

$$I_{\text{load}} = I_1 + I_{\text{reg}}$$

$$I_{\text{load}} = \left(1 + \frac{\text{R2}}{\text{R1}}\right) I_{\text{reg}}$$

TABLE 3-5. Lambda LAS 7000 Series 30 Amp Regulator IC Specifications

LAS 7000 SERIES

Regulator Performance Specifications

30 amp 400 watt positive regulator

General Description

The LAS 7200 Series of Power Hybrid Voltage Regulators is designed for applications requiring a well regulated output voltage for load current variations up to 30 amperes. A key feature of the LAS series of Power Hybrid Voltage Regulators is its construction. A high degree of thermal isolation between the heat generating power elements and the heat sensitive control and reference elements is achieved by the placing of the power section on the heat-dissipating base of the unit, and the control stage on the heat-dissipating upper surface. This thermal isolation results in extremely low thermal drift characteristics for changes in power levels. In addition, a unique thermal power limiting circuit is built into the power section of the unit for increased operational reliability. This reliability is accentuated by a demonstrated MTBF of 100,000 hrs.

PARAMETER	SYMBOL	CONDITIONS	MIN.	MAX.	UNITS
Input voltage to pin(1) (A)(H)	$V_{IN}(1)$		7.25	40.0	volts
Input voltage to pin(20) (A)(G)	$V_{IN}(20)$		12.3	40.0	volts
Output voltage	V_O		4.75	29.4	volts
Input-output differential (A)(F)	$(V_{IN}(1) - V_O)$		2.5	28.6	volts
Input-output differential (B)(F)	$(V_{IN}(20) - V_O)$		7.60	28.6	volts
Output current	I_O			30.0	amps
Standby current	$I_Q(1)$			40.0	mA
Standby current	$I_Q(20)$			7.0	mA
Power dissipation	P_D	Plate #1 @25°C		400	watts
Thermal resistance junction—Case 1	$\theta_j - C1$			0.44	°C/watt
Storage temperature	T_S		-55	+125	°C
Power transistor junction temperature	T_j			+200	°C
Regulation line (C)				0.016	$\%/\Delta V_{IN}$
Regulation load (D)				0.2	%
Programming resistance				1000 nominal	ohm/volt
Programming voltage				one/one	volt/volt
Temperature coefficient				0.015	%/°C
Ripple attenuation (E)		$V_{IN}(1)$ Minimum I_O Maximum	60		dB

(Courtesy of Lambda Electronics)

TABLE 3-5 *continued*

NOTES:

A. Separate DC input voltages for power circuit (pin 1) and control circuit pin (20).
B. Common input voltages for power circuit (pin 1) and control pin (20).
C. Io constant for entire input voltage range from [V_{IN}(1) & V_{IN}(20) min] to [V_{IN}(1) (20) max.]
D. V_{IN} constant for entire range from 0 to full load.
E. Ripple attenuation is 54 dB. min. for 24V, and 28V models.
F. Minimum input-output differential based on $T_j, \geqslant 25°C$.
G. For AC source to Pin 20 with source resistance less than 10 ohms, minimum VAC = 12V rms. For other conditions consult factory.
H. Maximum input voltage is 30V for LAS 7205 and 7206

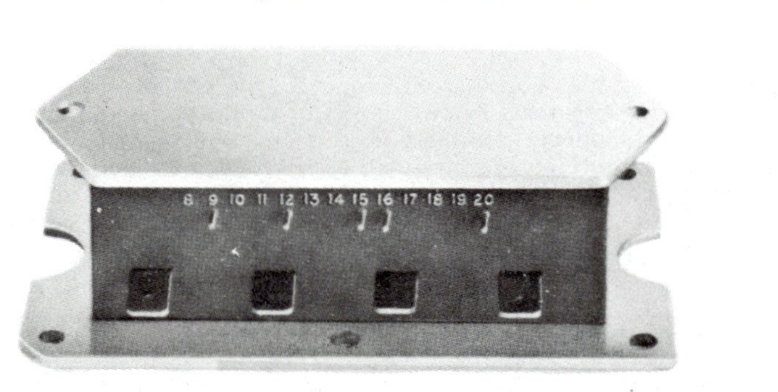

OUTLINE DRAWING

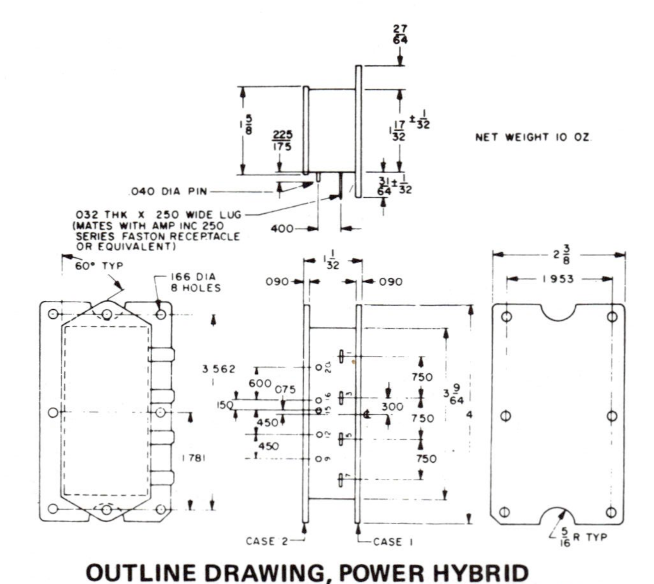

OUTLINE DRAWING, POWER HYBRID VOLTAGE REGULATOR, LAS 7000 SERIES

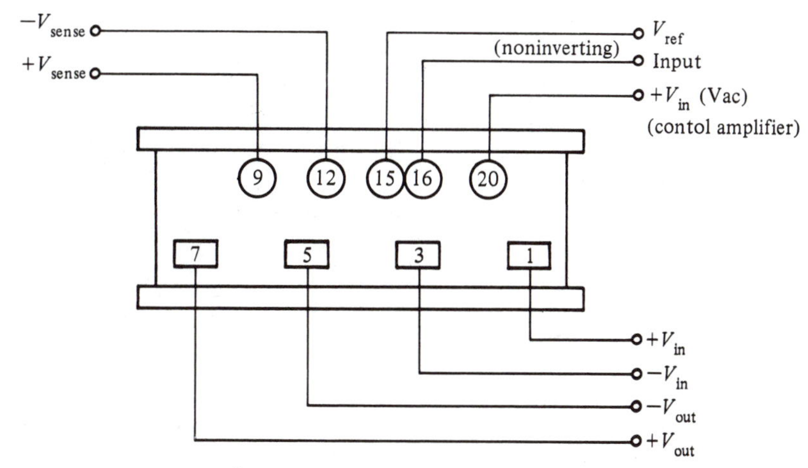

(a) 9-pin power hybrid voltage regulator

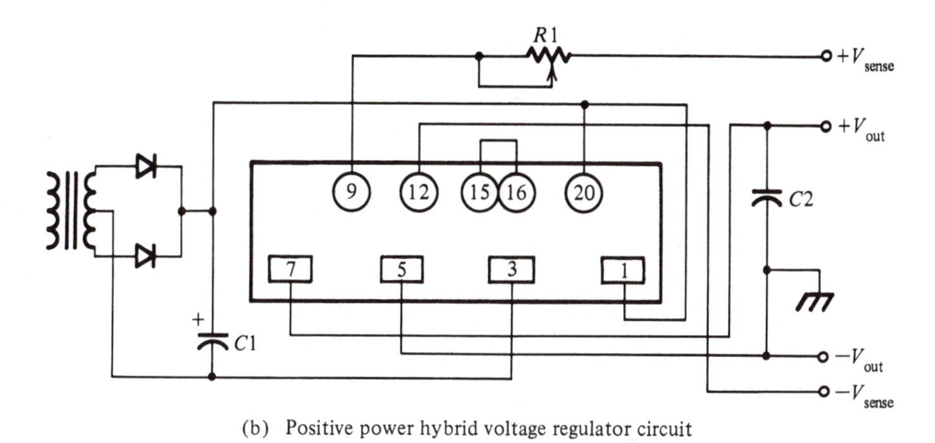

(b) Positive power hybrid voltage regulator circuit

Figure 3-24 LAS 7000 30 amp regulator (*Courtesy of Lambda Electronics*)

Should load current demand cause the regulator to go into thermal shutdown, I_{reg} will drop (to protect the regulator). Since the transistor current, I_1, is controlled by I_{reg}, this drop in I_{reg} will cause a similar drop in I_1. If the regulator and the transistor have the same θ_{JC}, and if the heat sink of the transistor is adequate, the transistor will be protected by the short circuit and thermal protection circuits of the regulator IC. The transistor's heat sink can be selected with exactly the same technique as used for the regulator IC. Also, don't forget to allow for the voltage drop across R2 and the diode when determining V_{min} for the filter capacitor selection.

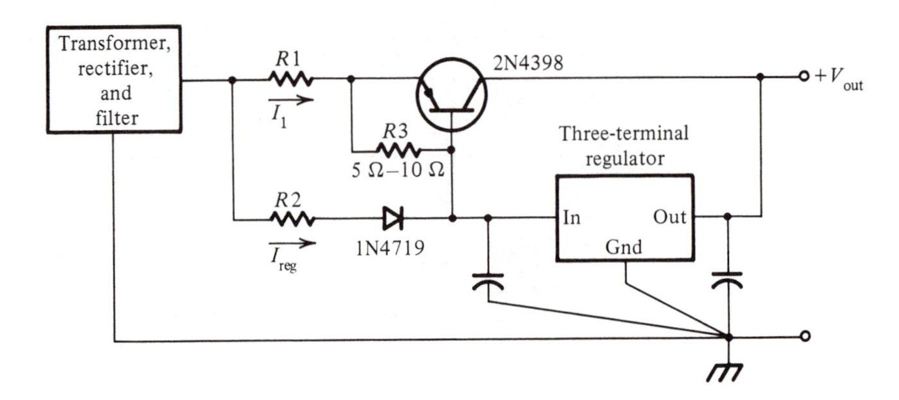

Figure 3-25 Current boosted three terminal regulator circuit (*Courtesy of National Semiconductor*)

3-4 THE 723 GENERAL PURPOSE REGULATOR

The three-terminal regulators of the previous section are convenient and relatively simple to use. However, this convenience is bought by the sacrifice of versatility. Current levels are set internally. You have no control over what will happen when the load is shorted. Different ICs are needed to provide positive and negative regulated voltages.

The 723 general purpose regulator can be set up for, or adjusted over, a wide range of both positive or negative regulated voltages. These can even be programmed remotely, or digitally. You can precisely determine how the regulator responds to a short circuit. Although the IC itself is inherently a low-current device, it can be boosted to provide 5 amps or more load current.

For this increased flexibility, you are going to have to work harder. The 723 requires several external components to function at all. Also, it is not nearly as forgiving as the three-terminal regulators. There is no on-chip thermal protection, nor are there short-circuit limits. You must provide these, or risk damaging the IC.

3-4.1 Basic Characteristics

A functional block diagram for the 723 regulator IC is shown in Fig. 3-26. There are two separate sections. The zener diode, constant current source, and reference amplifier produce a stable, fixed voltage of about 7 volts at the V_{ref} terminal. The constant current source forces the zener to operate at a particular *point* on its characteristic curve. This assures that the zener outputs a *fixed* voltage. The

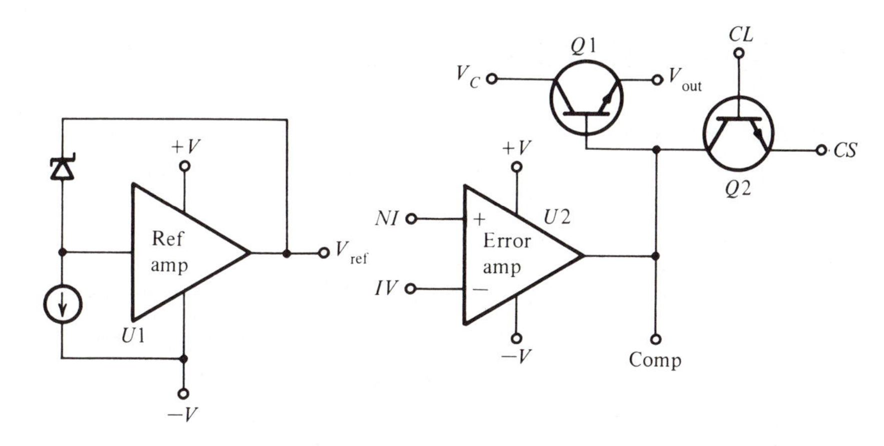

Figure 3-26 723 regulator IC functional block diagram

reference amplifier serves as a buffer, preventing any circuit attached to V_{ref} from loading down, and therefore shifting the voltage of, the zener diode.

The other section of the IC consists of the error amplifier, series pass transistor Q1, and the current limit transistor Q2. The error amplifier compares a sample of the output voltage (using negative feedback to ensure linear operation) to a reference voltage. An error signal, the difference between these two inputs, controls the conduction of Q1. When properly connected, Q2 can turn on at excessive load currents, stealing the base current of Q1, to turn the regulator off.

The primary specifications of the 723 are given in Table 3-6. There are several items you should notice. The maximum input voltage ($V+$ to $V-$) is 40 volts. However, the inverting and noninverting inputs of the error amp must never be more than 5 volts apart. Without an *external* pass transistor, the maximum possible output current is only 150 milliamps. The normal current may be much less than this. The input voltage must be at least 3 volts above the output voltage (not 2 volts, as with the three-terminal regulator).

You can use the 723 to build a simple positive low-voltage (2 V to 7 V) regulator. The schematic is shown in Fig. 3-27. This schematic is fine for building and testing a circuit, but circuit operation is easier to understand by looking at Fig. 3-28. The voltage from V_{ref} is typically 7.15 volts. This is divided down by R1 and R2. The voltage at the NI input of the error amplifier is

$$V_{NI} = V_{ref} \frac{R2}{R1 + R2}$$

TABLE 3-6a. 723 IC regulator schematic and connection diagram

LM723/LM723C Voltage Regulator

General Description

The LM723/LM723C is a voltage regulator designed primarily for series regulator applications. By itself, it will supply output currents up to 150 mA; but external transistors can be added to provide any desired load current. The circuit features extremely low standby current drain, and provision is made for either linear or foldback current limiting. Important characteristics are:

- 150 mA output current without external pass transistor
- Output currents in excess of 10A possible by adding external transistors
- Input voltage 40V max
- Output voltage adjustable from 2V to 37V
- Can be used as either a linear or a switching regulator.

The LM723/LM723C is also useful in a wide range of other applications such as a shunt regulator, a current regulator or a temperature controller.

The LM723C is identical to the LM723 except that the LM723C has its performance guaranteed over a 0°C to 70°C temperature range, instead of −55°C to +125°C.

Schematic and Connection Diagrams*

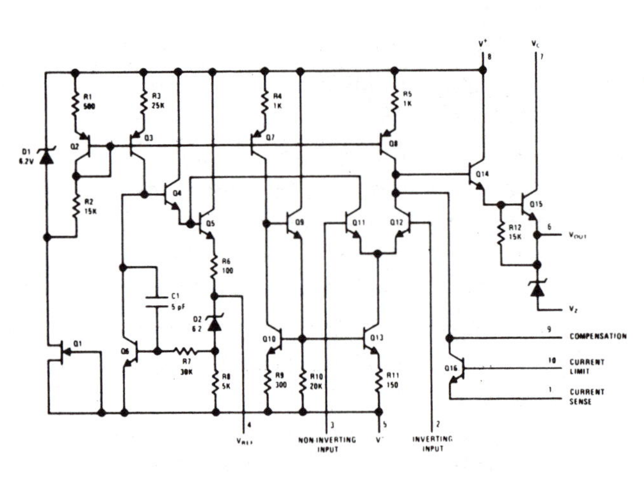

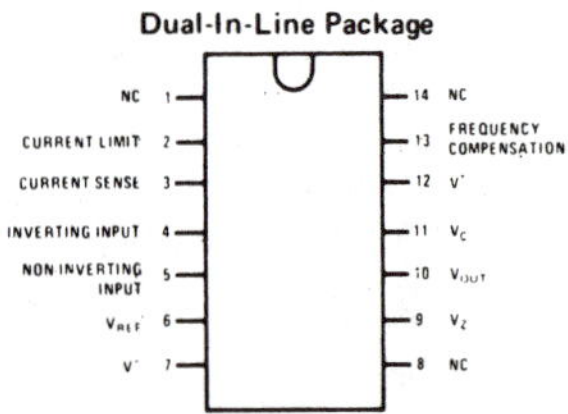

Order Number LM723CN
See NS Package N14A
Order Number LM723J or LM723CJ
See NS Package J14A

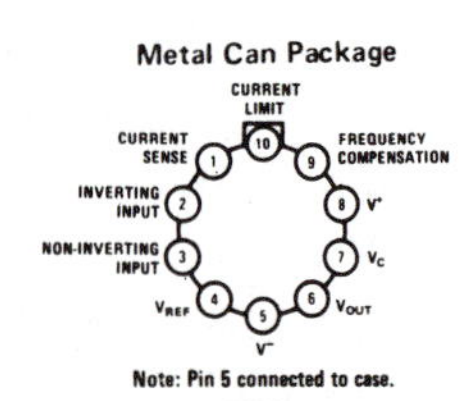

TOP VIEW

Order Number LM723H or LM723CH
See NS Package H10C

Equivalent Circuit*

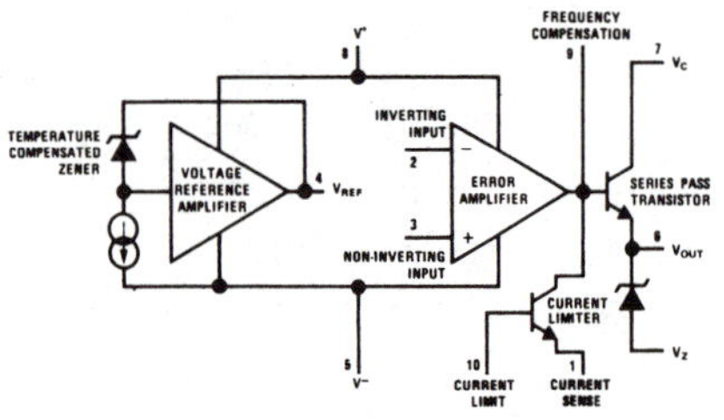

*Pin numbers refer to metal can package.

(Courtesy of National Semiconductor)

TABLE 3-6b. 723 IC regulator maximum ratings and electrical characteristics

Absolute Maximum Ratings

Pulse Voltage from V^+ to V^- (50 ms)	50V
Continuous Voltage from V^+ to V^-	40V
Input-Output Voltage Differential	40V
Maximum Amplifier Input Voltage (Either Input)	7.5V
Maximum Amplifier Input Voltage (Differential)	5V
Current from V_Z	25 mA
Current from V_{REF}	15 mA
Internal Power Dissipation Metal Can (Note 1)	800 mW
Cavity DIP (Note 1)	900 mW
Molded DIP (Note 1)	660 mW
Operating Temperature Range LM723	-55°C to +125°C
LM723C	0°C to +70°C
Storage Temperature Range Metal Can	-65°C to +150°C
DIP	-55°C to +125°C
Lead Temperature (Soldering, 10 sec)	300°C

Electrical Characteristics (Note 2)

PARAMETER	CONDITIONS	LM723			LM723C			UNITS
		MIN	TYP	MAX	MIN	TYP	MAX	
Line Regulation	V_{IN} = 12V to V_{IN} = 15V		.01	0.1		.01	0.1	% V_{OUT}
	-55°C ≤ T_A ≤ +125°C			0.3				% V_{OUT}
	0°C ≤ T_A ≤ +70°C						0.3	% V_{OUT}
	V_{IN} = 12V to V_{IN} = 40V		.02	0.2		0.1	0.5	% V_{OUT}
Load Regulation	I_L = 1 mA to I_L = 50 mA		.03	0.15		.03	0.2	% V_{OUT}
	-55°C ≤ T_A ≤ +125°C			0.6				% V_{OUT}
	0°C ≤ T_A ≤ = +70°C						0.6	% V_{OUT}
Ripple Rejection	f = 50 Hz to 10 kHz, C_{REF} = 0		74			74		dB
	f = 50 Hz to 10 kHz, C_{REF} = 5 μF		86			86		dB
Average Temperature Coefficient of Output Voltage	-55°C ≤ T_A ≤ +125°C		.002	.015				%/°C
	0°C ≤ T_A ≤ +70°C					.003	.015	%/°C
Short Circuit Current Limit	R_{SC} = 10Ω, V_{OUT} = 0		65			65		mA
Reference Voltage		6.95	7.15	7.35	6.80	7.15	7.50	V
Output Noise Voltage	BW = 100 Hz to 10 kHz, C_{REF} = 0		20			20		μVrms
	BW = 100 Hz to 10 kHz, C_{REF} = 5 μF		2.5			2.5		μVrms
Long Term Stability			0.1			0.1		%/1000 hrs
Standby Current Drain	I_L = 0, V_{IN} = 30V		1.3	3.5		1.3	4.0	mA
Input Voltage Range		9.5		40	9.5		40	V
Output Voltage Range		2.0		37	2.0		37	V
Input-Output Voltage Differential		3.0		38	3.0		38	V

Note 1: See derating curves for maximum power rating above 25°C.

Note 2: Unless otherwise specified, T_A = 25°C, V_{IN} = V^+ = V_C = 12V, V^- = 0, V_{OUT} = 5V, I_L = 1 mA, R_{SC} = 0, C_1 = 100 pF, C_{REF} = 0 and divider impedance as seen by error amplifier ≤ 10 kΩ connected as shown in Figure 1. Line and load regulation specifications are given for the condition of constant chip temperature. Temperature drifts must be taken into account separately for high dissipation conditions.

Note 3: L_1 is 40 turns of No. 20 enameled copper wire wound on Ferroxcube P36/22-3B7 pot core or equivalent with 0.009 in. air gap.

Note 4: Figures in parentheses may be used if R1/R2 divider is placed on opposite input of error amp.

Note 5: Replace R1/R2 in figures with divider shown in Figure 13.

Note 6: V^+ must be connected to a +3V or greater supply.

Note 7: For metal can applications where V_Z is required, an external 6.2 volt zener diode should be connected in series with V_{OUT}.

(Courtesy of National Semiconductor)

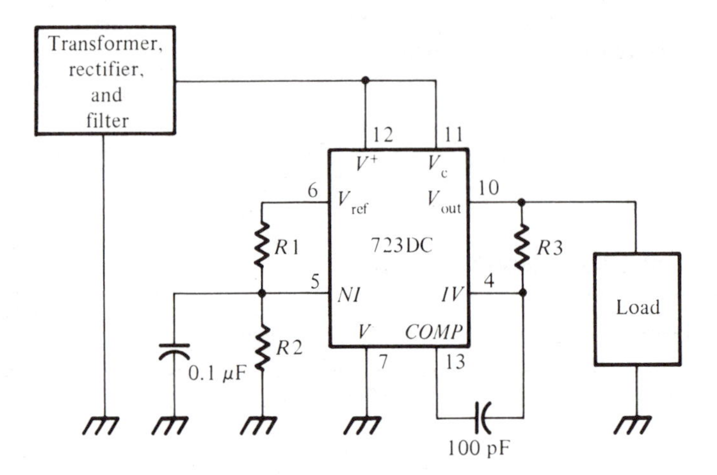

Figure 3-27 Basic low voltage 723 regulator

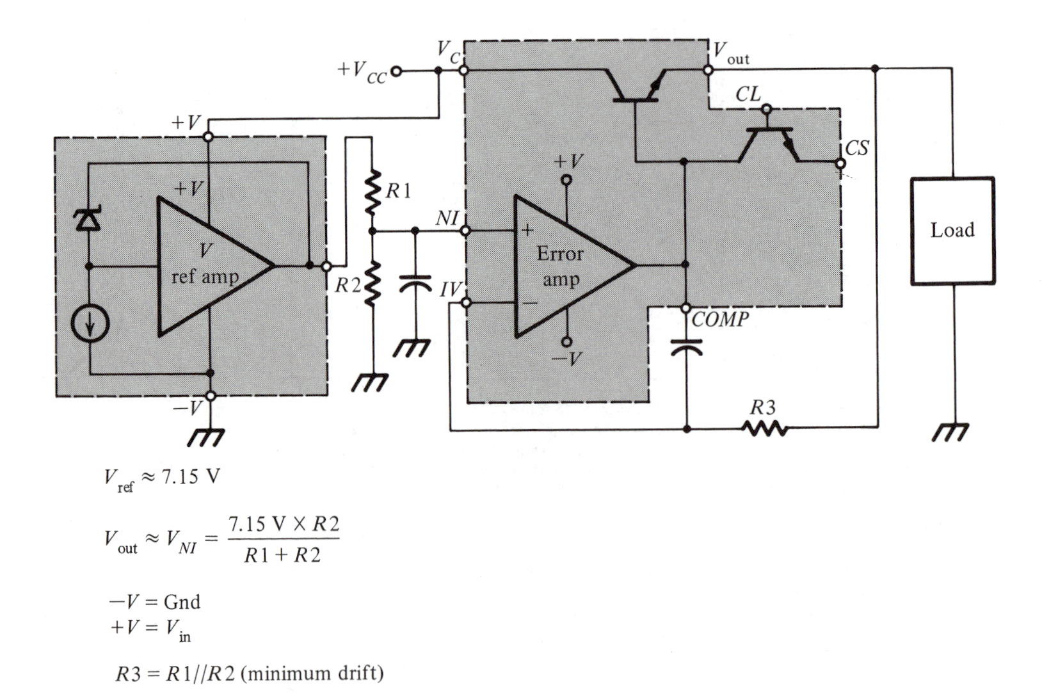

Figure 3-28 Functional block diagram for a low voltage regulator using the 723 IC

Any difference between this and the output voltage, which is fed directly back to the IV terminal, is amplified by the error amp. Pass transistor Q1 is driven to minimize any difference between the NI and IV inputs of the error amp. So, for all practical purposes, the output terminal (and IV terminal) is driven to the same potential as the NI terminal. The error amplifier/Q1 combination is working as a simple voltage follower:

$$V_{out} = V_{ref}\frac{R2}{R1 + R2} \tag{3-19}$$

If the output voltage drops, the inverting terminal of the error amp goes down. This forces the output of the error amp to go more positive (inverting). Pass transistor Q1 is driven on harder. Turning a transistor on harder lowers the voltage across it and drives more current into the load. More current driven into the load causes the load voltage to rise. Since the initial problem was a drop in load voltage, the drop has been compensated. A similar result occurs for a rise in load voltage, or changes of input voltage.

You can use the circuit in Fig. 3-29 to produce regulated output voltages above 7 volts. As with the low-voltage configuration, you may find the functional block diagram of Fig. 3-30 easier to follow.

The negative feedback for the error amplifier is divided by R1 and R2. The error amplifier/Q_1 combination now forms a noninverting amplifier with a voltage gain of

$$A_V = 1 + \frac{R1}{R2}$$

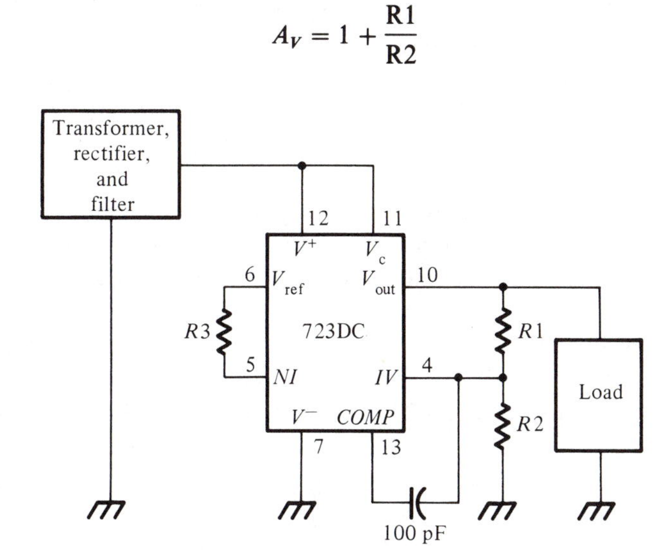

Figure 3-29 Basic high voltage 723 regulator

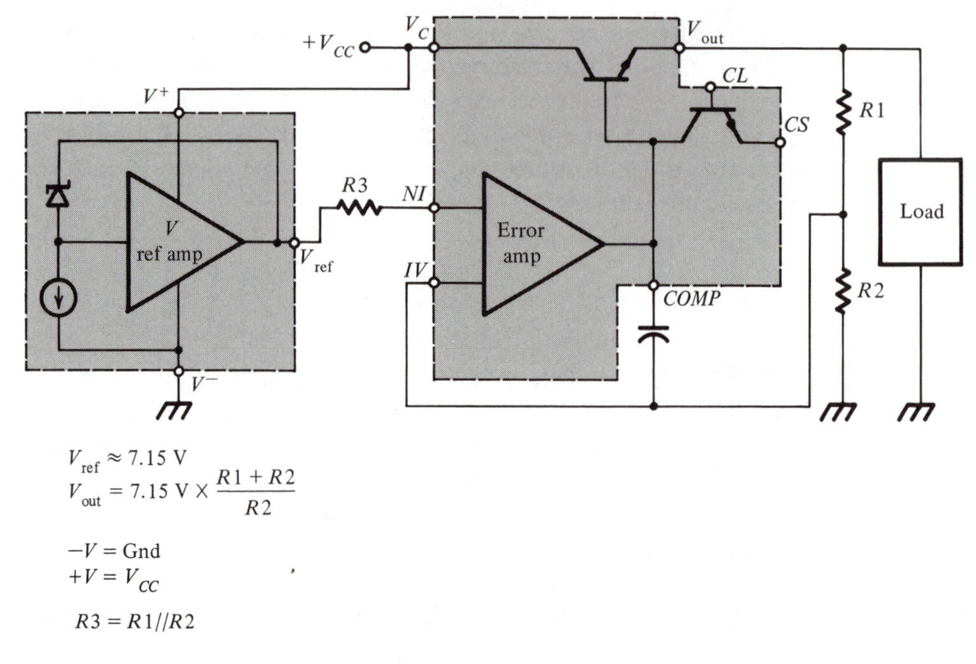

Figure 3-30 Functional block diagram for a high voltage regulator using a 723 IC

just like any other op amp. The input to this circuit comes directly from the reference voltage source, V_{ref}, which is typically 7.15 volts. So the output from the circuit in Figs. 3-29 and 3-30 is

$$V_{out} = 7.15\text{ V}\left(1 + \frac{\text{R1}}{\text{R2}}\right)$$

Remember that the regulator output voltage must be kept three volts below the minimum input from the filter (V_{min}). Also, the regulator IC is working hardest (dissipating most power) when V_{out} is adjusted to its lowest value.

3-4.2 Current Limiting

The circuits in Figs. 3-27 through 3-30 have no protection. As the load demands more current (under short-circuit conditions), the IC will try to provide it at a constant output voltage, getting hotter all the time. This will continue until the IC literally burns itself up.

You can protect the IC by using the current limit feature. The characteristic curve of a current-limited power supply is shown in Fig. 3-31. The output *voltage*

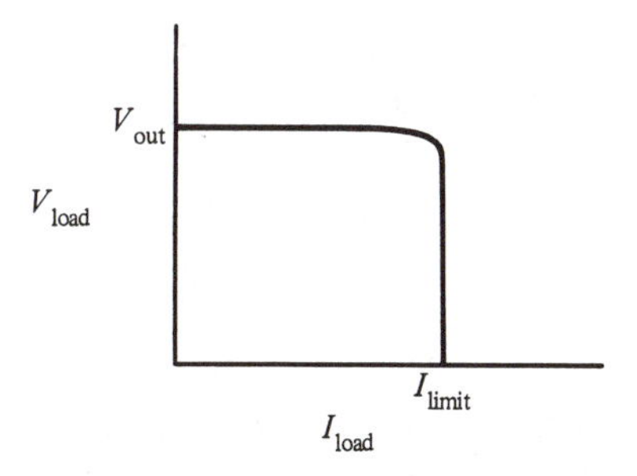

Figure 3-31 Characteristic curve for a current limited power supply

is held constant for any load current below the current limit. As current demand approaches the limit, the output voltage drops. The output voltage will continue to drop, as necessary, to keep the current at the current limit.

You can connect Q2 of 723, as shown in Fig. 3-32, to produce current limiting. The CL (current limit) terminal is connected to the output terminal. A small resistance resistor, R_{SC}, is connected between the output (CL) and the load. The emitter of Q2 (CS) is connected to the other side (load side) of R_{SC}.

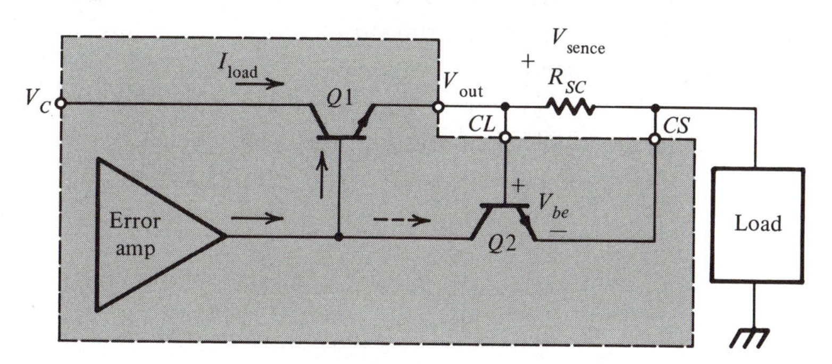

Figure 3-32 723 current limiting connection

The load current, passing through R_{SC}, develops a small voltage. This V_{sence} is applied directly across the base-emitter junction of Q2. When it reaches approximately 0.5 volt, Q2 begins to turn on. To turn on, Q2 pulls collector current from the error amplifier. But this current has been flowing into the base of Q1. With a drop of base current, the emitter current of Q1 (and therefore load current) decreases. However, if the load current decreases, V_{be} of Q2 drops. This causes less current to be pulled into the collector of Q2, away from the base of Q1. Transistor Q1 passes more current to the load. With more load current, V_{be} of Q2 goes up, turning Q2 back on. The cycle repeats itself. The net result of this cycling is that load current is held constant, at a level just large enough to produce a voltage across R_{SC} sufficient to turn on Q2. (Typically this voltage is about 0.5 V, but this varies from one IC to another.)

$$I_{limit} = \frac{V_{sence}}{R_{SC}} \approx \frac{0.5\ V}{R_{SC}} \tag{3-20}$$

Example 3-7

Design an adjustable voltage regulator (3 volts to 28 volts) with a short circuit limit of 60 milliamps at 30°C.

Solution

a. Transformer selection

Since the output must be able to go up to 28 volts, the minimum input value must be

$$V_{min} \geq V_{out\ max} + 3\ V$$

$$V_{min} \geq 28\ V + 3\ V = 31\ V$$

A 12.6 V transformer can output 17.8 volts maximum. Consequently, choose a 24-volt rms transformer.

b. Filter capacitor

You can see from Fig. 3-10 that a 60 milliamp load is quite small, so the low ripple equations must be used.

$$V_{pk} = 1.414 \times 24\ V - 1.4\ V = 32.5\ V \tag{3-2}$$

$$V_{min} = V_{pk} - \frac{I_{DC}}{fC} \tag{3-3}$$

Solving equation (3-3) for C, you obtain

$$C = \frac{I_{DC}}{f(V_{pk} - V_{min})}$$

$$C = \frac{60\ mA}{120(32.5\ V - 31\ V)} = 333\ \mu F$$

Choose $C = 350\ \mu F$

c. Regulator configuration

In order to obtain a low-voltage output of 3 volts, the reference voltage must be divided down, as it was in Fig. 3-27 by R1 and R2. However, to be able to reach 28 volts out, the error amplifier must also be given some gain, as with R1 and R2 of Fig. 3-29.

Finally, to provide short-circuit protection, R_{SC} and the CL and CS terminals must be connected. The schematic is shown in Fig. 3-33.

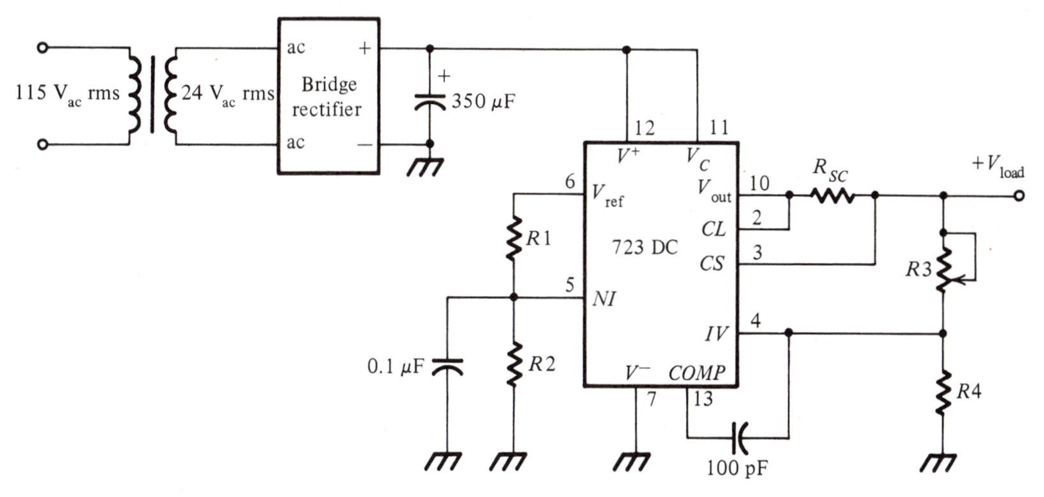

Figure 3-33 Schematic for Example 3-7

d. Regulator component calculation

Resistors R1 and R2 are calculated to set the NI terminal at 3 volts or less.

$$V_{NI} = 7.15\text{ V}\,\frac{R2}{R1 + R2}$$

When you solve this for R1, you obtain

$$R1 = \frac{R2}{V_{NI}}(7.15\text{ V} - V_{NI})$$

Since $\quad V_{NI} \approx 3$ volts

pick $\quad R2 = 3.3\text{ k}\Omega$

$$R1 = \frac{3.3\text{ k}\Omega}{3\text{ V}}(7.15\text{ V} - 3\text{ V})$$

$$R1 = 4.6\text{ k}\Omega$$

Pick $\quad R1 = 4.7\text{ k}\Omega$

To convert the 3 volts at the NI terminal to 28 volts at the output, R3 and R4 must be selected to give a gain of

$$\frac{28\text{ V}}{3\text{ V}} = 9.33 = 1 + \frac{\text{R3}}{\text{R4}}$$

or

$$\text{R4} = \frac{\text{R3}}{8.33}$$

Pick $\quad \text{R3} = 10\text{ k}\Omega$ potentiometer

$$\text{R4} = 1.2\text{ k}\Omega$$

Pick $\quad \text{R4} = 1.2\text{ k}\Omega$

The value of R_{SC} comes from equation (3-20):

$$I_{\text{limit}} = \frac{0.5\text{ V}}{\text{R}_{\text{SC}}}$$

or

$$\text{R}_{\text{SC}} = \frac{0.5\text{ V}}{I_{\text{limit}}} = \frac{0.5\text{ V}}{60\text{ mA}}$$

$$\text{R}_{\text{SC}} = 8.3\ \Omega$$

Pick $\quad \text{R}_{\text{SC}} = 8.6\ \Omega$

e. Will the regulator need a heat sink?

$$\theta_{JA\text{ max}} = \frac{T_J - T_A}{(V_{\text{DC}} - V_{\text{out min}})I_{\text{load}}} \tag{3-17}$$

The maximum junction temperature for the 723C is 125°C.

$$V_{\text{DC}} = V_{\text{pk}} - \tfrac{1}{2}V_{\text{ripple}}\text{ (p-p)}$$

$$= 32.5\text{ V} - \tfrac{1}{2}\frac{I_{\text{DC}}}{fC}$$

$$V_{\text{DC}} = 32.5\text{ V} - \tfrac{1}{2}\left(\frac{60\text{ mA}}{120 \times 350\ \mu\text{F}}\right) = 31.8\text{ V}$$

$$\theta_{JA\text{ max}} = \frac{125°\text{C} - 30°\text{C}}{(31.8\text{ V} - 0\text{ V})60\text{ mA}}$$

$$\theta_{JA\text{ max}} = 49.8°\text{C/W}$$

Notice that $V_{\text{out min}}$ has been set at zero volts, not 3 volts. When the output is short-circuited and R_{SC} produces limiting, the output drops to zero volts. The *entire* input voltage must be dropped across the regulator. This is the worst possible case for the IC.

$$\theta_{JA\,\text{max}} = 49.8°\text{C/W}$$

θ_{JA} is specified as 150°C/W, so a heat sink is definitely needed.

For the T0-100 metal case,

$$\theta_{JC} = 35°\text{C/W}$$

or

$$\theta_{SA_{\text{max}}} = \theta_{JA\,\text{max}} - \theta_{JC} \tag{3-15}$$

$$\theta_{SA_{\text{max}}} = 49.8°\text{C/W} - 35°\text{C/W} = 14.8°\text{C/W}$$

You must select a heat sink which will fit a T0-100 package with a thermal resistance below 14.8°C/W.

3-4.3 Current Foldback

You saw in Example 3-7 that when the load is short-circuited, maximum current flows, and the entire input voltage is dropped by the regulator. At this point ($I = I_{SC}$, $V_{\text{out}} = 0$) the regulator is working its hardest, yet nothing useful is being achieved, since the load is short-circuited. To protect the regulator you must severely limit the short-circuit current, but this places the same severe limit on the maximum load current before current limiting begins.

Current foldback, as shown in Fig. 3-34, offers an alternative. As current demand increases, the output voltage is held constant *until* a preset current level

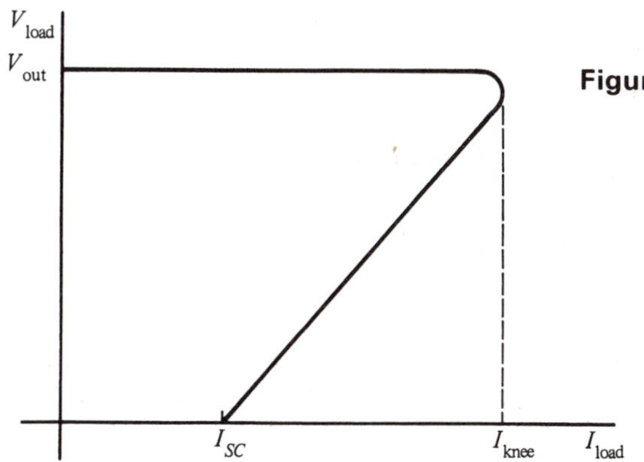

Figure 3-34 Current foldback characteristic curve

(I_{knee}) is reached. By increasing current demand above this level both output voltage and output current will decrease. This foldback allows your circuit to pass higher currents to the load, but to limit the short circuit current to a lower level.

How you can apply current foldback to a 723 IC regulator is shown in Fig. 3-35. The addition of R3 and R4, to divide the voltage fed to CL, is all that is necessary to achieve foldback.

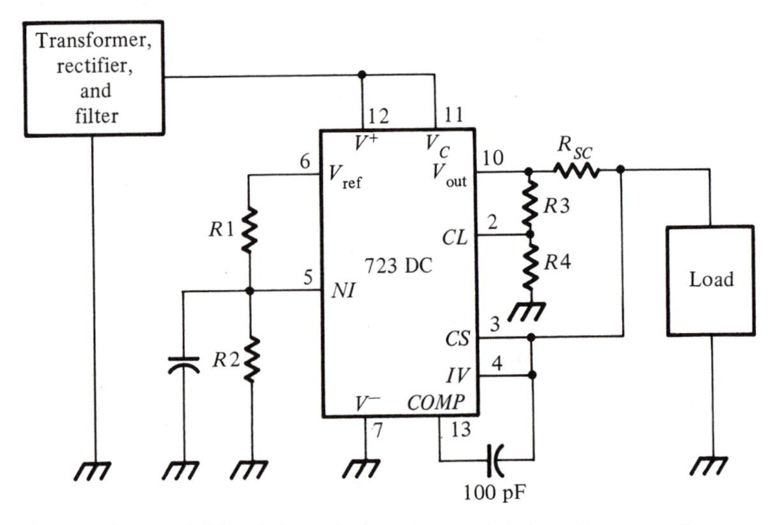

Figure 3-35 Current foldback installed on the basic low voltage regulator

The knee current is

$$I_K = \frac{(\mathrm{R3} + \mathrm{R4})(V_{out} + 0.5\ \mathrm{V})}{\mathrm{R_{SC}\,R4}} - \frac{V_{out}}{\mathrm{R_{SC}}} \tag{3-21a}$$

or

$$I_K = \frac{V_{out}\,\mathrm{R3}}{\mathrm{R_{SC}\,R4}} + \frac{0.5\ \mathrm{V}(\mathrm{R3} + \mathrm{R4})}{\mathrm{R_{SC}\,R4}} \tag{3-21b}$$

With foldback, the short-circuit limit is changed to

$$I_{SC} = \frac{(\mathrm{R3} + \mathrm{R4})(0.5\ \mathrm{V})}{\mathrm{R_{SC}\,R4}} \tag{3-22}$$

Usually, the I_{SC} and I_K requirements of a circuit are given and you must select component values. This requires that equations (3-21) and (3-22) be solved simultaneously. The results are

R4 = select arbitrarily (10k ohms is a good value)

$$R_{SC} = \frac{(0.5 \text{ V})(V_{reg})}{(I_{SC})(V_{reg} + 0.5 \text{ V}) - (I_K)(0.5 \text{ V})} \tag{3-23}$$

$$R3 = \frac{I_{SC}\, R4\, R_{SC}}{(0.5 \text{ V})} - R4 \tag{3-24}$$

Example 3-8

Design a foldback circuit for a 723 regulator to meet the following specifications:

$$I_K = 100 \text{ mA}$$

$$I_{SC} = 20 \text{ mA}$$

$$V_{reg} = 8 \text{ V}$$

Solution

1. Pick $R4 = 10 \text{ k}\Omega$

2. $$R_{SC} = \frac{(0.5 \text{ V})(V_{reg})}{(I_{SC})(V_{reg} + 0.5 \text{ V}) - (I_K)(0.5 \text{ V})} \tag{3-23}$$

$$R_{SC} = \frac{(0.5 \text{ V})(8 \text{ V})}{(20 \text{ mA})(8 \text{ V} + 0.5 \text{ V}) - (100 \text{ mA})(0.5 \text{ V})} = 33.3\ \Omega$$

Select $R_{SC} = 33\ \Omega$

3. $$R3 = \frac{I_{SC}\, R4\, R_{SC}}{0.5 \text{ V}} - R4$$

4. $$R3 = \frac{(20 \text{ mA})(10 \text{ k}\Omega)(33\ \Omega)}{0.5 \text{ V}} - 10 \text{ k}\Omega = 3.2 \text{ k}\Omega$$

Select $R3 = 3.3 \text{ k}\Omega$

Figure 3-36 shows how the current foldback resistors connect to the regulator's functional blocks.

Compare Fig. 3-36 (foldback) with Fig. 3-32 (simple current limiting). Resistors R3 and R4 divide the voltage developed across R_{SC} (V_{sence}) before applying it to the base of Q2. That is the only difference between the two (current foldback and current limiting) circuits.

The partial schematic of Fig. 3-37 will help you understand how foldback is achieved. The current limit transistor Q2 stays off until voltage developed across

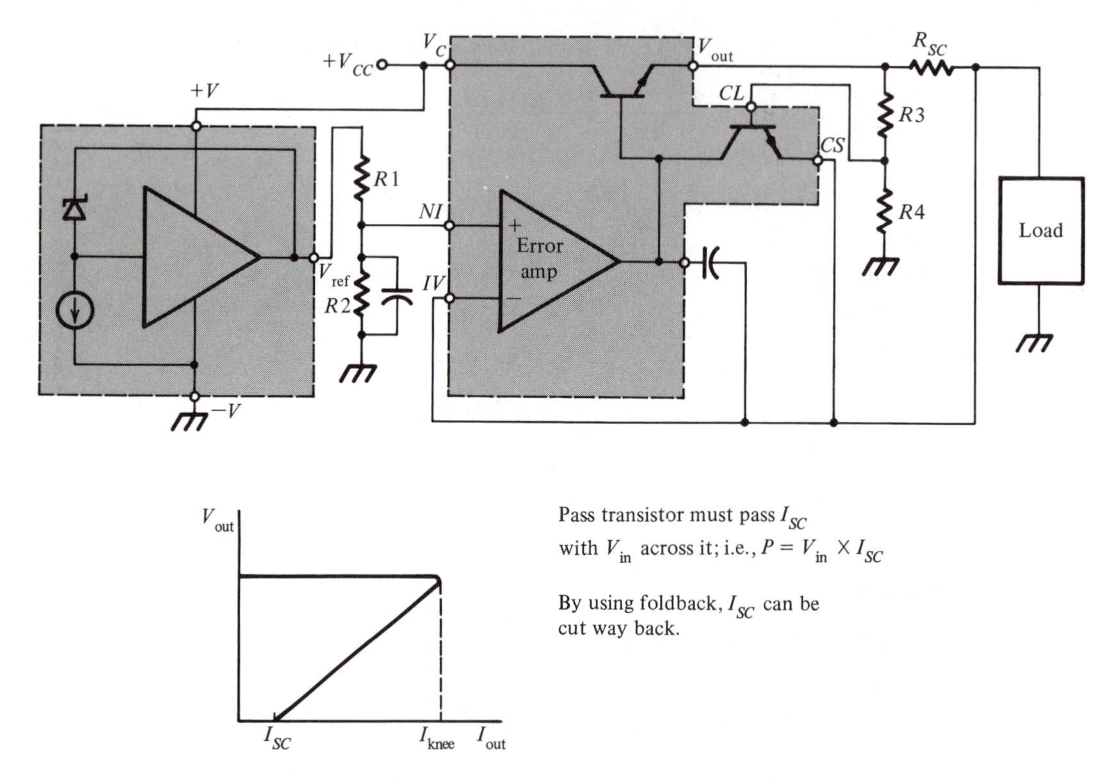

Figure 3-36 Functional block diagram of current foldback for the 723 IC

R_{SC} by the load current I is large enough to cause the base-emitter voltage of Q2 to be 0.5 volt. This means that the base potential (with respect to ground) must be 0.5 volt above the load (to ground) voltage (V_0).

Transistor Q2 then begins to turn on, stealing some of the base current from Q1. This, in turn, begins to turn Q1 off, which lowers I. A drop in current lowers both the voltage out of the regulator (Q1) and the voltage across the load. These two voltages will fall at approximately the same rate (because R_{SC} is normally much smaller than the load resistance). However, the base voltage of Q2 falls at a slower rate because of the voltage divider R3 and R4. That is, both Q2 emitter voltage (V_0) and Q2 base voltage drop, but the base drops at a *slower rate.*

The net result, then, is that the *difference* in potential between base and emitter of Q2 *increases.* This forces Q2 to conduct harder. It, in turn, steals more of Q1's base current, decreasing Q1's conduction further. This process continues (Q1 decreasing conduction, Q2 conducting harder) until $V_0 \approx 0$ V, and V1 is just large enough to keep 0.5 V across R4 and the base-emitter junction of Q2. This point is I_{SC} and was reached by lowering *both I* and V_0.

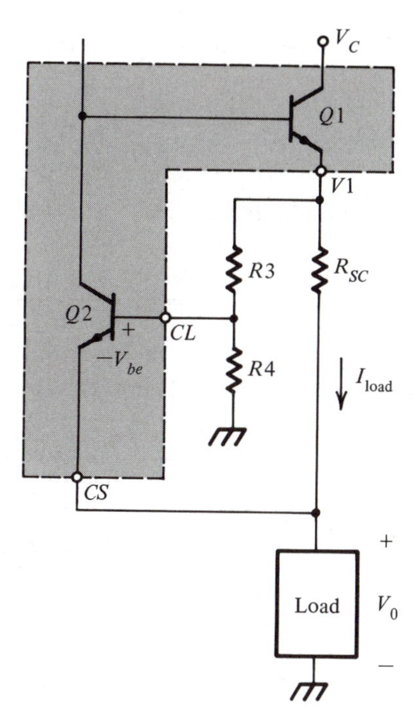

Figure 3-37 Current foldback partial schematic

You can now derive equations (3-21) and (3-22). Foldback starts when Q2 base voltage (voltage across R4) is 0.5 volt above the regulated output voltage:

$$V_{R_4} = V_{reg} + 0.5 \text{ V} \tag{3-25}$$

But this is produced by a voltage divider from V1:

$$V_{R_4} = \frac{\text{R4V1}}{\text{R3} + \text{R4}} \tag{3-26}$$

The relationship between V_{reg} and V1 is

$$\text{V1} = V_{reg} + I_K \text{R}_{SC} \tag{3-27}$$

Substitute equation (3-27) into equation (3-26):

$$\text{V}_{R4} = \frac{\text{R4}}{\text{R3} + \text{R4}} (V_{reg} + I_K \text{R}_{SC}) \tag{3-28}$$

Equation (3-28) can be algebraically manipulated to isolate $I_K R_{SC}$ on one side:

$$\frac{(R3 + R4)}{R4} V_{R4} - V_{reg} = I_K R_{SC}$$

Solving for I_K, you obtain

$$I_K = \frac{(R3 + R4)V_{R4}}{R_{SC} R4} - \frac{V_{reg}}{R_{SC}} \tag{3-29}$$

Finally, substitute equation (3-25) into equation (3-29) to eliminate V_{R4}:

$$I_K = \frac{(R3 + R4)(V_{reg} + 0.5 \text{ V})}{R_{SC} R4} - \frac{V_{reg}}{R_{SC}} \tag{3-21a}$$

When a full short circuit occurs, V_{reg} falls to zero volts:

$$V_{reg} = 0 \text{ V at } I_{SC}$$

Substituting this into equation (3-21a), you obtain

$$I_{SC} = \frac{(R3 + R4)(0.5 \text{ V})}{R_{SC} R4} \tag{3-22}$$

By adding two resistors, current foldback is achieved. Load current can be kept high under normal conditions, but folds back to a much lower level when a short-circuited load forces the regulator to drop the *entire* input voltage. So, using foldback, you can build a regulator with a much higher output current while still protecting the regulator against a short circuit.

3-4.4 Current Boosting

Even with proper use of current foldback, the maximum current that the 723 IC regulator itself can provide is 140 milliamps. For many applications, this just is not enough. You must current-boost the 723 regulator.

A current-boosted low-voltage regulator is shown in Fig. 3-38. You can also add a boost transistor to the high voltage regulator (Fig. 3-29) and the adjustable regulator (Fig. 3-33). Current foldback could also be used.

The block diagram of Fig. 3-39 illustrates how the pass transistor works with the functional blocks of the regulator IC. The collector current of the pass transistor is fed from the capacitive filter. The output current from the IC (from the $+V_{out}$ terminal) drives the base of the pass transistor. This base current is multi-

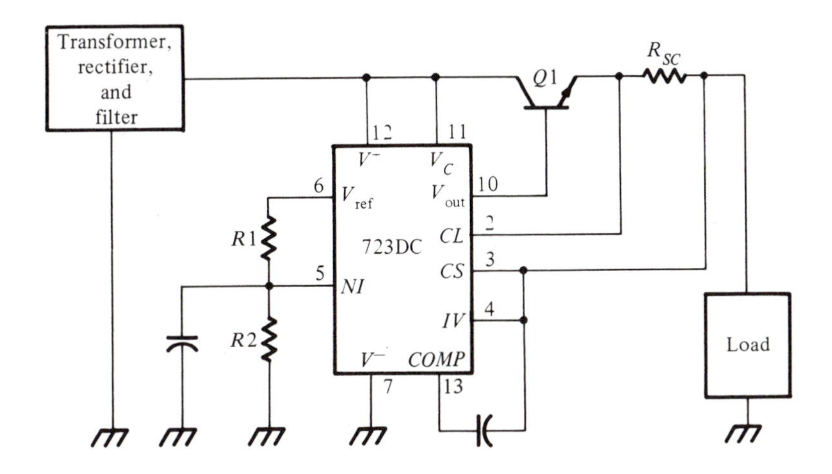

Figure 3-38 Current boosted low voltage regulator

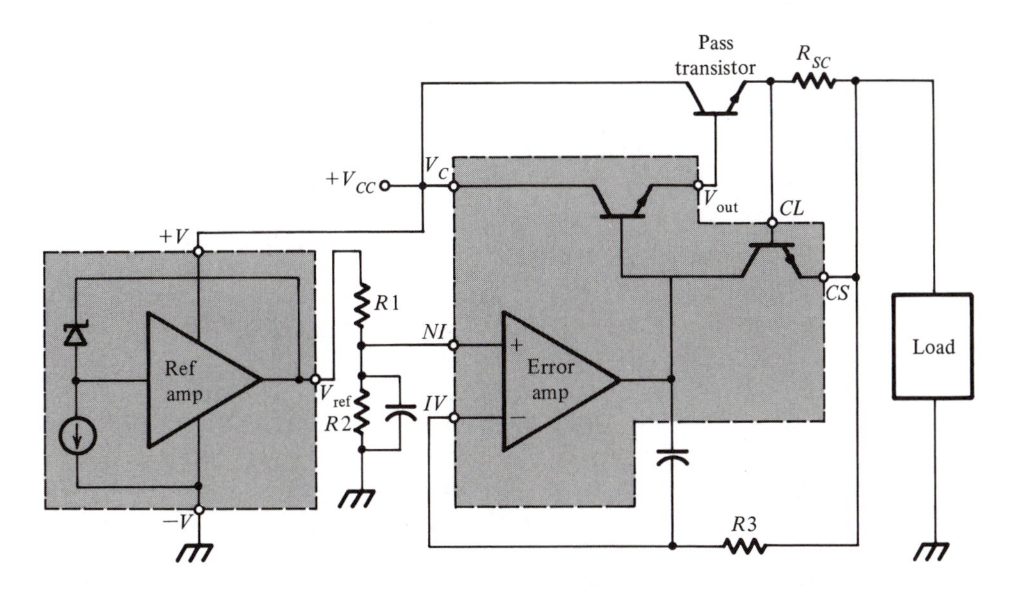

Figure 3-39 Functional block diagram of a current boosted 723 regulator

plied by the beta of the pass transistor. Consequently, the 723 must provide only the base current. Since the pass transistor and R_{SC} are enclosed in the negative feedback loop of the error amplifier, offsets, nonlinearities, and other voltage drops are automatically compensated for, keeping the *load* voltage regulated.

All of the equations developed for regulated voltage, short-circuit current, and knee current still hold for the current-boosted regulator. In addition,

$$I_{\text{load}} = \beta_{\text{pass transistor}} \times I_{\text{out 723}}$$

$$P_{\text{pass transistor}} = (V_{\text{in}} - V_{\text{load}})I_{\text{load}}$$

Heat sink requirements for the pass transistor are calculated just as you calculate them for the current-boosted simple op amp regulator. Look back at Example 3-3. Keep in mind, as you do, that under short-circuit conditions, V_{load} drops to zero volts, causing the pass transistor's power dissipation to go up.

3-4.5 Negative Voltage Regulation

The 723 regulator IC can also be used to produce regulated negative voltages. The basic schematic is shown in Fig. 3-40. There are several fundamental differences between the positive and negative regulator circuits. The $V+$ and V_c terminals are connected to ground (not to the input), and the $V-$ terminal is tied to

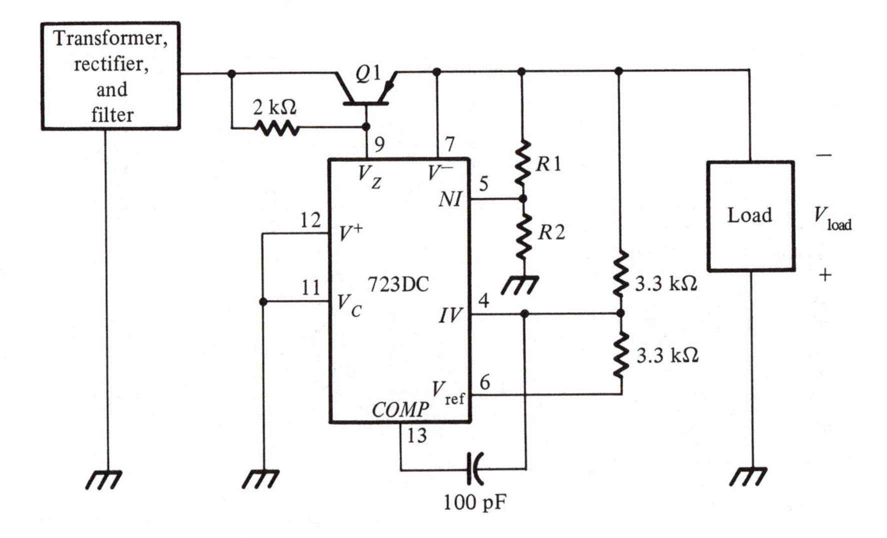

Figure 3-40 Basic negative regulator

the *regulated* negative voltage. A PNP external pass transistor *must* be used.

A new terminal, V_z, is used to drive the pass transistor (rather than V_{out}). Feedback from the output is provided to *both NI* and *IV* inputs to the error amp with R1 < R2. Finally, the reference signal divider is tied to the negative regulated output (not grounded).

The functional diagram is shown in Fig. 3-41. Load current flows from

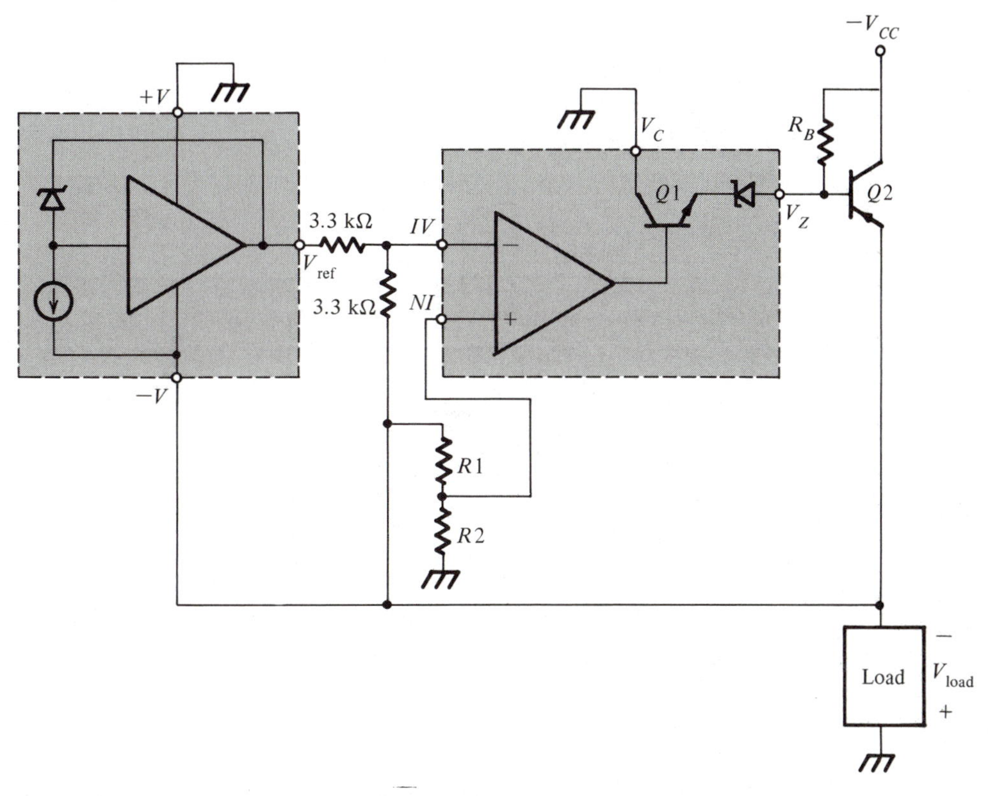

Figure 3-41 Functional diagram of 723 IC regulator as a negative voltage regulator

ground, through the load, and into the pass transistor. Current from the error amp controls Q1 conduction. Current flows from ground, through Q2, through the zener (which shifts voltage levels, assuring an adequate negative output voltage), and through R_B. The voltage across R_B, established by current from Q1, sets the voltage at the base of the external pass transistor. This, in turn, controls the conduction of the pass transistor, and thereby controls the load voltage.

Any variation of that load voltage is applied to both inputs of the error amp. Also, the $V-$ terminal (which provides the reference for the V_{ref} signal) is tied to the load, *not ground*. The V_{ref} signal will be held 7 volts more positive than the load voltage, but will change as the load voltage changes. How this circuit regulates the load voltage is best seen with a numerical example.

Example 3-9

For R1 = 3.47 kΩ and R2 = 8.45 kΩ, calculate V_{ref}, IV, NI, and $NI - IV$, and describe circuit operation for:

a. $V_L = -12$ V
b. $V_L = -11$ V
c. $V_L = -13$ V

Solution

a.

$$V_{\text{load}} = -12 \text{ V}$$

$$V_{\text{ref}} = V_{\text{load}} + 7 \text{ V} = -5 \text{ V}$$

$$IV = V_{\text{ref}} + \frac{(V_{\text{load}} - V_{\text{ref}})}{2} = -5 \text{ V} + \frac{(-12 \text{ V} + 5 \text{ V})}{2} = -8.5 \text{ V}$$

$$NI = \frac{V_{\text{load}} \times 8.45 \text{ k}\Omega}{8.45 \text{ k}\Omega + 3.47 \text{ k}\Omega} = -8.5 \text{ V}$$

$$NI - IV = 0$$

This is the condition necessary for a stable negative feedback system. Remember that an op amp with negative feedback alters its output in such a way as to minimize the differences in potential between its inputs ($NI - IV$).

Consequently, -12 V is the load voltage to which this circuit regulates.

b. $V_{\text{load}} = -11$ V (The load voltage has gone more positive by 1 volt.)

$$V_{\text{ref}} = V_{\text{load}} + 7 \text{ V} = -4 \text{ V}$$

$$IV = V_{\text{ref}} + \frac{(V_{\text{load}} - V_{\text{ref}})}{2}$$

$$= -4 \text{ V} + \frac{(-11 \text{ V} + 4 \text{ V})}{2} = -7.5 \text{ V}$$

$$NI = \frac{V_{\text{load}} \times 8.45 \text{ k}\Omega}{8.45 \text{ k}\Omega + 3.47 \text{ k}\Omega} = -7.8 \text{ V}$$

$$NI - IV = -0.3 \text{ V}$$

A negative difference in potential at the input of the error amp will drive its output more negative. Placing a more negative voltage on the base of Q1 (a NPN transistor) causes it to conduct less. This drives less current through the base resistor of Q2, allowing its base to become more negative. Since Q2 is a PNP transistor, it will conduct harder, passing more current from the load. More current through the load produces a larger (more negative) load voltage. The regulator is driving the output voltage back toward the stable −12 V level.

c. $V_{load} = -13$ V (The load voltage has gone more negative by 1 volt.)

$$V_{ref} = V_{load} + 7\text{ V} = -6\text{ V}$$

$$IV = V_{ref} + \frac{(V_{load} - V_{ref})}{2}$$

$$= -6\text{ V} + \frac{(-13\text{ V} + 6\text{ V})}{2} = -9.5\text{ V}$$

$$NI = \frac{V_{load} \times 8.45\text{ k}\Omega}{8.45\text{ k}\Omega + 3.47\text{ k}\Omega} = -9.2\text{ V}$$

$$NI - IV = +0.3\text{ V}$$

A positive difference in potential at the input of the error amp will drive its output more positive (less negative). Placing a less negative voltage on the base of Q1 causes it to conduct harder. This drives more current through the base resistor of Q2, driving its base less negative (further from $-V_{cc}$). Transistor Q2 conducts less. Less load current means less voltage dropped across the load. Again, the regulator is driving the output back toward −12 V.

For the negative regulator circuit of Figs. 3-40 and 3-41 the load voltage will be regulated at

$$V_{load} = -3.5\text{ V}\left(1 + \frac{R2}{R1}\right)$$

However, the smallest output voltage is −6 volts. Also, you should have noticed from the schematic that even though an external boost transistor is available to allow higher load currents, there is no current limiting or current foldback incorporated. The *CL* and *CS* cannot simply be connected because of the reversed load current flow. The polarity of V_{sence} would be backwards, never turning on the short-circuit limit transistor.

3-5 TRACKING REGULATORS

Many applications require several different voltage power supplies. One solution is to build several independent regulators. However, very often it is important that all of these supply voltages track. That is, if one of the supply voltages goes up two percent, it is best if *all* of the supply voltages go up the same amount. You can do this by adding an op amp to the adjustable three-terminal regulator of Fig. 3-23.

3-5.1 Op Amp/Three-terminal Regulator

A simple tracking power supply is shown in Fig. 3-42. It consists of a center tapped transformer, rectifier, and dual filtering. This is followed by an adjustable voltage regulator to provide the positive regulated output voltage. The op amp is configured as an inverting amplifier, with a gain of -1. The negative output voltage is produced by this op amp, from $+V_{out}$. Any change in the positive output voltage is inverted by the op amp and appears at $-V_{out}$. This is tracking.

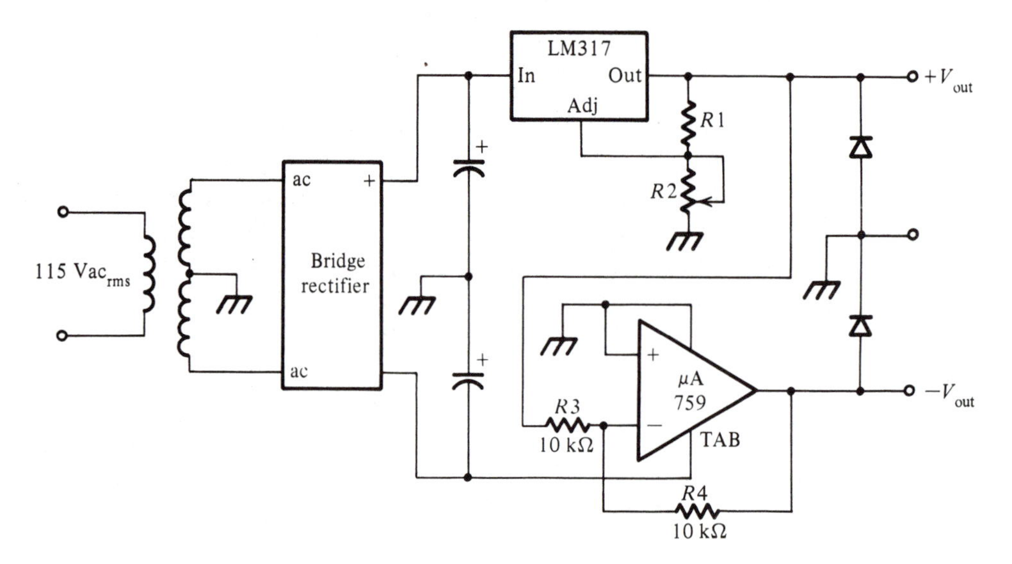

Figure 3-42 Simple tracking power supply

Two adjustments are possible. Changing R2 alters $+V_{out}$. This will cause a similar change in $-V_{out}$. Changing the relationship between R3 and R4 changes the gain of the inverting amp. Altering this gain will change the relationship between $+V_{out}$ and $-V_{out}$. If the gain were -1, a $+V_{out}$ of $+6$ volts would cause a $-V_{out}$ of -6 volts. If you change the gain to -2 (R3 = 10 kΩ, R4 = 20 kΩ), a $+V_{out}$ of $+6$ volts produces a $-V_{out}$ of -12 volts. You can expand this technique to provide multiple positive and negative regulated output voltages, all at different levels, all of which track the single adjustable regulator IC.

There are several cautions you should observe. The op amp selected is capable of high output currents (at least 325 mA short circuit). If you are forced to use standard, lower current op amps, then you must provide a pass transistor (as was discussed in section 3-2). Since most op amps do not provide internal thermal limiting (as many regulator ICs do), you must be sure to provide adequate heat sinking. This is to protect the op amp during short circuits. The heat sink calculations of section 3-3.3 apply. Finally, remember to keep the op amp out of saturation. This restricts you to an output voltage which is about two volts smaller than the supply voltage.

The circuit in figure 3-43 provides current limiting and thermal shut down for the negative as well as the positive output voltage. The positive regulated voltage is produced with a LM317 adjustable positive regulator IC, as it was in Fig. 3-42.

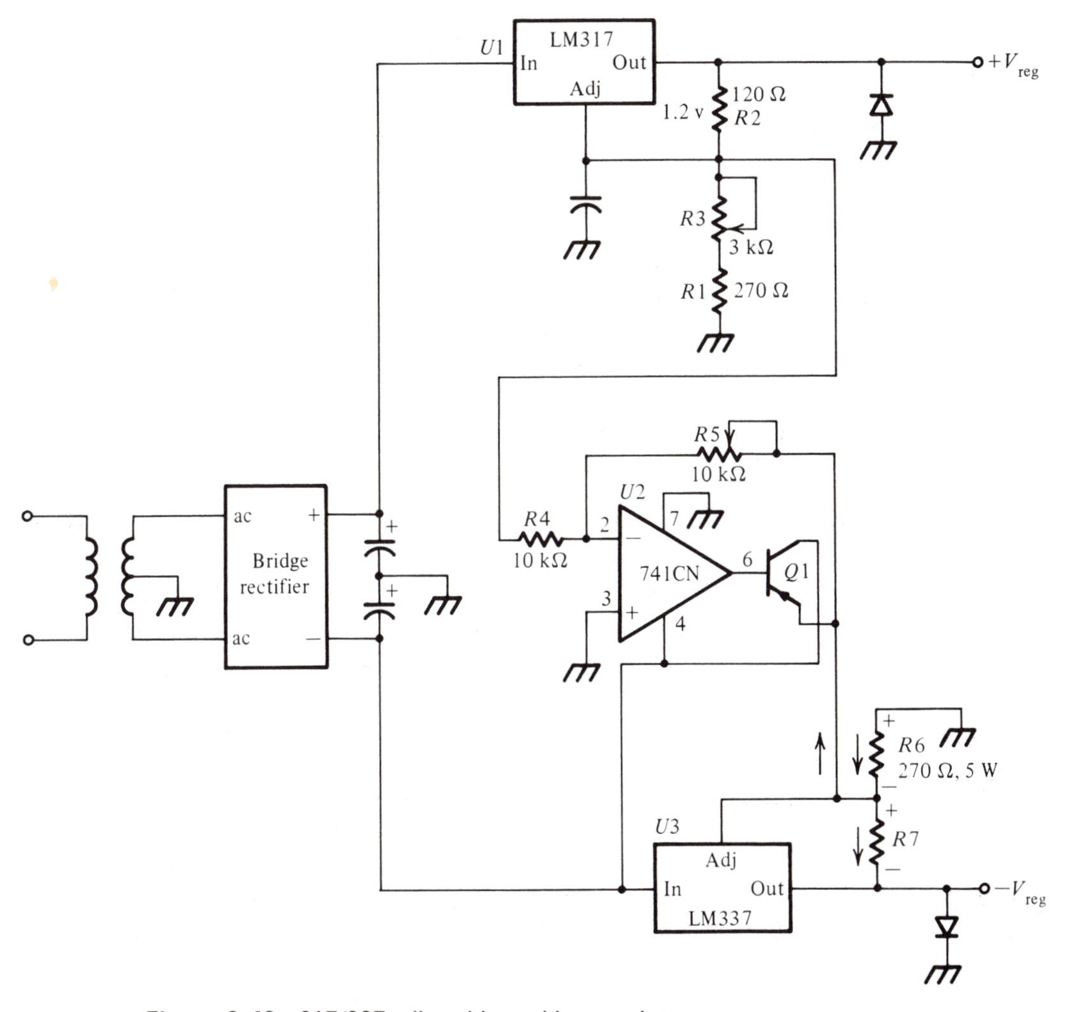

Figure 3-43 317/337 adjustable tracking regulator

The regulated positive voltage is always 1.2 V more positive than the voltage across R3 (on the adjust pin). This voltage, across R3, is inverted by the inverting amplifier U2 (and associated components) and drives the adjust pin of U3. The LM337 (U3) is the negative compliment of the LM317. Its output will be 1.2 V more negative than its adjust pin.

With R3 adjusted all the way down, the 10 mA produced by the 1.2 V across R2 causes a 2.7 V potential on U1's adjust pin, sending the output to 3.9 V (1.2 V + 2.7 V). The +2.7 V across R1 is inverted to −2.7 by U2 (and associated components). Negative 2.7 V across R6 produces 10 mA through R6 and R7, biasing U3 on and producing 10 mA · (270 Ω + 120 Ω) = −3.9 V regulated negative output.

Increasing the resistance of R3 causes the 10 mA from R2 to produce a larger voltage across R3. The positive regulated voltage increases, staying 1.2 V above the voltage across R3. The inverting amplifier drives the adjust pin of U3 proportionally more negative. In turn, the negative regulated output voltage goes more negative, staying 1.2 V more negative than its adjust pin. Ten milliamps of the current flowing through R6 goes through R7 (producing the 1.2 V drop). The remainder of that current must be sunk by Q1.

3-5.2 Dedicated Tracking Regulator ICs

The regulator/op amp technique of producing a dual tracking power supply is simple. However, it requires two IC packages and possibly an additional pass transistor. Also, there is no way to set the short-circuit current limit.

The MC1468/1568 is a dual tracking regulator IC. It allows output voltage adjustment between ±8 volts and ±20 volts. The short-circuit current limit can be adjusted with an external resistor. The basic regulator schematic is shown in Fig. 3-44. Connection and operation are similar to the 723. Dual, unregulated voltages are connected from the filters between V_{CC} and V_{EE}. Resistor R_{SC} is placed in series with the load current (as with the 723). The sense terminals are also connected as you connect the 723 sense terminals. The short circuit current is limited to

$$I_{SC} = \frac{0.6\ \text{V}}{R_{SC}}$$

The output voltages are nominally ±15 volts. To adjust these voltages, a potentiometer can be used to sample and feed back positive voltage to the V_{adj} terminal. This is shown in Fig. 3-45.

$$R2 = \frac{R1(1\ \text{k}\Omega)(7.28\ \text{V})}{(1\ \text{k}\Omega)(V_{out} - 7.28\ \text{V}) - .68\ \text{V}\ R1}$$

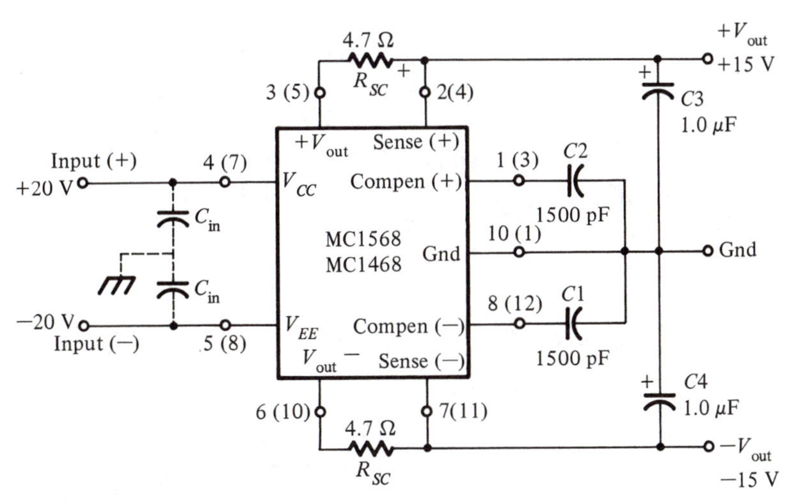

C1 and C2 should be located as close to the device as possible. A 0.1 μF ceramic capacitor (C_{in}) may be required on the input lines if the device is located an appreciable distance from the rectifier filter capacitors. C3 and C4 may be increased to improve load transient response and to reduce the output noise voltage. At low temperature operation, it may be necessary to bypass C4 with a 0.1 μF ceramic disc capacitor.

Figure 3-44 Basic tracking regulator using a MC1568 (*Courtesy of Motorola Semiconductor Products*)

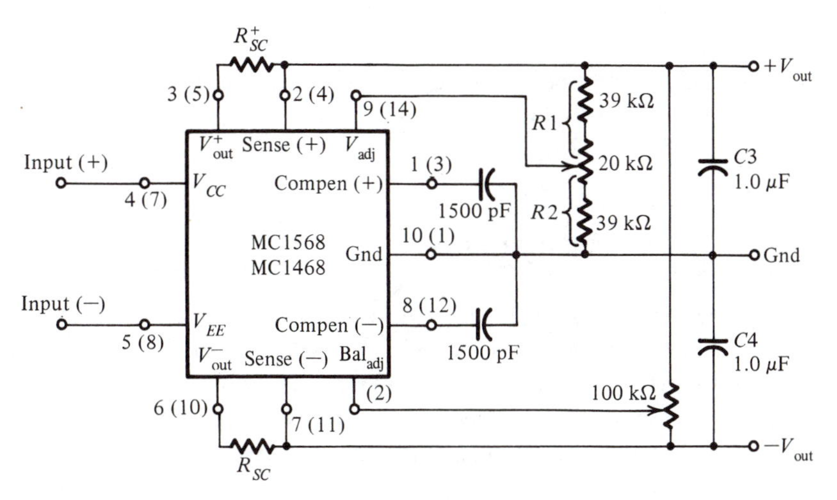

Balance adjust available in MC1568L, MC1468L ceramic dual in line package only.

Figure 3-45 Adjustable output tracking voltage (*Courtesy of Motorola Semiconductor Products*)

Also, like the 723 regulator, the MC1568/1468 IC has a limited output current capacity. This can be boosted with external pass transistors. A ±1.5 amp regulator schematic is given in Fig. 3-46. Note the similarities between this circuit and the 723 boosted regulator of Fig. 3-38. You must apply the heat sink calculations for both the pass transistor and the IC. The maximum ratings and electrical characteristics for the MC1568/1468 are given in Table 3-7.

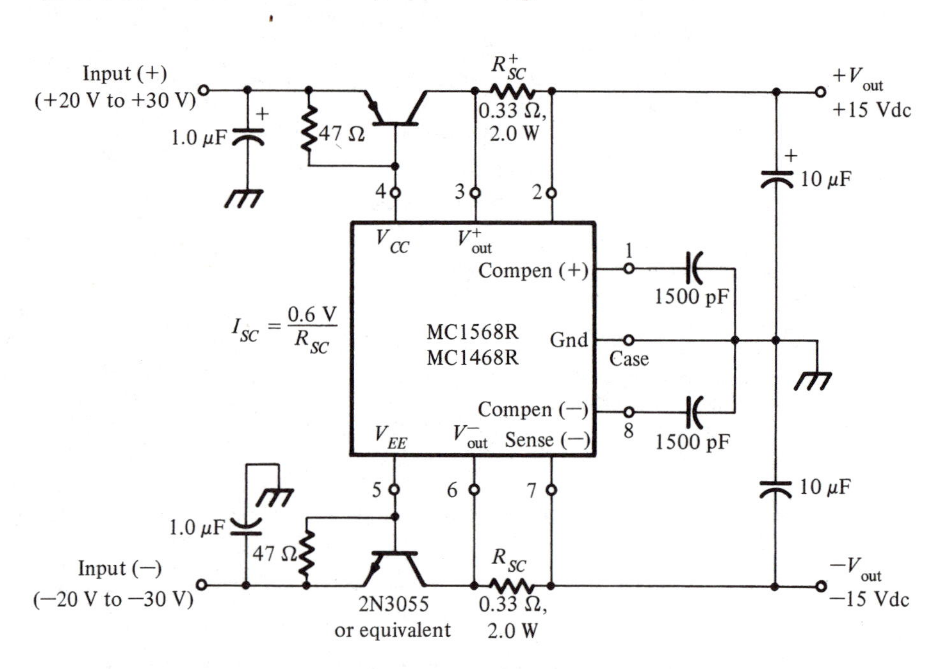

Figure 3-46 Boosted output tracking regulator (*Courtesy of Motorola Semiconductor Products*)

3-6 SWITCHING POWER SUPPLY

The regulated power supplies discussed so far are often referred to as *linear supplies*, since the pass transistors operate in their linear region. This mode of operation has several disadvantages. The power stepdown transformer is the largest, most expensive component of the supply. This is caused by the relatively low line frequency (60 Hz). The low line frequency also requires large filter capacitors to decrease the ripple. Finally, operating the pass transistor in the active region forces it to dissipate quite a bit of power. These three factors combine to lower the efficiency of the power supply, while increasing its bulk and cost.

Switching power supplies overcome these problems. Switchers rely on pulse width modulation to control the average value of the output voltage. This is

TABLE 3-7. MC1468, MC1568 dual regulator IC maximum ratings and electrical characteristics

MAXIMUM RATINGS (T_C = +25°C unless otherwise noted.)

Rating	Symbol	Value			Unit
Input Voltage	V_{CC}, $\lvert V_{EE}\rvert$	30			Vdc
Peak Load Current	I_{PK}	100			mA
Power Dissipation and Thermal Characteristics		G Package	R Package	L Package	
T_A = +25°C	P_D	0.8	2.4	1.0	Watts
Derate above T_A = +25°C	$1/\theta_{JA}$	6.6	28.5	10	mW/°C
Thermal Resistance, Junction to Air	θ_{JA}	150	35	100	°C/W
T_C = +25°C	P_D	2.1	9.0	2.5	Watts
Derate above T_C = +25°C	$1/\theta_{JC}$	14	61	20	mW/°C
Thermal Resistance, Junction to Case	θ_{JC}	70	17	50	°C/W
Storage Junction Temperature Range	T_J, T_{stg}	-65 to +175			°C
Minimum Short-Circuit Resistance	$R_{SC}(min)$	4.0			Ohms

OPERATING TEMPERATURE RANGE

		Symbol	Value	Unit
Ambient Temperature		T_A		°C
	MC1468		0 to +70	
	MC1568		-55 to +125	

ELECTRICAL CHARACTERISTICS (V_{CC} = +20 V, V_{EE} = -20 V, C1 = C2 = 1500 pF, C3 = C4 = 1.0 μF, R_{SC}^+ = R_{SC}^- = 4.0 Ω, I_L^+ = I_L^- = 0, T_C = +25°C unless otherwise noted.) (See Figure 1.)

		MC1568			MC1468			
Characteristic	**Symbol***	**Min**	**Typ**	**Max**	**Min**	**Typ**	**Max**	**Unit**
Output Voltage	V_O	±14.8	±15	±15.2	±14.5	±15	±15.5	Vdc
Input Voltage	V_{in}	–	–	±30	–	–	±30	Vdc
Input-Output Voltage Differential	$\lvert V_{in} - V_O\rvert$	2.0	–	–	2.0	–	–	Vdc
Output Voltage Balance	V_{Bal}	–	±50	±150	–	±50	±300	mV
Line Regulation Voltage	Reg_{in}							mV
(V_{in} = 18 V to 30 V)		–	–	10	–	–	10	
(T_{low} ① to T_{high} ②)		–	–	20	–	–	20	
Load Regulation Voltage	Reg_L							mV
(I_L = 0 to 50 mA, T_J = constant)		–	–	10	–	–	10	
(T_A = T_{low} to T_{high})		–	–	30	–	–	30	
Output Voltage Range	V_{OR}							Vdc
L Package (See Figure 4.)		±8.0	–	±20	±8.0	–	±20	
R and G Packages (See Figures 2 and 13.)		±14.5	–	±20	±14.5	–	±20	
Ripple Rejection (f = 120 Hz)	RR	–	75	–	–	75	–	dB
Output Voltage Temperature Stability (T_{low} to T_{high})	$\lvert TS_{V_O}\rvert$	–	0.3	1.0	–	0.3	1.0	%
Short-Circuit Current Limit (R_{SC} = 10 ohms)	I_{SC}	–	60	–	–	60	–	mA
Output Noise Voltage (BW = 100 Hz - 10 kHz)	V_N	–	100	–	–	100	–	μV(RMS)
Positive Standby Current (V_{in} = +30 V)	I_B^+	–	2.4	4.0	–	2.4	4.0	mA
Negative Standby Current (V_{in} = -30 V)	I_B^-	–	1.0	3.0	–	1.0	3.0	mA
Long-Term Stability	$\Delta V_O/\Delta t$	–	0.2	–	–	0.2	–	%/k Hr

① T_{low} = 0°C for MC1468
= -55°C for MC1568

② T_{high} = +70°C for MC1468
= +125°C for MC1568

(Courtesy of Motorola Semiconductors)

illustrated in Fig. 3-47. The average value of a repetitive pulse waveform depends on the area under the waveform curve. Varying the duty cycle (ON time per period) changes the average value of the voltage proportionally.

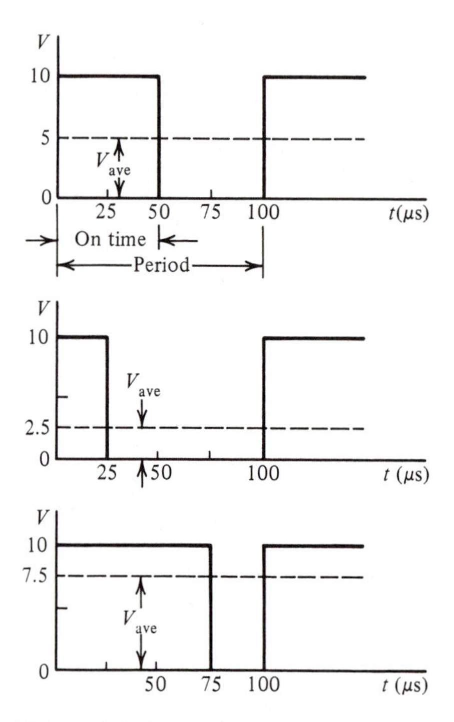

Figure 3-47 Pulse width modulation and average value

A switching power supply is shown in Fig. 3-48. The bridge rectifier and capacitor filters are connected directly to the AC line. Thermistor Rt° limits high initial capacitor charge currents. Transistors Q1 and Q2 are alternately switched off and on at 20 kHz. Since these transistors are either hard off (no collector current) or hard on ($V_{ce} \approx 0.2$ V) they dissipate very little power. The transistors drive the primary of the main power transformer. The secondary is full wave center-tapped rectified. This square wave is then filtered to yield its average value with minimum ripple at the output.

Regulation is provided by the pulse width modulator IC chip. It varies the time that it drives each of the transistors on. By keeping the period of the drive signals constant, and varying the turn-on time, pulse width modulation is achieved. As shown in Fig. 3-47, varying the pulse width changes the average value.

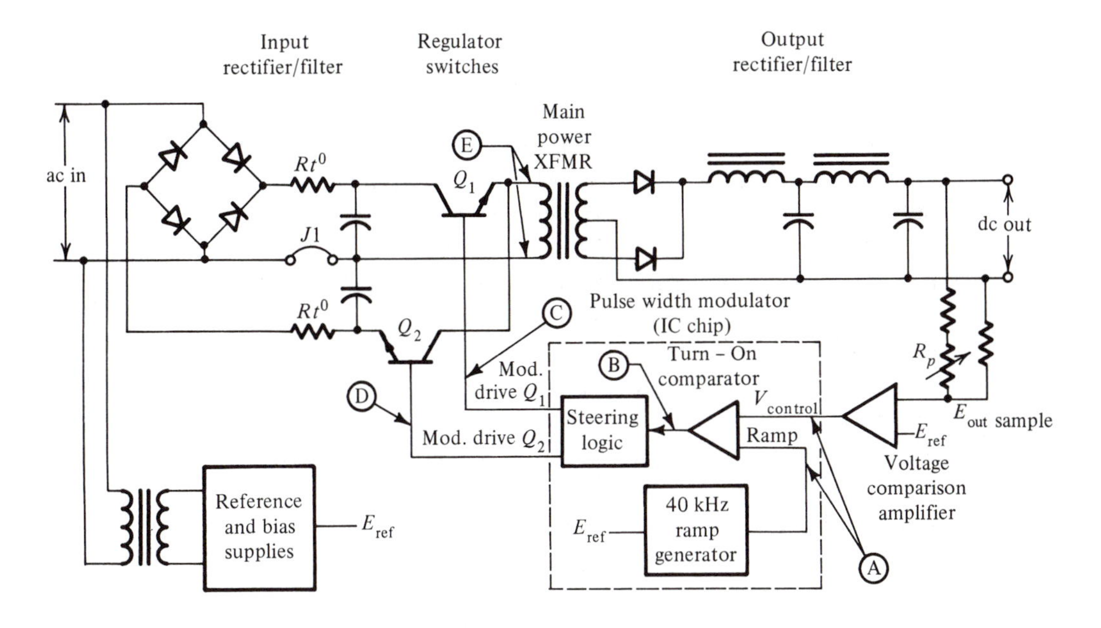

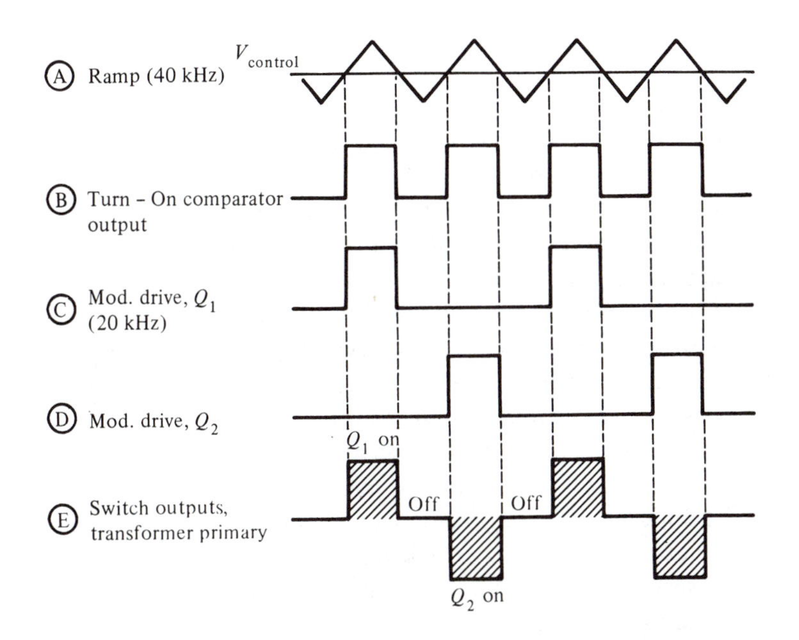

Figure 3-48 Push-pull switching power supply (*Courtesy of Hewlett Packard*)

Should the output voltage rise, the output from the voltage comparison amplifier rises. The intersection of this level and the internally generated ramp occurs later. This decreases the width out of the turn-on comparator, and therefore decreases the pulse width from the transistors driving the main power transformer. Dropping the pulse width lowers the average value of the output. The output rise, which started this process, has been compensated for.

The Motorola MC3420/3520 is a pulse width modulator IC chip. Its equivalent circuit and typical application are shown in Fig. 3-49. Further operation and application information is available from the Literature Distribution Center, P.O. Box 20912, Phoenix, Arizona.

There are several cautions and limitations of switching power supplies. Since the rectifier is tied directly to the line voltage, the rectifiers, capacitors, and switching transistors must be able to withstand the full peak line voltage (168 V for 115 V AC rms line). Input AC surge current protection (Rt°) must be provided to prevent the uncharged capacitors from shorting out the line when initially turned on. Increasing the frequency of the regulator IC will increase the noise coupled into associated circuits and will decrease the effectiveness of the capacitors, transistors, rectifiers, and inductors. Decreasing the switching frequency, however, requires an increase in the value, size, and cost of the transformer, capacitors, and inductors.

SUMMARY

This chapter has described the use of analog integrated circuits as power supply regulators. A simple power supply uses a stepdown transformer, bridge rectifier, filter capacitor, and op amp based regulator. Under light load conditions (< 10 percent ripple) there is a group of equations [equations (3-1) through (3-4)] which allow you to analyze or design a filtered supply. With higher ripple, a set of experimental curves (Figs. 3-7 through 3-10) should be used.

A zener diode can be used to provide a degree of regulation, but the addition of an op amp significantly improves regulations and even allows you to adjust the regulated output. A series (pass) transistor can be added to the op amp to boost the load current by the transistor's beta. You must be careful to consider the transistor's power dissipation and heat sink requirements.

There are several series of three-terminal IC regulators. These combine the reference zener, error op amp, and pass transistor with thermal shutdown, short-circuit limiting and high ripple rejection. Filter capacitance should be kept as low as possible while keeping the input, unregulated voltage two volts above the regulated output. This decreases cost, size, and power dissipation requirements. Component selection was illustrated in Example 3-6. These three-terminal reg-

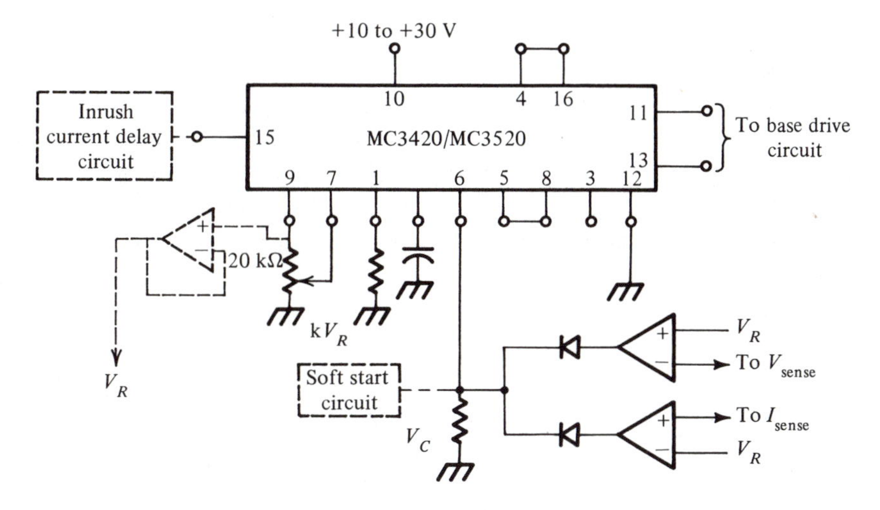

(a) Typical application

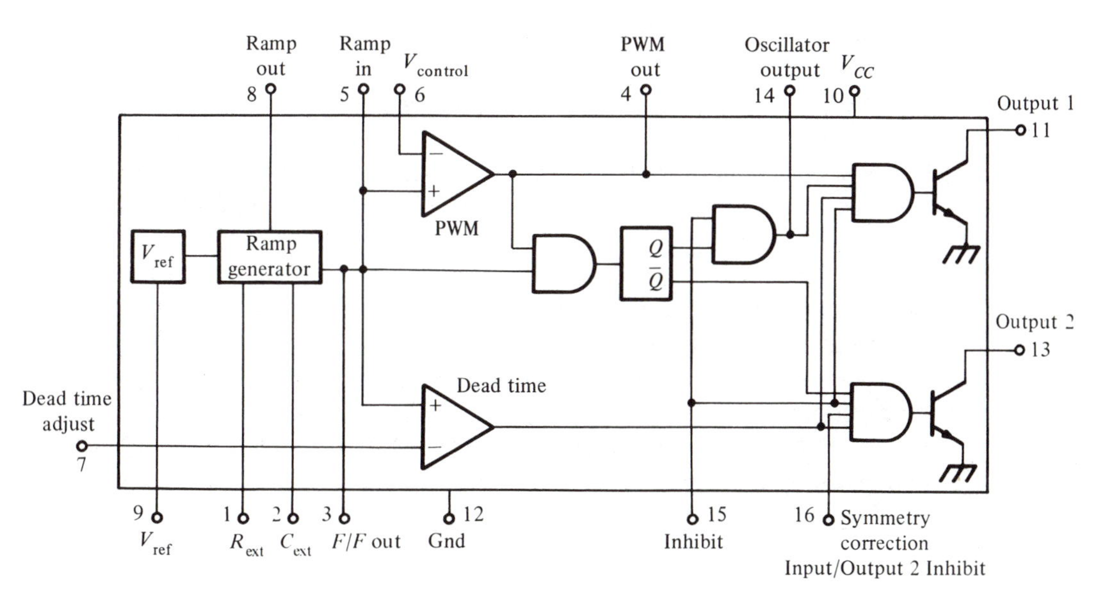

(b) Equivalent circuit

Figure 3-49 Equivalent circuit and application of MC3420/3520 switch mode regulator control circuit (*Courtesy of Motorola Semiconductor Products*)

ulators can be combined to provide a dual power supply, though protection diodes should be added.

Although most three-terminal regulators provide a single, fixed voltage, the ground potential can be floated to allow you to adjust the output voltage. The LM317 is optimized to work as an adjustable three-terminal regulator.

Initially, three-terminal regulators were limited to load currents below one amp. However, should a circuit board require more current, there are 3 A and 5 A monolithic IC regulators and 5 A, 10 A, 20 A, and 30 A hybrid IC regulators available.

The 723 regulator IC allows you to set output voltage, short circuit, and current foldback points by the selection of external components. The basic regulator can be current-boosted with an external pass transistor and can even be configured to produce a negative, regulated voltage.

Tracking regulators provide several regulated output voltages, each specifically related to the others, so small variations will not disturb the symmetry of a set of supplies. Also, adjustment of a single control tunes all supplies. The simplest tracking regulator consists of a single, adjustable positive regulator followed by an inverting power op amp. There is also a single-chip, dual tracking regulator IC which allows you to set output voltages, and short-circuit limit. Component selection is similar to the 723 regulator.

The switching power supply allows a decrease in size, and cost, without performance degradation. Instead of operating the pass transistors in their linear region, they are switched on and off at 20 kHz or faster. The output level is controlled by varying the pulse width of the switching waveform. Operating at this frequency allows the use of smaller transformers, capacitors, and inductors. Switchers are one of the most progressive power supply designs and will be used more and more widely as the price of switching transistors with both high power and high voltage ratings drops.

These first three chapters have assumed ideal op amp characteristics. Under many circumstances this will work adequately. However, as the input voltage level drops to the millivolt range, source impedance rises above 100 kΩ, source frequency rises above 10 kHz, or rise time falls into the microsecond range, the actual performance of the transistors within the IC op amp significantly alters the circuit's performance. In Chapter 4 you will see how the DC and AC characteristics of the op amp affect its performance. Limitations and compensation techniques will be discussed.

PROBLEMS

3-1 For the circuit in Fig. 3-50, let $V_{sec} = 6.3\ V_{AC\ rms}$, $C = 1500\ \mu F$, and $I_{load} = 73$ mA. Calculate V_{pk}, V_{min}, V_{ripple}(p-p), and V_{DC}.

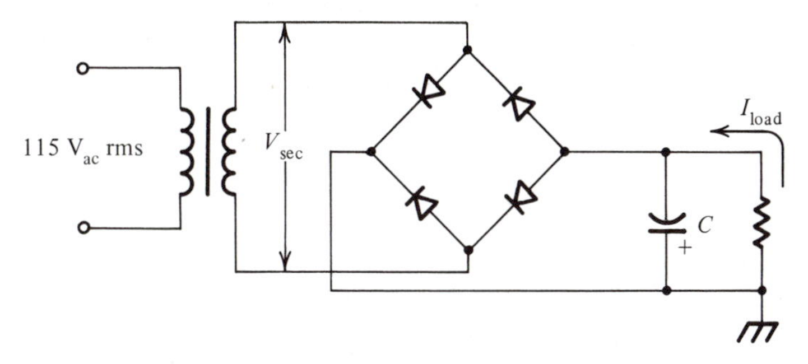

Figure 3-50 Schematic for Problem 3-1

3-2 Draw the schematic for and determine component values for a capacitive filtered, unregulated power supply which will meet or exceed the following:

$$I_{load} = 120\ \text{mA}$$

$$V_{pk} < 9\ \text{V}$$

$$V_{min} > 5\ \text{V}$$

$$V_{ripple}(\text{p-p}) < 2\ \text{V}$$

3-3 For the circuit in Fig. 3-50, let $V_{sec} = 12.6\ V_{AC\ rms}$, $C = 470\ \mu F$, $I_{load} = 500$ mA. Determine V_{pk}, V_{min}, V_{ripple}(p-p), and V_{DC}.

3-4 Calculate the values of components necessary to build a capacitive filtered, unregulated power supply. Specify the smallest standard component values which will meet the following: $I_{load} = 1$ A, $V_{min} \geq 7$ V, $V_{ripple}(\text{p-p}) \leq 8$ V, $V_{pk} \leq 30$ V.

3-5 For the circuit in Fig. 3-51, (p. 138) calculate the following: V_{load}, $V_{out\ op\ amp}$, $I_{out\ op\ amp}$, $V_{DC\ in}$, $P_{transistor}$.

3-6 Determine the component values necessary to build a $+5\ V_{DC}$ op amp/zener/transistor-type regulator for a load current of 120 mA. Draw the complete power supply schematic.

3-7 For the circuit in Fig. 3-52, (p. 138) do the following:

a. Assume $I_L = 50$ mA

b. Calculate V_{DC}(cap) and T_J.

c. If T_J is less than the maximum allowable T_J, assume a higher I_L and repeat step b.

d. If T_J is above the maximum allowable T_J, assume a lower I_L and repeat step b.

e. Continue the above iterative steps until you have found a T_J that is within 10 percent of $T_{J\,max}$. This is the maximum load current before the regulator goes into thermal shutdown.

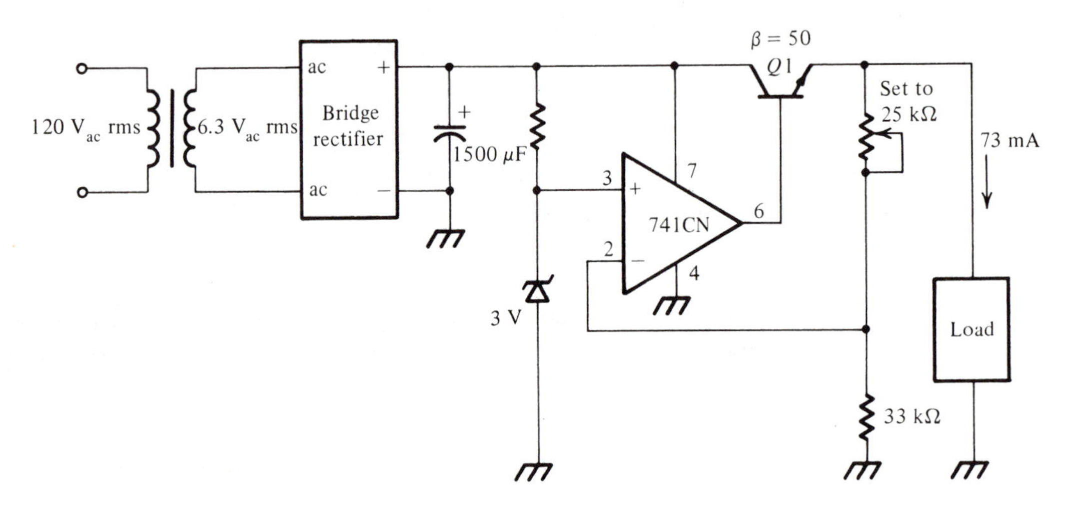

Figure 3-51 Schematic for problem 3-5

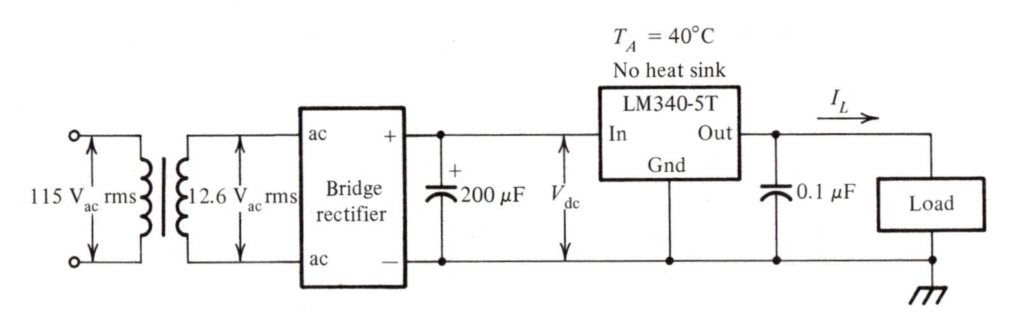

Figure 3-52 Schematic for problem 3-7

3-8 Design a solid state power supply, using a full-wave bridge rectifier and three-terminal integrated regulator to meet or exceed the following:

$$V_L = 8 \text{ V regulated}$$

$$I_L \leq 800 \text{ mA}$$

$$V_{\text{ripple load}}\,(\text{p-p}) \leq 0.1 \text{ V}$$

$$T_A \leq 60°\text{C}$$

a. Draw the schematic diagram.
b. Determine the AC input to the rectifier.
c. Determine the allowable input ripple to the regulator.
d. Determine the input filter capacitance required.
e. Determine the appropriate DC input voltage to the regulator.
f. Select the diode ratings.
g. Select the regulator and heat sink (if necessary).

3-9 Repeat Problem 3-8, using an LM317 with output voltage adjustable from 3 V to 15 V.

Answer Problems 3-10 through 3-20 for the 723CN (DIP).

3-10 What is the highest continuous input voltage?
3-11 What is the largest difference in potential allowable between the two inputs of the error amplifier?
3-12 What is the largest load that can be applied to the V_{ref}?
3-13 What is the maximum
a. Junction temperature?
b. Ambient temperature?
3-14 Does the 723 have better or worse ripple rejection than the LM 340-5T?
3-15 What is V_{ref}?
3-16 What is a good value for C_{comp}?
3-17 What is the minimum input voltage?
3-18 For $V_{\text{out}} = 8$ V, what is the best value of $V_{\text{min in}}$? Explain.
3-19 What is the thermal resistance, junction to ambient, for the bare package (no heat sink)?
3-20 For the circuit in Fig. 3-53, calculate V_{DC} and T_J starting at $I = 10$ mA and incrementing in 10 mA until the maximum junction temperature is exceeded. This will give you the allowable load current for this specific configuration with $V_{\text{out}} = 5$ V and $T_A = 40°$C. (See p. 140.)
3-21 What would happen to the maximum allowable load current if the regulator were adjusted to output 8 volts? Explain.

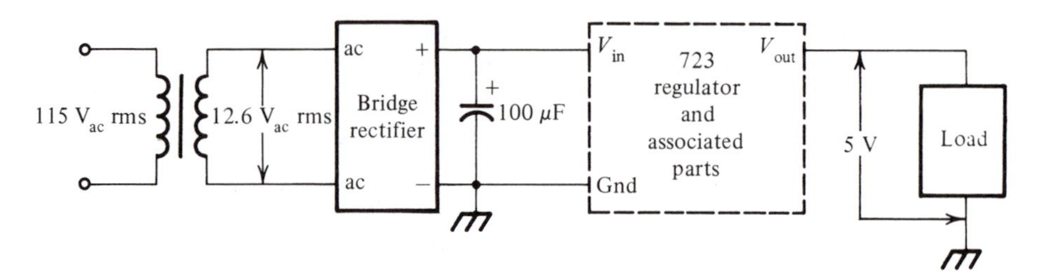

Figure 3-53 Schematic for problem 3-20

3-22 Design the regulator section of the circuit in Fig. 3-53 for a ±10 percent adjustable 5 volt regulator with current limiting. Set the current limiting for a level which would protect the package indefinitely from a short circuit. Include DIP pin numbers.

3-23 Redesign the regulator section of the circuit above for a knee (foldback) current 20 percent below that calculated in Problem 3-20 and a short-circuit limit 20 percent of that calculated in Problem 3-22. (Pick R4 = 10 kΩ; solve for R_{SC} and R3.)

3-24 Draw the schematic of the circuit which would provide $I_L = 1$ A, current limited, with an output voltage of 4 volts, using a 723. No calculations or component values are necessary. (CAREFULLY! Do not just copy from a manual.)

3-25 For the circuit in Fig. 3-54, calculate all output voltages if:

a. R = 470 Ω

b. R = 2200 Ω

3-26 Design a tracking regulator using an LM317 and op amps to output +5 V, ±12 V all adjustable ±10 percent with a single adjustment. Select a heat sink (if necessary) for the 317 for $I_{load} = 100$ mA, $V_{DC} = \pm 18$ V, $T_A = 30$°C.

3-27 For the circuit in Fig. 3-55, (p. 142) calculate the:

a. Regulated output voltage.

b. Short-circuit current limit.

c. Power dissipated by the pass transistor at $I_{load} = 0.9\ I_{short\ circuit\ limit}$

3-28 Design a dual tracking regulator, using the MC1468 to meet or exceed the following:

a. $V_{out} = \pm 12$ V

b. Adjustable output voltage range: at least ±10 percent.

c. Short-circuit current limit: adjustable—50 mA–500 mA

3-29 Draw the full schematic of a switching regulator, using a MC3420 for the pulse width modulator IC. Do not calculate component values.

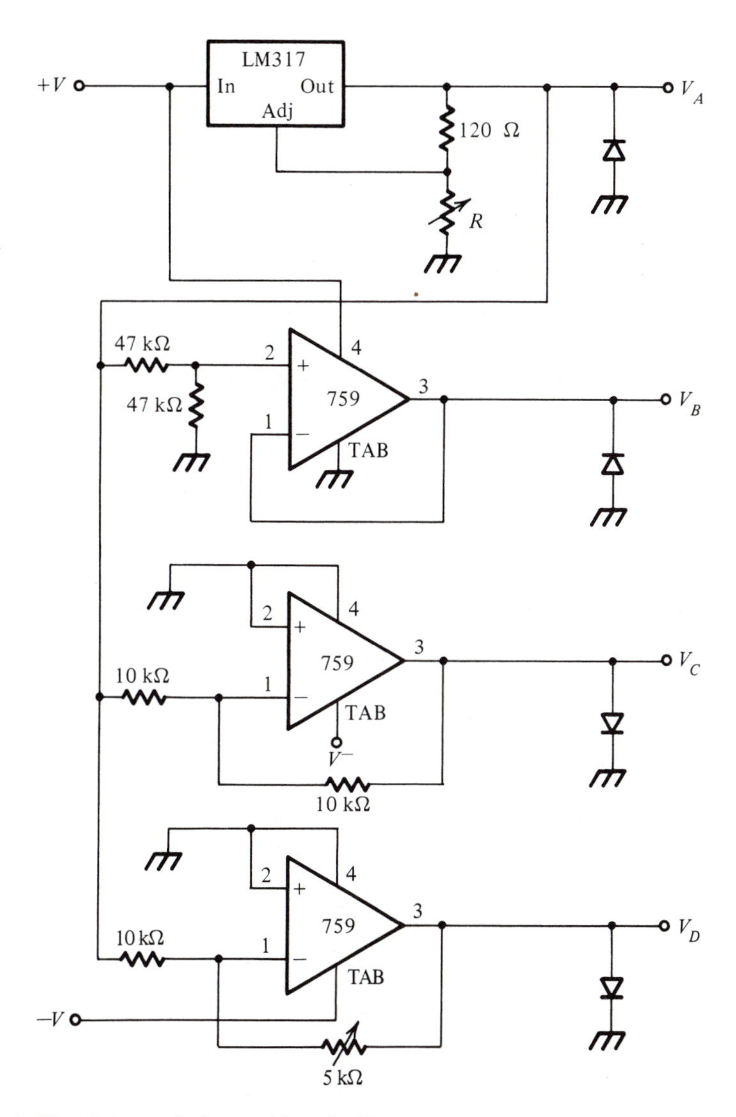

Figure 3-54 Schematic for problem 3-25

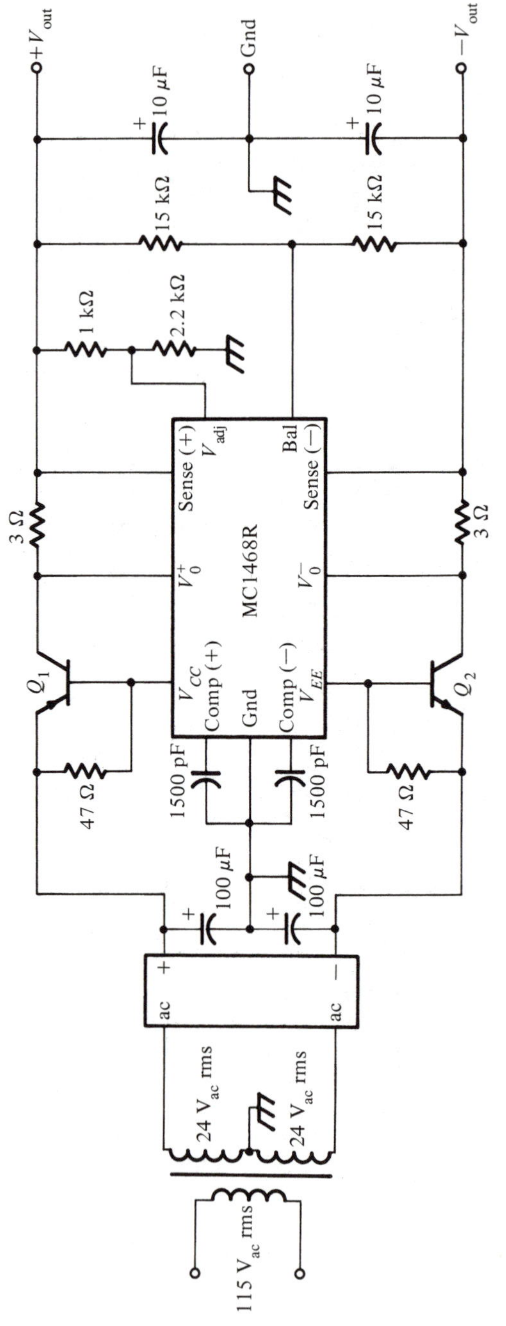

Figure 3-55 Schematic for problem 3-27

CHAPTER 4

Operational Amplifier Characteristics

In the previous chapters you have used an ideal op amp; the DC and AC characteristics have been ignored. For temporary circuits with rather loose requirements, these characteristics may be ignorable. However, you must be much more careful when building prototype or production grade circuits. These circuits are expected to exhibit stable operation over long time intervals and wide variations in temperature. They may have to respond identically to DC, audio, and radio frequencies. High-speed transients may need to be transmitted. Input levels may be in or below the millivolt range from signal sources with high output impedances. To meet these requirements, you must choose an op amp whose DC and AC characteristics are adequate. In this chapter you will learn the performance limitations imposed by the op amp DC and AC characteristics and appropriate compensation techniques.

OBJECTIVES

After studying this chapter you should be able to do the following:

1. Define each of the following and explain the effect each has on the operation of the amplifier:
   a. Bias currents
   b. Offset currents
   c. Offset voltage
   d. Drift
   e. Unity gain bandwidth (gain bandwidth product)
   f. Rise time
   g. Slew rate
   h. Full power response
   i. Noise

2. Draw a schematic and explain compensation techniques.
3. List four ways to decrease internally generated noise in an op amp.
4. Explain the application and effects of single capacitor frequency compensation, and feed forward compensation.

4-1 DC CHARACTERISTICS

The ideal op amp draws no current from the source driving it. Both the inverting and the noninverting inputs look and respond identically. The circuit response does not vary with temperature. Real op amps do not work this way. Current is taken from the source into the op amp's inputs. There are slight differences in the way the two inputs respond to current and voltage. A real op amp will shift its operation with temperature. These nonideal DC characteristics and compensation techniques are described in this section.

4-1.1 Bias Currents

The op amp's input is a differential amplifier. It may be made of bipolar junction transistors, as in Fig. 4-1(a) or field effect transistors, as in Fig. 4-1(b). In either case, these transistors must be biased, and this takes current. For the bipolar transistors this bias current, supplied by the external circuit, is the base bias current, biasing the input transistors into their linear region. The bias current for the FET input op amp is actually the leakage current across the reverse-biased gate to channel junction.

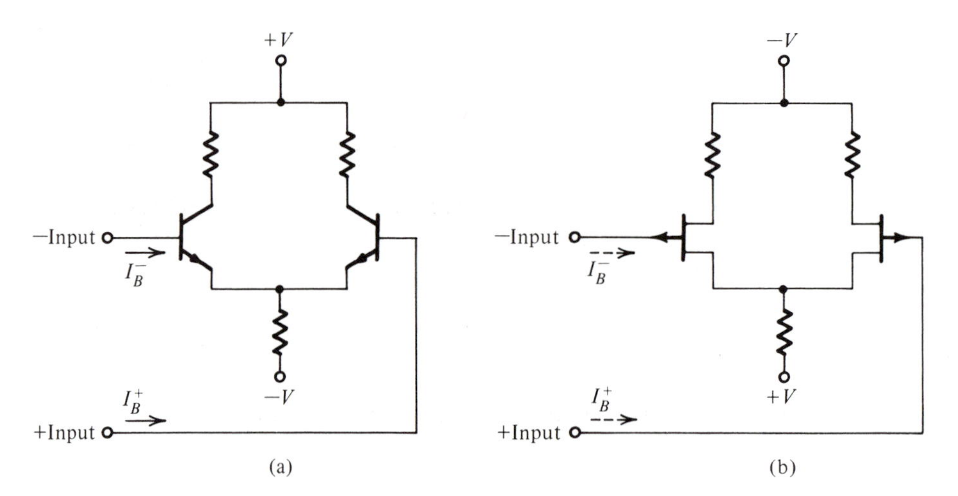

Figure 4-1 Op amp input circuits

Manufacturers specify input bias current as the average of the bias current at each input.

$$I_B = \frac{I_B^+ + I_B^-}{2} \tag{4-1}$$

For the 741, a bipolar input op amp, the bias current is 500 nanoamps or less. A FET input op amp will have bias currents as low as 50 picoamps at room temperature.

Can such small current levels really have much effect? Figure 4-2 shows a basic inverting amplifier. When e_{in} is set to zero volts, the output, e_o, should also be zero volts. However, both the inverting and the noninverting inputs are drawing bias current from the external circuit. Current I_B^+ comes directly from ground, and produces no voltage. Current I_B^- comes entirely through the feedback resistor (since both ends of R_i are at zero volts, no current can flow through it). Therefore, bias current I_B^- produces a voltage drop across R_F. Since the left end of R_F is held at virtual ground, the right end of R_F must be driven to $(I_B^-)(R_F)$ by the op amp's output. So, instead of an output voltage of zero volts,

$$e_o = (I_B^-)(R_F)$$

For the 741 and a 1 MΩ feedback resistor,

$$e_o = 500 \text{ nA} \times 1 \text{ M}\Omega$$

$$e_o = 500 \text{ mV} \quad \text{or} \quad 0.5 \text{ V}$$

The output is driven to 500 mV, with zero input, because of the bias currents. In applications where signal levels are measured in millivolts, or even hundreds of millivolts, this is totally unacceptable.

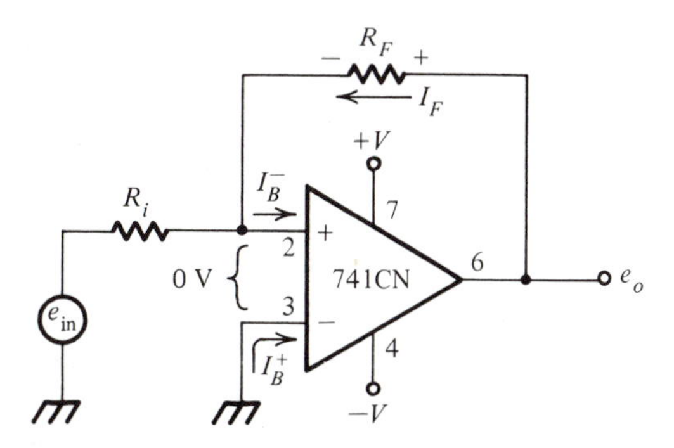

Figure 4-2 Basic inverting amplifier with bias currents

This effect can be compensated for, as shown in Fig. 4-3. A compensating resistor has been added between ground and the noninverting input. Current I_B^+ flowing through this resistor develops a voltage V_1. Summing the voltages around the input-feedback-output loop, you obtain

$$-V_1 + 0 + V_F - e_o = 0$$

or

$$e_o = V_F - V_1$$

If the proper size R_{comp} is selected, V_1 will cancel V_F and the output will go to zero.

The value of R_{comp} is derived below:

$$V_1 = I_B^+ R_{comp}$$

or

$$I_B^+ = \frac{V_1}{R_{comp}}$$

$$I_{R_i} = \frac{V_1}{R_i}$$

$$I_{R_F} = \frac{V_F}{R_F}$$

For compensation,

$$V_F = V_1$$

$$I_{R_F} = \frac{V_1}{R_F}$$

The current I_B^- is the sum of I_{R_F} and I_{R_i}

$$I_B^- = I_{R_F} + I_{R_i}$$

$$I_B^- = \frac{V_1}{R_F} + \frac{V_1}{R_i}$$

$$I_B^- = V_1\left(\frac{R_i + R_F}{R_i R_F}\right)$$

Figure 4-3 Bias current compensation

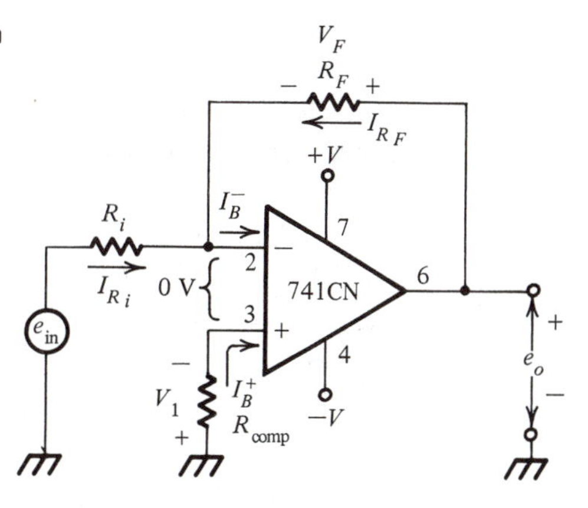

Assuming

$$I_B^- = I_B^+$$

$$V_1\left(\frac{R_i + R_F}{R_F\,R_i}\right) = \frac{V_1}{R_{comp}}$$

$$R_{comp} = \frac{R_F\,R_i}{R_i + R_F}$$

$$R_{comp} = R_F\,//\,R_i$$

So to compensate for bias currents, the compensating resistor should equal the parallel combination of the resistors tied to the other input.

To compensate for bias currents, set the total resistance seen from the noninverting input to ground, equal to the total resistance seen from the inverting input to ground.

4-1.2 Offset Current

Bias current compensation will work *if* both bias currents are equal. Since the input transistors cannot be made identical, there will always be some small difference between I_B^+ and I_B^-. This difference is called the offset current.

$$I_{OS} = |\,I_B^+ - I_B^-\,| \tag{4-2}$$

The absolute value sign indicates that there is no way to predict which of the bias currents will be larger. Offset current for the 741 general purpose (bipolar transistor) op amp will always be less than 200 nanoamps. For a typical FET op amp, offset is 10 picoamps or less.

Even with bias current compensation, offset current will produce an output voltage when the input voltage is zero. In Fig. 4-3,

$$V_1 = (I_B^+)(\mathrm{R_{comp}})$$

$$I_{\mathrm{R_i}} = \frac{V_1}{\mathrm{R_i}} = \frac{(I_B^+)(\mathrm{R_{comp}})}{\mathrm{R_i}}$$

$$I_{\mathrm{R_F}} = (I_B^-) - (I_{\mathrm{R_i}}) = (I_B^-) - \frac{(I_B^+)(\mathrm{R_{comp}})}{\mathrm{R_i}}$$

Summing the input-feedback-output loop gives

$$e_o = I_{\mathrm{R_F}}\mathrm{R_F} - V_1$$

$$e_o = I_{\mathrm{R_F}}\mathrm{R_F} - (I_B^+)(\mathrm{R_{comp}})$$

Substituting for $I_{\mathrm{R_F}}$, you obtain

$$e_o = \left[(I_B^-) - \frac{(I_B^+)\mathrm{R_{comp}}}{\mathrm{R_i}}\right]\mathrm{R_F} - (I_B^+)(\mathrm{R_{comp}})$$

After several steps of algebraic manipulation

$$e_o = \mathrm{R_F}[(I_B^-) - (I_B^+)]$$

$$e_o = \mathrm{R_F}I_{\mathrm{OS}} \tag{4-3}$$

So even with bias current compensation, if the feedback resistor were 1 MΩ, a 741 op amp could have an output as large as

$$e_0 = 1\ \mathrm{M\Omega} \times 200\ \mathrm{nA}$$

$$e_0 = 200\ \mathrm{mV}$$

with a zero input voltage.

For equation (4-3) you can see that:

To minimize the effects of offset current, keep the feedback resistance small.

Unfortunately, to obtain both high input impedance ($\mathrm{R_i}$ large) and reasonable gain, the feedback resistance must be very big. The T feedback network in Fig. 4-4 will allow you to have large feedback resistance.

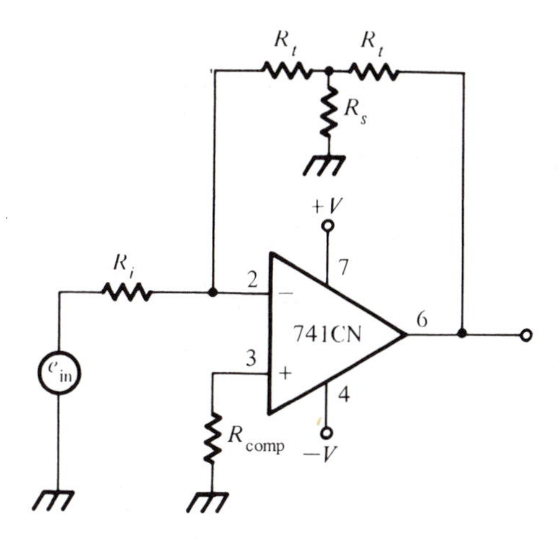

Figure 4-4 Inverting amplifier with *T* feedback network

The T network provides a feedback signal as if the network were a single feedback resistor:

$$R_F = \frac{R_t^2 + 2R_t R_s}{R_s} \tag{4-4}$$

To design a T network, first pick

$$R_t \ll \frac{R_F}{2} \tag{4-5}$$

then calculate

$$R_s = \frac{R_t^2}{R_F - 2R_t} \tag{4-6}$$

Example 4-1

Design an inverting amplifier using a 741C op amp, with a gain of -10, an input impedance of 10 MΩ.

Solution

The circuit schematic of Fig. 4-4 will be used.
To set $Z_{IN} = 10\ M\Omega$, pick $R_i = 10\ M\Omega$
Since $A_v = -(R_F/R_i)$,

$$R_F = -A_v R_i$$

$$R_F = -(-10)(10\ M\Omega)$$

$$R_F = 100\ M\Omega$$

Pick $$R_t = 47\ k\Omega$$

$$R_s = \frac{R_t^2}{R_F - 2R_t}$$

$$R_s = \frac{(47\ k\Omega)^2}{100\ M\Omega - 2(47\ k\Omega)}$$

$$R_s = 22\ \Omega$$

Proper selection of R_{comp} will eliminate the effects of bias currents. However, the *T* feedback network has virtually no effect on output voltage caused by offset current.

$$V_{out} = I_B^-\left(2R_t + \frac{R_t^2}{R_S}\right) - I_B^+\left(\frac{R_C R_t^2}{R_i R_S} + \frac{2R_C R_t}{R_i} + \frac{R_C R_t}{R_S} + R_C\right) \tag{4-7}$$

For example 4-1, the proper value of R_{comp} is

$$R_{comp} = R_C = 46.8\ k\Omega$$

Assuming

$$I_B^+ = 80\ nA \quad \text{and} \quad I_B^- = 60\ nA$$

substitution into equation (4-7) gives

$$V_{out} = -2.0\ V$$

But this is the same exact result you would have gotten using a 10 MΩ input resistor and a 100 MΩ feedback resistor and an offset current of 20 nA.

4-1.3 Offset Voltage

Even though you use bias current compensation and minimize the size of feedback resistance (or resistance to ground for the T network), a zero input to your amplifier may not produce a zero output. This effect is called offset voltage. It is represented schematically as a small voltage source placed in series with the noninverting input terminal, as shown in Fig. 4-5. Keep in mind that this is an *effect*, not a real voltage source. For the general-purpose 741C, the offset voltage will be typically ± 2 millivolts.

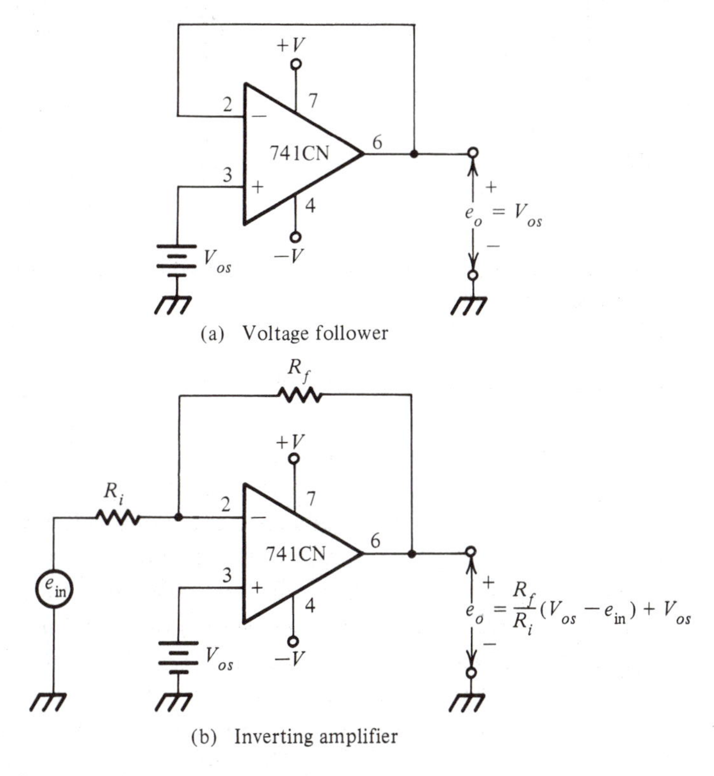

Figure 4-5 Offset voltage effect on amplifier outputs

This offset voltage is amplified by the circuit gain along with the signal. Also, since the offset voltage may vary quite a bit from one IC to another, even among devices of the same type number, standard, fixed compensation is difficult.

Many, but not all, op amps provide offset compensation pins. The use of these pins is specific to the particular op amp. So you must refer to the manufacturer's specifications when using the provided offset null or trim connections. Fig. 4-6 gives the connections for the 741C and the 318.

Use of the provided null offset pins, when it works, is the best way to compensate for offset voltage. However, there are times when the op amp that you are using does not have offset null pins, or when they do not give you the

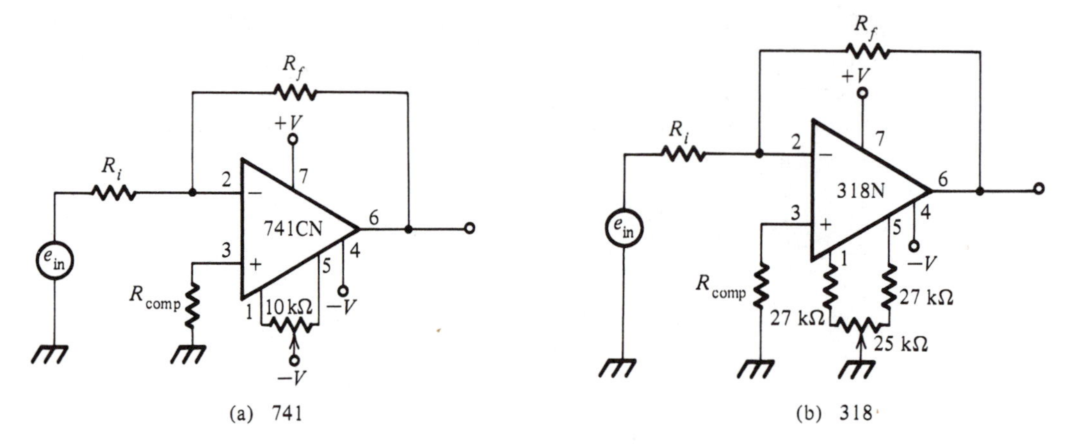

Figure 4-6 Inverting amplifiers using offset voltage null pins (*Courtesy of National Semiconductor*)

range necessary to eliminate the offset. External balancing techniques are then needed.

Several are shown in Fig. 4-7. All these techniques shown use an adjustable voltage. This voltage is divided to a very small range and is then fed to one of the op amp inputs. These external balancing techniques allow you to set the range of the offset adjustment. Also, compensation is independent of the op amp chosen (so changing the op amp has little effect on the circuit configuration).

There are several precautions you should observe. First, keep the adjustment range small. It is very difficult to null out 10 millivolts of offset when the adjustment range is ± 1 volt. Second, do not try to balance offset by turning your amplifier into a summer, with one input to give DC adjustment. This could lower the effective resistance between the inverting input and ground, which *increases* the effect of offset and noise voltages on the output. Finally, the external balance circuit of the differential amplifier [Fig. 4-7(c)], could lower the common mode rejection of the circuit. Resistor R3 should be kept at least 100 times larger than R4.

Proper adjustment of the offset balancing circuitry, either internal or external, will compensate for the offset voltage and for any effects of bias currents and offset currents you could not eliminate otherwise. To set up for this adjustment, make sure that the bias current resistor is installed, feedback resistance is minimized, and all signal sources are off (or replaced by a resistor equal to their output impedance). While monitoring the output with an instrument which is accurate in the millivolt range, adjust the balance network to produce zero volts at the op amp output. Do not change the setting of this control during normal circuit operation.

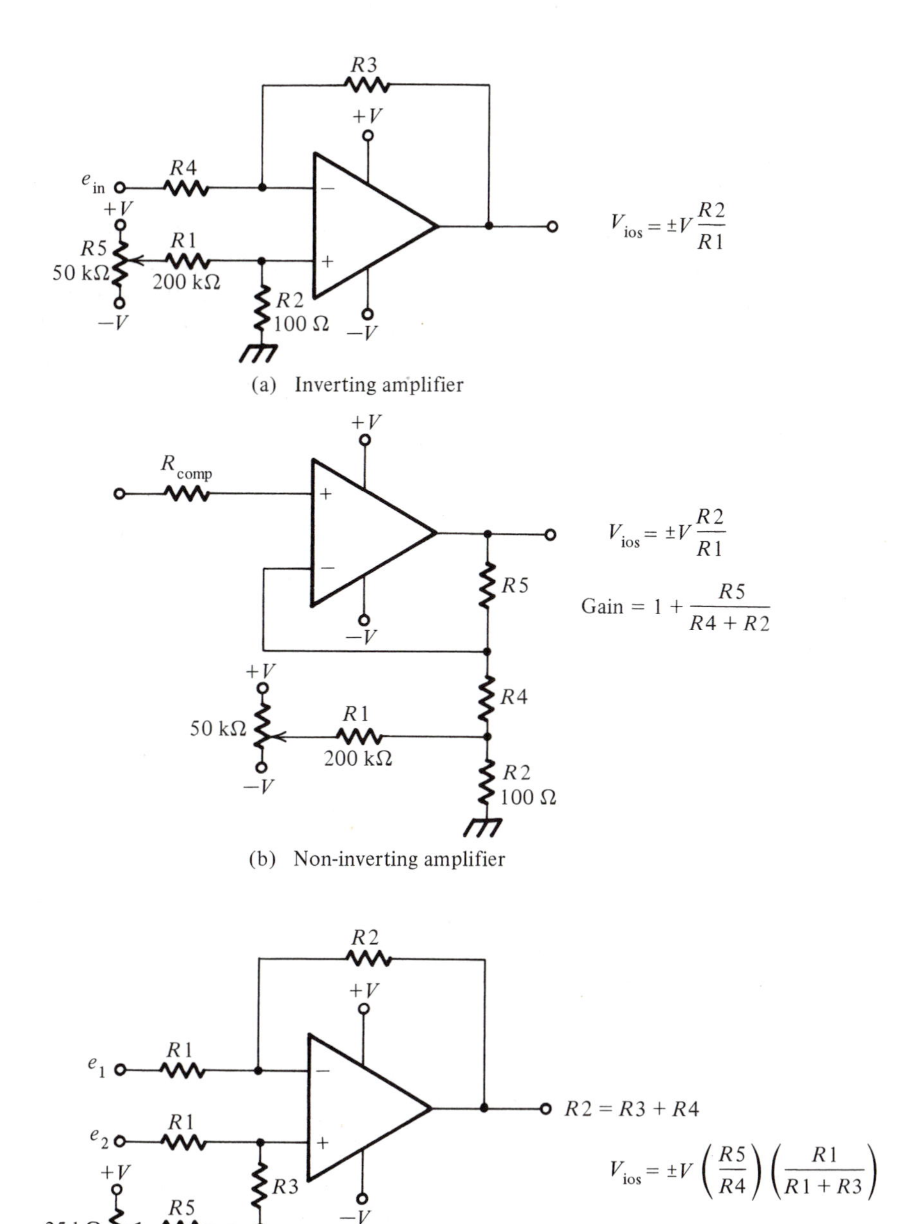

Figure 4-7 External balancing techniques (*Courtesy of National Semiconductor*)

Follow proper procedures (above) **when adjusting offset voltage balance circuits.**

4-1.4 Characteristic Measurement

Usually, bias currents, offset current and offset voltage are too small to be accurately measured directly. However, their effect on the output voltage can be pronounced because of the circuit's gain. You can use this apparent problem to more easily measure these characteristics.

Measurement of bias currents is shown in Fig. 4-8. Bias current at one input is forced to flow through a large resistance. Since the bias current at the other input does not flow through a resistor, a compensating voltage is not produced.

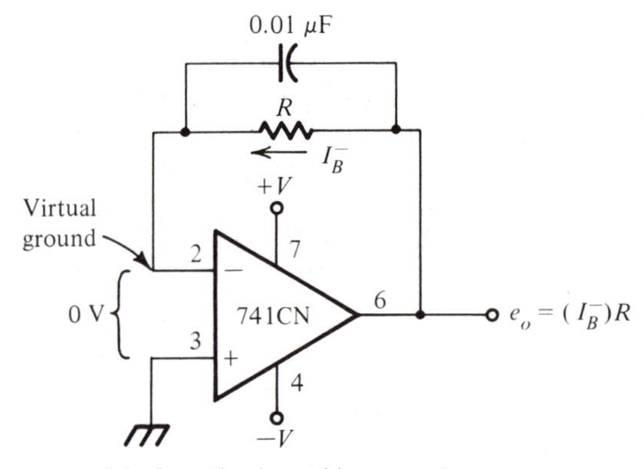

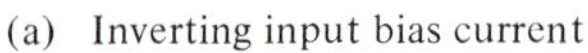

(a) Inverting input bias current

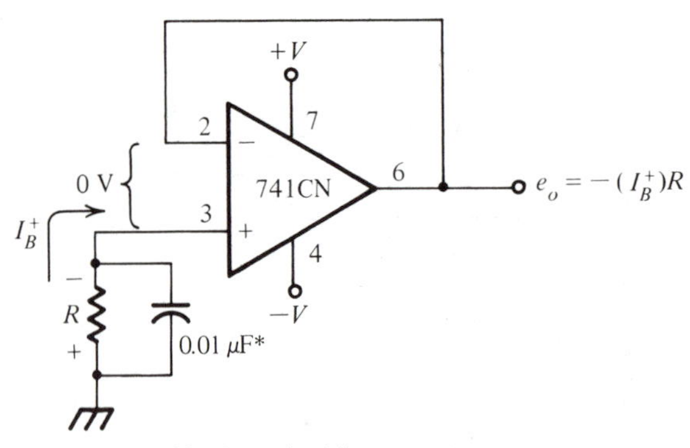

(b) Noninverting bias current

*Minimize noise

Figure 4-8 Measurement of bias currents

The output voltage is the result of the bias current at one input only. By selecting a large (MΩ) resistor, the output voltage (and therefore bias current) can be accurately and precisely measured. You should select R large enough that the output produced by the bias currents is 500 millivolts or larger. Otherwise, the offset voltage can not be ignored. Note that current flowing *into* the input is defined as positive. Be careful not to miss the output voltage polarity (shown in Fig. 4-8).

A simple extension of Fig. 4-8 allows you to measure offset current. This is shown in Fig. 4-9. With both resistors set *exactly* equal, the effects of bias currents cancel. Any voltage at the output is caused by offset current and offset voltage. By selecting R large, the effects of the offset current can be made so much larger than the few millivolts of offset voltage that V_{os} can be ignored.

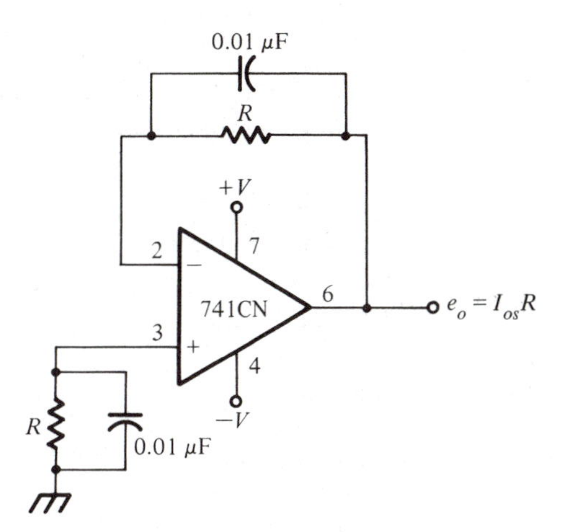

Figure 4-9 Offset current measurement

Offset voltage can be amplified as shown in Fig. 4-10. Bias currents are compensated for by R_{comp}. The output is an amplified version of the input (offset voltage) plus a term dependent on offset current times feedback resistance. To accurately measure the offset voltage, then, you should make R_f small, while keeping a good amplification factor (R_f/R_i). The values shown in Fig. 4-10 would typically give 98 percent or better accuracy.

4-1.5 Drift

Bias currents, offset current, and offset voltage change with temperature. You may have carefully nulled a circuit shortly after turning on power, with an ambient temperature of 25°C. Later, you return to find that the temperature has risen to 35°C, and the circuit is no longer nulled. This is called *drift*.

Figure 4-10 Offset voltage measurement

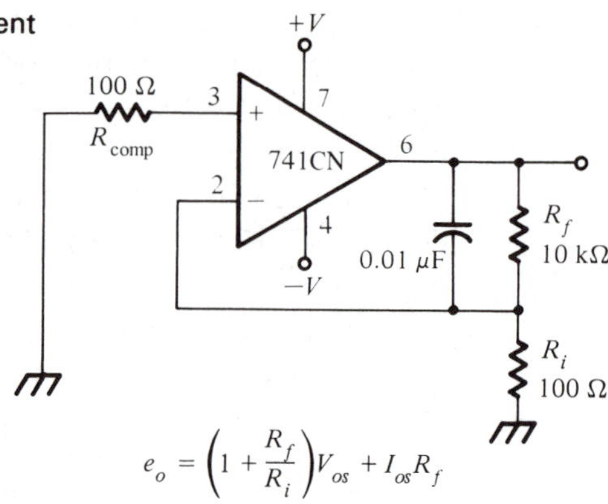

Often offset current drift is expressed in nA/°C, while offset voltage drift is expressed in mV/°C. These indicate the *change* in offset for each degree Celsius *change* in temperature.

Example 4-2

If a noninverting amplifier with a gain of 100 is nulled at 25°C, what will happen to the *output* voltage if the temperature rises to 50°C for an offset voltage drift of 0.15 mV/°C?

Solution

$$0.15 \text{ mV/°C} \times (50\text{°C} - 25\text{°C}) = 3.75 \text{ mV}$$

The *input* offset will drift from the nulled condition by 3.75 mV. Since this is an input change, the output voltage will change by

$$e_O = v_{OS} \times A$$

$$e_O = 3.75 \text{ mV} \times 100 = 375 \text{ mV}$$

This could represent a very major shift at the output.

The numbers specified for drift are not necessarily constant across the temperature range of the op amp. A drift of 0.15 mV/°C at 25°C may be 0.25 mV/°C or −0.05 mV/°C at 75°C. Read the specification carefully. If a plot of offset versus temperature is given, it is far better to use it.

There are very few circuit techniques that you can use to minimize the effects of drift. Careful PC board layout, to keep op amps away from sources of heat (power supply regulators and power transistors), and forced air cooling will help stabilize the ambient temperature. The T feedback network in Fig. 4-4 is

$[1 + (R_t/R_s)]$ times more sensitive to drift than is a conventional inverting amplifier. Consequently, avoid the T feedback network when drift is a concern. The only other way to minimize the effects of drift is to buy op amps with very low drift specifications.

Drift can drastically alter the null of a circuit. Anticipate this temperature-sensitive effect when selecting the op amp.

4-2 AC CHARACTERISTICS

Although the op amp is used in some industrial circuits as strictly a DC amplifier, many op amp applications are AC. Bias currents, offset currents, offset voltage, and drift all affect the steady-state (DC) response of the op amp. How the op amp's output responds to *changes* in its input must also be considered. The effects of small signal sinusoidal ($<1\ V_{PK}$) and large signal or transient inputs ($>1\ V_{PK}$) on the op amp's output are presented in this section.

4-2.1 Gain Bandwidth Product

For the ideal op amp, you were told to assume that the open loop gain was arbitrarily large, and that this did not vary significantly with frequency. Neither of these assumptions is valid for a real op amp.

The frequency response curve of the 741C is given in Fig. 4-11. The DC and very low frequency open loop gain of the 741C is about 200K, which is generally large enough to be considered arbitrarily large. However, as frequency increases, the gain drops proportionally. The open loop gain falls until it reaches 1 at a frequency of 1 MHz. This is a decrease of 20 dB/decade, or 6 dB/octave. Expressed another way, *increasing* the frequency by a factor of 10 decreases the gain by 10. At any point along the curve, the product of the gain and the frequency (or bandwidth) is a constant 1 MHz.

Example 4-3

Determine the gain and gain bandwidth product at 10 Hz, 1 kHz, 100 kHz and 1 MHz.

Solution

At 10 Hz, $A_{OL} = 100$ k $\quad$ GBW = 100 k × 10 Hz = 1 MHz

At 1 kHz, $A_{OL} = 1$ k $\quad$ GBW = 1 k × 1 kHz = 1 MHz

At 100 kHz, $A_{OL} = 10$ $\quad$ GBW = 10 × 100 kHz = 1 MHz

At 1 MHz, $A_{OL} = 1$ $\quad$ GBW = 1 × 1 MHz = 1 MHz

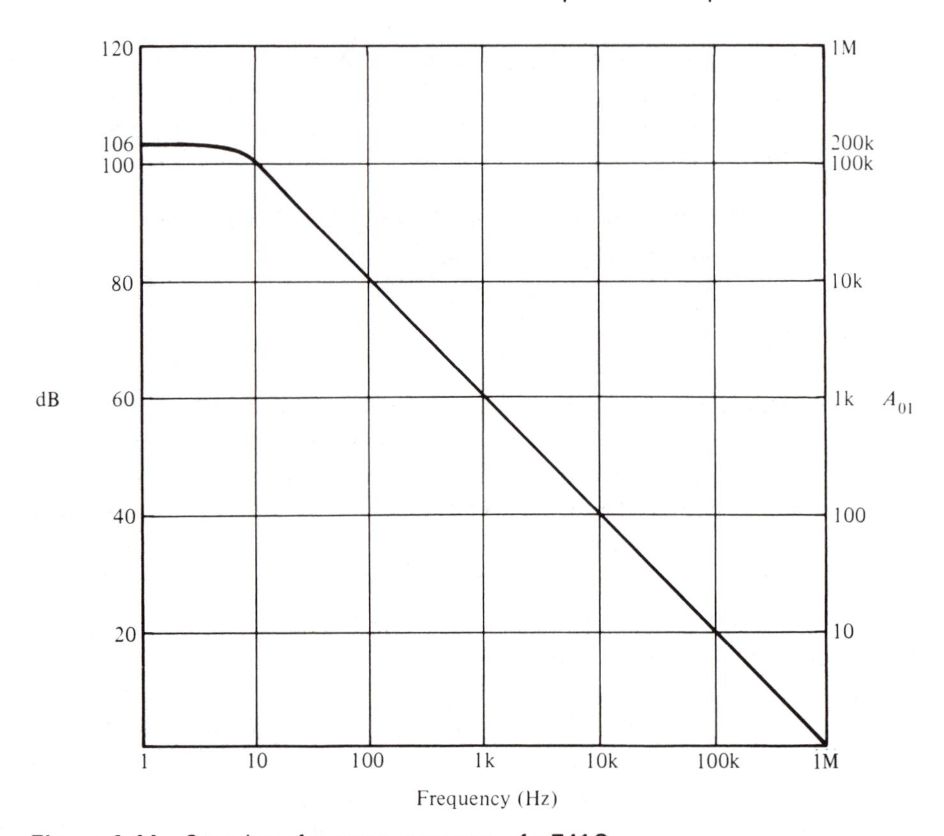

Figure 4-11 Open loop frequency response of a 741C

Initially, you might think that this drastic roll-off of gain at high frequencies would be a disadvantage. However, very often, it is intentionally built into the op amp by installing a simple low pass RC filter (see Fig. 4-12). This reduced gain ensures that high frequency noise will not be large enough when fed back into the input through parasitic capacitance to cause oscillations. A stable amplifier is ensured this way, but only by sacrificing frequency response.

This reduction in open loop gain will also severely limit the high frequency gain when negative feedback is used. This is illustrated in Fig. 4-13 for an amplifier with a closed loop gain of 100 (40 dB). At low frequencies, the open loop gain is so much larger than the closed loop gain that the circuit's performance is not affected. But as the open loop gain falls closer to the closed loop gain, the closed loop gain begins to suffer, and falls below that predicted by ideal theory. When the open loop gain coincides with the low-frequency closed loop gain (100 in this example), the actual closed loop gain has fallen to 0.707 of its low frequency level.

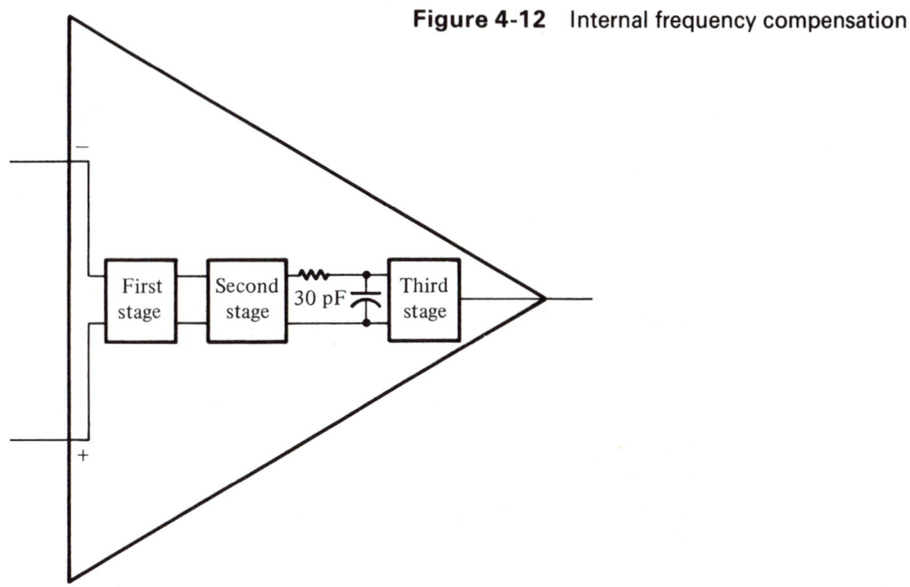

Figure 4-12 Internal frequency compensation

This is the high frequency cutoff f_H, and defines the small signal bandwidth of the amplifier.

$$f_H = \frac{\text{GBW}}{A_{\text{CL}}} \tag{4-8}$$

where

GBW = gain bandwidth product (1 MHz for the 741C).

A_{CL} = circuit's closed loop gain (100 in Fig. 4-13).

Above this high frequency cutoff, the closed loop gain closely follows the fall of the open loop gain.

Example 4-4

What is the maximum closed-loop gain from a 741C-based circuit with a cutoff of 20 kHz (upper audio frequency limit)?

Solution

$$f_H = \frac{\text{GBW}}{A_{\text{CL}}}$$

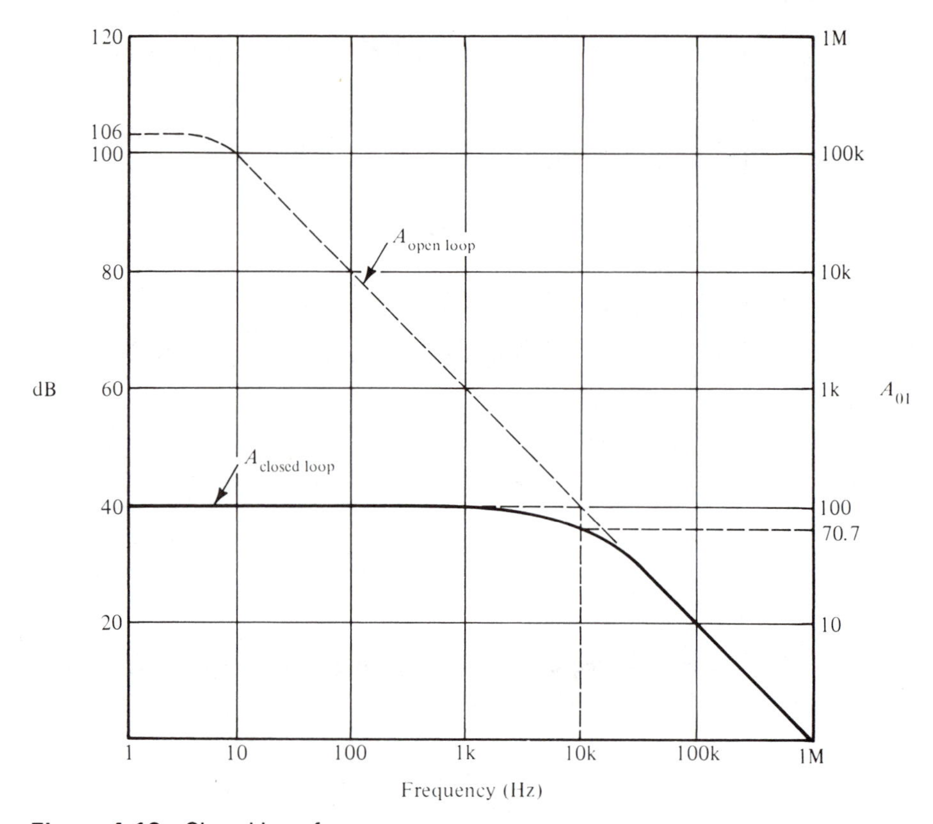

Figure 4-13 Closed loop frequency response

or

$$A_{\text{CL}} = \frac{\text{GBW}}{f_H}$$

$$A_{\text{CL}} = \frac{1\ \text{MHz}}{20\ \text{kHz}} = 50$$

So to use a 741C in an audio amplifier you must limit its closed loop gain to 50 or less.

Closely examine Fig. 4-13 and you will see that even at frequencies below f_H, the closed loop gain has fallen significantly below the 100 predicted by ideal theory. At f_H, the gain predicted assuming an ideal op amp is 30 percent above that you would actually get. This amount of error is hard to accept. How high a frequency can be used before the gain drops 1 percent from the ideal? Ten percent from the ideal? Knowing these two frequencies would give better guidance than f_H.

$$f_{1\%} = \tfrac{1}{7} f_H \tag{4-9}$$

$$f_{10\%} = \tfrac{1}{2} f_H \tag{4-10}$$

Equation (4-9) gives a frequency $f_{1\%}$ at and below which the actual circuit gain will be within one percent of that predicted assuming an ideal op amp. Equation (4-10) gives a frequency $f_{10\%}$ at and below which the actual gain will deviate less than 10 percent.

Example 4-5

Using a 741C, find:

a. $f_{1\%}$ for $A_{CL} = 200$
b. $f_{10\%}$ for $A_{CL} = 200$
c. f_H for $A_{CL} = 200$
d. Actual gain at $f = 20$ kHz

Solutions

a. $$f_{1\%} = \tfrac{1}{7} f_H = \frac{1}{7}\frac{\text{GBW}}{A_{CL}}$$

$$f_{1\%} = \frac{\tfrac{1}{7} \times 1\ \text{MHz}}{200} = 714\ \text{Hz}$$

If you want the gain to be within 1 percent of 200, do not use this circuit above 714 Hz.

b. $$f_{10\%} = \tfrac{1}{2} f_H = \frac{1}{2}\frac{1\ \text{MHz}}{200}$$

$$f_{10\%} = 2.5\ \text{kHz}$$

If you can tolerate up to 10 percent deviation from the 200 gain calculated, the circuit may be used up to 2.5 kHz.

c. $$f_H = \frac{\text{GBW}}{A_{CL}} = 5\ \text{kHz}$$

The small signal bandwidth is 5 kHz.

d. $A_{CL} = \dfrac{\text{GBW}}{f}$

$$A_{CL} = \frac{1\text{ MHz}}{20\text{ kHz}} = 50$$

At the upper end of the audio frequency range, an amplifier designed for a gain of 200 using a 741C will actually have a gain of only 50, an error of 75 percent.

So the frequency response can prove to be much more restrictive than the 1 MHz gain bandwidth product may at first seem to indicate.

Calculate $f_{1\%}$ and $f_{10\%}$ for all amplifiers you design.

There are two solutions to this restriction. The simplest is to select an op amp with a higher gain bandwidth product. Careful though! Op amps must be frequency compensated to prevent spontaneous oscillations. Uncompensated op amps, such as the 301, have much higher gain bandwidth products than does the 741C. However, you *must* add external compensating components to eliminate oscillations, and these reduce the gain bandwidth product. The 318 op amp is internally compensated and has a gain bandwidth product of 15 MHz. Using a 318 will extend the frequency response of your circuits.

An alternative to buying more expensive higher frequency op amps is to build the circuit with several stages, each stage with a lower gain (and therefore higher frequency response). This is illustrated in Example 4-6.

Example 4-6

Design a circuit with a gain of 125 ± 10 percent over the full audio frequency band, and a 560 Ω input impedance.

Solution

$$f_{10\%} = 0.1\,f_H = \frac{1}{2}\,\frac{\text{GBW}}{A_{CL}}$$

or

$$\text{GBW} = \frac{f_{10\%} \times A_{CL}}{\frac{1}{2}}$$

$$\text{GBW} = \frac{20\text{ kHz} \times 125}{\frac{1}{2}} = 5\text{ MHz}$$

a. One solution is to use a 318 op amp, as shown in Fig. 4-14.

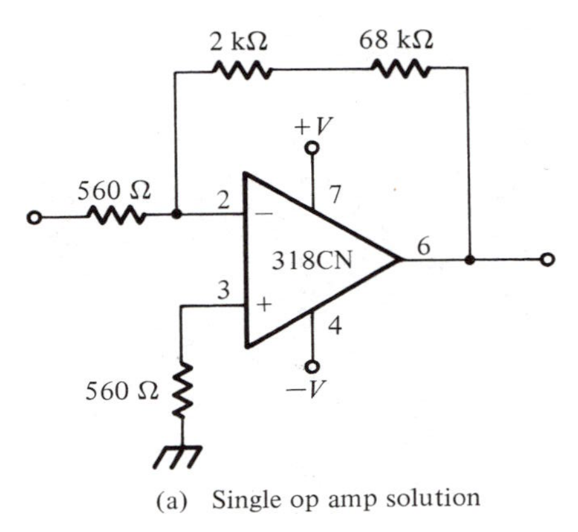

(a) Single op amp solution

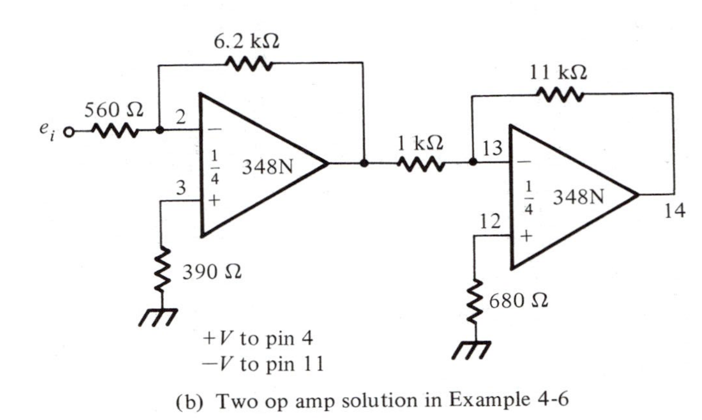

(b) Two op amp solution in Example 4-6

Figure 4-14 Schematics for Example 4-6

$$R_i = Z_{IN} = 560\ \Omega$$

$$R_f = R_i \times A = 560\ \Omega \times 125 = 70\ k\Omega$$

Choose a 2 kΩ resistor in series with a 68 kΩ resistor.

b. Another solution is to use two cascaded amplifiers, each with a closed loop gain of about 11. Then for each separate amplifier

$$GBW = \frac{20\ kHz \times 11}{\frac{1}{2}} = 440\ kHz$$

and a 348 (four 741C op amps in a single package) can be used. This is shown in Fig. 4-14(b).

4-2.2 Rise Time

For small signal ($V_{peak} < 1$ V) sinusoidal signals, the bandwidth gives all of the frequency response information you will need. However, for small signal, rectangular waves, the rise time is a more critical parameter.

Rise time is illustrated in Fig. 4-15. It is the time required for the output signal to rise from 10 percent of the amplitude to 90 percent of the amplitude. The faster the op amp is, the shorter the rise time, and the more like an ideal rectangular wave the output signal appears.

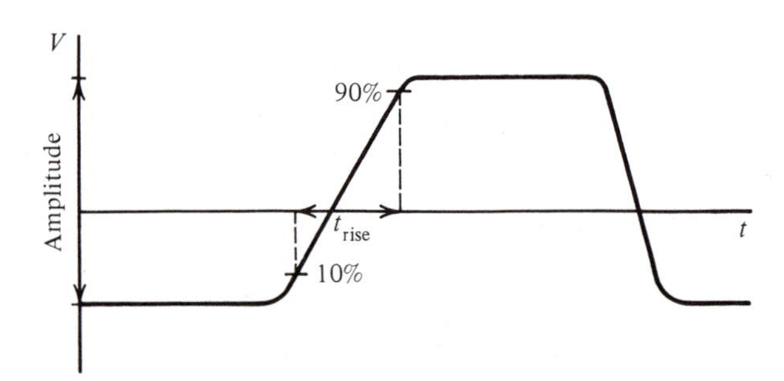

Figure 4-15 Rise time definition

For *small signals*, $V_{pk} < 1$ volt, the rise time is directly related to the gain bandwidth.

$$t_{rise} = \frac{0.35}{GBW} \tag{4-11}$$

This is caused by the transistor propagation delays, the time it takes a signal to alter the conduction of the transistors within the op amp.

Often op amp manufacturers will specify the rise time, but will not give the gain bandwidth product. You can calculate the gain bandwidth product from equation (4-11).

$$\text{GBW} = \frac{0.35}{t_{\text{rise}}} \tag{4-12}$$

4-2.3 Slew Rate

Gain bandwidth and rise time are both characteristics of small signals, where the peak output voltage is less than one volt. For many front-end applications this may be adequate, but most power or driver stages must output larger signals.

For large signal outputs, $V_{\text{peak}} > 1$ volt, the op amp's speed is limited by the slew rate. Slew rate is defined as the maximum rate at which the output voltage can rise.

$$SR = \left.\frac{dv}{dt}\right|_{\text{max}} \tag{4-13}$$

The rate at which the voltage can change is primarily limited by the frequency compensation capacitor, illustrated in Fig. 4-12 (which may be either internally provided, or connected externally by you). The voltage across a capacitor cannot change instantaneously, but is governed by the size of the capacitor and the amount of current available to change the capacitor.

$$\frac{dv_{\text{capacitor}}}{dt} = \frac{I}{C}$$

or

$$SR = \left.\frac{dv}{dt}\right|_{\text{max}} = \frac{I_{\text{max}}}{C}$$

For the 741C the maximum internal capacitor charging current is limited to about 15 μA. So the slew rate of a 741C is

$$SR = \left.\frac{dv}{dt}\right|_{\text{max}} = \frac{I_{\text{max}}}{C} = \frac{15\ \mu\text{A}}{30\ \text{pF}} = 0.5\ \text{V}/\mu\text{s}$$

The units of slew rate are volts per μs. Independent of all other considerations (GBW, t_r, A_{OL}, A_{CL}, input, etc.), it will take at least 2 μs for the output to change one volt, or twenty μs for the output voltage to change by 10 volts.

Example 4-7

A square wave with negligible rise time and a peak-to-peak amplitude of 500 mV must be amplified to a peak-to-peak amplitude of 3 volts, with a rise time of 4 μs or less.

a. Can a 741 be used?

b. Can a 318 be used?

Solution

Since the output has a peak amplitude of greater than 1 volt, the slew rate is the limiting factor, not the gain bandwidth product, or the rise time (t_r).

The required slew rate is

$$SR_{\text{needed}} = \frac{\Delta V}{\Delta t} = \frac{.8 \times 3\ \text{V}}{4\ \mu\text{s}} = 0.6\ \text{V}/\mu\text{s}$$

The op amp must have a slew rate equal or faster than 0.6 $V/\mu s$.

a. The 741 slew rate is 0.5 $V/\mu s$. It is too slow and cannot be used.

b. The 318 slew rate is 50 $V/\mu s$. It is more than fast enough. The rise time using a 318 op amp will be

$$SR = \frac{\Delta V}{\Delta t}$$

or

$$\Delta t = \frac{\Delta V}{SR} = \frac{.8 \times 3\ \text{V}}{50\ \text{V}/\mu\text{s}} = 48\ \text{ns}$$

4-2.4 Full Power Response

Slew rate limits the response speed to all large signal waveshapes. The effects on a square wave are easily calculated (as in Example 4-7). However, deriving the slew rate limiting of a sine wave requires a bit more work.

$$SR = \frac{dv}{dt}$$

$$v = E_{pk} \sin \omega t$$

$$\omega = 2\pi f$$

$$SR = \frac{d(E_{pk} \sin \omega t)}{dt}$$

$$SR = \omega E_{pk} \cos \omega t$$

The maximum (dv/dt) is the maximum slope, which occurs at $t = 0$ for a sine wave.

$$SR_{max} = \omega E_{pk} \cos \omega(0) = \omega_{max} E_{pk}$$

$$\omega_{max} = \frac{SR_{max}}{E_{pk}}$$

$$f_{max} = \frac{SR_{max}}{2\pi E_{pk}} \tag{4-14}$$

This, f_{max}, is called the *full power response*. It is the maximum frequency of a large amplitude sine wave that the op amp can output without distortion. Notice that this is entirely separate from the gain bandwidth product (which limits the output frequency because of a drop in gain). Above the full power response frequency, the op amp cannot charge the compensation capacitor fast enough to cause the output signal to swing to E_{pk}, before the input begins lowering the signal.

Example 4-8

It is necessary to amplify a 20 mV RMS sine wave, 15 kHz, to 6 volts RMS, ± 10 percent.

a. Can a 741 be used?

b. Can a 318 be used?

Solution

The amplifier must produce a closed loop gain of

$$A_{CL} = \frac{v_O}{v_i} = \frac{6\ V_{RMS}}{20\ mV_{RMS}} = 300 \text{ at } 15 \text{ kHz}$$

The op amp's slew rate must equal or exceed

$$f_{max} = \frac{SR_{max}}{2\pi E_{pk}}$$

or

$$SR_{max} = 2\pi E_{pk} f_{max}$$

$$SR_{max} = 2\pi(6 \text{ V} \times 1.414)(15 \text{ kHz})$$

$$SR_{max} = 0.8 \text{ V}/\mu\text{s}$$

a. The 741 has a slew rate of 0.5V/μs. It is too slow and would distort the sine wave output.

b. The 318 has a slew rate of 50 V/μs. It is fast enough to produce a 6 V_{RMS} sine wave output at 15 kHz.

Does the 318 have an adequate gain bandwidth product?

$$f_{10\%} = .1 f_H = .1 \frac{\text{GBW}}{A_{CL}}$$

or

$$\text{GBW} = \frac{f_{10\%} \times A_{CL}}{0.1}$$

$$\text{GBW} = \frac{15 \text{ kHz} \times 300}{0.1} = 45 \text{ MHz}$$

The GBW of the 318 is only 15 MHz. So even though the 318 will not be slew rate limited, it does not have a large enough gain bandwidth to give a gain of 300 ($\pm$10 percent) at 15 kHz.

You can solve this problem by cascading two 318 amplifiers, one with a gain of 10 and the other with a gain of 30. Neither will be slew rate limited, as has been shown.

$$\text{GBW (gain of 10)} = \frac{15 \text{ kHz} \times 10}{0.1} = 1.5 \text{ MHz}$$

$$\text{GBW (gain of 30)} = \frac{15 \text{ kHz} \times 30}{0.1} = 4.5 \text{ MHz}$$

The 318 has a large enough gain bandwidth to operate both amplifiers.

An amplifier must meet both the gain bandwidth and the slew rate requirements to operate properly at a given frequency. Check both.

4-2.5 Noise

Any unwanted signals at the output of an amplifier circuit are often referred to as *noise.* Under this broad definition, variation of e_{out} caused by bias currents, offset current, and offset voltage are an example of DC noise. You have seen how to minimize these effects in section 4-1.

Alternating current noise can be divided into two major classes; that caused by external effects (or interference), and that caused by effects internal to the op amp IC.

Internally generated noise is modeled as a voltage source and as a current source (see Fig. 4-16). The noise *voltage* is primarily Johnson or *thermal noise.* It is directly proportional to the temperature of the IC chip, series circuit resistance, and the IC's bandwidth.

$$e_n(T, R, f_H)$$

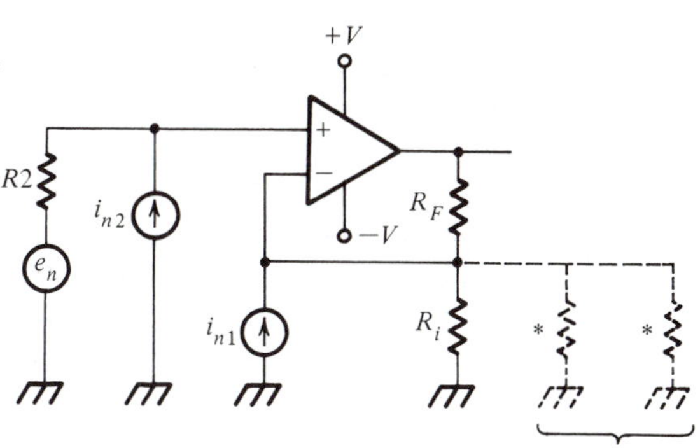

Figure 4-16 Noise current and voltage

The noise *current* is often referred to as *shot noise.* The exact number of electrons flowing past a point at any *instant* is random. The *average* rate is the DC currents. This random variation of the currents, about the I_{DC} average, is the noise current. It is directly proportional to the DC current and the IC's bandwidth.

$$i_N(I_{DC}, f_H)$$

Manufacturers' specifications of noise voltage and current are done at stated temperature, circuit resistance, supply voltage, and frequency. If your circuit is not operating at these precise conditions, the specified noise voltage and current may not apply.

To lower the noise voltage and current present in your IC, keep the supply voltage as low as practical. This will decrease the power dissipated by the IC and therefore its temperature. This will also lower internal bias currents, lowering the noise current. Keep the circuit's bandwidth as low as possible. (Do not use a 318 if a 741 will work.) This will drop the high frequency noise harmonics.

You can also lower the effect of the noise you cannot eliminate.

$$e_0 = \left(1 + \frac{R_f}{R_i}\right) \sqrt{(R2 i_{n2})^2 + \left(\frac{R_f R_i i_{n1}}{R_f + R_i}\right)^2 + e_n^2} \quad * \text{ (4-15)}$$

Equation (4-15) gives the output voltage produced by the noise sources in Fig. 4-16. The noise term (the square root factor) is multiplied by the gain of a noninverting amplifier. So noise can be viewed as a signal injected into the noninverting terminal. Reducing R_F will reduce this gain, and therefore the effect of the noise. But this also will drop the gain of the amplifier for any signal you want. One solution is to place a small capacitor around R_F (3 pF to 30 pF). This will lower the high-frequency (noise) feedback impedance, but will not affect the lower-frequency (signal) gain.

However, do not allow any shunting capacitance around R_i. Capacitance here would lower Z_i (R_i in parallel with the capacitance), thereby raising the high-frequency (noise) gain. Also, inverting summers effectively place each of their input resistors in parallel. This also lowers R_i, raising the noise gain. Use inverting summers only when necessary and then carefully limit the number of inputs.

Finally, keep the input and source impedance low. This will allow you to drop *both* R_i and R2. The square root factor, [equation (4-15)] then will decrease more than the gain factor will increase.

For noise reduction,

Use all DC compensation techniques. Keep $\pm V$ supply low. Keep IC cool. Keep gain and R_F low. Place a capacitor across R_F but not R_i. Avoid the inverting summer. Lower R_i and R2 (source impedances).

4-2.6 Frequency Compensation Techniques

High gain at high frequency jeopardizes amplifier stability. It is probable that uncompensated amplifiers will provide enough gain and phase shift for signals above the 100 kHz range to be fed back to the input through parasitic capacitance and cause oscillations. This oscillation can occur spontaneously even though no input signal is applied. The 741 and 318 op amps eliminate this problem by using a metal-oxide capacitor built within the IC. The circuit configuration is shown in Fig. 4-12.

---

* Gerald G. Graeme, Gene E. Tobey, Lawrence P. Hueleman, *Operational Amplifiers Design and Applications* (New York, N.Y.: McGraw Hill Book Co., 1971), p. 439.

Although this works well, internal compensation does not allow you any control over the op amp's frequency response. The 709 and 301 op amps have no internal frequency compensation capacitor. Instead, frequency compensation terminals are provided, and you *must* connect compensation capacitors yourself. Failure to connect these external compensation capacitors will practically guarantee that the op amp will oscillate. However, the op amp's frequency response can be tailored to your needs by the value and type of compensation network you connect.

The 709, one of the first integrated circuit op amps, requires two external capacitors and a resistor for proper compensation. This is shown in Fig. 4-17.

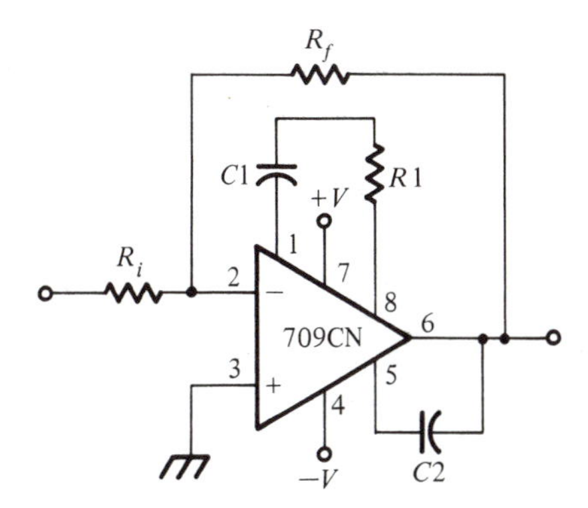

Figure 4-17 709 frequency compensation circuit

The effects of various sets of compensation component values on the op amp's performance in a closed-loop circuit are shown in Fig. 4-18. Increasing the compensation capacitors' sizes will improve the circuit's stability, but will also decrease bandwidth and full power response (output voltage swing). This is not necessarily bad. Remember, noise voltage and current are both proportional to bandwidth. You should always restrict the op amp's bandwidth as much as signal amplification requirements allow, to keep noise low. In fact, using C1 = 0.05 μF and C2 = 2000 pF will give 20 dB more stability, dropping output swing and bandwidth below those shown in Fig. 4-18. This could be handy if you are using the 709 as strictly a DC amplifier.

The 101/201/301 is a later generation IC op amp. Several types of compensation are possible. The simplest, shown in Fig. 4-19, requires a single capacitor connected between two IC pins. The minimum value of compensation capacitances is dependent on the resistor feedback network.

$$C1 = \frac{R_i \times 30\text{ pF}}{R_i + R_f}$$

You should keep this in mind when selecting R_i and R_f. Since the same equation

Curve A: $C1 = 10$ pF, $R1 = 0$, $C2 = 3$ pF
Curve B: $C1 = 100$ pF, $R1 = 1.5$ kΩ, $C2 = 3$ pF
Curve C: $C1 = 500$ pF, $R1 = 1.5$ kΩ, $C2 = 20$ pF
Curve D: $C1 = 5000$ pF, $R1 = 1.5$ kΩ, $C2 = 200$ pF

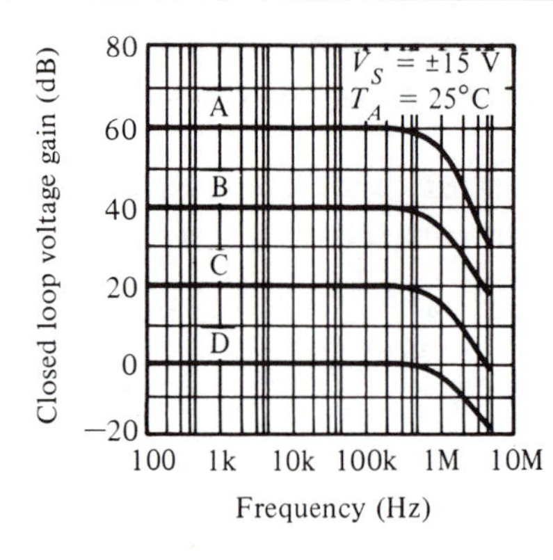

(a) Frequency responses for various closed-loop gains

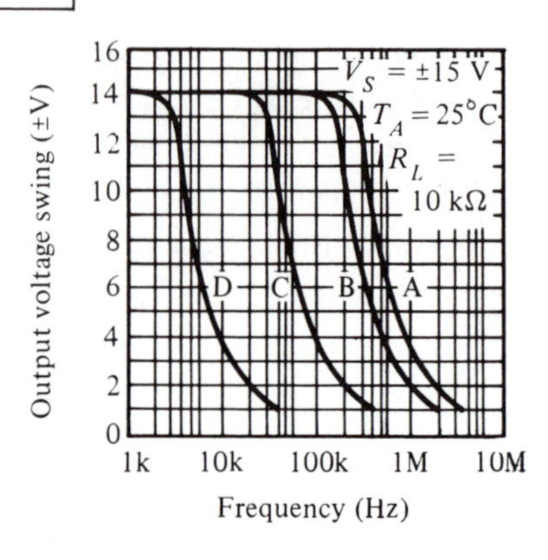

(b) Output voltage swing as a function of frequency

Figure 4-18 709 frequency response and output voltage characteristics (*Courtesy of National Semiconductor*)

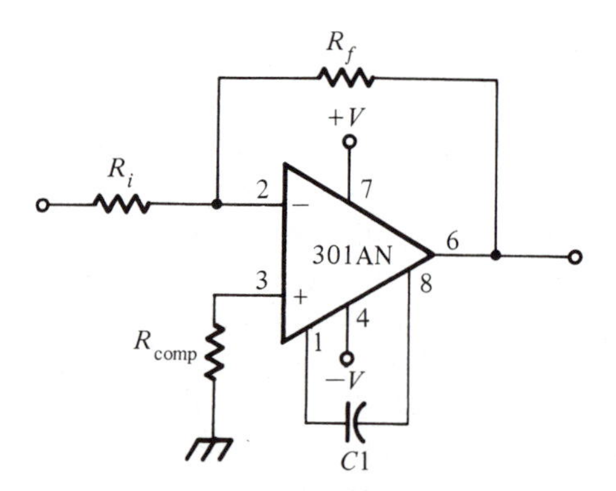

Figure 4-19 301 single capacitor compensation

holds for a noninverting amp, using a noninverting amplifier will allow you to lower R_i and C1 without sacrificing the input impedance. The effects of a 3 pF and a 30 pF compensation capacitor on open-loop frequency response, and output voltage swing are shown in Fig. 4-20. As with the 709, larger compensation capacitance can be used to lower the frequency response and output voltage swing further.

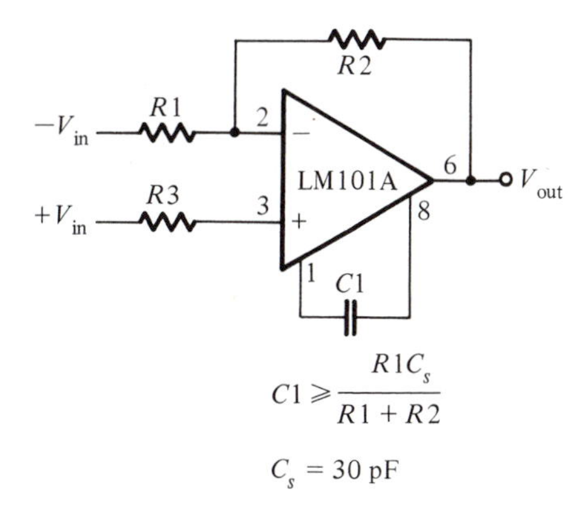

(a) Single pole compensation
Pin connections shown are for metal can

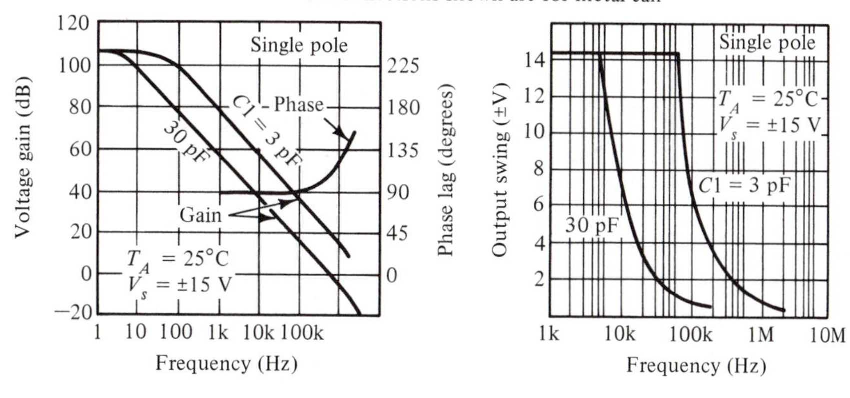

(b) Open loop frequency response (c) Large signal frequency response

Figure 4-20 301 frequency response with a single compensation capacitor (*Courtesy of National Semiconductor*)

Using a single 30 picofarad compensation capacitor on a 301 op amp drops the gain bandwidth to 1 MHz and restricts the full power response to 6 kHz. Using a two-capacitor compensation network allows you to more than double the full power response, for good signal output, while keeping the gain bandwidth at 1 MHz for good noise suppression and stability. The schematic is shown in Fig. 4-21. At low frequencies the reactance of C2 is high, feeding back only a small signal. This has very little effect on gain. But as frequency increases, the reactance of C2 drops. More of the output is fed back into the compensation terminal, lowering gain. A comparison of single capacitor and two-capacitor compensated frequency response is given in Fig. 4-22. Notice the boosted low-frequency response of the capacitor technique. The open loop frequency response and full signal output response curves for the two-capacitor compensation network are given in Fig. 4-22. Full 14 volt peak-to-peak output is available at 20

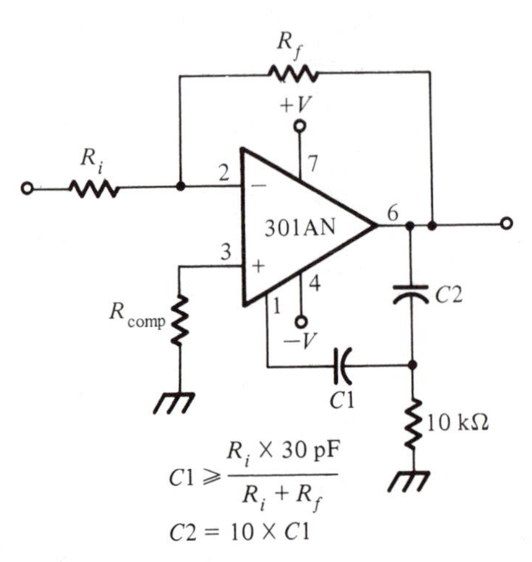

Figure 4-21 Two capacitor compensation (*Courtesy of National Semiconductor*)

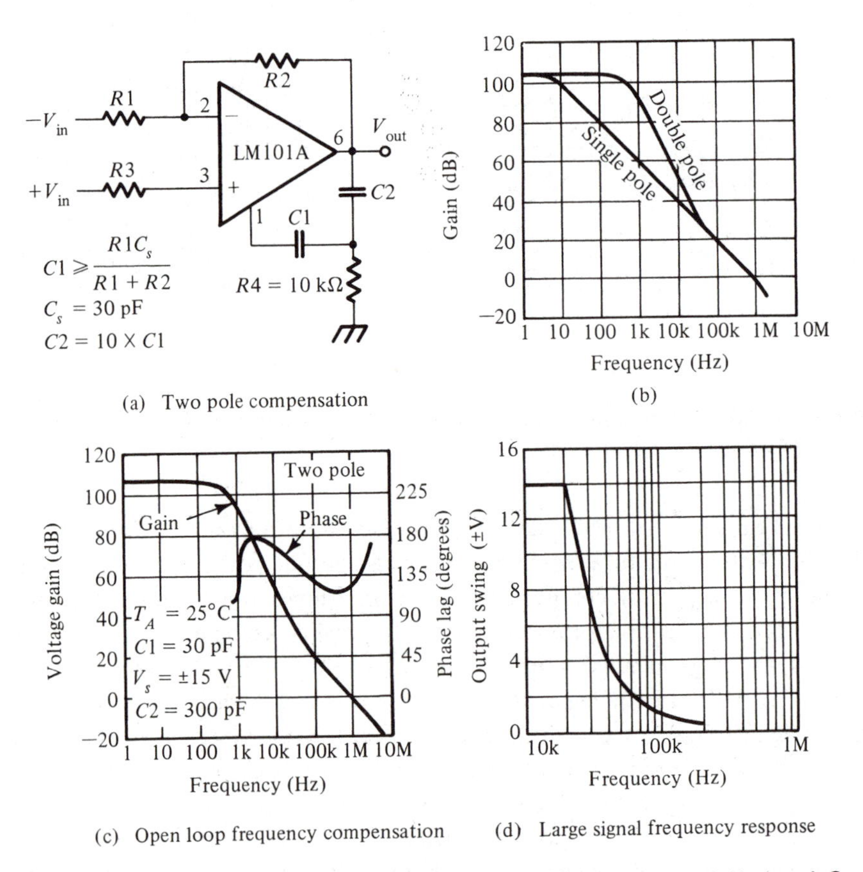

Figure 4-22 Two capacitor compensation responses (*Courtesy of National Semiconductor*)

kHz, while the gain bandwidth is still restricted to 1 MHz. For single-capacitor compensation (Fig. 4-20), to lower gain bandwidth to 1 MHz, a 30 pF capacitor is needed. This, however, restricts the output swing to less than 4 volts at 20 kHz.

Two capacitor compensation of the 301 allows low gain bandwidth (for stability) and high full power response.

In order to provide the proper DC level, a lateral PNP transistor must be included in the input stage of an op amp. This is shown in Fig. 4-23. Such transistors have poor frequency response and excessive phase shift at high frequencies. It is largely this transistor which limits the stable frequency response of the entire circuit. Without it, however, proper DC operation could not be established.

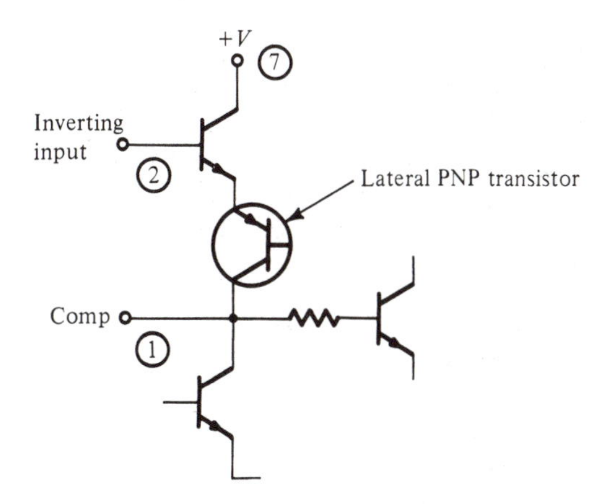

Figure 4-23 Partial schematic of op amp input showing lateral PNP transistor

Significant improvements in frequency response and slew rate can be obtained by feeding the high frequency signals forward, through an external capacitor, around this slow PNP transistor. The circuit schematic for this feed-forward compensation is given in Fig. 4-24. A 150 pF feed-forward capacitor allows high-frequency signals to bypass the input stage and its slow PNP transistor, entering the op amp through the compensation pin. Lower frequency and DC signals, however, are forced into the input stage for proper level shifting by the PNP transistor. Capacitor C2, around the gain setting resistor R_f, forces the circuit's gain to fall (at 20 dB per decade, with a corner frequency of f_0). This ensures stable operation at high frequencies. This technique allows a closed loop $f_{1\%}$ (1% gain error) gain bandwidth of 100 kHz ($f_0 = 3$ MHz) and a full power response (for $\pm V_s = \pm 15$ V) in excess of 150 kHz. The frequency response curves for a feed-forward compensated 301 op amp are shown in Fig. 4-25. Compare them

Figure 4-24 Feedforward frequency compensation (*Courtesy of National Semiconductor*)

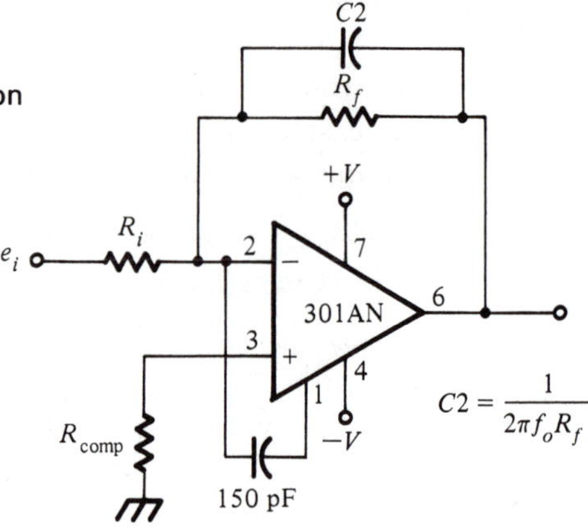

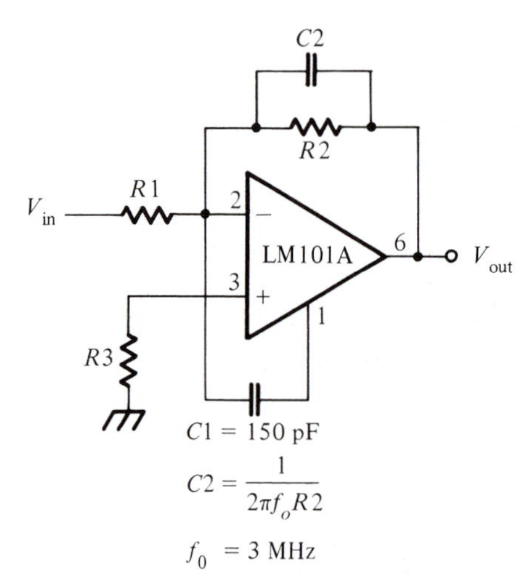

(a) Feed forward compensation

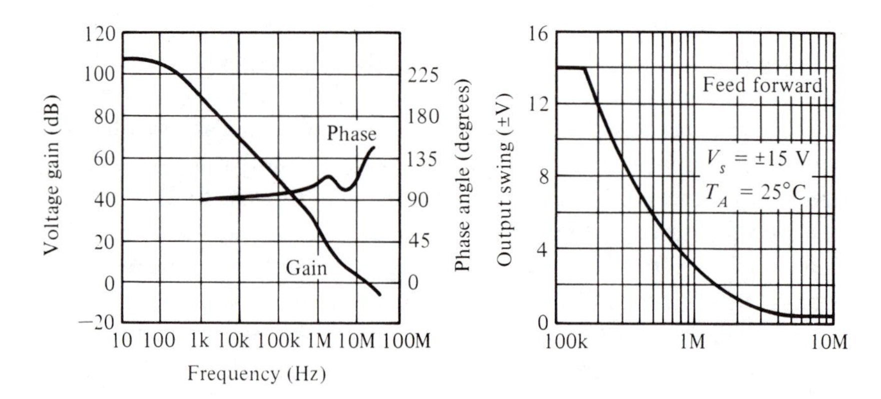

(b) Open loop frequency response

(c) Large signal frequency response

Figure 4-25 Feedforward frequency response (*Courtesy of National Semiconductor*)

carefully with those of the single capacitor and two capacitor compensation techniques. You will see that feed-forward compensation gives far better open loop, large signal, and pulse responses. Figure 4-26 gives a summary of compensation techniques' effects on gain bandwidth. Full power effects are summarized in Fig. 4-27 (p. 178).

Bypassing the slow PNP input stage gives feed-forward compensated op amps better open loop, large signal, and pulse response than either of the other compensation techniques.

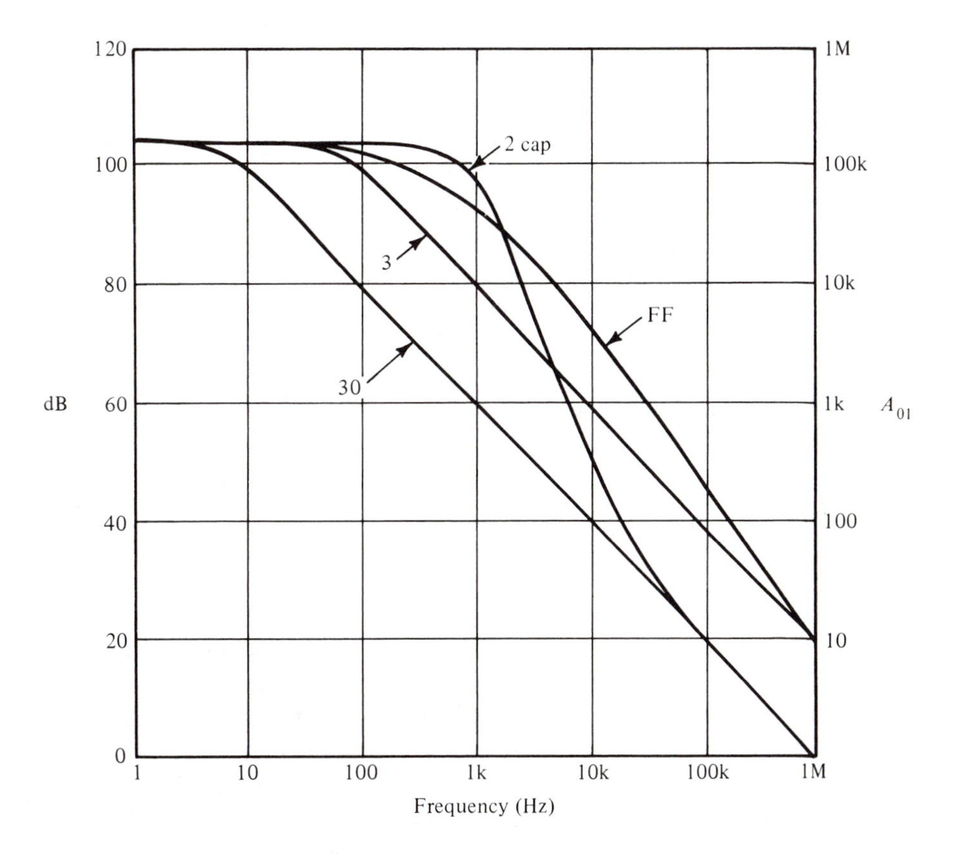

Figure 4-26 Summary of external compensation techniques on gain bandwidth

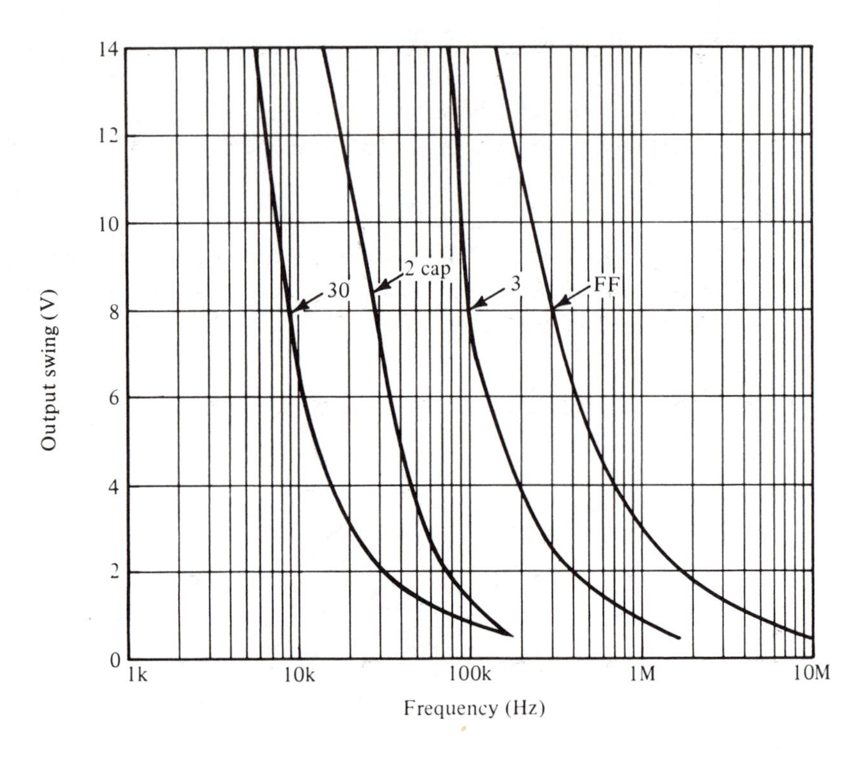

Figure 4-27 Summary of external compensation effects on the full power response

SUMMARY

This chapter has presented the nonideal DC and AC characteristics of operational amplifiers. Compensation techniques for each were also given. Bias current is the DC current necessary to flow into the inputs of an op amp to turn it on. The effects of bias currents can be eliminated by placing a resistor in the noninverting input lead which equals the resistance in the inverting input lead. Offset current is the difference between the two bias currents. Its effect on the output voltage can be minimized by keeping the value of the negative feedback resistance small. Offset voltage is modeled as a signal source at the noninverting input. Its effect can be eliminated only by proper nulling, using either internally supplied pins or external null circuits. Drift describes the variation of offsets with temperature. Layout, power dissipation, and ambient temperature all must be considered if drift is critical.

To prevent high frequency oscillations, the gain of an op amp must be reduced at high frequencies. The gain bandwidth product is the open-loop gain times frequency and is a constant. So, as frequency rises, the open-loop gain falls off. For closed-loop gain errors of less than ten percent, the closed-loop gain times frequency must be one-tenth of the open-loop gain bandwidth product. Closed-loop gain errors of less than one percent require a closed-loop gain bandwidth product one hundred times smaller than the manufacturers' stated open-loop gain bandwidth.

Large signal outputs ($e_o > 1$ V) are limited by the speed at which the op amp can charge compensation capacitance. This is defined as the slew rate, and is expressed in volts per microsecond. The limits that slew rate places on the op amp's response to a pulse are easily seen. However, slew rate limiting of a sine wave output is a bit more complex, and is defined as the full power (or large signal) frequency response. An amplifier may be gain bandwidth limited (small signal), or slew rate or full power response limited (large signals). You should check both.

Internally generated AC noise results from thermal effects and random current variations. Lowering supply voltage and package temperature will minimize noise. Bypassing the negative feedback resistor with a small capacitor reduces the effects of high frequency noise. Low source impedances will also help lower noise effects.

Some op amps do not limit the high frequency gain with an internal capacitor. Such uncompensated op amps must have external capacitance added to prevent oscillations. Single capacitor compensation is easy to use and allows you to select the open loop gain bandwidth. However, two capacitor compensation gives more audio frequency gain with a steeper roll-off. This allows a wider full power response while still strongly limiting high frequencies (oscillations). Feed-forward compensation bypasses the slow, input PNP transistor to give superior small signal, large signal, and pulse response.

The op amps discussed so far have required dual power supplies. For many applications, this is either inconvenient or highly undesirable. In the next chapter you will see how to operate a standard op amp and op amps specifically designed for such operation from a single power supply. Requirements and limitations will be pointed out. Also, the current differencing amp (CDA) or Norton amp will be presented. Its specifications, operational characteristics, biasing requirements, and circuit application will be discussed. The Norton amp's lower price, four amplifiers within a single package, and operation from a single power supply will often allow you to cut circuit cost, complexity, and size.

PROBLEMS

4-1 For the circuit in Fig. 4-28, when the signal source is set to zero volts, the bias current I_B^- flows entirely through R_f. None flows through R_i. Explain why.

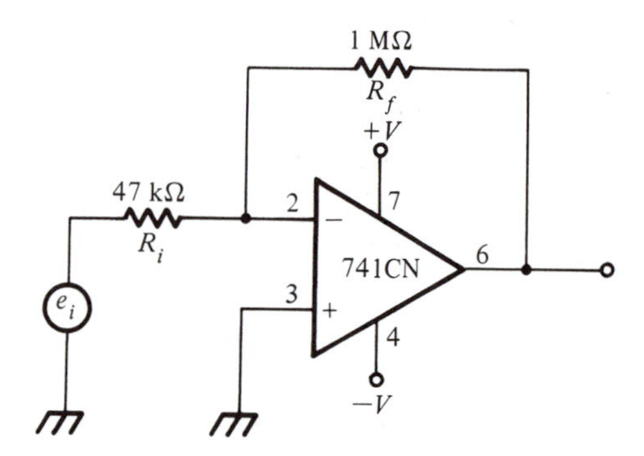

Figure 4-28 Schematic for problems 4-1 and 4-2

4-2 For the circuit in Fig. 4-28, with the signal source set to zero volts, calculate the output voltage produced by a bias current of 80 nA.

4-3 For the circuit in Fig. 4-29, with the signal source set to zero volts, calculate the output voltage produced by a bias current of 100 nA.

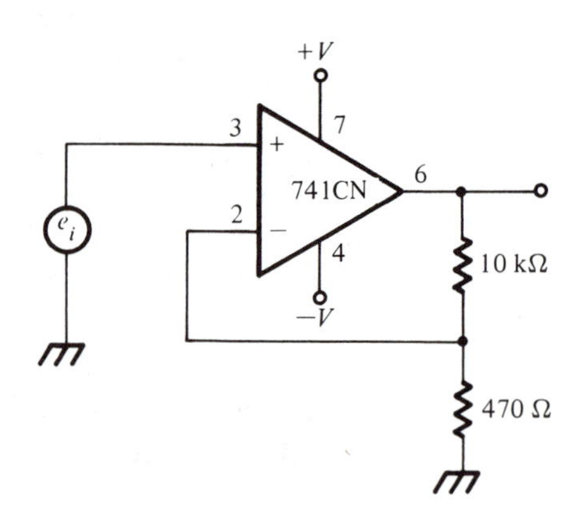

Figure 4-29 Schematic for problem 4-3

4-4 Calculate the components necessary to compensate the circuits in Figs. 4-28 and 4-29 for bias currents. Draw the schematics of the compensated circuits.

4-5 For the circuit in Fig. 4-30, calculate

a. $V_{\text{R COMP}}$ **d.** I_{R_f}
b. V_{R_i} **e.** V_{R_f}
c. I_{R_i} **f.** e_o

Figure 4-30 Schematic for problems 4-5, 4-6 and 4-7

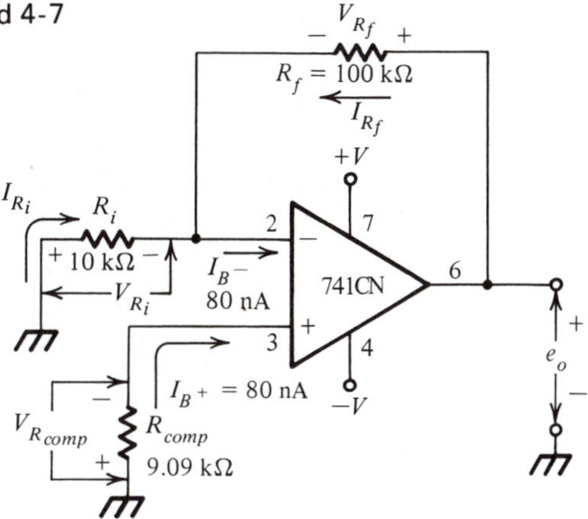

4-6 Repeat Problems 4-5 with $I_B^+ = 80$ nA and $I_B^- = 60$ nA. ($I_{OS} = 20$ nA).

4-7 How can you modify the circuit in Fig. 4-30 to reduce the effect of the offset current (Problem 4-6) by a factor of 10?

a. Draw the schematic.

b. Repeat the calculations of Problem 4-6 on your modified circuit to prove that the offset current effect on e_o has been dropped by 10.

4-8 For the circuit in Fig. 4-31, calculate

a. Z_{IN}

b. A_v

c. e_o with $e_i = 0$ $I_B^+ = 80$ nA, $I_B^- = 60$ nA

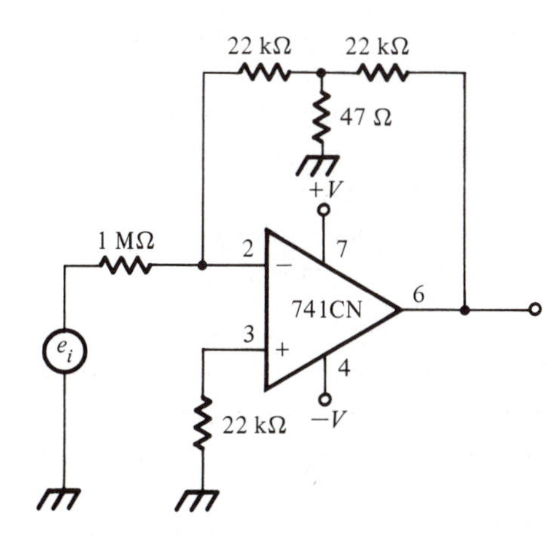

Figure 4-31 Schematic for problem 4-8

4-9 Design an inverting amplifier using a 741C op amp with a gain of -22, an input impedance of 4.7 MΩ, and an output error voltage due to typical offset current of less than 20 mV.

4-10 Refer to the specification of the 741C amp. How much can the output voltage be changed by using the provided offset compensation pins?

4-11 Design an inverting amplifier to meet all of the following:

a. $Z_{IN} = 10\ k\Omega$
b. $A_v = -25$
c. Input offset voltage adjustment range of ± 100 mV.

4-12 Design a noninverting amplifier to meet the following:

a. $A_v = 48$
b. Output offset voltage adjustment range of ± 150 mV.

4-13 Design a difference amplifier to implement the following:

a. $e_o = 10(e_2 - e_1)$
b. Output offset voltage adjustment range of ± 50 mV.

4-14 Draw the schematic of the circuits needed to allow you to measure I_B^+ and I_B^- (two separate circuits) of a 741C with three digits of precision with a digital multimeter on the 1.999 volt range. Give all component values and the output voltage to bias current conversion equation.

4-15 Draw the schematic of the circuit which will allow you to measure the offset current of a 741C with three digits of precision on a digital multimeter (1.999 volt range). Give all component values and output voltage to offset current conversion equation.

4-16 **a.** Draw the schematic of the circuit which will allow you to measure the offset voltage (while minimizing the offset current effects) of a 741C with three digits of precision on a digital multimeter (1.999 volt range).
b. Give all component values and the output voltage to offset voltage conversion equation.
c. For *your* circuit, if I_{OS} has been measured to be 20 nA, and e_o is 469 mV, calculate the offset voltage.

4-17 The rise time of the 741C is specified as typically 0.35 μs. Calculate the

a. Open loop gain bandwidth.
b. Open loop gain at 20 Hz, 15 kHz, 650 kHz.
c. Small signal bandwidth for a closed loop gain of 80.
d. Maximum frequency for a closed loop gain of 80 with 1 percent gain error.
e. Maximum frequency for a closed loop gain of 110 with 10 percent gain error.

4-18 What must be the open loop gain bandwidth product of an op amp used in an amplifier with a gain of 50, gain error of 10 percent, at 20 kHz?

4-19 How could you use several 741C op amps to build the circuit described in Problem 4-18?

4-20 A 741C is used to build a comparator. How long will it take the output pulse to rise from 0.8 volt to 2.4 volts?

4-21 Design a comparator using an op amp (to be used as the input stage of a digital counter). The output should switch from -3 volts to $+3$ volts in 150 ns, whenever the input crosses ground, going positive. Specify the op amp's minimum required slew rate.

4-22 Calculate the full power response (frequency) of a 741C which must output a 2 V RMS sine wave.

4-23 A circuit is needed to amplify a 50 mV RMS 10 kHz input, outputting 3 volts RMS with a gain error of 1 percent. Calculate the

a. Closed loop gain.
b. Required open loop gain bandwidth product of the op amp.
c. Required slew rate of the op amp.

4-24 **a.** List three ways to decrease the noise signals generated within an op amp.
b. List two ways to decrease the *effect* of internally generated noise signals you cannot eliminate.

4-25 Explain why adding inputs to an inverting summer increases the effect of the noise at the circuit's output.

4-26 Determine the open loop gain of a 301 op amp at 10 kHz for the following external compensation techniques:

a. Single 30 pF capacitor.
b. Single 3 pF capacitor.
c. Single 3000 pF capacitor.
d. Two capacitors, as shown in Fig. 4-22.
e. Feed forward, as shown in Figs. 4-24 and 4-25.

4-27 What is the full power response (frequency) of a 301 using the external compensation techniques in Problem 4-26 for a 6 volt peak output voltage?

CHAPTER 5

Single Supply Amplifiers

The operational amplifier circuits you have seen so far have been supplied with bipolar (±V) power. This was done to allow the circuit to amplify DC. The output could go positive or negative with no significant distortion as it crossed zero volts. This feature is important in low frequency and DC circuits such as industrial instrumentation and control systems.

However, a very large portion of analog electronics deals strictly with an AC signal. For these applications the precise DC level of that AC signal is not important. Audio amplifiers (entertainment systems, telephone system, intercoms), active filters, oscillators, rf communications circuitry, and AC instruments all traditionally operate from a single power supply. The DC level is blocked while the signal is passed between amplifier stages, using RC coupling.

In this chapter you will learn how to use an op amp, for AC amplification, biased from a single power supply. Also, the Norton amp, a single supply current differencing amp, will be covered. Using these techniques for single supply amplifiers, you will be able to build AC amplifiers more cheaply and compactly, by eliminating the negative power supply.

OBJECTIVES

After studying this chapter you should be able to do the following:

1. Explain why the output of an op amp must be offset from ground when powered from a single supply.
2. Draw the schematic for an inverting amplifier circuit, and a noninverting amplifier circuit using a single supply op amp.
3. Calculate component values necessary to bias the circuits above (Objective 2) for class A operation.
4. Describe the characteristics of the Norton amp (current differencing amp) as compared with the operational amplifier.

5. Analyze a given circuit or design a circuit to properly bias the Norton amp's output.
6. For each of these circuits, built with Norton amps:
   inverting amplifier
   inverting summer
   noninverting amplifier
   noninverting summer
   difference amplifier
   Do the following:
   a. Draw the schematic.
   b. Analyze a given circuit for Z_i, A_v, V_{DC} out, f_c.
   c. Design a circuit to meet given specifications.
   d. Derive the equation for the output voltage.
7. Draw the schematic for and describe the operation of a Norton amp comparator with a reference voltage and a two level Norton amp comparator.

5-1 OPERATING THE OP AMP FROM A SINGLE SUPPLY

There is no earth ground connection required on an op amp. It simply requires a certain minimum difference in potential between the $+V$ and $-V$ supply pins. This could be a difference of 24 volts, with all signals and instruments referenced at half this difference (that is, ± 12 volts). However, the op amp will perform equally as well with the same 24 volt supply if circuit common was defined as the most negative potential. (That is, connect $+24$ volts to the $+V$ pin and ground to the $-V$ pin.) This would allow you to build a single regulated (24 V) power supply, rather than two (± 12 V) supplies. One problem arises, however. If the source inputs a signal which should cause the op amp's output to swing negative, the output can only go down to ground. The negative portion of the signal will be clipped. This is illustrated in Fig. 5-1.

5-1.1 Inverting Amplifier

This problem can be solved by adding a DC level to the output. Offsetting the op amp's output by $+12$ V DC (*half of the supply voltage*) will remove the clipping, as shown in Fig. 5-2.

The simplest way to add this DC offset to the output of an inverting amplifier is shown in Fig. 5-3. The 324 consists of four independent op amps in a single package. Although they will work well from ($\pm V$) split supplies, their operation has been optimized for single supply biasing (as low as $+5$ V).

Resistors R_f and R_i form the traditional inverting amp configuration, giving a gain of $-(R_f/R_i)$ to the AC input signal e_i.

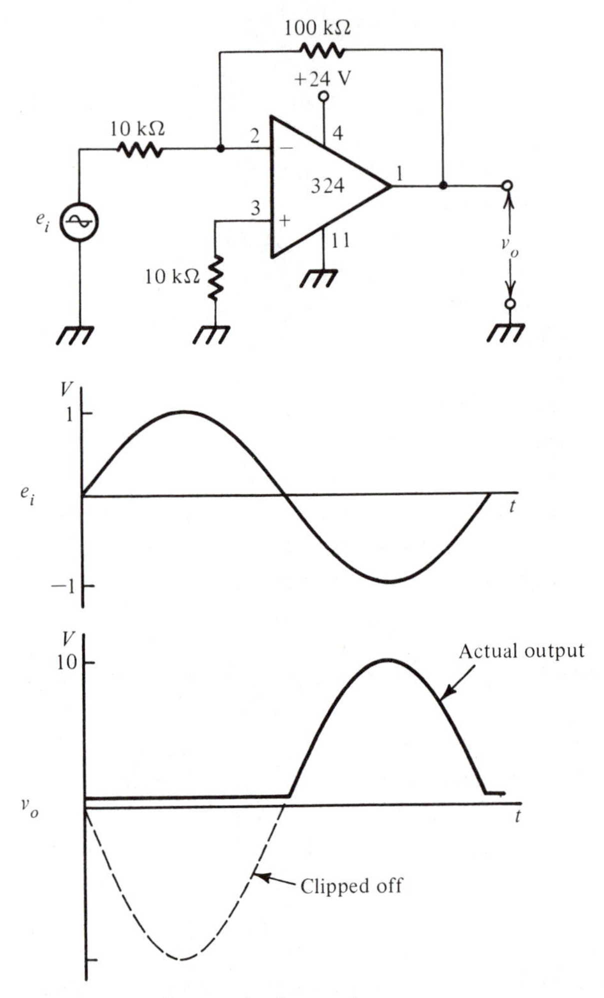

Figure 5-1 Clipped output from a single supply op amp

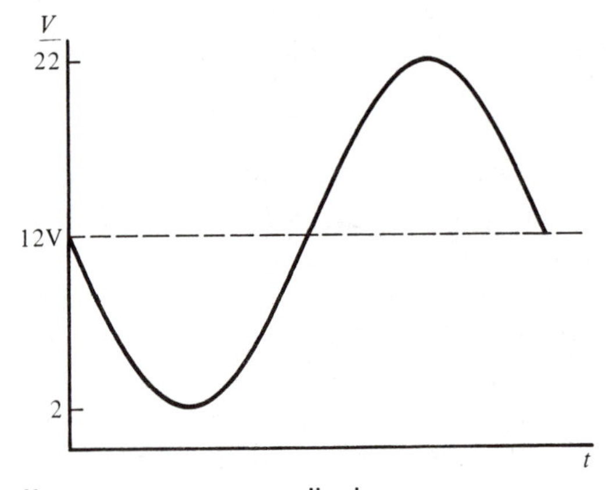

Figure 5-2 DC offset to remove output clipping

Figure 5-3 Biased inverting amplifier

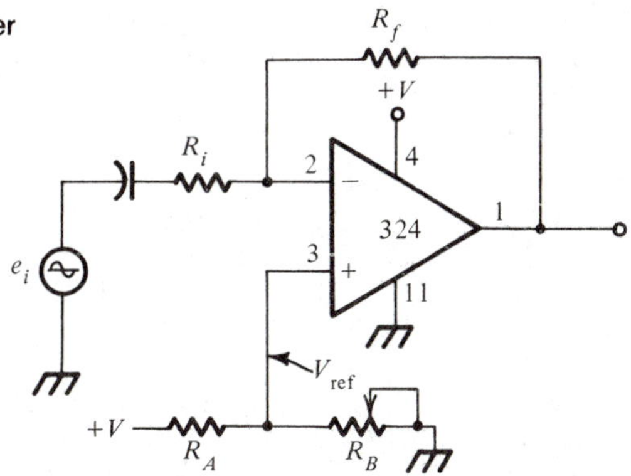

Instead of connecting the noninverting input to ground, as was done in Fig. 5-1, it has been tied to a positive reference voltage. This V_{ref} is easily obtained from the positive power supply with a voltage divider, R_A and R_B. As long as R_A and R_B are small (100 kΩ) compared to the input impedance of the op amp, the reference voltage will be stable (not loaded down).

The key to DC operation of the circuit in Fig. 5-3 is the input blocking capacitor. To the DC bias voltage on the noninverting input, V_{ref}, the capacitor appears as an open circuit. The resulting circuit is shown in Fig. 5-4. For the DC voltage, the circuit is actually a voltage follower. With negligible I_B^-, there is no voltage drop across R_f. The output DC equals the voltage at the inverting input (100 percent negative feedback), which is the reference voltage at the noninverting input.

$$V_{DC} = V_{ref} \tag{5-1}$$

Remember that this relationship holds because the input capacitor removes R_i from the DC circuit, dropping DC circuit gain to 1 for the reference voltage.

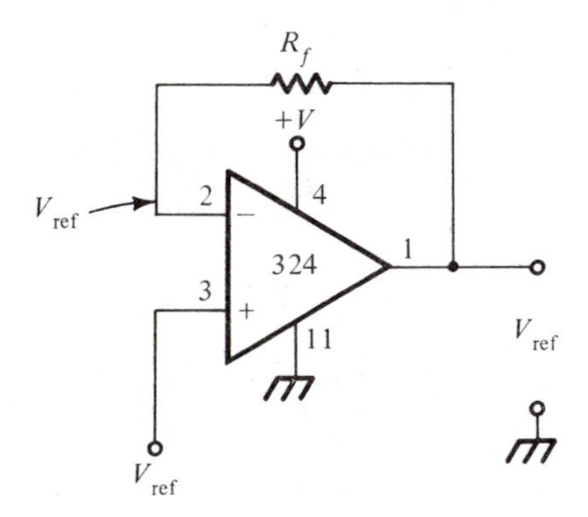

Figure 5-4 V_{ref} as input to noninverting amp

Example 5-1

Calculate the AC and DC output voltage for the circuit in Fig. 5-3 if:

$$+V = 12\text{ V}$$

$$R_f = 22\text{ k}\Omega$$

$$R_i = 3.3\text{ k}\Omega$$

$$R_A = 100\text{ k}\Omega$$

$$R_B = 41.8\text{ k}\Omega$$

$$e_i = 200\text{ mV peak, centered around ground}$$

Solution

AC:
$$A_v = -\frac{R_f}{R_i} = \frac{22\text{ k}\Omega}{3.3\text{ k}\Omega} = -6.67$$

$$v_O = A_v \times e_i = -6.67 \times 200\text{ mV peak}$$

$$v_O = 1.33\text{ V peak (180 deg phase shift)}$$

DC:
$$V_{DC} = V_{ref}$$

$$V_{ref} = \frac{R_B}{R_A + R_B}(+V)$$

$$V_{ref} = \frac{41.8\text{ k}\Omega}{100\text{ k}\Omega + 41.8\text{ k}\Omega}(12\text{ V})$$

$$V_{ref} = 3.54\text{ V}$$

The composite output is shown in Fig. 5-5.

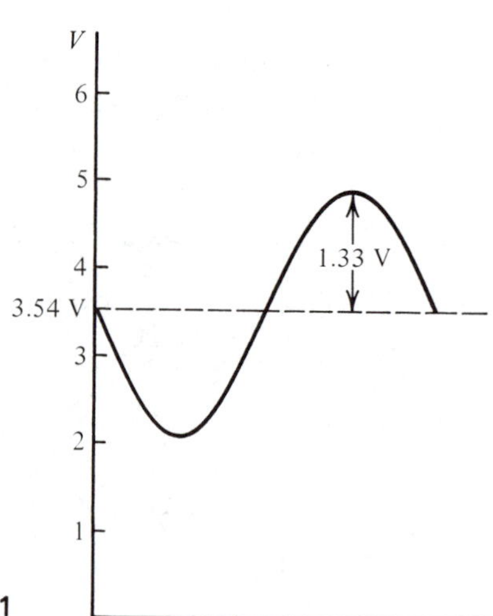

Figure 5-5 Composite output for Example 5-1

Example 5-2

What would happen if the input to the circuit in Example 5-1 were increased to 700 mV peak?

Solution

AC: $v_O = A_v \times e_i = -6.67 \times 700 \text{ mV peak}$

$v_O = 4.67 \text{ V peak (180 deg phase shift)}$

DC: There is no change. $V_{DC} = 3.54$ V

The composite output is given in Fig. 5-6. Notice that the output peak signal is larger than the DC level. This means that a portion of the negative-going half-cycle will be clipped. The biasing must be altered. The best output DC level is

$$V_{DC} \approx \tfrac{1}{2} V_{\text{supply}} \tag{5-2}$$

This will allow maximum signal swing in both directions, before clipping (reaching the saturation voltage).

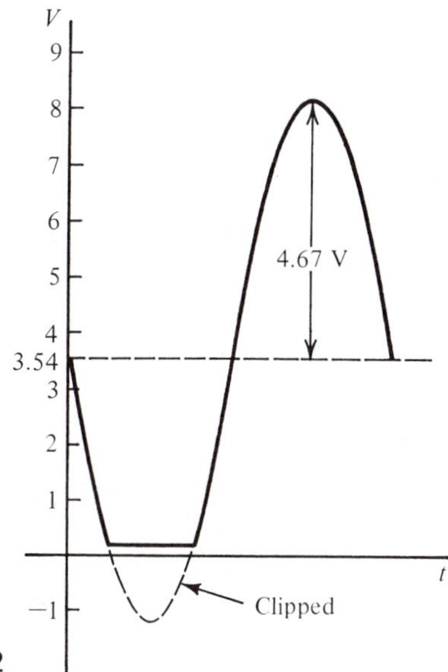

Figure 5-6 Output from circuit for Example 5-2

Example 5-3

How must the circuit of Example 5-1 be changed for best output biasing? Calculate needed component values.

Solution

With a supply voltage of +12 volts the output DC voltage should be

$$V_{DC} = \tfrac{1}{2} V_{supply} = 6 \text{ V}$$

This is done by setting

$$R_A = R_B$$

To bias an inverting op amp circuit, powered from a single supply, remember

Couple the input AC into the amp with a capacitor. Insert the desired DC output voltage at the noninverting input. $V_{DC} = \tfrac{1}{2} V_{supply}$ is often optimum.

5-1.2 Noninverting Amplifier

A noninverting amplifier can also be built from an op amp running from a single supply. The schematic is given in Fig. 5-7a.

The basic noninverting amplifier has been modified by the addition of R_A, R_B, C_i and C_f.

$$v_{AC} = \left(1 + \frac{R_f}{R_i}\right) e_i \tag{5-3a}$$

$$V_{DC} = \frac{R_B}{R_A + R_B} (+V) \tag{5-3b}$$

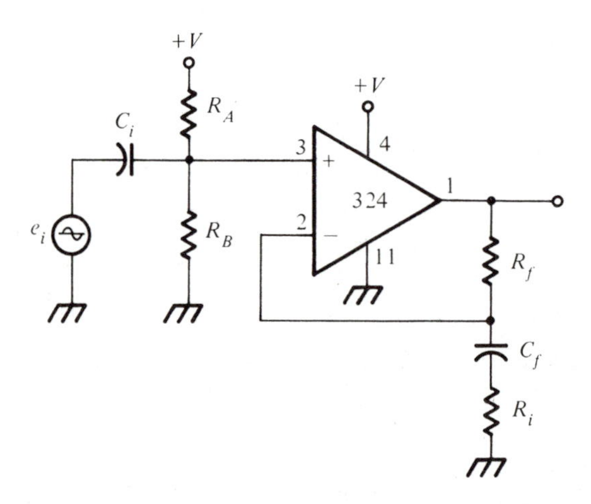

Figure 5-7a Full schematic of noninverting single supply op amp

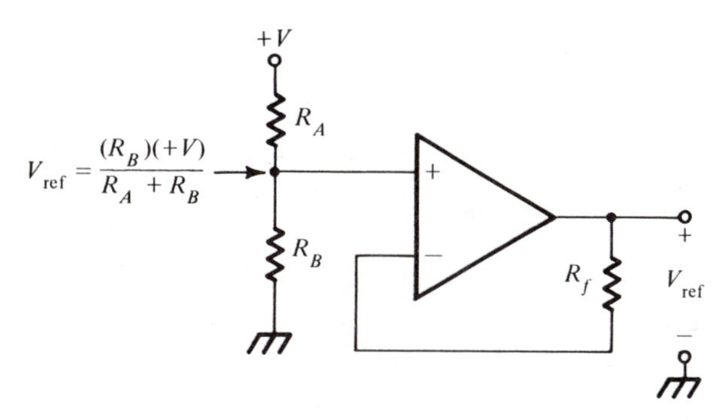

Figure 5-7b DC circuit of noninverting single supply op amp

The DC circuit is shown in Fig. 5-7(b). Resistors R_A and R_B form a voltage divider, placing V_{ref} on the noninverting input. The capacitor, C_f, appears as an open to the DC. This removes R_i from the circuit, converting it to a voltage follower. This is similar to the inverting circuit of Fig. 5-4. Consequently, the output DC is just the DC at the noninverting input (if it is assumed that there is no loss across R_f caused by I_B^-).

The capacitor, C_f, looks like a short to AC. This gives a gain of

$$A_v = 1 + \frac{R_f}{R_i}$$

to the AC voltage at the noninverting input.

The input impedance is set by R_A and R_B.

$$Z_i = \frac{R_A R_B}{R_A + R_B}$$

since R_A and R_B appear in parallel to the AC input.

Example 5-4

Design a noninverting amplifier, using a single +5 volt supply, with an AC gain of 15 and $Z_i > 100$ kΩ. Make sure that the output is properly biased.

Solution

To properly set the DC bias, the output DC should equal half of the supply.

$$V_{DC} = \frac{R_B}{R_A + R_B}(5\text{ V}) = 2.5\text{ V}$$

This is done by setting

$$R_A = R_B$$

To make

$$Z_i = \frac{R_A R_B}{R_A + R_B} > 100\ k\Omega$$

pick

$$R_A = R_B = 220\ k\Omega$$

making

$$Z_i = 110\ k\Omega$$

$$v_{AC} = \left(1 + \frac{R_f}{R_i}\right)e_i$$

$$A_{AC} = \left(1 + \frac{R_f}{R_i}\right)$$

$$15 = \left(1 + \frac{R_f}{R_i}\right)$$

$$\frac{R_f}{R_i} = 14$$

$$R_i = \frac{R_f}{14}$$

Pick

$$R_f = 2.2\ k\Omega$$

$$R_i = \frac{2.2\ k\Omega}{14} = 157\ \Omega$$

The capacitor value depends on circuit resistance and signal frequency, as you will see in the next section.

To bias a noninverting op amp circuit, powered from a single supply, use

A DC voltage divider at the noninverting input.

If you need only one or two op amps biased from a single supply, the techniques of Figs. 5-3 and 5-7 are adequate. For more extensive circuits requiring many op amps, the separate stage capacitors and biasing resistors can be replaced with a single supply splitter circuit.

This splitter circuit divides the +V supply in half, producing a common or $V_{ref} = \frac{1}{2}(+V)$. You can then use this common to replace ground. By RC coupling into and out of this common referenced (rather than ground referenced) circuit, your biasing problems are solved.

Figure 5-8(a) is the schematic of a four stage amplifier biased from dual supplies. The same circuit is shown in Fig. 5-8(b) using a single supply and

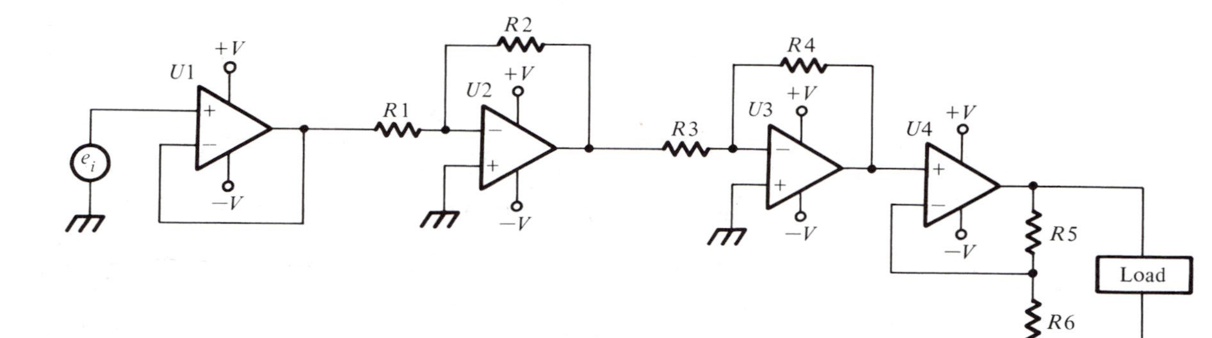

Figure 5-8a Four stage amplifier biased from dual supplies

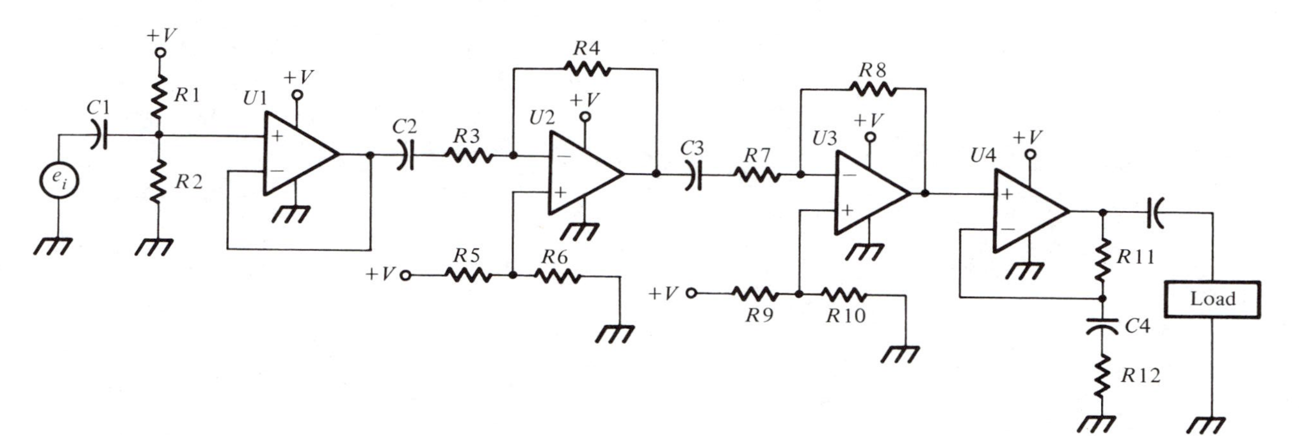

Figure 5-8b Four stage amplifier biased from single supply using individual stage biasing

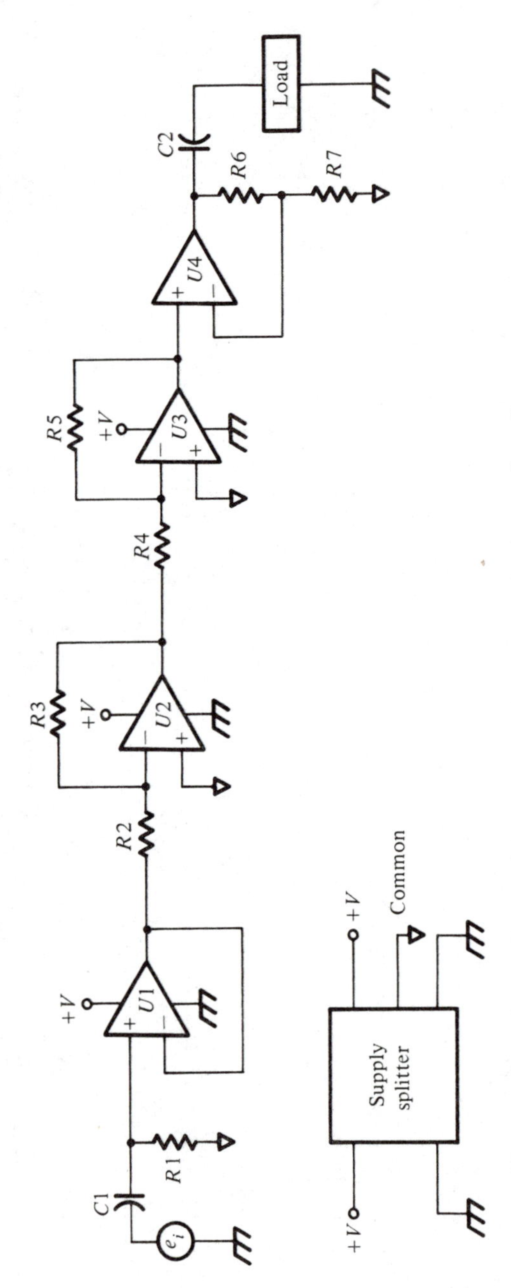

Figure 5-8c Four stage amplifier biased from single supply using a supply splitter

individual stage biasing and coupling. Finally, Fig. 5-8(c) is the same single supply amplifier circuit using a supply splitter and common referencing. Individual stage biasing is certainly the most complex (and expensive). Use of a supply splitter to produce a circuit common gives a circuit which is almost as simple as the dual supply circuit.

The simplest form of supply splitter is shown in Fig. 5-9(a). It consists of a voltage divider with shunt capacitance. The 1 kΩ resistors evenly split +V establishing the common or bias reference potential. The two capacitors ensure good AC grounding.

If you only draw I_B^+ from the common connection, the circuit in Fig. 5-9(a) works well. However, for larger loads, an active splitter may be needed. This is shown in Fig. 5-9(b). The two input resistors split the supply (+V), setting the op amp's noninverting input to $\frac{1}{2}(+V)$. The negative feedback line (inverting input) is connected to the emitters of the boost transistors. Circuit performance is best analyzed in three steps.

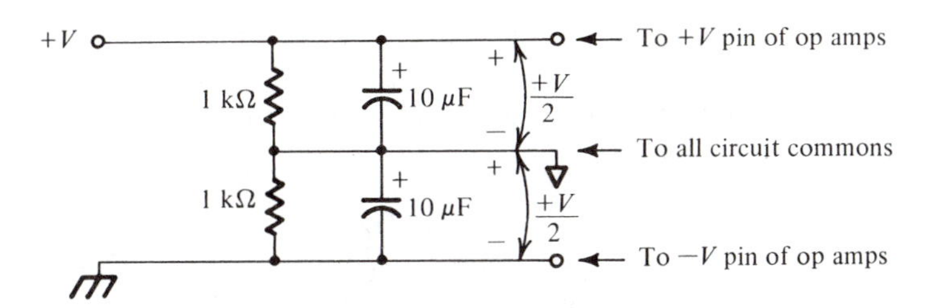

Figure 5-9a Simple supply splitter

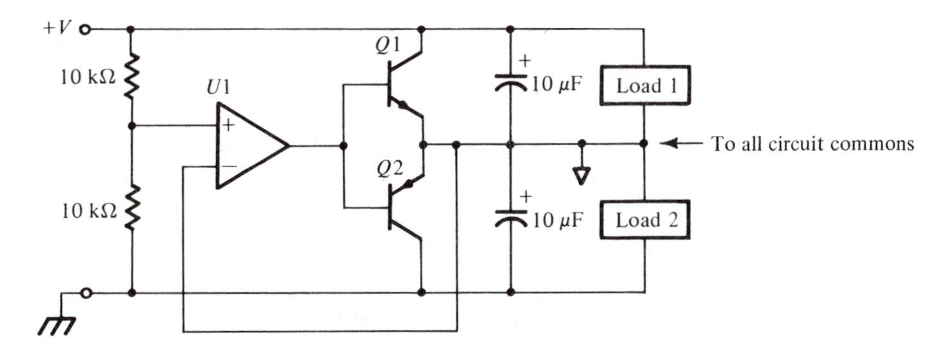

Figure 5-9b Active supply splitter

Example 5-5

For the circuit in Fig. 5-9(b), calculate current through each load, current through each transistor, and the output voltage of the op amp for $+V = 24$ V and

a. Load 1 = Load 2 = 400 Ω
b. Load 1 = 400 Ω Load 2 = 600 Ω
c. Load 1 = 400 Ω Load 2 = 250 Ω

Solution

a. With the loads equal, they evenly split $+V$, placing $+12$ V at the common and on the inverting input of the op amp. The input resistors also evenly split $+V$. The op amp outputs $+12$ volts, keeping both transistors off.

$$I_{\text{Load 1}} = I_{\text{Load 2}} = \frac{12\text{ V}}{400\ \Omega} = 30\text{ mA}$$

b. Increasing Load 2 to 600 Ω would cause the voltage at the common to *try* to increase to

$$\frac{600\ \Omega}{400\ \Omega + 600\ \Omega}(24\text{ V}) = +14.4\text{ V}$$

But this increase is fed to the inverting input of the op amp, driving it above the $+12$ V on the noninverting input. The output of the op amp heads negative (less positive). When it reaches 11.4 V, Q2 is turned on. Some of the current from Load 1 is now diverted around Load 2 and through Q2. Transistor Q2 is driven on hard enough to balance currents, dropping the voltage across Load 2 to 12 V. This puts the op amp's inputs at the same potential, and the circuit stabilizes.

$$I_{\text{Load 1}} = \frac{12\text{ V}}{400\ \Omega} = 30\text{ mA}$$

$$I_{\text{Load 2}} = \frac{12\text{ V}}{600\ \Omega} = 20\text{ mA}$$

All the Load 1 current which does not flow through Load 2 must go through Q2.

$$I_{Q2} = I_{\text{Load 1}} - I_{\text{Load 2}} = 10\text{ mA}$$

Transistor Q2 is on, Q1 is off. This is caused by an op amp output of 11.4 V.

c. Decreasing Load 2 to 250 Ω would cause the voltage at the common to *try* to decrease to

$$\frac{250\ \Omega}{400\ \Omega + 250\ \Omega}(24\ \text{V}) = +9.2\ \text{V}$$

But this decrease is fed to the inverting input of the op amp, driving it below the +12 V on the noninverting input. The output of the op amp increases. Transistor Q2 is turned off. When U1 output reaches 12.6 V, Q1 turns on. Additional current is driven through Q1 into Load 2. This increases the voltage across Load 2 (i.e., at the common) to +12 V, stabilizing the circuit.

$$I_{\text{Load 1}} = \frac{12\ \text{V}}{400\ \Omega} = 30\ \text{mA}$$

$$I_{\text{Load 2}} = \frac{12\ \text{V}}{250\ \Omega} = 48\ \text{mA}$$

All of the Load 2 current which did not come from Load 1 must be provided by Q1.

$$I_{Q1} = I_{\text{Load 2}} - I_{\text{Load 1}} = 18\ \text{mA}$$

Transistor Q1 is on, Q2 is off. This is caused by an op amp output of 12.6 V.

The active supply splitter's op amp shifts its output voltage to whatever level is necessary to turn on the proper boost transistor. That transistor then sources (Q1) or sinks (Q2) the current necessary to cause the unbalanced load resistances to drop the same voltage; in effect evenly splitting the single supply voltage.

Use a supply splitter circuit to simplify multi-op amp single supply circuits.

5-1.3 RC Coupling

These single supply circuits are for AC amplification. This signal must be coupled from one stage to another, or to the load. However, the output DC voltage should be blocked. Otherwise, signal distortion, severe loading or saturation of the following stages may result.

The simplest way to pass the AC signal, while blocking its DC level is with a RC coupler, shown in Fig. 5-10(a). It may be connected at the input of an amplifier, or at the output.

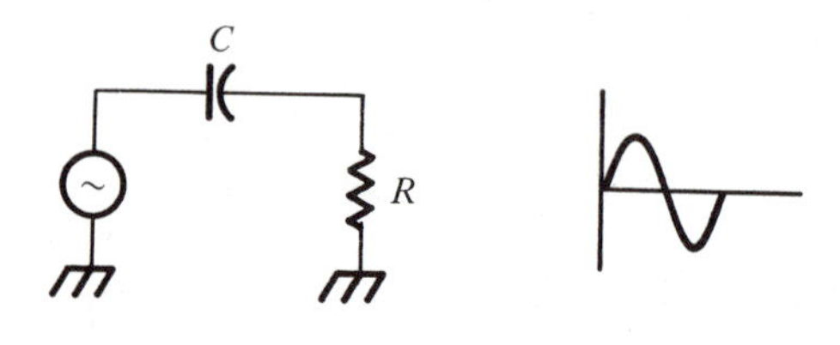

(a) Simple *RC* coupler

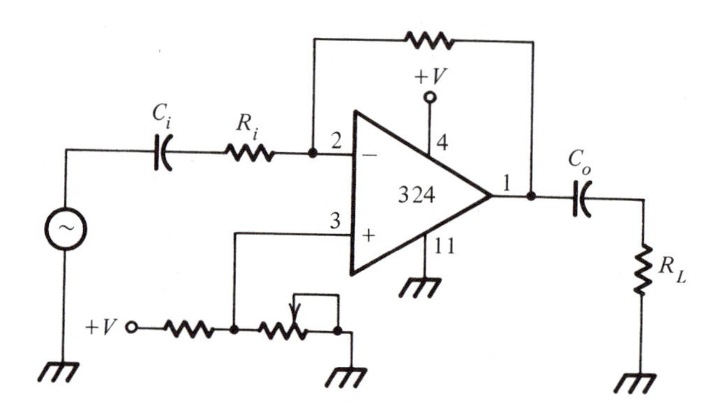

(b) Coupling into and out of an inverting amplifier

Figure 5-10 RC coupler

The RC coupler is a single-pole, high pass filter. Its frequency response is shown in Fig. 5-11. Direct current (zero hertz) is blocked completely. Low frequency signals are strongly attenuated, while high frequency signals pass without any attenuation. At the critical frequency only 70.7 percent of the signal input to the RC coupler will actually be passed through.

$$f_c = \frac{1}{2\pi RC} \tag{5-6}$$

where R is the input impedance of the amplifier if the coupler is going into an amp or where R is the load impedance if the coupler is between the amplifier's output and the load.

You *do not* want to operate at the critical frequency (f_c). Thirty percent of the signal is lost across the coupler at f_c. Select a critical frequency which is far below the lowest frequency of the signal. This is shown as f_o in Fig. 5-11.

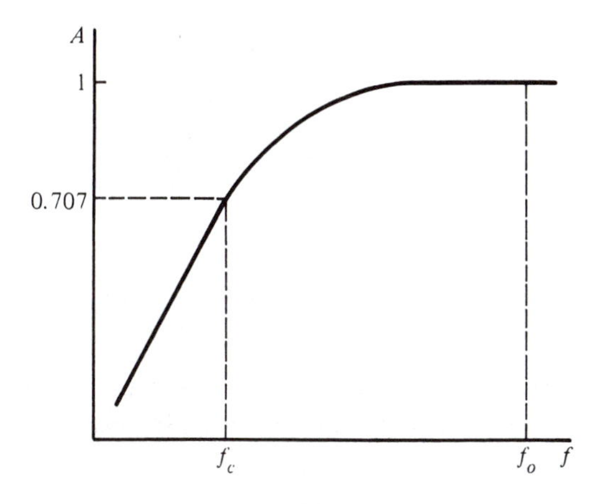

Figure 5-11 Frequency response of the RC coupler

Select:

$$f_c \leq 0.1 f_o \tag{5-7}$$

where f_o is the lowest operating frequency.

Example 5-6

For the circuit in Fig. 5-10(b), calculate values for C_i and C_o if

$$f_o \geq 20 \text{ Hz}$$

$$R_i = 220 \text{ k}\Omega$$

$$R_L = 600 \ \Omega$$

Solution

$$f_o \geq 20 \text{ Hz}$$

$$f_c \leq 0.1 f_o = 2 \text{ Hz}$$

$$f_c = \frac{1}{2\pi RC}$$

or

$$C = \frac{1}{2\pi R f_c}$$

$$C_i = \frac{1}{2\pi \times 220 \text{ k}\Omega \times 2 \text{ Hz}} = 0.362 \ \mu\text{F}$$

Pick $C_i = 0.47\ \mu\text{F}$

$$C_o = \frac{1}{2\pi \times 600 \times 2\ \text{Hz}} = 133\ \mu\text{F}$$

Pick $C_o = 200\ \mu\text{F}$

By selecting standard capacitance values *above* those calculated, the critical frequency is moved even further from the minimum operating frequency.

RC couple signals into and out of single supply amplifiers. Use a critical frequency of one-tenth the operating frequency.

5-2 NORTON AMPLIFIER

Much of the internal complexity of the operational amplifier's integrated circuit is not needed if you plan to operate entirely from a single power supply. Also, many operational amplifiers, designed for split power supply operation, have a limited output voltage swing and a restricted input voltage range when operated from a single supply. Finally, some of the performance specifications of the op amp could be reduced without seriously affecting AC circuit performance. With these facts in mind, the Norton amplifier IC, or current differencing amplifier IC, was designed for use in AC amplifier circuits operating from a single power supply.

5-2.1 Characteristics

The schematic symbol for a single Norton amp is given in Fig. 5-12. The arrow pointing toward the noninverting input is the major distinguishing feature. It signifies that the Norton amp is a *current* driven device. No power supply connections are shown, because usually two or four Norton amps are located in a single IC package and share supply and ground pins. The package layout for the LM 3900, a popular low-frequency Norton amp and the LM 359, a high-speed amp, are given in Fig. 5-13.

The Norton or current differencing amp outputs a voltage proportional to the difference in input *currents*. For linear circuits, this difference in input currents is kept very small by negative feedback. These input currents flow into the bases of the two input transistors (see Fig. 5-14) Q2 and Q3. Since the emitters of both transistors are tied to ground, the voltage at the inputs is fixed at about 0.6 V and will not vary significantly.

These input characteristics are very different from those of the operational amplifier. A comparison is given in Table 5-1. Consider it carefully. Your success in using the Norton amp is very dependent on understanding how to drive it. The

Figure 5-12 Norton amplifier schematic symbol

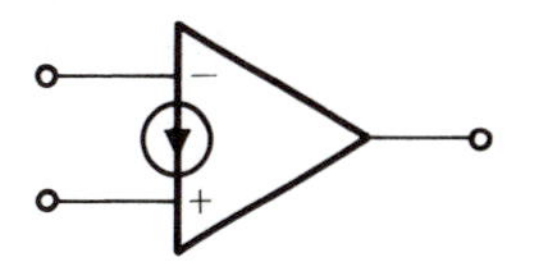

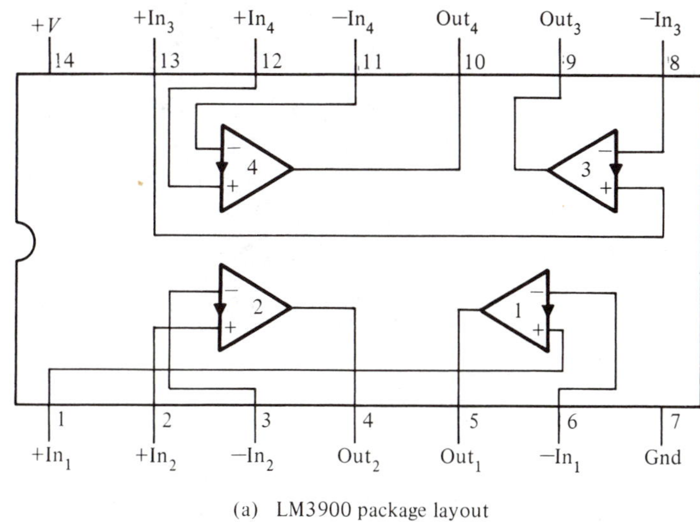

(a) LM3900 package layout

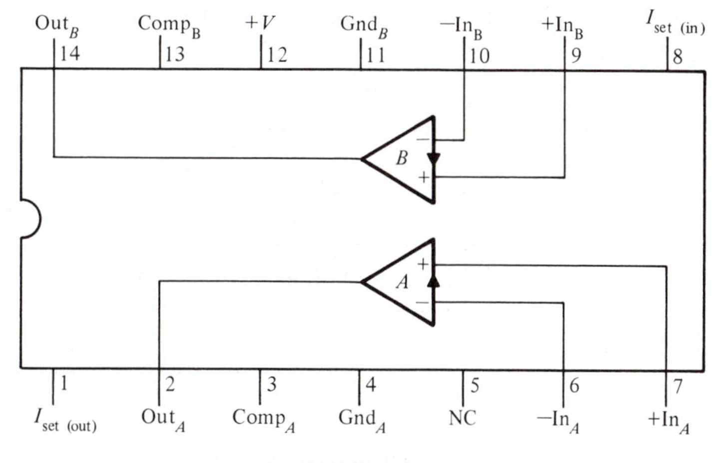

(b) LM359 package layout

Figure 5-13 Norton amp package layouts

Figure 5-14 Norton amp schematic (*Courtesy of National Semiconductor*)

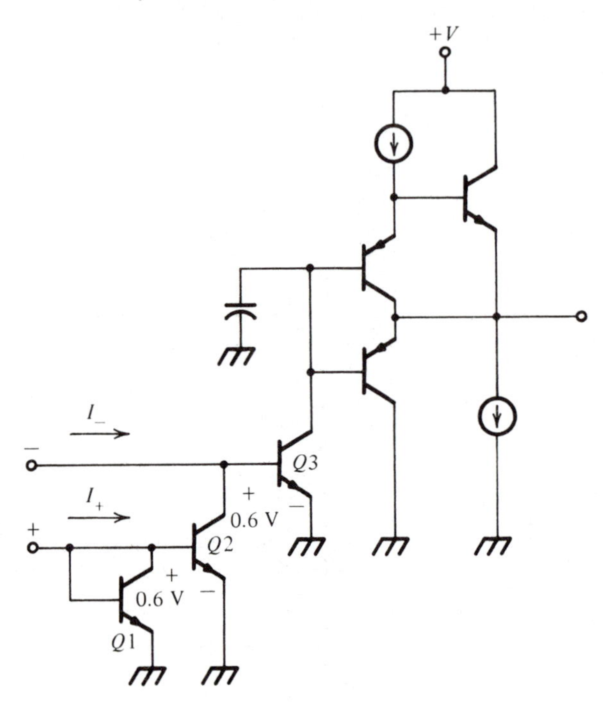

op amp, a voltage-driven device, requires no significant input current. Negative feedback is used to keep the two inputs at the same (though not necessarily zero) potential. However, the current differencing amp is *current-driven.* The input potentials are almost zero (~0.6 V), and negative feedback is used to keep the two input *currents* equal.

TABLE 5-1. Op amp and Norton amp input characteristics.

	Op Amp	*Norton Amp*
Drive	Voltage	Current
Voltages	$v_- = v_+$	$v_- \approx 0; v_+ \approx 0$
Currents	$i_+ \approx 0; i_- \approx 0$	$i_+ = i_-$

For a Norton amp, in a linear application (negative feedback):

$$v_- \approx 0 \tag{5-8a}$$

$$v_+ \approx 0 \tag{5-8b}$$

$$i_- = i_+ \tag{5-9}$$

5-2.2 Biasing the Norton Amplifier

Since the Norton amp is to be used as an AC amplifier (operating from a single power supply), its output must be biased to a DC level equal to half of the power supply. This allows composite output signals such as you saw in Figs. 5-2 and 5-5.

You can establish the output DC level as shown in Fig. 5-15. First, notice how much simpler this is than the biasing circuitry you had to add to an op amp. Since the Norton amp is current driven, *series resistors must always be connected to each input*. Current into the noninverting input is established through R_B by the power supply $+V$.

$$i_+ = \frac{+V}{R_B}$$

Negative feedback forces the two input currents to be equal [equation (5-9)].

$$i_- = i_+ = \frac{+V}{R_B} \tag{5-10}$$

Since the inverting input is at ground [equation (5-8)], the output voltage is just the voltage dropped across R_f by i_-.

$$V_o = i_- R_f \tag{5-11}$$

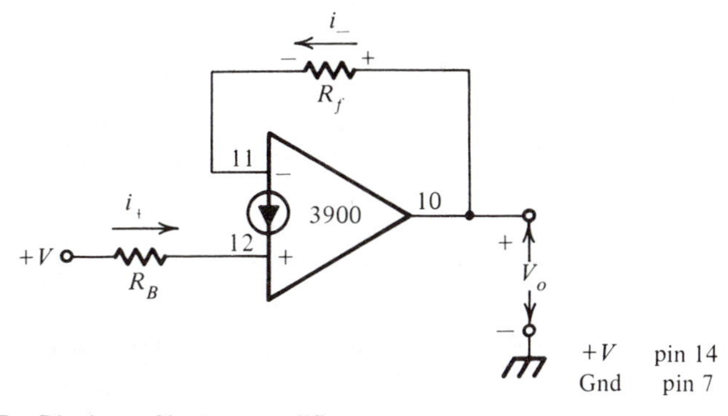

Figure 5-15 Biasing a Norton amplifier

Substituting equation (5-10) into equation (5-11), you obtain

$$V_o = \frac{+V}{R_B} R_f$$

or

$$V_o = \frac{R_f}{R_B}(+V) \tag{5-12}$$

The DC output level is the input bias voltage $(+V)$ times a DC gain (R_f/R_B). To set the output at half of the supply voltage,

$$R_B = 2R_f \tag{5-13}$$

To bias the output of a Norton amp at half of the supply voltage,

Connect a series resistor between the power supply and the noninverting input. Make it twice the size of the feedback resistor.

5-2.3 Inverting Norton Amplifier

The schematic for an inverting amplifier using a Norton amp is given in Fig. 5-16. It looks very much like an inverting amplifier made with an op amp. Capacitors C_i and C_o couple the AC signals in and out of the amplifier, while blocking DC. Resistors R_f and R_i set the AC gain. Biasing is achieved with $+V$ through R_B to the noninverting input.

Like the inverting amplifier using an op amp,

$$Z_i = R_i \tag{5-14}$$

$$A_{AC} = -\frac{R_f}{R_i} \tag{5-15}$$

Select

$$R_B = 2R_f \tag{5-13}$$

for optimum biasing.

There are two precautions that you should observe. First, as pointed out in section 5-1.3, be sure to set the critical frequencies of the input and output couplers at one-tenth of the operating frequency. Second, the output impedance of the 3900 is typically 8 kΩ. Output current is typically 5 mA or less. Care must be taken to prevent loading. This is illustrated in Fig. 5-17. Both R_L and R_f load

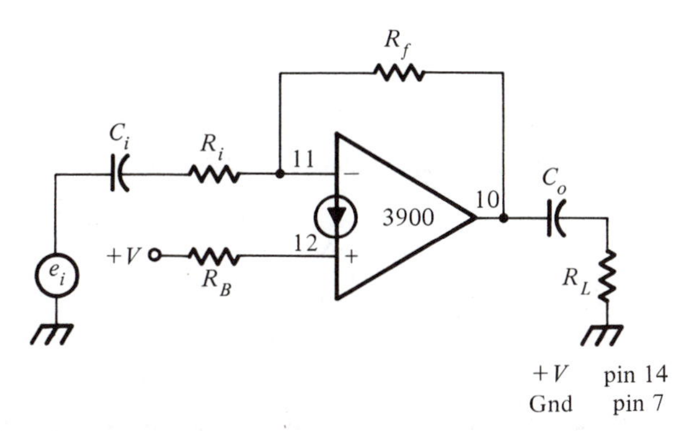

Figure 5-16 Inverting Norton amplifier

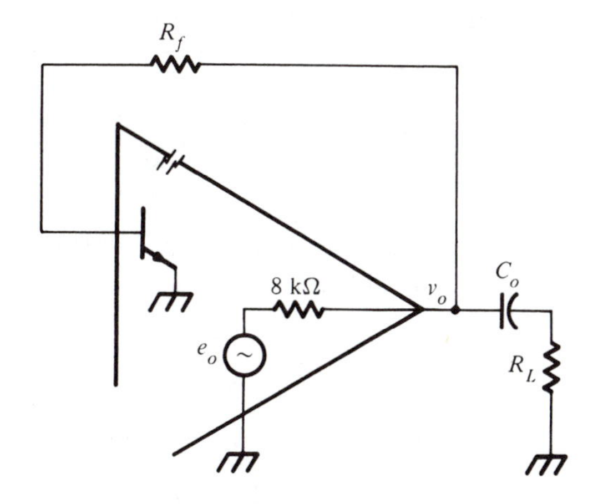

Figure 5-17 Output loading of Norton amp

the output. Keep both R_f and R_L as high as practical to minimize this effect. You were told to keep R_f low for an op amp to minimize output offset voltages. However, since you are intentionally driving the output to half the DC supply, a large R_f will cause no significant trouble.

The derivation of the output voltage as a function of supply voltage and input signal is given below. Refer to Fig. 5-18. All loading effects are assumed to be negligible.

$$v_- = 0 \text{ V} \tag{5-8a}$$

$$v_+ = 0 \text{ V} \tag{5-8b}$$

$$i_+ = i_- \tag{5-9}$$

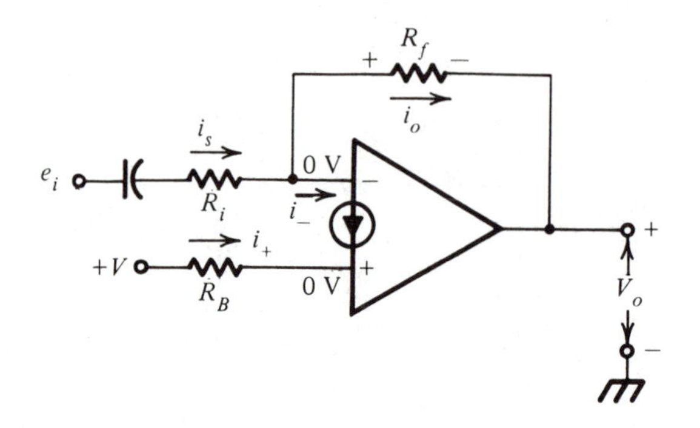

Figure 5-18 Derivation of output voltage for the inverting Norton amp

The input signal current i_s is that current driven by the input voltage signal e_i through R_i. Since $v_- = 0$ V,

$$i_s = \frac{e_i}{R_i}$$

At the inverting input i_s splits, part flows into the Norton amp (i_-) and part flows through the feedback resistor R_f.

$$i_o = i_s - i_-$$

The output voltage is the voltage developed by i_0 flowing through R_f.

$$V_0 = -i_0 R_f$$

Note the polarity.

$$V_0 = -(i_s - i_-)R_f$$

$$V_0 = -\left(\frac{e_i}{R_i} - i_-\right)R_f$$

$$i_- = i_+ = \frac{+V}{R_B} \tag{5-10}$$

$$V_0 = -\left(\frac{e_i}{R_i} - \frac{+V}{R_B}\right)R_f$$

$$V_0 = \frac{+VR_f}{R_B} - \frac{e_i R_f}{R_i}$$

$$V_0 = \frac{R_f}{R_B}(+V) - \frac{R_f}{R_i} e_i \tag{5-16}$$

The first term on the right of equation (5-16) is the DC output level. It is the same as developed in section 5-2.2, equation (5-12). The second term on the right side of equation (5-16) is the AC output.

$$v_{AC} = -\frac{R_f}{R_i} e_i$$

or

$$A_{AC} = -\frac{R_f}{R_i}$$

which is the same for equation (5-15).

Example 5-7

For the circuit in Fig. 5-16, with

$$+V = 5 \text{ V}$$
$$C_i = 0.1\ \mu\text{F}$$
$$R_i = 22\ \text{k}\Omega$$
$$R_f = 470\ \text{k}\Omega$$
$$R_B = 1\ \text{M}\Omega$$
$$e_i = 75 \text{ mV p-p}$$
$$C_0 \text{ and } R_L \text{ not connected}$$

calculate the following:

a. V_{DC} out
b. v_{AC} out
c. Draw the composite output signal
d. f_c
e. Z_i

Solutions

a.

$$V_{DC}\ \text{out} = \frac{R_f}{R_B}(+V) \tag{5-12}$$

$$V_{DC}\ \text{out} = \frac{470\ \text{k}\Omega}{1\ \text{M}\Omega}(5\ \text{V}) = 2.35\ \text{V}_{DC}$$

b.

$$A_{AC} = -\frac{R_f}{R_i} \tag{5-15}$$

$$A_{AC} = -\frac{470\ \text{k}\Omega}{22\ \text{k}\Omega} = -21.4$$

$$v_{AC} = A_{AC} \times e_i$$

$$v_{oAC} = -21.4 \times 75\ \text{mV p-p} = 1.61\ \text{V p-p}$$

c. See Fig. 5-19

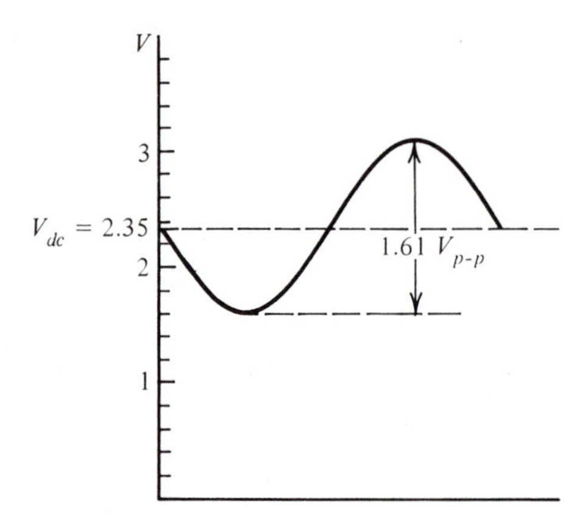

Figure 5-19 Composite output for Example 5-6

d.

$$f_c = \frac{1}{2\pi C_i R_i} \tag{5-6}$$

$$f_c = \frac{1}{2\pi \times 0.1\ \mu\text{F} \times 22\ \text{k}\Omega} = 72\ \text{Hz}$$

Note that this is the *cutoff* frequency. The gain will have fallen by 30 percent (0.707) at that frequency. For no signal attenuation by $R_i C_i$, you should operate the amplifier at or above approximately 720 Hz.

e.

$$Z_i = R_i = 22\ \text{k}\Omega \tag{5-14}$$

5-2.4 Inverting Summer

The inverting amplifier of Fig. 5-16 can be converted into an inverting summer. This is done by adding more input voltages, each with its own series input resistor to convert the voltage into an input current. This is shown in Fig. 5-20.

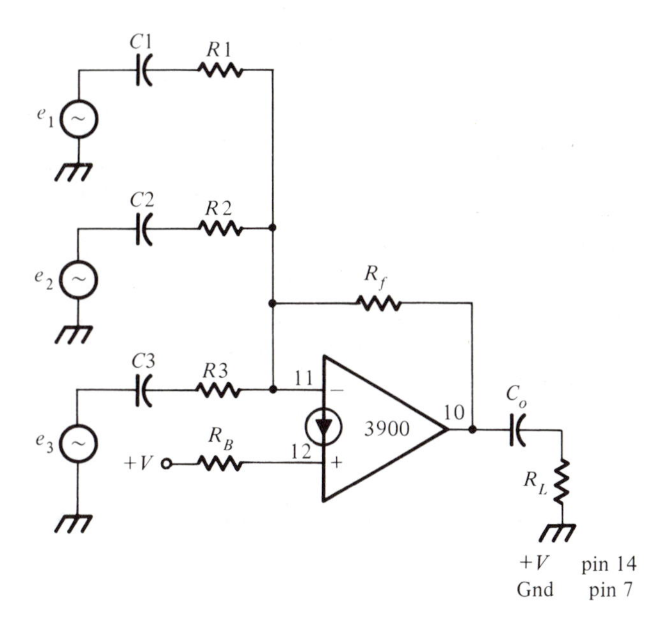

Figure 5-20 Inverting Norton summer

For optimum output signal swing, the output should be biased to half of the supply voltage. As with the inverting Norton amplifier, this is done by making

$$R_B = 2R_f \tag{5-13}$$

Also, as with the inverting Norton amplifier, each input voltage source sees its series resistor as the input impedance.

$$Z1 = R1$$

$$Z2 = R2$$

$$Z3 = R3$$

$$\vdots$$

The output voltage equation is very similar to equation (5-16).

$$v_0 = \frac{R_f}{R_B}(+\text{V}) - \frac{R_f}{\text{R1}} e_1 - \frac{R_f}{\text{R2}} e_2 - \frac{R_f}{\text{R3}} e_3$$

or

$$v_0 = \frac{R_f}{R_B}(+\text{V}) - R_f\left(\frac{e_1}{\text{R1}} + \frac{e_2}{\text{R2}} + \frac{e_3}{\text{R3}}\right) \tag{5-17}$$

The first term on the right side of the equation is the DC bias level. The feedback resistor controls the effect of each of the inputs equally. However, each input source's effect is also independently controlled by its own input resistor. The derivation of equation (5-17) follows that outlined for the inverting Norton amp.

Example 5-8

Design a four channel audio mixer using a Norton amplifier. The input impedance of each channel must be at least 10 kΩ. The contribution of each channel to the output should be adjustable by a factor of 20. There must be an overall volume control as well. The low frequency *cutoff* is 20 Hz or lower. It will drive a 10 kΩ load.

Solution

The circuit configuration is shown in Fig. 5-21. Each input channel is identical. To ensure that the input impedance never falls below 10 kΩ,

$$R1 = 10\ k\Omega$$

Each channel affects the output according to

$$\frac{R_f}{R_1 + R_2}$$

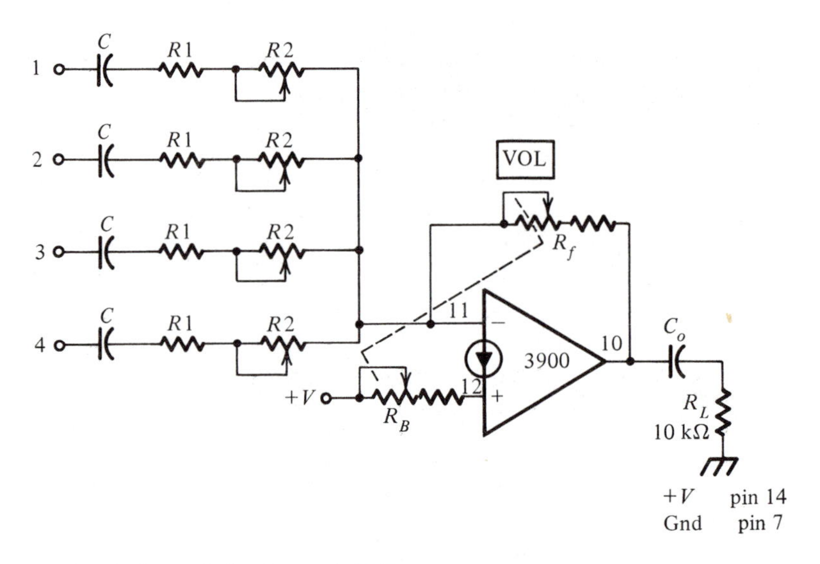

Figure 5-21 Schematic for Example 5-8

where R2 is variable. To ensure an adjustment factor of 20,

$$R2 = 19R1$$

so that adjusting R2 from zero to full resistance will lower the output by a factor of 20.

$$R2 = 200\ \text{k}\Omega \text{ potentiometer}$$

Cutoff frequency must be 20 hertz or below.

$$f_c = \frac{1}{2\pi R_i C_i}$$

Since R_i varies from 10 kΩ to 210 kΩ, using $R_i = 10$ kΩ will assure a 20 Hz cutoff when R2 is set to zero Ω, and a lower cutoff as R2 is increased.

$$C_i = \frac{1}{2\pi \times 10\ \text{k}\Omega \times 20\ \text{Hz}} = 0.8\ \mu\text{F}$$

$$C_i = 1\ \mu\text{F} = C_0$$

The feedback resistor has been made a potentiometer and labeled VOL. This is the overall volume control. Varying it will directly and equally alter the effect on the output of each input. Selecting $R_f = 100$ kΩ gives a gain of -10 when the mixer controls (R2) are set to zero and a gain of -0.5 when the mixer controls are set to full resistance.

The biasing resistor R_B is mechanically ganged to R_f.

$$V_{\text{DC out}} = \frac{R_f}{R_B}(+\text{V})$$

As the volume control (R_f) is changed, R_B changes proportionally, to prevent any variation in the output bias level.

For the inverting Norton amplifier and summer,

The input impedance is set by R_i the gain by $-R_f/R_i$ and the output bias by R_f/R_B.

5-2.5 Noninverting Norton Amplifier

To produce a noninverting Norton amplifier, you simply move the input source, with its coupling capacitor and series resistor to the noninverting input. This is shown in Fig. 5-22. Biasing is still achieved with current from $+V$ through R_B into the noninverting input. Negative feedback *current* still flows through R_f into the inverting input.

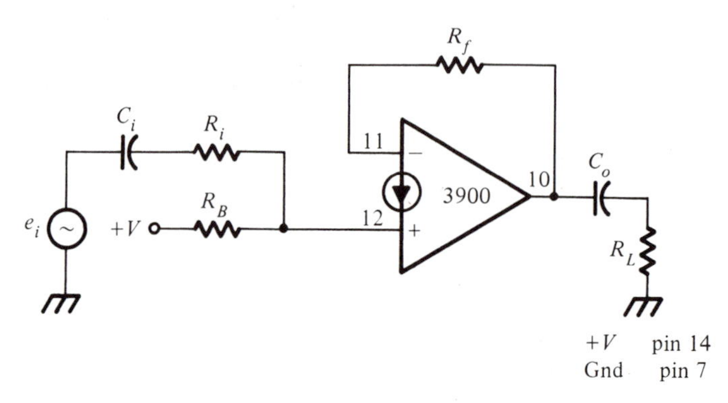

Figure 5-22 Noninverting Norton amplifier

The characteristics of the noninverting Norton amplifier are much closer to those of the inverting Norton amplifier than to the noninverting op amp circuit. The input impedance, seen by the source, is

$$Z_i = \mathrm{R}_i$$

since the noninverting input is just a base-emitter drop from ground. The signal gain is

$$A_{\mathrm{AC}} = \frac{\mathrm{R_f}}{\mathrm{R}_i}$$

Notice that this gives the same magnitude as the inverting amplifier. There is no $1 + \mathrm{R_f}/\mathrm{R}_i$ in the gain of a noninverting Norton amplifier.

You must observe the same precautions for the noninverting Norton amplifier as you did with the inverting Norton amplifier. Be sure to pick the coupling capacitors to give a cutoff frequency well below (0.1) the operating frequency. Also, beware of loading down the output with both R_L and $\mathrm{R_f}$.

The derivation of the output voltage as a function of supply voltage and input signal is given below. Refer to Fig. 5-23.

$$v_- = 0 \text{ V} \tag{5-8a}$$

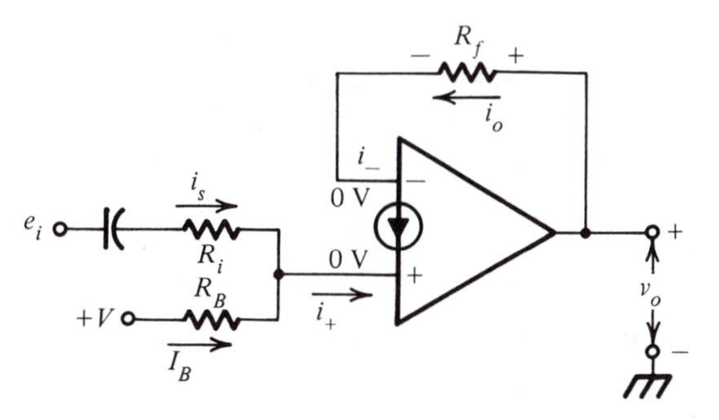

Figure 5-23 Derivation of output voltage for the noninverting Norton amp

$$v_+ = 0 \text{ V} \tag{5-8b}$$

$$i_+ = i_- \tag{5-9}$$

The input signal i_s is the current driven by the input voltage signal, e_i through R_i. Since $v_+ = 0$ V,

$$i_s = \frac{e_i}{R_i} \quad \text{and} \quad I_B = \frac{+V}{R_B}$$

At the noninverting input, the signal current adds with the bias current to produce i_+.

$$i_+ = I_B + i_s$$

But since $\quad i_+ = i_-$

and since $\quad i_- = i_O$

$$i_O = I_B + i_s$$

The output voltage is the voltage developed by i_O flowing through R_f

$$v_O = i_O R_f$$

$$v_O = (I_B + i_s)R_f$$

$$v_O = \left(\frac{+V}{R_B} + \frac{e_i}{R_i}\right)R_f$$

$$v_O = \frac{R_f}{R_B}(+V) + \frac{R_f}{R_i}e_i \tag{5-18}$$

Note the similarity to equation (5-16).

The noninverting Norton amplifier is injecting its signal current into a node in such a way as to add to the bias current to form the output current. The signal current for the inverting Norton amplifier is subtracted from the bias current.

Example 5-9

Design an AC amplifier which can give either an inverted or noninverted output, at the push of a button. The gain must be a constant 10 (in either case) and the input impedance constant at 100 kΩ (for either case). The lowest *operating* frequency is 50 Hz.

Solution

The circuit configuration is shown in Fig. 5-24. With S1 in the Out position the circuit is an inverting Norton amplifier, with

$$Z_i = R_i$$

$$A_{AC} = -\frac{R_f}{R_i}$$

$$V_{OUT\ DC} = \frac{R_f}{R_B}(+V)$$

Moving S1 to the in position changes the circuit to a noninverting Norton amplifier. Input impedance, gain magnitude, and biasing, however, do not change.

$$Z_i = R_i = 100\ \text{k}\Omega$$

$$A_{AC} = \frac{R_f}{R_i}$$

$$R_f = A_{AC} \times R_i = 10 \times 100\ \text{k}\Omega$$

$$R_f = 1\ \text{M}\Omega$$

To set the output biasing properly,

$$R_B = 2R_f = 2\ \text{M}\Omega$$

$$f_c = 0.1 f_0 = 0.1 \times 50\ \text{Hz} = 5\ \text{Hz}$$

$$C_i = \frac{1}{2\pi f_c \times R_i} = \frac{1}{2\pi \times 5\ \text{Hz} \times 100\ \text{k}\Omega}$$

$$C_i = 0.32\ \mu\text{F}$$

$$\text{Choose } C_i = 0.47\ \mu\text{F}$$

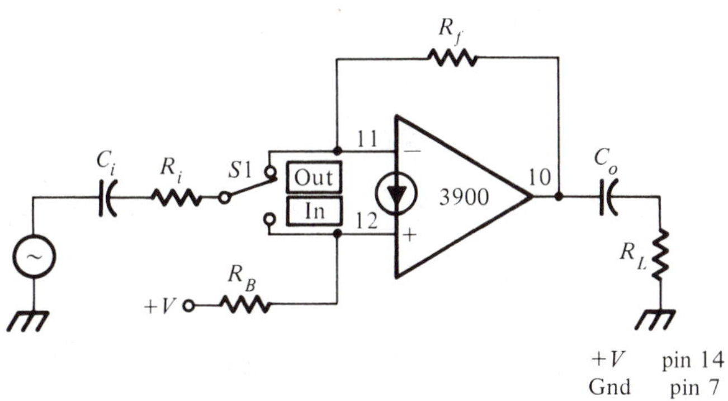

Figure 5-24 Schematic for Example 5-9

A noninverting Norton amplifier has the same characteristics (except for phase shift), equations, and precautions as the inverting Norton amplifier.

5-2.6 Noninverting Norton Summer

The noninverting Norton summer is very similar to the inverting Norton summer. All input voltages are fed through coupling capacitors and series input resistors, to the noninverting input of the Norton amplifier. This is illustrated in Fig. 5-25. As with all of the other Norton amplifier circuits, the input impedance seen by a signal source is its series resistor. The output voltage is

$$v_O = \frac{\mathrm{R_f}}{\mathrm{R_B}}(+V) + \frac{\mathrm{R_f}}{\mathrm{R1}}e_1 + \frac{\mathrm{R_f}}{\mathrm{R2}}e_2 + \frac{\mathrm{R_f}}{\mathrm{R3}}e_3$$

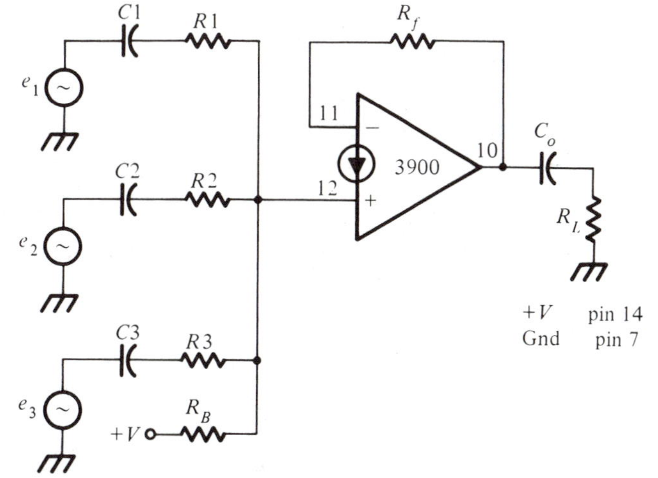

Figure 5-25 Noninverting Norton summer

or

$$v_O = \frac{R_f}{R_B}(+V) + R_f\left(\frac{e_1}{R1} + \frac{e_2}{R2} + \frac{e_3}{R3}\right) \quad (5\text{-}18)$$

Compare this relationship to the two-input, noninverting summer, made from an op amp (given below):

$$v_O = \left(1 + \frac{R3}{R4}\right)\left(e_1 \frac{R2}{R1 + R2} + e_2 \frac{R1}{R1 + R2}\right)$$

The Norton noninverting summer gives a much simpler relationship than does the op amp version. Not only does this make your math easier, but adjustments to the circuit are not nearly so interdependent.

5-2.7 Norton Difference Amplifier

A subtractor or difference amplifier can be made by using a Norton amp. The schematic is given in Fig. 5-26

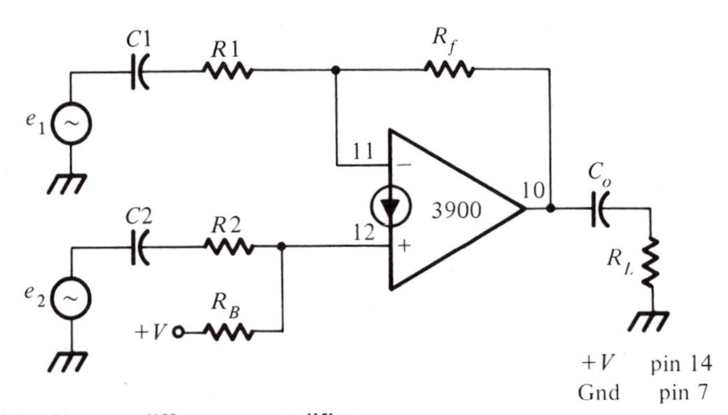

Figure 5-26 Norton difference amplifier

As with all of the other Norton amplifier circuits you have seen, the input impedance seen by each source is its series resistor.

$$Z_{i1} = R1$$

$$Z_{i2} = R2$$

The output voltage depends on the bias circuit, e_1 and e_2.

$$v_O = \frac{R_f}{R_B}(+V) + \frac{R_f}{R2}e_2 - \frac{R_f}{R1}e_1$$

or

$$v_O = \frac{R_f}{R_B}(+V) + R_f\left(\frac{e_2}{R2} - \frac{e_1}{R1}\right) \tag{5-19}$$

For simplicity, usually

$$R1 = R2 = R$$

$$v_O = \frac{R_f}{R_B}(+V) + \frac{R_f}{R}(e_2 - e_1)$$

The first term on the right of this equation is just the DC output bias, so

$$v_{AC} = \frac{R_f}{R}(e_2 - e_1) \tag{5-20}$$

As with other Norton amplifier circuits, do not forget to make R_B twice the size of R_f to provide proper biasing. Also, output loading caused by R_L and R_f must be kept in mind. This AC difference amplifier is subject to the same critical and operating frequency considerations previously discussed.

The derivation of the output voltage as a function of supply voltage and input signals is given below. Refer to Fig. 5-27.

$$v_- = 0 \text{ V} \tag{5-8a}$$

$$v_+ = 0 \text{ V} \tag{5-8b}$$

$$i_+ = i_- \tag{5-9}$$

$$i_1 = \frac{e_1}{R1}$$

$$i_2 = \frac{e_2}{R2}$$

$$I_B = \frac{+V}{R_B}$$

$$i_+ = I_B + i_2$$

$$i_+ = \frac{+V}{R_B} + \frac{e_2}{R2} \tag{5-21}$$

$$i_- = i_1 + i_O$$

or

$$i_O = i_- - i_1$$

$$i_O = i_- - \frac{e_1}{R1}$$

But since $\quad i_+ = i_-$

$$i_O = i_+ - \frac{e_1}{R1} \tag{5-22}$$

Substituting equation (5-21) into equation (5-22), you obtain

$$i_O = \frac{+V}{R_B} + \frac{e_2}{R2} - \frac{e_1}{R1}$$

$$v_O = i_O R_f$$

$$v_O = \left(\frac{+V}{R_B} + \frac{e_2}{R2} - \frac{e_1}{R1}\right) R_f$$

or

$$v_O = \frac{R_f}{R_B}(+V) + \frac{R_f}{R2} e_2 - \frac{R_f}{R1} e_1$$

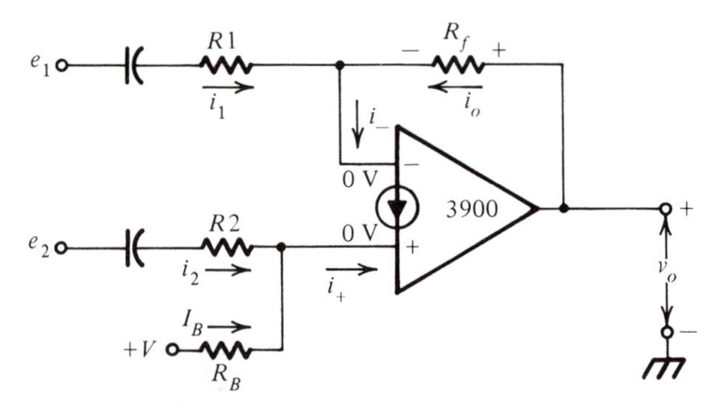

Figure 5-27 Derivation of output voltage for the Norton difference amp

5-2.8 Norton Comparator

The Norton amplifier can be used to build a comparator. Its schematic is given in Fig. 5-28. There are several differences you should notice. There is no biasing resistor or source. The output is to be at either ground or $+V_{SAT}$. There is no need for class A biasing. Also, comparators do not use negative feedback, so R_f has been removed.

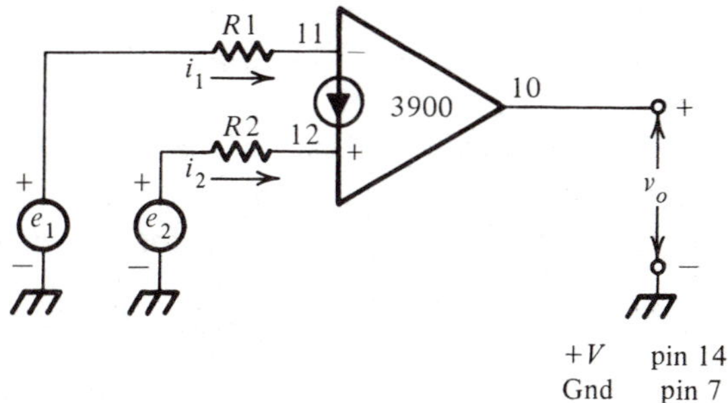

Figure 5-28 Norton comparator

Since the Norton amp is a current driven device, comparison is made between the two input currents, i_1 and i_2. When

$$i_2 > i_1$$

then

$$v_O = +V_{SAT} = (+V) - 0.8 \text{ V}$$

That is, if the current into the noninverting input terminal is greater than the current into the inverting input terminal, the output voltage goes to positive saturation. Positive saturation is typically about 0.8 volt below the positive power supply voltage. On the other hand, if

$$i_1 > i_2$$

then

$$v_O \approx 0.2 \text{ V}$$

When current into the inverting input exceeds current into the noninverting input, the output goes to negative saturation. For the Norton amp, this is the collector-emitter voltage drop of a saturated transistor, 0.2 volt.

There is one important consideration to note. There must be *some* current into both inputs to assure proper operation. If either of the inputs, e_1 or e_2, is set to zero (ground), the output voltage may not be properly driven when current is applied by the other source.

With this in mind, there are two ways to use the Norton comparator. These are shown in Figs. 5-29(a) and (b). A reference voltage, V_{ref}, is established with $+V$ and the voltage divider R1 and R2. For proper operation, this voltage must be at least 0.7 V to ensure that the input transistor is turned on.

$$V_{ref} \geq 0.7 \text{ V} \tag{5-23}$$

This voltage is then converted into a reference current by the series resistor R. This resistor should have at least ten times the resistance of R1 or R2, to ensure that the voltage divider and V_{ref} are not loaded down by the Norton amp's input.

$$\text{R} \gg \text{R1 or R2} \tag{5-24}$$

By setting the two input resistors to the same value, you can make direct comparison between the input voltage, e_i, and the reference voltage. For the noninverting comparator of Fig. 5-29(a),

$$v_O = +V_{SAT}$$

when

$$i_2 > i_1$$

But

$$i_1 = \frac{V_{ref}}{\text{R}} \text{ and } i_2 = \frac{e_i}{\text{R}}$$

$$\frac{e_i}{\text{R}} > \frac{V_{ref}}{\text{R}}$$

So when

$$e_i > V_{ref}$$

then

$$v_O = +V_{SAT}$$

and, conversely, when

$$e_i < V_{ref}$$

then

$$v_O = 0.2 \text{ V}$$

The inverting reference comparator in Fig. 5-29(b) gives the opposite output.

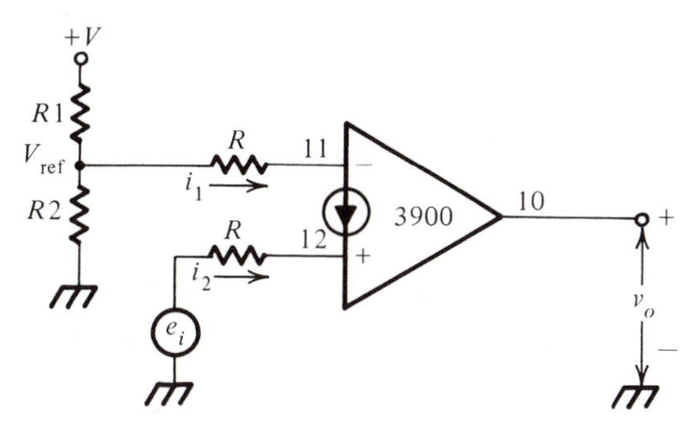

(a) Noninverting referenced Norton comparator

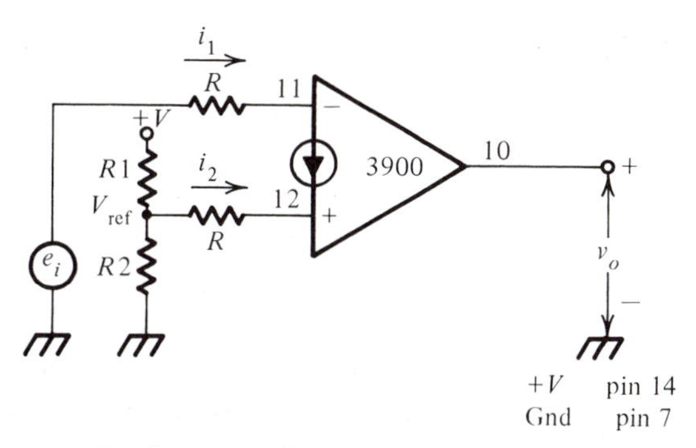

(b) Inverting referenced Norton comparator

Figure 5-29 Referenced Norton comparators

Example 5-10

a. Draw the output from the circuit in Fig. 5-29a, with

$$R1 = 100\ k\Omega$$

$$R2 = 33\ k\Omega$$

$$+V = 5\ V$$

$$R = 1\ M\Omega$$

and e_i as shown in Fig. 5-30.

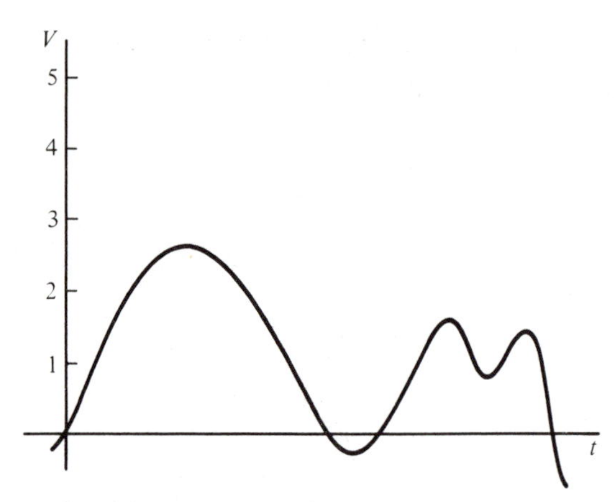

Figure 5-30 Input signal for Example 5-10

Solution

$$V_{\text{ref}} = \frac{\text{R2} \times (+V)}{\text{R1} + \text{R2}}$$

$$V_{\text{ref}} = \frac{33\ \text{k}\Omega \times 5\ \text{V}}{100\ \text{k}\Omega + 33\ \text{k}\Omega} = 1.24\ \text{V}$$

When the input signal exceeds 1.24 volts, the output goes to $+V_{\text{SAT}}$. When the input signal is less than 1.24 volts, the output goes to 0.2 volt.

This is shown in Fig. 5-31(a). The input signal voltage may go negative without damaging the Norton amplifier, as long as current is limited.

b. Repeat part a for the inverting referenced Norton comparator of Fig. 5-29(b).

Solution

An inverting referenced comparator will have the following relationship:

$$e_i > V_{\text{ref}} \Rightarrow v_O = 0.2\ \text{V}$$

$$e_i < V_{\text{ref}} \Rightarrow v_O = +V_{\text{SAT}}$$

This is shown in Fig. 5-31(b). It is 180 degrees out of phase with (or the logical complement of) the result from the noninverting comparator.

There are three factors you should consider. First, the input impedance is set by R. However, so is the sensitivity (i.e., how much bigger does one input voltage

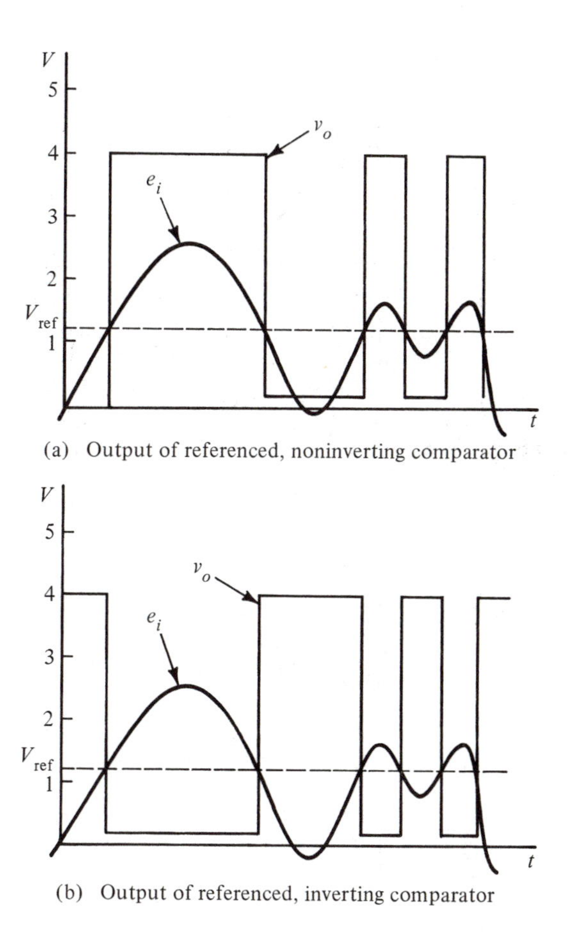

(a) Output of referenced, noninverting comparator

(b) Output of referenced, inverting comparator

Figure 5-31 Solutions for Example 5-10

have to be than the reference to cause switching?). To improve sensitivity you should keep R small. However, not only does this lower the input impedance, but it also increases the loading on V_{ref} and the voltage divider. A tradeoff is necessary. Second, switching on the positive-going edge may not occur at exactly the same level (V_1) as switching on the negative-going edge (V_2). This is illustrated in Fig. 5-32. Though usually fairly small, this difference in switch levels, or hysteresis, can cause you problems in a precision comparator. Finally, as mentioned before, there must be some reference current for proper operation. You may not set the reference voltage to ground.

You can cause the Norton comparator to trigger at two separate levels, establishing the hysteresis *you* want. This is shown in Fig. 5-33. Notice that positive (not negative) feedback has been added through R_f.

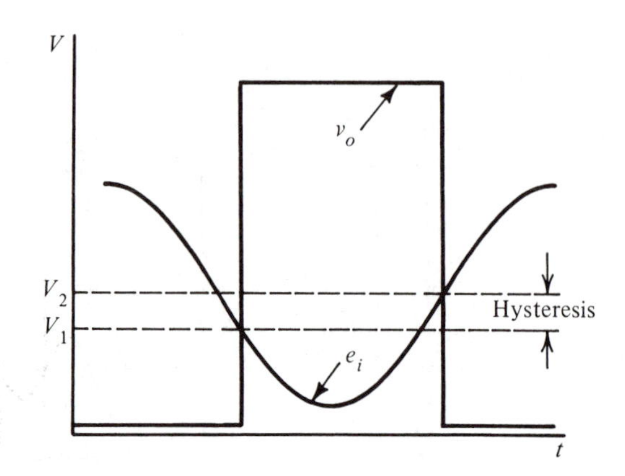

Figure 5-32 Hysteresis in an inverting, referenced Norton comparator

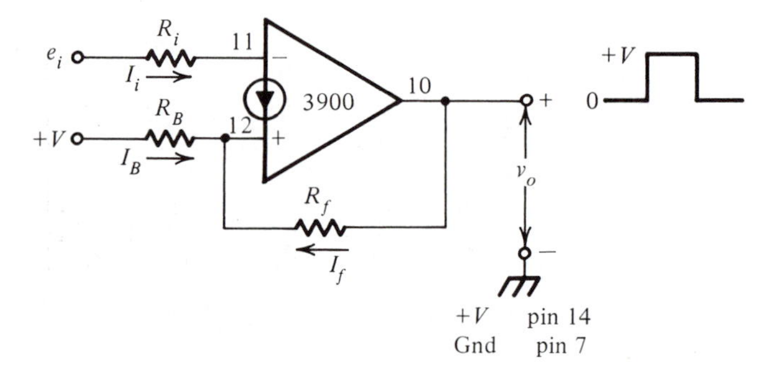

Figure 5-33 Norton comparator with two trigger points (hysteresis)

When current I_i is greater than the current flowing into the noninverting input, the output voltage is low (almost ground). However, when I_i falls below the current into the noninverting input, the output swings high (almost $+V$).

To determine V_{LT}, the lower trigger point, notice that before it is reached, e_i is large, forcing a large I_i into the amplifier, keeping the output at ground.

$$v_o \cong 0$$

With both ends of R_f at ground,

$$I_f = 0$$

So the only current flowing into the noninverting input is

$$I_B = \frac{+V}{R_B}$$

Switching will occur when the input voltage falls low enough to drop I_i to just below I_B. The value of e_i, at that point, is the lower trigger point V_{LT}.

$$V_{LT} = e_i = I_i R_i$$

but

$$I_i = I_B = \frac{+V}{R_B}$$

$$V_{LT} = \frac{R_i}{R_B}(+V) \tag{5-25}$$

When the lower trigger point is crossed, the output goes to approximately $+V$. Current is now forced through R_f.

$$I_f = \frac{+V}{R_f}$$

This sums with I_B to flow into the noninverting input. Consequently, the current into this noninverting input has increased significantly. For the output voltage to switch again, the current into the inverting input must exceed this new, larger noninverting input current. When the input voltage e_i is large enough to cause I_i to equal (or slightly exceed) $I_B + I_f$, the output will switch. This is V_{UT}, the upper trigger point.

$$V_{UT} = e_i = I_i R_i$$

but

$$I_i = I_B + I_f$$

$$V_{UT} = (I_B + I_f)R_i$$

$$V_{UT} = \left(\frac{+V}{R_B} + \frac{+V}{R_f}\right)R_i$$

$$V_{UT} = R_i\left[\frac{R_f(+V) + R_B(+V)}{R_B R_f}\right]$$

$$V_{UT} = \frac{R_i(+V)(R_B + R_f)}{R_B R_f} \tag{5-26}$$

To design a circuit, given $+V$, V_{LT} and V_{UT}, you first pick a large value of R_B. Then solve equation (5-25) for R_i. Equation (5-26) can then be solved for R_f.

$$R_f = \frac{R_i R_B(+V)}{V_{UT} R_B - (+V)R_i} \tag{5-27}$$

Finally, you must be careful to ensure that

$$V_{LT} \geq 0.7 \text{ V}$$

since it is, after all, just a reference voltage. Also,

$$V_{UT} < V_{SAT} = +V - 0.8 \text{ V}$$

The biased Norton comparator is shown in Fig. 5-34. Bias currents are provided through R1 and R2 for the inverting input and through R3 for the noninverting input. For noninverting operation, the signal voltage is coupled through a capacitor and series resistor to the noninverting input. To make an inverting comparator, just remove R_i from the noninverting input and connect it to the inverting input.

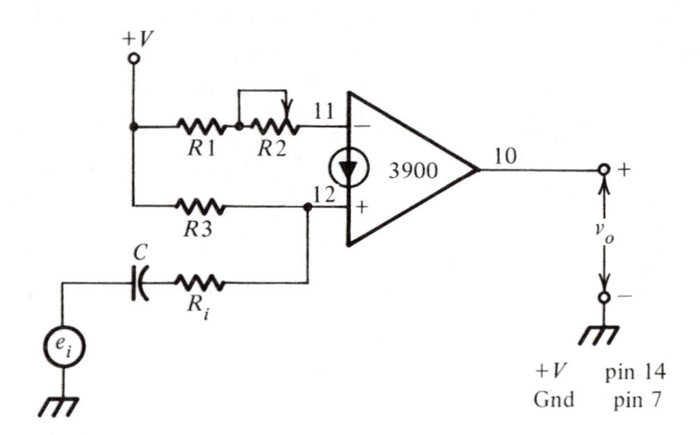

Figure 5-34 Biased Norton comparator

As with the other Norton comparators,

$$v_O = +V_{SAT}$$

when

$$i_+ > i_-$$

But

$$i_+ = I_{R3} + i_{R_i}$$

$$I_{R3} = \frac{+V}{R3}$$

$$i_{R_i} = \frac{e_i}{R_i}$$

$$i_- = \frac{+V}{R1 + R2}$$

so that

$$v_O = +V_{SAT}$$

when

$$\frac{+V}{R3} + \frac{e_i}{R_i} > \frac{+V}{R1 + R2}$$

$$\frac{e_i}{R_i} > \frac{+V}{R1 + R2} - \frac{+V}{R3}$$

Switching occurs when

$$e_i = \frac{(+V)R_i}{R1 + R2} - \frac{(+V)R_i}{R3} \tag{5-28}$$

This then is the switching level.

When

$$R1 + R2 < R3$$

the switching level is positive, when

$$R1 + R2 = R3$$

the switching level is zero, when

$$R1 + R2 > R3$$

the switching level is negative.

Solving equation (5-28) for R1 + R2

$$R1 + R2 = \frac{(+V)R_i R3}{(+V)R_i + R3e_i} \tag{5-29}$$

Resistor R_i affects both the input impedance and the switching sensitivity discussed earlier. Capacitor C with R_i sets the critical frequency, also discussed earlier. It (f_c) should be selected at one-tenth of the operating frequency. Resistor R1 is chosen to limit the input current (to protect the Norton amp) when R2 is shorted. Several kilohms or larger are adequate to protect the 3900.

Inverting and noninverting Norton comparators can be made. Either a reference voltage or *two* bias connections are required.

SUMMARY

Single supply, AC amplifiers can be built with op amps. However, a DC voltage ($\sim \frac{1}{2}V_{supply}$) must be induced at the output to allow the output signal to be driven both up and down by the input. For the inverting and the noninverting amplifiers this is done by applying a reference voltage to the noninverting input (from a voltage divider or supply splitter).

You must couple the AC signal into and out of the amplifier while blocking the DC levels. This is done with a RC coupler. Be sure to keep the critical frequency of that coupler well below the lowest operating frequency of the amplifier.

The Norton amplifier is optimized for single supply operations, and for many applications provides attractive performance and density to cost ratios. Its output voltage is determined by the input *currents*, and the input terminals are at approximately zero volts.

Although this is quite different from an op amp, inverting, noninverting, and difference amplifiers and summers as well as comparators can be made from a Norton amp. For all linear Norton amps, biasing is accomplished with a single series resistor. The AC gain is always R_f/R_i, and $Z_i = R_i$, for all configurations.

You have seen in these chapters how to build and power signal amplifiers. The op amp and Norton amp allowed you to do this much more easily, more precisely, and more cheaply than you could have with discrete components. These signals being fed into your IC amplifiers must be generated. In the following chapter, you will apply the characteristics of op amps and Norton amps to produce square wave, triangle wave, and sine wave generators. Also special-purpose signal and function generator ICs will be presented.

PROBLEMS

5-1 **a.** Carefully draw the output you would obtain from the circuit in Fig. 5-35.
b. Explain why the output is shaped as it is.
c. Modify the schematic of Fig. 5-35 to improve the quality of the output signal.

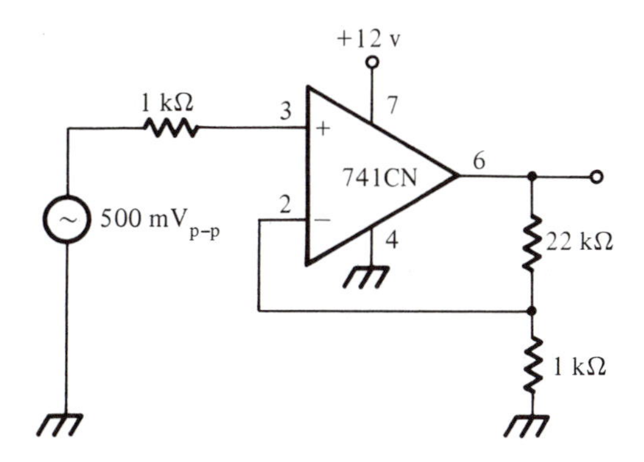

Figure 5-35 Schematic for problem 5-1

5-2 For the circuit in figure 5-3, with

$$e_i = 100 \text{ mV p-p}$$
$$R_i = 10 \text{ k}\Omega$$
$$R_f = 330 \text{ k}\Omega$$
$$R_A = 100 \text{ k}\Omega$$
$$R_B = 120 \text{ k}\Omega$$
$$+V = 12 \text{ V}$$

calculate the following:
a. Output AC signal peak-to-peak voltage.
b. V_{ref}.
c. Output DC voltage.
d. Draw the composite output voltage.
e. Input impedance.

5-3 Design an inverting amplifier, using a 324 op amp operating from a single +5 volt supply. The amplifier must have an input impedance of at least 20 kΩ, and a gain of −15. Make sure that the output is properly biased.

5-4 For the circuit in Fig. 5-7a with

$$e_i = 100 \text{ mV p-p}$$

$$R_A = 200 \text{ k}\Omega$$

$$R_B = 100 \text{ k}\Omega$$

$$R_f = 100 \text{ k}\Omega$$

$$R_i = 47 \text{ k}\Omega$$

$$+V = 12 \text{ V}$$

calculate the following:

a. Output AC signal peak-to-peak voltage.

b. V_{ref}.

c. Output DC voltage.

d. Draw the composite output voltage.

e. Input impedance seen by the voltage source e_i.

5-5 Design a noninverting amplifier, using a 324 op amp operating from a single +5 volt supply. The amplifier must have an input impedance of at least 20 kΩ, and a gain of 15. Make sure that the output is properly biased. (This is a repeat of Problem 5-3 for a noninverting amp.)

5-6 Describe the major difference between an op amp and a Norton amp. What do these different characteristics mean about input voltages, input currents, biasing circuit configuration, and feedback circuit configuration?

5-7 For the circuit in Fig. 5-15, calculate i_+, i_-, V_o, given:

a. $+V = +12$ V, $R_f = 180$ kΩ, $R_B = 1$ MΩ

b. $+V = +12$ V, $R_f = 470$ kΩ, $R_B = 1$ MΩ

c. $+V = +12$ V, $R_f = 1$ MΩ $R_B = 1$ MΩ

d. $+V = +5$ V, $R_f = 1$ MΩ, $R_B = 2.2$ MΩ

5-8 For the circuit in Fig. 5-16, with

$$e_i = 100 \text{ mV p-p}$$

$$R_i = 10 \text{ k}\Omega$$

$$R_f = 330 \text{ k}\Omega$$

$$R_B = 980 \text{ k}\Omega$$

$$C_i = 0.1\ \mu\text{F}$$

$$C_o = 0.047\ \mu\text{F}$$

$$R_L = 47 \text{ k}\Omega$$

$$+V = 12 \text{ V}$$

a. Calculate the following:

1. Output AC signal peak-to-peak voltage.
2. Output DC voltage.
3. Input impedance.
4. Input and output couplers' critical frequency.
5. Lowest operating frequency at which the amplifier should be used.

b. Draw the composite output voltage.

5-9 Design an inverting amplifier using a 3900 Norton amp operating from a single +5 volt supply. The amplifier must have an input impedance of at least 20 kΩ, a gain of −15, and an operating frequency of 1 kHz. It will drive a 100 kΩ load. Make sure that the output is properly biased. (This is a repeat of Problem 5-3 for a Norton amp.)

5-10 The circuit in Fig. 5-20 is to be used as a mixer:

$$+V = 30\ \text{V}$$

$$\text{R1} = 25\ \text{k}\Omega$$

$$\text{R2} = 47\ \text{k}\Omega$$

$$\text{R3} = 18\ \text{k}\Omega$$

$$\text{R}_\text{f} = 100\ \text{k}\Omega$$

$$\text{R}_\text{B} = 220\ \text{k}\Omega$$

If each input contributes 2 V RMS to the output, what are the three input signal levels (e_1, e_2, e_3)?

5-11 Design an inverting mixer using a Norton amp. It is to have eight channels. Each channel must have an input impedance of at least 100 kΩ, and an adjustment range of at least a factor of 100. With any channel's control giving maximum contribution to the output, the overall gain should be 5. The low frequency cutoff is 30 Hz or lower. It will drive a 22 kΩ load. Make sure that biasing does not vary with changes in overall gain.

5-12 For the circuit in Fig. 5-22, with

$$e_i = 100\ \text{mV p-p}$$

$$\text{R}_i = 22\ \text{k}\Omega$$

$$\text{R}_\text{f} = 680\ \text{k}\Omega$$

$$\text{R}_\text{B} = 1.2\ \text{M}\Omega$$

$$C_i = 0.047\ \mu\text{F}$$

$$C_\text{o} = 0.022\ \mu\text{F}$$

$$\text{R}_L = 91\ \text{k}\Omega$$

$$+V = 12\ \text{V}$$

a. Calculate the following:
  1. Output AC signal peak-to-peak voltage.
  2. Output DC voltage.
  3. Input impedance.
  4. Input and output couplers' critical frequency.
  5. Lowest operating frequency at which the amplifier should be used.

b. Draw the composite output voltage.

5-13 Design a noninverting amplifier using a Norton amp operating from a single +5 volt supply. The amplifier must have an input impedance of at least 20 kΩ, a gain of 15, and an operating frequency of 1 kHz. It will drive a 100 kΩ load. Make sure that the output is properly biased. (This is a repeat of Problem 5-9 for a noninverting amp.)

5-14 For the circuit in Fig. 5-25,

$$e_1 = 10 \text{ mV RMS}$$

$$e_2 = 47 \text{ mV RMS}$$

$$e_3 = 27 \text{ mV RMS}$$

$$+V = 30 \text{ V}$$

$$\text{R}_\text{f} = 1 \text{ M}\Omega$$

Calculate the required values of R1, R2, R3, and R_B so that each input will have a 2.5 V RMS contribution to the output signal, and so that the output is biased to the proper DC level.

5-15 Rework Problem 5-11 for a noninverting mixer.

5-16 For the circuit in Fig. 5-26,

$$e_1 = e_2 = 100 \text{ mV RMS}, \quad 60 \text{ Hz}$$

$$\text{C1} = \text{C2} = C_o = 1\ \mu\text{F}$$

$$\text{R1} = \text{R2} = 10 \text{ k}\Omega$$

$$\text{R}_\text{f} = 220 \text{ k}\Omega$$

$$\text{R}_\text{B} = 470 \text{ k}\Omega$$

$$\text{R}_\text{L} = 100 \text{ k}\Omega$$

$$+V = 12 \text{ V}$$

calculate and draw the composite output voltage.

5-17 **a.** Repeat Problem 5-16 with

$$e_1 = 10 \text{ mV RMS} \qquad 2 \text{ kHz}$$

$$e_2 = 10 \text{ mV RMS} \qquad 2 \text{ kHz}$$

e_2 is 180 degrees out of phase with e_1.

b. Discuss the response of the difference amplifier to the common mode noise of Problem 5-16 as compared with its response to the differential (180 degrees out of phase) signal of Problem 5-17.

5-18 For the circuits in Figs. 5-29(a) and (b),

$$R1 = 10 \text{ k}\Omega$$

$$R2 = 1.8 \text{ k}\Omega$$

$$R = 1 \text{ M}\Omega$$

$$+V = 15 \text{ V}$$

Draw the output from each in response to the input signal shown in Fig. 5-36.

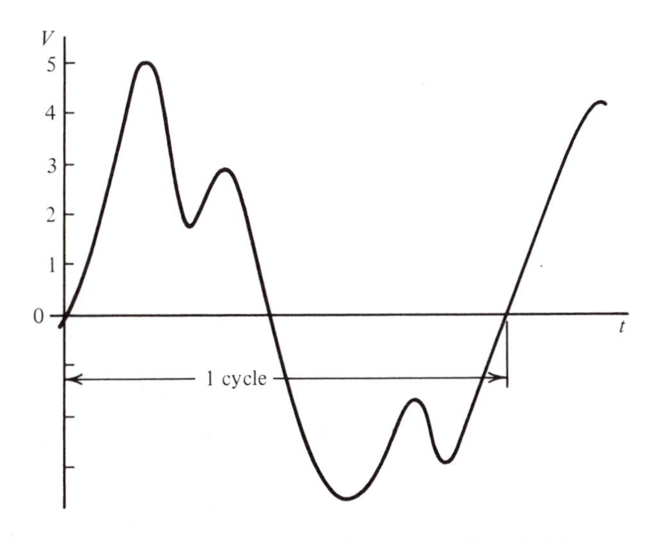

Figure 5-36 Input signal for problems 5-18, 5-19, 5-20 and 5-21

5-19 Design an inverting referenced comparator and a noninverting referenced comparator which will give only one output cycle for the one input cycle of Fig. 5-36.

5-20 For the circuit in Fig. 5-33,

$$R_i = 37\ k\Omega$$

$$R_B = 220\ k\Omega$$

$$R_f = 920\ k\Omega$$

$$+V = 12\ V$$

e_i given in Fig. 5-36

draw the output.

5-21 Modify the circuit of Problem 5-20 so that it switches at two separate levels, and outputs only one cycle for the one input cycle of Fig. 5-36.

5-22 For the circuit in Fig. 5-34, calculate the trigger level:

$$+V = 5\ V,\ R1 = R2 = 100\ k\Omega,\ R_i = 100\ k\Omega, \text{ and}$$

a. R3 = 470 kΩ
b. R3 = 200 kΩ
c. R3 = 100 kΩ

5-23 For the circuit in Fig. 5-37, select values for R1, R2, R3, R4, R5, and R6 so that:

$e_i > 2$ V CR1 ON

$e_i < -2$ V CR2 ON

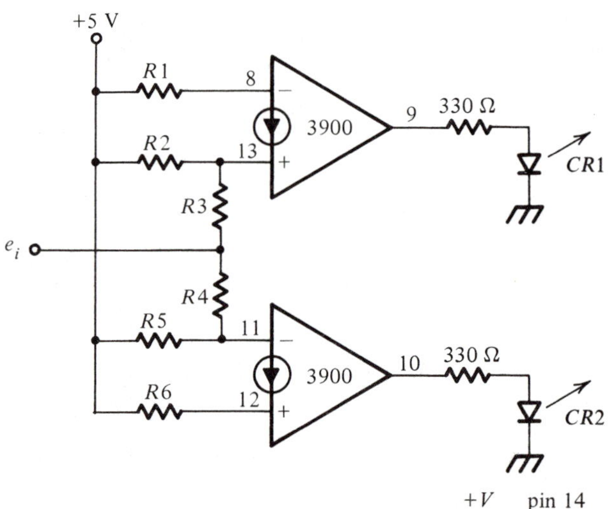

Figure 5-37 Schematic for problem 5-23

CHAPTER 6

Waveform Generators

During the previous chapters you have seen how to condition signals with op amps and Norton amp amplifiers. However, these signals must be produced, system control commands must be generated, and measurements and timing must be performed. Those three functions (signal generation, command generation, and measurement and timing) depend heavily on precision waveform generation.

In this chapter you will see how to produce square waves, ramps, and sine waves. These can come from circuits based around op amps, Norton amps, or special purpose integrated circuits. Circuits to produce all three waveshapes from each of these ICs will be covered.

OBJECTIVES

After studying this chapter you should be able to do the items listed below for each of the following circuits:

Circuits

Op amp square wave generator
Norton amp square wave generator
555 astable multivibrator
2240 long interval timer
Op amp ramp (triangle) generator
Norton amp ramp (triangle) generator
Sawtooth generator
Op amp Wien-bridge oscillator
Norton amp Wien-bridge oscillator
Op amp twin T oscillator
Norton amp twin T oscillator
Multi-op amp function generator
8038 function generator IC

Tasks:

a. Draw the schematic.
b. Qualitatively and quantitatively describe circuit operation.

c. Calculate component values necessary to meet given frequency and amplitude specifications.

d. Discuss advantages, disadvantages, limitations, and precautions of the given circuit and compare with others which produce the same waveshape.

6-1 SQUARE WAVE GENERATORS

Square waves are the simplest waveforms to produce. They are used in system control (both analog and digital), tone generation, and instrumentation, and can be used to produce both ramps and sine waves. In this section you will see square wave generators made with op amps, Norton amps and special timer ICs.

6-1.1 Op Amp Square Wave Generator

Most square wave generators are based on the charge and discharge of a capacitor. When charging through a resistor toward a fixed voltage, the voltage across the capacitor, as a function of time, is

$$e(t) = A(1 - e^{-t/RC}) + V_o \tag{6-1}$$

where

A is the total *difference* in potential through which the capacitor tries to charge.

V_0 is the charge on the capacitor at $t = 0$.

Equation (6-1) will form the basis for all of the square wave frequency calculations.

The only other element you need is a comparator which detects *two* levels. This is shown in Fig. 6-1. Since there is no negative feedback, the output will be at either $\pm V_{SAT}$. A portion of this output is divided by R1 and R2 and fed back to the noninverting input.

$$V_{ref} = \frac{R2}{R1 + R2}(\pm V_{SAT})$$

Let

$$\beta = \frac{R2}{R1 + R2}$$

Then

$$V_{ref} = \beta(\pm V_{SAT})$$

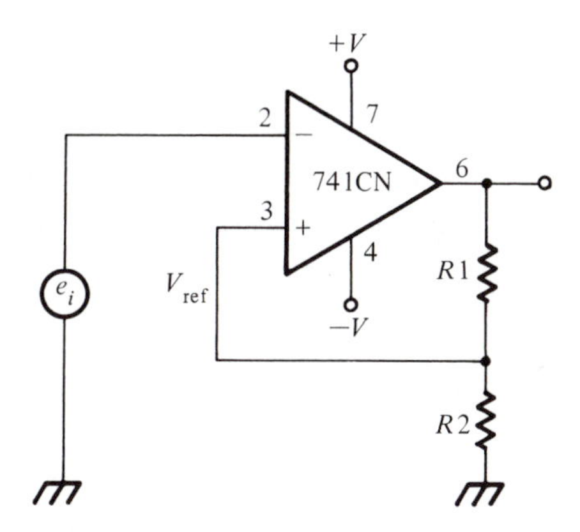

Figure 6-1 Comparator with two levels or trigger points

When the output is at $+V_{SAT}$, the reference level is at βV_{SAT}, a *positive* voltage. When the output is forced to $-V_{SAT}$, the reference level switches to $-\beta V_{SAT}$, a *negative* voltage. This circuit, then, provides comparison at two different levels, depending on the condition of its own output.

To build a square wave generator, you allow the output of the comparator to drive a RC circuit [described by equation (6-1)] and use the voltage across the capacitor as the input. This is shown in Fig. 6-2. When the output is at $+V_{SAT}$, current flows from that output, through R, charging C positively. The voltage at the noninverting input is held at $+\beta V_{SAT}$ by R1 and R2. This condition continues as the charge on C rises, until it has just exceeded $+\beta V_{SAT}$, the reference voltage.

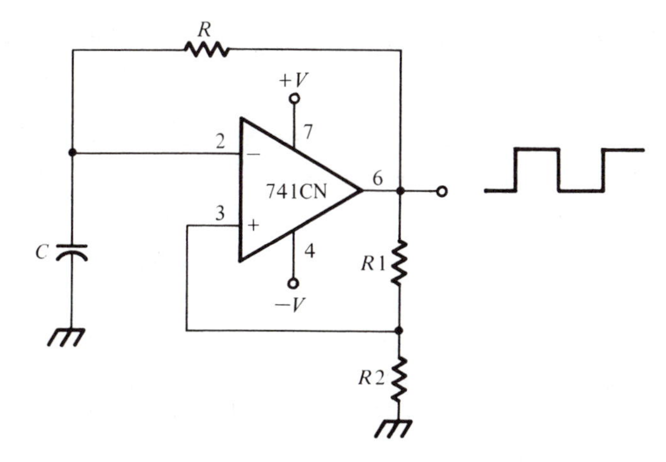

Figure 6-2 Simple op amp square wave generator

When the voltage at the inverting input exceeds the reference voltage at the noninverting input, the output of the comparator is driven to $-V_{SAT}$. At this instant, the voltage on the capacitor is $+\beta V_{SAT}$. It begins to discharge through R, and then charge toward $-V_{SAT}$. When the output voltage switched to $-V_{SAT}$, the reference voltage (on the noninverting input) switched to $-\beta V_{SAT}$. The capacitor charges more and more negatively until its voltage just exceeds $-\beta V_{SAT}$. The output switches back to $+V_{SAT}$. The cycle repeats itself. The waveforms are shown in Fig. 6-3.

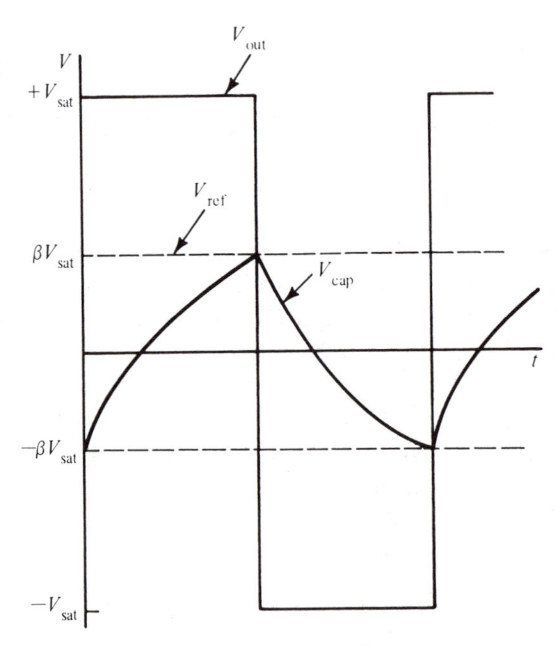

Figure 6-3 Op amp square wave generator waveforms

The frequency is determined by the time it takes the capacitor to charge from $-\beta V_{SAT}$ to $+\beta V_{SAT}$ and vice versa. To determine the time required to charge from $-\beta V_{SAT}$ to $+\beta V_{SAT}$,

$$e(t) = A(1 - e^{-t/RC}) + V_o$$

where A is the total difference in potential through which the capacitor tries to charge. Since the capacitor begins its charge at $-\beta V_{SAT}$, and would charge to $+V_{SAT}$ (if the comparator did not switch), the total difference is

$$A = V_{SAT} + \beta V_{SAT} = (1 + \beta)V_{SAT}$$

Equation (6-1) becomes

$$e(t) = (1 + \beta)V_{SAT}(1 - e^{-t/RC}) - \beta V_{SAT}$$

Switching occurs when the voltage across the capacitor, $e(t)$, reaches βV_{SAT}.

$$\beta V_{SAT} = (1 + \beta)V_{SAT}(1 - e^{-t/RC}) - \beta V_{SAT}$$

With a bit of perseverence and algebraic manipulation, this can be solved for t:

$$t = RC \ln \frac{1 + \beta}{1 - \beta}$$

This is the time to charge from $-\beta V_{SAT}$ to $+\beta V_{SAT}$, which is only half of the period (T).

$$T = 2t = 2 \text{ RC} \ln \frac{1 + \beta}{1 - \beta}$$

This relationship can be R1 = R2. Then simplified if you choose

$$\beta = 0.5$$

$$T = 2 \text{ RC} \ln 3$$

$$R2 = 0.90 \text{ R1}$$

This choice of β causes the log term to equal one, so for

$$T = 2.2 \text{ RC} \tag{6-2}$$

$$f = \frac{1}{2.2 \text{ RC}} \tag{6-3}$$

The output swings from $+V_{SAT}$ to $-V_{SAT}$, so

$$V_{out} \text{ peak-peak} = 2V_{SAT}$$

You could alter the output amplitude by varying the power supply voltage, but this is rarely a good idea. A better technique is shown in Fig. 6-4. When the output of the op amp goes to $+V_{SAT}$, CR1 forward biases, and CR2 breaks down

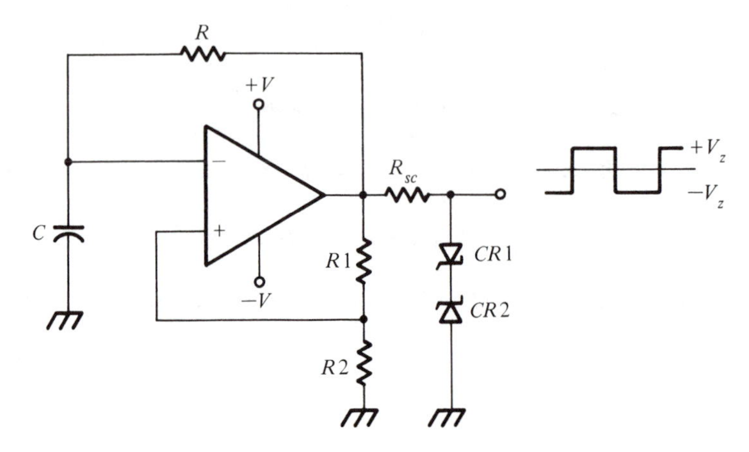

Figure 6-4 Amplitude limiting with zener diodes

at its zener voltage. The opposite occurs on the negative half cycle. The output voltage is regulated to $\pm V_Z$ by the zener diodes.

$$V_{out} \text{ peak-peak} = 2V_Z$$

Resistor R_{SC} limits the current drawn from the op amp to

$$I_{SC} = \frac{V_{SAT} - V_Z}{R_{SC}}$$

It should be chosen to keep from exceeding the op amp's maximum output current. If you use an op amp which is internally protected against short circuits (such as the 741), R_{SC} is not needed.

This circuit works reasonably well at audio frequencies. To lower the frequency, you must raise *R*, *C*, or both. However, since *C* charges between $\pm\beta V_{SAT}$, it cannot be electrolytic. As you increase *R*, to lower the frequency, the current available to charge the capacitor falls. At high levels of *R*, a significant amount of the current flowing through *R* is I_B, and does not charge *C* at all. Pulse width and frequency begin to drastically deviate from theory.

Slew rate is one of the major limiting factors on high frequency response. Also, there is a delay time in pulling the op amp out of saturation. This time is different for coming out of $+V_{SAT}$ and coming out of $-V_{SAT}$. Consequently, at higher frequencies, even if you are not slew rate limited, these delay times can alter the pulse width and lower the frequency significantly. The use of zeners tied *directly* to the output will help keep the op amp out of saturation, but will force it into short-circuit current limiting.

Should you replace a 741 (0.5 V/μs) with a 318 (50 V/μs) to improve high frequency response, you must watch for two things. The higher slew rate gives you much better rise times. However, these large voltage transients couple high frequency harmonic noise onto the power supplies and output lines. Be sure to decouple the power supply pins by placing a 0.1 μF ceramic capacitor from each power pin to ground. Also, placing a 1 kΩ load, and/or an 80 pF capacitor, from the output of the 318 to ground should absorb enough energy to clean up the output.

6-1.2 Norton Amp Square Wave Generator

You saw in the previous section that a square wave generator can be made by feeding a two-level comparator with an RC network. This comparator can be a Norton amp. A two-level Norton amp comparator was discussed in Chapter 5. It is shown, again, in Fig. 6-5 along with the equations for lower and upper trigger points.

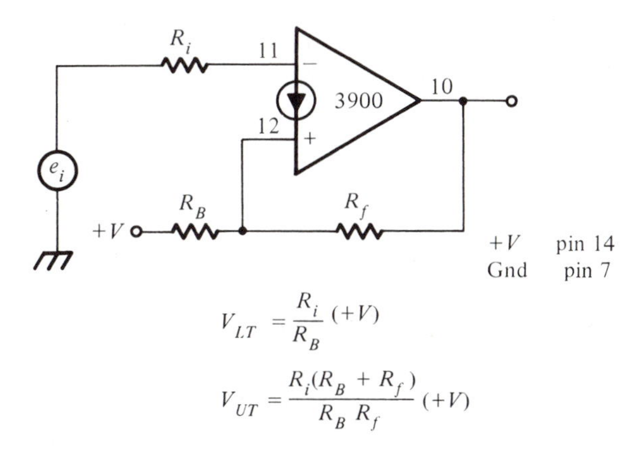

$$V_{LT} = \frac{R_i}{R_B}(+V)$$

$$V_{UT} = \frac{R_i(R_B + R_f)}{R_B R_f}(+V)$$

Figure 6-5 Two level Norton amp comparator

A Norton amp square wave generator is shown in Fig. 6-6. Initially, with a low charge on the capacitor, there will be only a small I_i, since

$$I_i = \frac{V_c}{R_i}$$

Current I_B flowing into the noninverting input forces the output to the positive supply voltage. This output voltage will produce two currents, I_{charge} through R to charge the capacitor, and I_f through R_f. Current I_f sums with I_B and flows into the noninverting input,

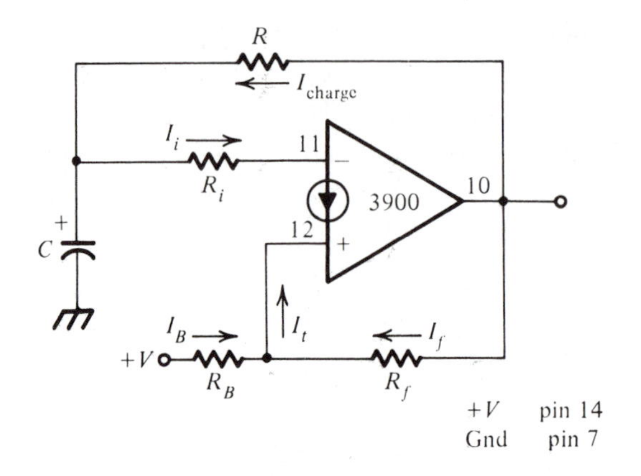

Figure 6-6 Norton amp square wave generator

$$I_t = I_B + I_f$$

This condition continues as the capacitor charges. As this charge increases, I_i increases. Eventually, the voltage across the capacitor just exceeds the upper trigger point (V_{UT}). At this point,

$$I_i > I_t$$

or

$$I_i > I_B + I_f$$

The output of the Norton amp is driven to ground. This ends the positive half-cycle.

With the output voltage at ground, I_f falls to zero. So

$$I_t = I_B$$

and the current into the noninverting input drops. Also, the capacitor has a charge of V_{UT}. It now discharges through R. This condition continues as the voltage across the capacitor drops.

As the voltage drops, I_i also decreases. When the voltage on the capacitor is just below the lower trigger point (V_{LT}),

$$I_i < I_t$$

or

$$I_i < I_B$$

The output switches, again, to $+V$. This ends the negative half-cycle.

To summarize, during the positive half-cycle, the capacitor charges from a voltage of V_{LT} through R until it reaches V_{UT}. At this point $I_i > I_B + I_f$. The output switches to ground and I_t drops to I_B. The capacitor now discharges from V_{UT} toward V_{LT}. When the capacitor voltage reaches V_{LT}, I_i has fallen to just below I_B. The output switches again. The output switches between ground and $+V$ (actually between 0.2 V and $+V - 0.8$ V) while the capacitor charges and discharges between V_{LT} and V_{UT}.

These charge and discharge times are controlled by R, C, V_{LT}, and V_{UT}. Although any V_{LT} and V_{UT} could be chosen, it is convenient to select:

$$V_{LT} = \tfrac{1}{3}(+V) \tag{6-4}$$

$$V_{UT} = \tfrac{2}{3}(+V) \tag{6-5}$$

This is done by selecting

$$R_B = R_f$$

$$R_i = \tfrac{1}{3}R_B$$

You should substitute these values into the trigger level equations given in Fig. 6-5 to verify equations (6-4) and (6-5).

The time it takes the capacitor to charge to $\frac{1}{3}(+V)$ from zero volts is

$$e(t) = A(1 - e^{-t/RC}) + V_o$$

$$\tfrac{1}{3}(+V) = +V(1 - e^{-t/RC})$$

$$\tfrac{1}{3} = (1 - e^{-t/RC})$$

$$e^{-t/RC} = \tfrac{2}{3}$$

$$t/RC = -\ln(\tfrac{2}{3})$$

$$t = -RC\ln(\tfrac{2}{3})$$

$$t_{1/3} = 0.405RC$$

To charge from zero volts to $\frac{2}{3}(+V)$:

$$\tfrac{2}{3}(+V) = +V(1 - e^{-t/RC})$$

$$\tfrac{2}{3} = (1 - e^{-t/RC})$$

$$e^{-t/RC} = \tfrac{1}{3}$$

$$t/RC = -\ln(\tfrac{1}{3})$$

$$t = -RC\ln(\tfrac{1}{3})$$

$$t_{2/3} = 1.09\ RC$$

The time to charge from $\frac{1}{3}(+V)$ to $\frac{2}{3}(+V)$ is

$$t = t_{2/3} - t_{1/3} = 0.69\text{RC}$$

This is the time of the positive half-cycle. The negative half-cycle time equals the positive half-cycle time. So the period of the square wave is

$$T = 2t = 1.38\text{RC} \sim 1.4\text{RC}$$

or

$$f = \frac{1}{1.4\text{RC}} \tag{6-6}$$

Equation (6-6) is valid for the Norton amp square wave generator with

$$\text{R}_i = \tfrac{1}{3}\text{R}_\text{B}$$

$$\text{R}_f = \text{R}_\text{B}$$

There are several conditions you must be aware of. In Fig. 6-6, I_i subtracts from the charging current. This was not accounted for in equation (6-6). To keep this effect small, set

$$\text{R}_i > 10\text{R}$$

so that

$$I_i \ll I_{\text{charge}}$$

Second, the amplitude is

$$V_{\text{out}}\ \text{peak-peak} = +V$$

If you want to lower this, a series resistor and *single* zener can be used. In this case

$$V_{\text{out}}\ \text{peak-peak} = V_Z$$

Be sure to limit output current to a safe level with the series resistor.

Since the capacitor charges and discharges between $\frac{1}{3}(+V)$ and $\frac{2}{3}(+V)$, no charge reversal is required. So, you may use an electrolytic capacitor to lower the frequency. However, bias current (I_B^- and I_B^+) must be at least 0.2 μA. This requirement places an upper limit on the size of R_B, R_f, R_i, and R.

$$R_B \leq \frac{+V}{0.2\ \mu A}$$

The match between the two inputs degrades at bias currents below 10 μA, which could affect the symmetry of the signal. So

$$R_B \leq \frac{+V}{10\ \mu A}$$

is a more conservative limit.

As with the op amp square wave generator, the high frequency limit is dependent on the slew rate and delay times in coming out of saturation.

6-1.3 The 555 Timer

The 555 timer is an integrated circuit specifically designed to perform signal generation and timing functions. It allows a great deal of versatility. It will operate from a wide range of power supplies, sinking or sourcing 200 mA of load current. Proper selection of only a few external components will allow timing intervals of several minutes or frequencies as high as a few MHz. The block diagram for the 555 timer is given in Fig. 6-7.

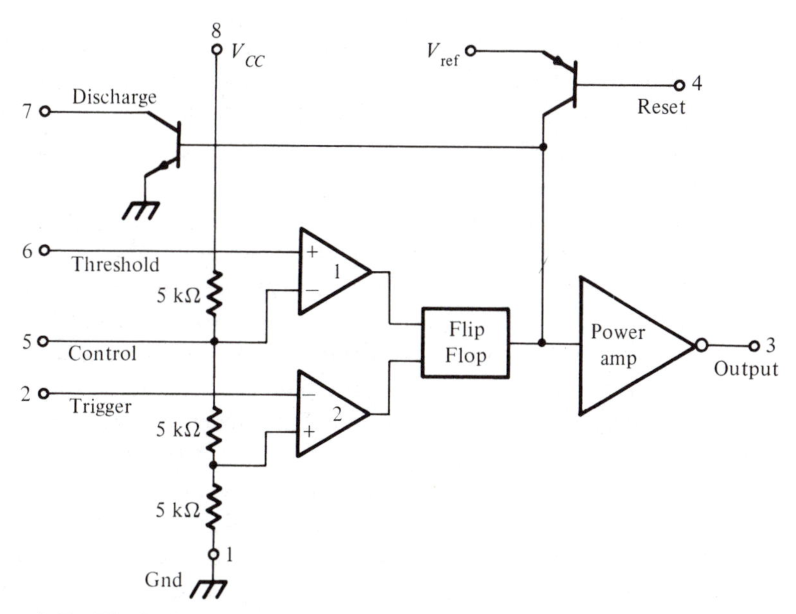

Figure 6-7 Block diagram of a 555 timer IC

The core of the 555 timer is a voltage divider and two comparators. The reference voltage to comparator 1 is $\frac{2}{3}V_{cc}$, while the reference voltage to comparator 2 is $\frac{1}{3}V_{cc}$. Whenever the voltage on the *threshold* input exceeds $\frac{2}{3}V_{cc}$, comparator 1 forces the flip-flop to output a high. A high from the flip-flop saturates the discharge transistor and forces the output from the power amplifier to go low. These conditions will persist until comparator 2 triggers the flip-flop. Even if the voltage at the threshold input falls below $\frac{2}{3}V_{cc}$, comparator 1 cannot cause the flip-flop to change again. Comparator 1 can only force the flip-flop's output high.

To change the output of the flip-flop to low, the voltage at the *trigger* input must fall below $\frac{1}{3}V_{cc}$. When this occurs, comparator 2 triggers the flip-flop, forcing its output low. This low from the flip-flop turns the discharge transistor off and forces the power amplifier to output a high. These conditions will continue, independent of the voltage on the *trigger* input. Comparator 2 can only cause the flip-flop to output a low.

In summary, to force the output of the 555 timer low, the voltage on the *threshold* input must exceed $\frac{2}{3}V_{cc}$. This also turns the discharge transistor *on*. To force the output of the timer high, the voltage on the *trigger* input must fall below $\frac{1}{3}V_{cc}$. This also turns the discharge transistor *off*.

A voltage may be applied to the control input to change the levels at which the switching occurs. When not in use, a 0.01 μF capacitor should be connected between pin 5 and ground to prevent noise coupled onto this pin from causing false triggering.

Connecting the *reset* (pin 4) to a logic low will place a high on the output of the flip-flop. The discharge transistor will go *on* and the power amplifier will output a low. This condition will continue until *reset* is taken high. This allows synchronization or resetting of the circuit's operation. When not in use, *reset* should be tied to V_{cc}.

An astable multivibrator can be produced by adding two resistors and a capacitor to the basic timer IC. This is shown in Fig. 6-8.

When power is first applied, the capacitor is discharged. This places the *trigger* below $\frac{1}{3}V_{cc}$. The output goes high and the discharge transistor goes *off*.

The capacitor charges through R_A and R_B (solid line) until its voltage exceeds $\frac{2}{3}V_{cc}$. This voltage on the *threshold* input causes the output to drop to a low and the discharge transistor to turn *on*. The capacitor now discharges (dotted line) through R_B and the discharge transistor (inside the IC) to ground. Current also flows through R_A into the discharge transistor. Resistors R_A and R_B must be large enough to limit this current and prevent damage to the discharge transistor.

As the capacitor discharges, its voltage falls. When the voltage on the *trigger* input falls below $\frac{1}{3}V_{cc}$, the output goes high again and the discharge transistor goes *off*. This cycle repeats itself for as long as power is applied.

The length of time that the output remains high is the time for the capacitor to charge from $\frac{1}{3}V_{cc}$ to $\frac{2}{3}V_{cc}$.

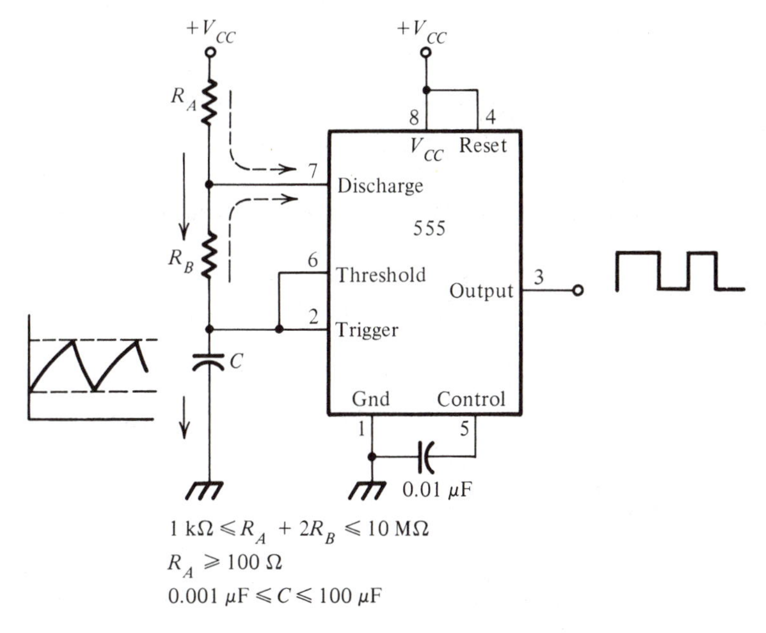

Figure 6-8 Basic square wave generator

$$V(t) = A(1 - e^{-t/\mathrm{RC}}) \quad \text{charge of an RC circuit}$$

For the time it takes the circuit to charge from 0 to $\frac{2}{3}$ V_{cc},

$$\tfrac{2}{3}V_{cc} = V_{cc}\,(1 - e^{-t/\mathrm{RC}})$$

$$\tfrac{2}{3} = 1 - e^{-t/\mathrm{RC}}$$

$$e^{-t/\mathrm{RC}} = \tfrac{1}{3}$$

$$-t/\mathrm{RC} = \ln\,(\tfrac{1}{3}) = -1.09$$

$$t = 1.09\mathrm{RC}$$

For the time it takes the circuit to charge from 0 to $\frac{1}{3}$ V_{cc},

$$\tfrac{1}{3}V_{cc} = V_{cc}\,(1 - e^{-t/\mathrm{RC}})$$

$$\tfrac{1}{3} = 1 - e^{-t/\mathrm{RC}}$$

$$e^{-t/\mathrm{RC}} = \tfrac{2}{3}$$

$$-t/\mathrm{RC} = \ln\,(\tfrac{2}{3}) = -0.405$$

$$t = 0.405\mathrm{RC}$$

So the time to charge from $\frac{1}{3}V_{cc}$ to $\frac{2}{3}V_{cc}$ is

$$t_{high} = 1.09RC - 0.405RC = 0.69RC$$

$$t_{high} = 0.69(R_A + R_B)C$$

The output is low while the capacitor discharges from $\frac{2}{3}V_{cc}$ to $\frac{1}{3}V_{cc}$.

$$V(t) = Ae^{-t/RC}$$

$$\tfrac{1}{3}V_{cc} = \tfrac{2}{3}V_{cc}\, e^{-t/RC}$$

$$\tfrac{1}{2} = e^{-t/RC}$$

$$-t/RC = \ln\left(\tfrac{1}{2}\right) = -0.69$$

$$t = 0.69RC$$

$$t_{low} = 0.69R_B C$$

Notice that both R_A and R_B are in the charge path, but only R_B is in the discharge path.

$$T = t_{high} + t_{low} = 0.69(R_A + R_B)C + 0.69R_B C$$

$$T = 0.69(R_A + 2R_B)C$$

$$f = \frac{1}{T} = \frac{1.45}{(R_A + 2R_B)C} \tag{6-8}$$

With this circuit configuration, it is impossible to have a duty cycle of 50 percent or less, since $t_{high} = 0.69(R_A + R_B)C > t_{low} = 0.69R_B C$. However, the circuit in Fig. 6-9 will allow the duty cycle to be set at practically any level.

During the charge portion of the cycle, diode CR1 forward biases, effectively shorting out R_B:

$$t_{high} = 0.69R_A C$$

However, during the discharge portion of the cycle, the diode is reverse biased:

$$t_{low} = 0.69R_B C$$

$$T = t_{high} + t_{low} = 0.69(R_A + R_B)C$$

$$f = \frac{1.45}{(R_A + R_B)C} \tag{6-9}$$

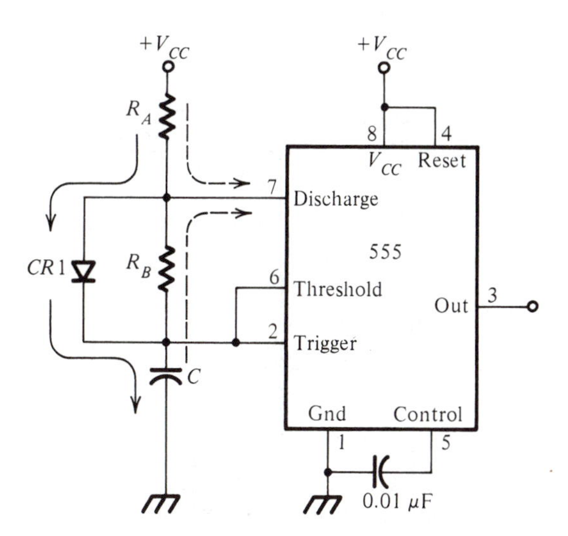

Figure 6-9 Adjustable duty cycle rectangular wave generator

Resistors R_A and R_B could be variable to allow adjustment of frequency and pulse width. However, a series fixed resistance of at least 100 Ω should be added to each R_A and R_B. This will limit peak current to the discharge transistor should the rheostat be adjusted to its minimum value.

Many timing problems call for a monostable operation. The 555 timer can easily be configured for that mode. Figure 6-10 is the schematic of a basic monostable multivibrator. When power is applied, the timer can "come up" in either output state. If the output is initially low, the discharge transistor will be *on*, preventing C_T from charging. This is the stable state. However, if the output is initially *high*, C_T will charge through R_T to 2/3 V_{cc}. When this occurs, the output goes low, the discharge transistor comes *on*, and the stable state is entered. Therefore, shortly after power is applied, the monostable enters its stable state: output *low* and the discharge transistor *on*.

Bringing the voltage on the *trigger* input below 1/3 V_{cc} will cause the monostable to flip to its unstable state. The output goes *high* and the discharge transistor goes *off*. The capacitor now begins to charge from 0 volts toward V_{cc}. The circuit operation will continue like this until the voltage on the *threshold* input exceeds 2/3 V_{cc}. At that time, *if the trigger input voltage is high*, the output will be sent low and the discharge transistor will short out the capacitor.

Should the *trigger* input pulse still be below 1/3 V_{cc} when the *threshold* input voltage exceeds 2/3 V_{cc}, the flip-flop would receive commands to set the output both high and low. Unpredictable and possibly damaging conditions would result. Therefore, t_{in} must be kept less than the output pulse width.

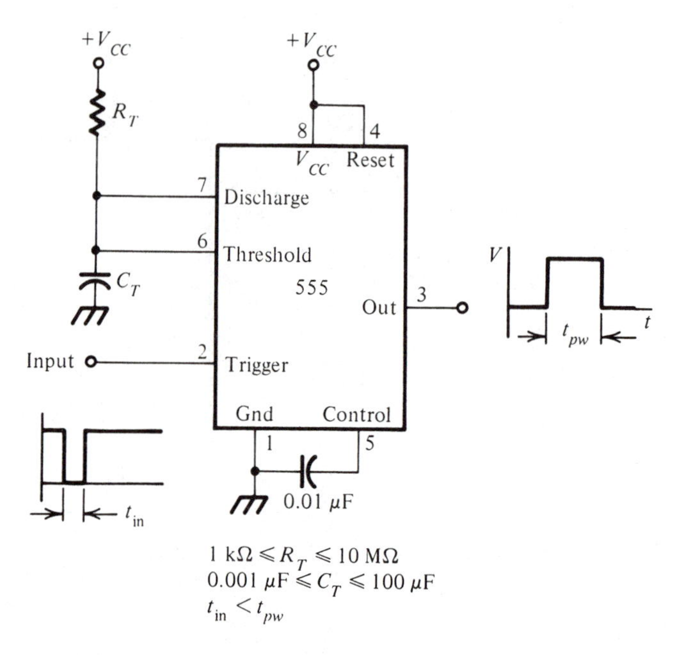

Figure 6-10 Basic monostable multivibrator using a 555 timer IC

The width of the output pulse is determined by the time required for the capacitor to charge from 0 to 2/3 V_{cc}.

$$V(t) = A(1 - e^{-t/RC})$$

$$\tfrac{2}{3} V_{cc} = V_{cc}(1 - e^{-t/RC})$$

$$\tfrac{2}{3} = 1 - e^{-t/RC}$$

$$e^{-t/RC} = \tfrac{1}{3}$$

$$-t/RC = \ln\left(\tfrac{1}{3}\right) = -1.09$$

$$t_{pw} = 1.09 R_T C_T \tag{6-10}$$

A differentiator and positive limiter can be added to the *trigger* input of the monostable, as shown in Fig. 6-11. The differentiator shapes the negative transition into a sharp negative-going spike. The diode (positive limiter) ensures that the positive transition does not significantly exceed the power supply voltage, and damage the IC. The differentiator is returned to V_{cc} rather than ground to ensure that the *trigger* input is held high when there is no pulse present.

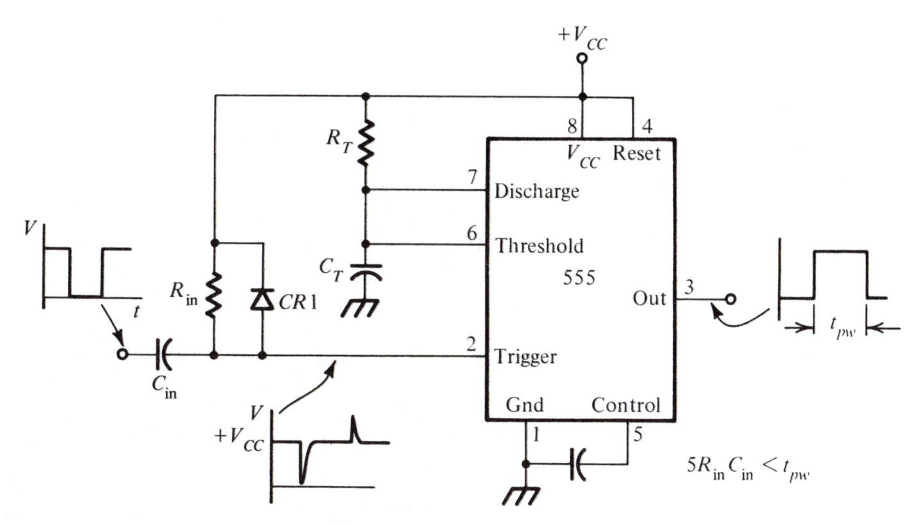

Figure 6-11 Trigger pulse shaping for 555 monostable multivibrator

6-1.4 Long Interval Timer

By driving a binary counter with the output of a 555 timer, you can build several very interesting new circuits. The timing interval can be drastically lengthened, and can be switch-selected (or digitally set). With similar ease, you can digitally synthesize a wide variety of frequencies. You can also generate many pulse patterns, staircase waveforms, waves synchronized with an external signal, and even build an analog-to-digital converter.

The XR 2240 programmable timer/counter consists of a 555-type timer, an eight-bit binary counter, and some additional control logic. This is all integrated in a single monolithic chip. The internal schematic is given in Fig. 6-12, and the block diagram in Fig. 6-13.

Comparators U1 and U2 form the heart of the timer, as they do in the 555. However, the comparison reference levels are set differently. With this setting, the charge or discharge time is set to one-half time constant,

$$t = RC/2$$

a bit simpler to work with than the timing equation for a 555.

The flip-flop U3 is controlled by both of the comparators, as it is in the 555. It can also be flipped (or flopped) by the control logic. Transistor Q1 is the discharge transistor, and Q2 is the power amplifier. The output (collector) of this

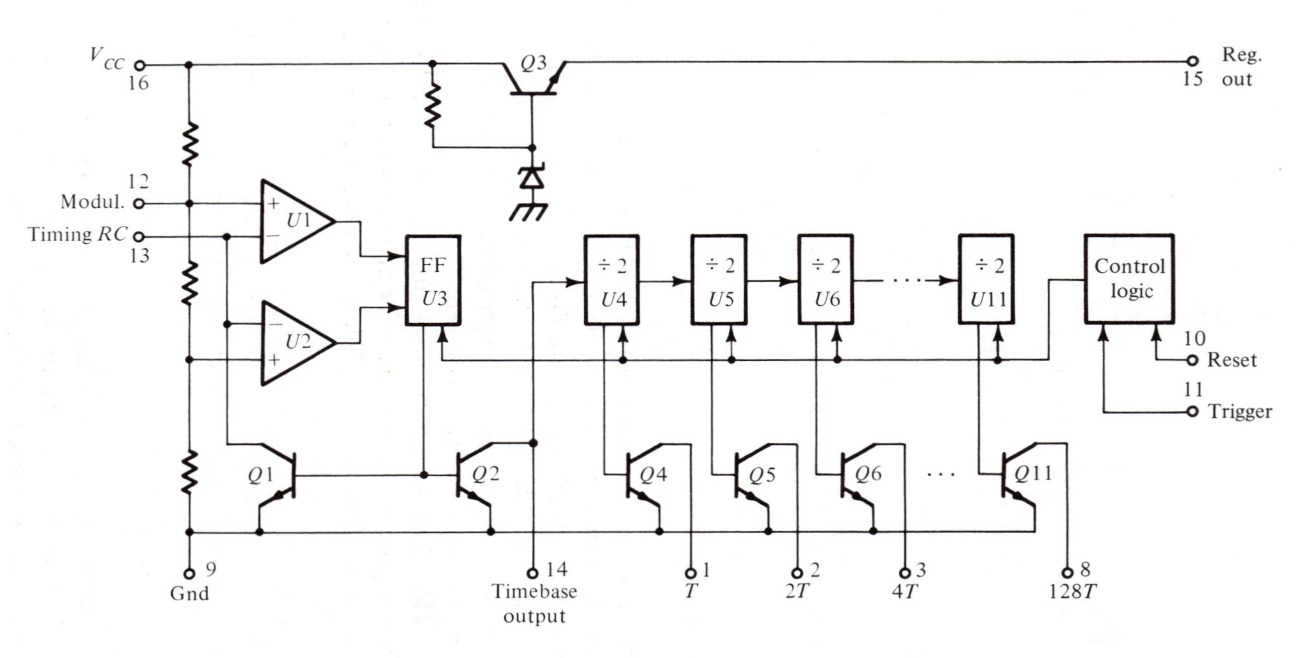

Figure 6-12 XR 2240 Internal schematic (*Courtesy of EXAR Corporation*)

Figure 6-13 XR 2240 block diagram

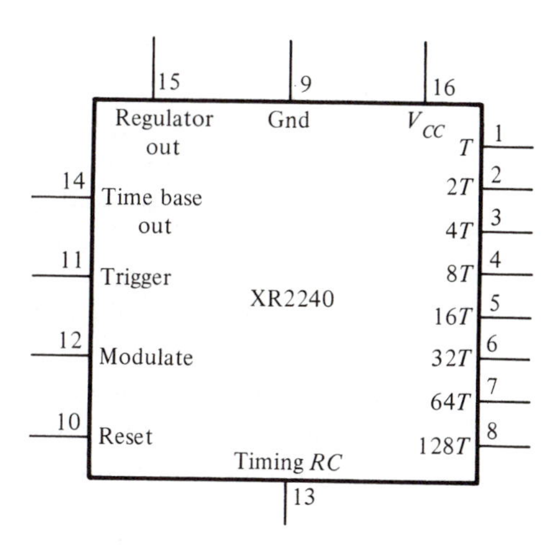

power amplifier is wired directly to the input of the binary counters. However, also notice that there is no bias (or pull-up) resistor for this transistor power amp. You must provide one externally.

The output of each binary counter drives the next in the chain, so it takes two pulses from Q2 to produce one out of U4; two from U4 to produce one out of U5; and so forth. This means that the pulse width out at each stage will be twice that of the previous stage. Consequently,

$$T = 2t = \text{RC} \qquad \text{PIN 1}$$

$$2T = 2\text{RC} \qquad \text{PIN 2}$$

$$4T = 4\text{RC} \qquad \text{PIN 3}$$

$$\vdots$$

$$128T = 128\text{RC} \qquad \text{PIN 8}$$

The output pins are the open collectors of transistors. Instead of outputting a high voltage, or a low voltage, these open collector outputs will either be open to ground (transistor off) or short to ground (transistor on), just like switches. To convert an open or short to ground to a voltage, you must connect a pull-up resistor (~ 10 kΩ) between that output pin and the supply voltage.

The control logic drives the flip-flop and counters to the proper states when stimulated by either a *positive-going* reset or trigger pulse. These pulses must be at least 1.4 V above ground, with a width of at least three μs. When a trigger pulse is applied, all outputs initially go *low*, and the 555 timer portion of the IC begins to oscillate, flipping and flopping the binary counters. All subsequent trigger pulses are ignored and the timer continues to run until a reset pulse is applied. The reset pulse turns off the 555 timer and sends all outputs *high* (open to ground). When power is initially applied, the 2240 should "come up" in this reset condition. Should a trigger and a reset pulse be applied simultaneously, the trigger pulse overrides, and the 2240 begins to oscillate.

A simple application of the 2240 is as an astable multivibrator. This is shown in Fig. 6-14. All outputs are pulled up to the supply voltage by 10 kΩ pull-up resistors. Basic timing is provided by the charge and discharge of C through R, connected to pin 13. Since the circuit is to free-run, there is no need for a reset pulse, so reset pin 10 has been tied to ground. Noise is shorted to ground through the 0.01 μF capacitor, keeping the modulation pin 12 stable. The 22 kΩ resistor between pin 15 (regulator out) and pin 14 (time base) properly biases the 555 type timer's power amplifier output stage. This ensures that a voltage signal will be properly driven into the counters.

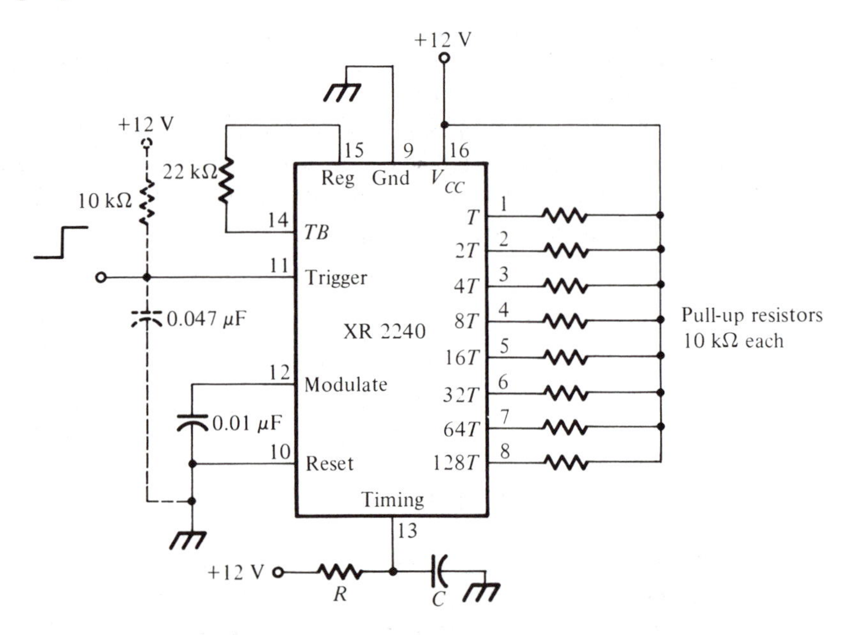

Figure 6-14 XR 2240 astable multivibrator

To start the oscillation, you increase the voltage on the trigger pin above 1.4 volts. This can be done with an external pulse, applied after the power has been turned on and chip operation has stabilized. Or you can connect an RC time delay as shown in the dotted lines. When power is first applied, the trigger pin is held at ground by the uncharged capacitor, allowing time for the IC to stabilize. As time goes on, the capacitor charges, bringing the trigger input above 1.4 V, and oscillations begin.

There are eight outputs available, each at half the frequency (or twice the period) of the one before it. Remember, to get a voltage output, you must provide a pull-up resistor. The relationship of the first six outputs is shown in Fig. 6-15.

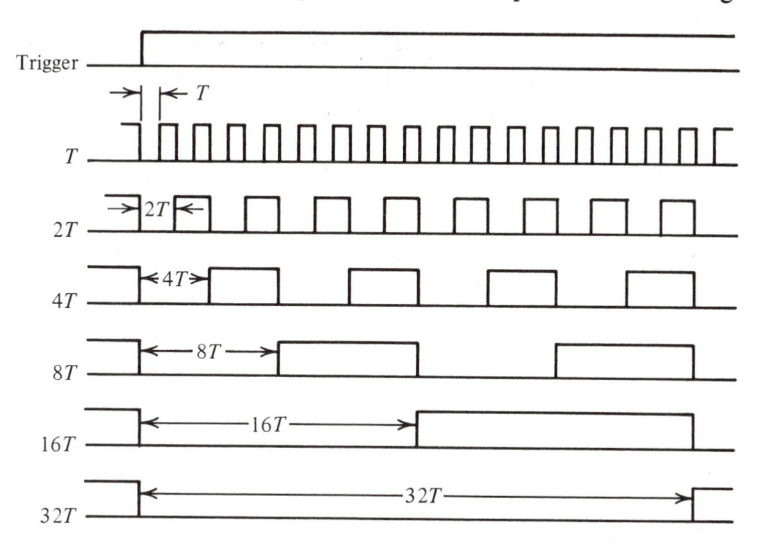

Figure 6-15 Binary outputs from a 2240 astable multivibrator

Actually, you do not have to provide a pull-up resistor at every output pin. If you only wanted an output which was 128T, then you need only use pin 8. Another interesting use of the outputs comes from their open collector configuration. You can connect two or more output pins together, then through a single pull-up resistor to the supply. This is shown in Fig. 6-16. Since each output presents either an open or a short to the ground, shorting outputs together simply puts switches to ground in parallel. Whenever *any* pin connected to the output is low, the output is shorted to ground. Only when *all* pins connected to the output are open (high), will the output go high.

Example 6-1

a. Draw the output waveshape if pins 1, 2, and 4 are wired together in Fig. 6-16.

b. What is the positive pulse width?

Solution

The output will be low until *all* pins connected to the output go open (high). Analysis of the circuit, then, is just a matter of examining the timing diagram to determine when pins 1, 2, and 4 all go high. The result is shown in Fig.

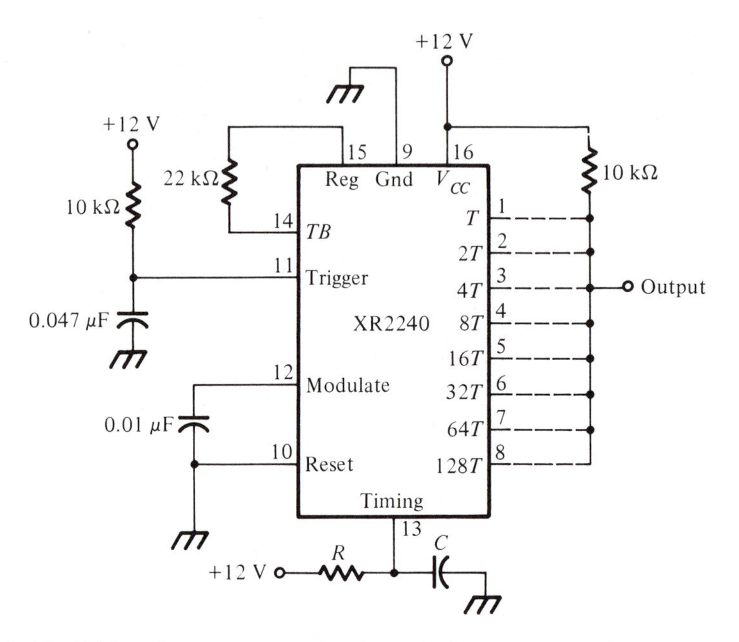

Figure 6-16 Wiring the output together for varied output waveshapes

6-17. The positive pulse width is equal to the narrowest pulse width attached to the output.

This technique can be used to generate a large variety of binary patterns. It can also be used to produce an astable multivibrator with switch-selected (or digitally controlled) frequency.

This switch selectable frequency synthesizer is shown in Fig. 6-18. The reset pin (10) is connected through a current-limiting 100 kΩ resistor to the output. Whenever the output goes high, the 2240 will be reset. The trigger pin (11) is connected, through an RC circuit, also to the output.

When power is applied, the 2240 resets itself, sending all outputs high. Capacitor C2 begins to charge, through R2. When the voltage across C2 has reached 1.4 volts, the IC is triggered and begins to oscillate. The output immediately goes low, and stays low until all pins connected to the output are high simultaneously. (During this time C2 fully discharges.) To this point, the circuit works just as the circuit of Fig. 6-16 (Example 6-1).

When the outputs all go high (open), the reset pin is pulled high. The 2240 is reset and the output is held high. This condition continues until a trigger is received. Capacitor C2 begins to charge through R2 when the output has gone high. When this voltage reaches 1.4 volts the IC is triggered, and the cycle repeats

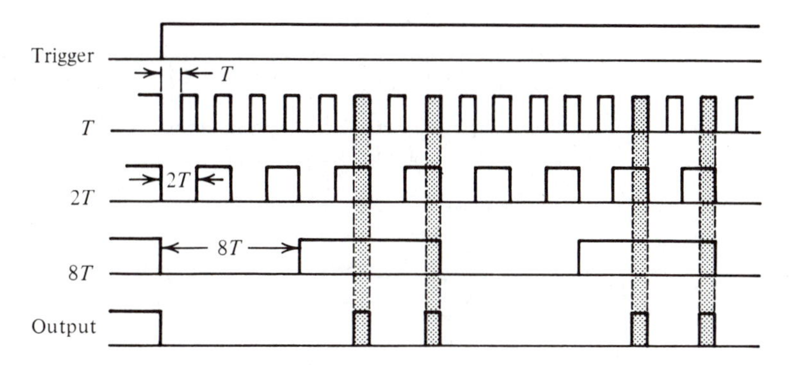

Figure 6-17 Pulse pattern for Example 6-1

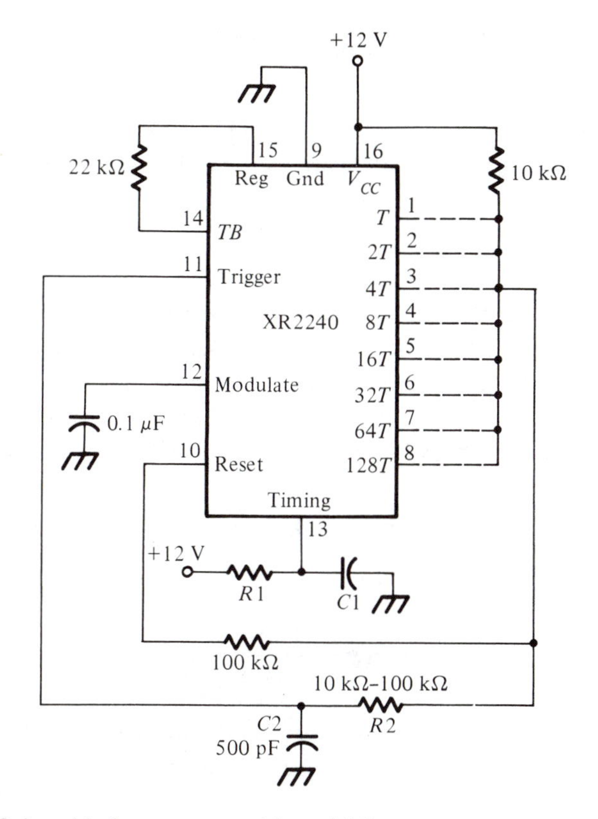

Figure 6-18 Selectable frequency astable multivibrator

itself. The output is illustrated in Fig. 6-19, if it is assumed that pin 1(T), pin 2($2T$), and pin 4($8T$) are wired to the pull-up resistor.

The total period depends on two factors. The time the output is low is the

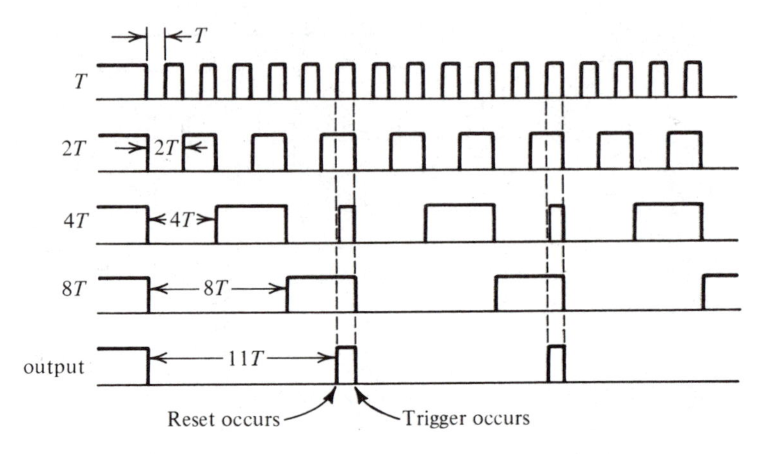

Figure 6-19 Frequency synthesizer output (pins 1, 2 and 4 wired to the output)

sum of the times for each of the pins tied to the output. In Fig. 6-19, pin 1(1T), pin 2(2T), and pin 4(8T) are used. So the time low is

$$1T + 2T + 8T = 11T$$

Once the reset occurs (after 11T in Fig. 6-19), all outputs are driven high and held high until a trigger pulse occurs. How long this takes depends on how long it takes C2 to charge to 1.4 volts through R2. In Fig. 6-19 this time has been shown as a little longer than T. However, R2 can be adjusted for stable operation, adjusting this positive pulse width time to T. If you do this, then:

$$\text{Period} = \text{time output is low} + \text{time output is high}$$

$$\text{Period} = NT + T = (N + 1)T$$

$$\text{Frequency} = \frac{1}{(N + 1)T}$$

N = sum of the times for each pin tied to the output
T = RC (the timing components)

Example 6-2

Design a frequency synthesizer using a 2240 to produce 50 kHz, 20 kHz, 10.3 kHz, 5 kHz, 400 Hz. Indicate which switches must be closed to obtain each of these frequencies.

Solution

The schematic in Fig. 6-18 will work. T is set to 10 μs (the shortest allowed according to the manufacturer's specifications) by selecting:

$$R = 1\ \text{k}\Omega \qquad C = 0.01\ \mu\text{F}$$

By properly adjusting or selecting R2, the positive pulse width can also be adjusted to T(10 μs). Then

$$f = \frac{1}{(N+1)T}$$

or

$$N = \frac{1}{fT} - 1$$

For

$$f = 50\ \text{kHz}$$

$$N = \frac{1}{(50\ \text{kHz})(10\ \mu\text{s})} - 1 = 1$$

So tie only T (pin 1) to the output. The same procedure is followed for each of the other required frequencies. The results are listed in Table 6-1.

TABLE 6-1. Solutions for example 6-2

Frequency (Hz)	*N*	*Pins*
50 k	1	T (pin 1)
20 k	4	$4T$ (pin 3)
10 k	9	$8T$ (pin 4), $1T$ (pin 1)
5 k	19	$16T$ (pin 5), $2T$ (pin 2), $1T$ (pin 1)
400	249	$128T$ (pin 8), $64T$ (pin 7), $32T$ (pin 6) $16T$ (pin 5), $8T$ (pin 4), $1T$ (pin 1)

One problem, however, arises. The output is asymmetric. No matter what the period, the positive pulse width is fixed at 10 μs. At lower frequencies, this may be impractically narrow; so narrow that you could not see it on a scope. You can solve this problem by using the output to trigger (clock) a toggle flip-flop. Each cycle the flip-flop would be either flipped or flopped. The output of the flip-flop would be a symmetric square wave at half the frequency of the pulses coming out of the 2240.

You can also build a programmable timer (or monostable multivibrator) by slightly modifying the circuit in Fig. 6-18. This is shown in Fig. 6-20. You must supply the trigger. Once triggered, the 2240 oscillates until the output goes high. This high, coupled to the reset pin, resets the 2240, sending all outputs high. The circuit then waits for another trigger pulse. There is a delay of

$$t_{\text{delay}} = NT$$

between the time the trigger pulse goes high and when the output goes high.

Since the trigger pulse will override a reset pulse, you must make sure that the voltage on *trigger*, pin 11, is low when the reset occurs. If the trigger signal does not return to zero on its own, you can add a differentiator (capacitor and resistor) and negative limiter (diode), as shown in the dotted lines.

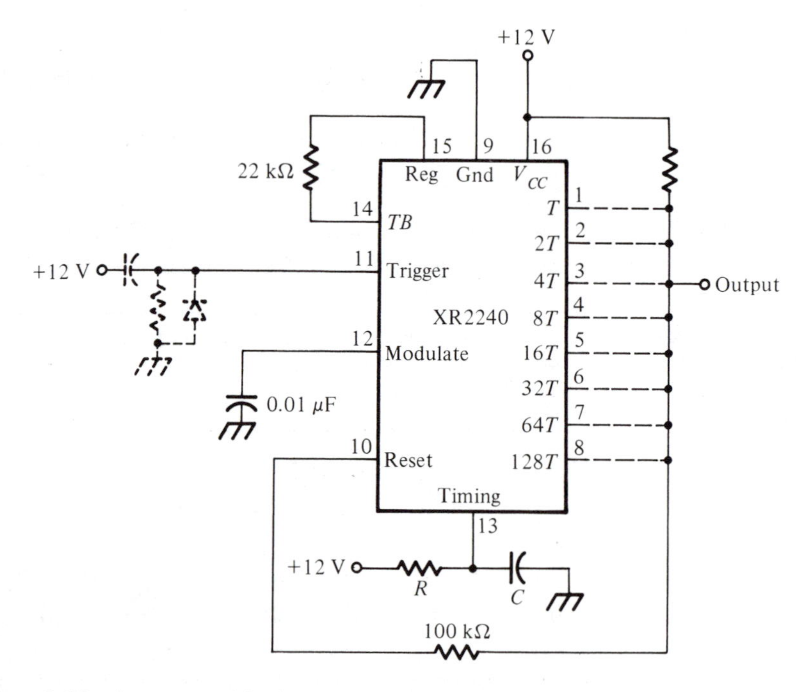

Figure 6-20 Programmable timer (monostable multivibrator)

Example 6-3

a. Design a programmable timer with a resolution of 20 seconds.

b. What is the maximum delay?
c. How must the switches be set to produce a delay of one hour?

Solution

a. The circuit in Fig. 6-20 will work. To get a resolution of 20 seconds,

$$T = RC = 20 \text{ s}$$

Pick

$$C = 100\ \mu\text{F}$$

$$R = \frac{20 \text{ s}}{100\ \mu\text{F}} = 200 \text{ k}\Omega$$

b. A delay of $255T$ is possible (all switches closed).

$$\text{Maximum time} = 255T = 255 \times 20 \text{ s}$$

$$\text{Maximum time} = 5100 \text{ s} = 85 \text{ minutes}$$

c. To get a delay of one hour,

$$N = \frac{60 \text{ min} \times 60 \text{ sec/min}}{20 \text{ sec}}$$

$$N = 180$$

Wire $128T$ (pin 8), $32T$ (pin 6), $16T$ (pin 5), and $4T$ (pin 3) together to produce a delay of $180T$.

6-2 RAMP GENERATORS

Ramps, triangle waves, and sawtooth waves, are used extensively in sweeping (i.e., causing a variable to change at a linear rate) in many industrial, measurement, and signal and data conversion circuits. You depend on a ramp generator every time you use an oscilloscope to move the beam linearly across the screen. Sine waves can also be derived from triangle waves. Op amps and Norton amps can be used to produce these important timing and driving waveforms.

6-2.1 Op Amp Triangle Wave Generator

Square waves were produced by using an RC circuit and a two-level comparator. These same two components will produce a triangle wave if you rearrange them a bit.

The voltage across a capacitor, when charging toward a fixed *voltage*, is exponential. However, the output from a triangle wave generator varies *linearly* with time. These two waveshapes are compared in Fig. 6-21. Clearly, charging the capacitor with a fixed voltage will not produce a triangle wave.

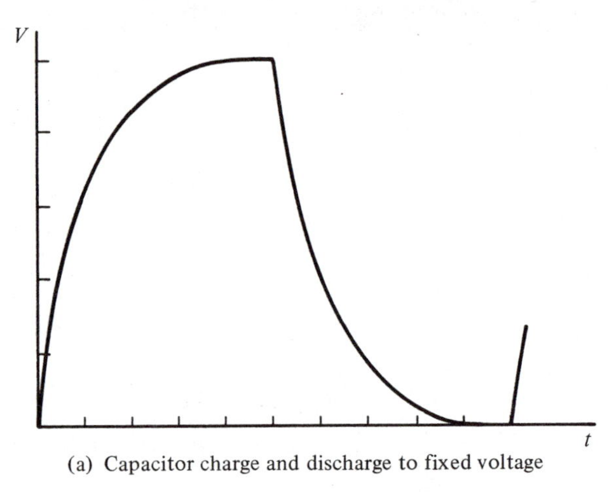

(a) Capacitor charge and discharge to fixed voltage

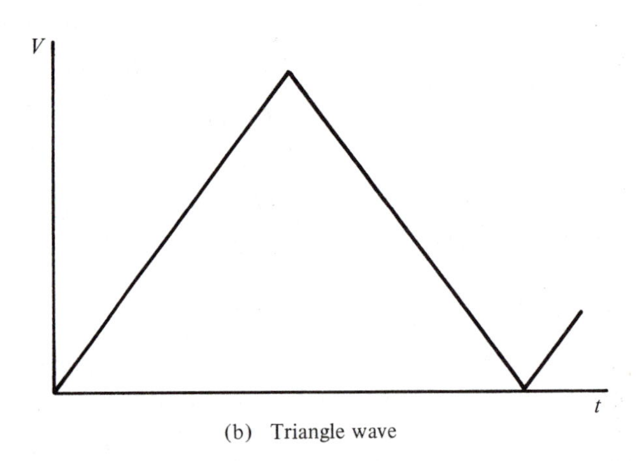

(b) Triangle wave

Figure 6-21 Capacitor charge waveforms

There are two relationships involving capacitors you should recall. First, the charge stored on a capacitor is determined by the current which has flowed into the capacitor and the length of time that current has flowed.

$$Q = It \tag{6-11}$$

Second, the voltage across a capacitor is directly proportional to the charge and inversely proportional to the value of the capacitance.

$$V = \frac{Q}{C} \tag{6-12}$$

Substituting equation (6-11) into equation (6-12), you obtain

$$V = \frac{I}{C}t \tag{6-13a}$$

You can see from equation (6-13a) that the voltage across the capacitor is directly proportional to the time that current has been flowing into the capacitor. By making the current a constant, the voltage will change *linearly* with time, exactly what you need to produce a signal like Fig. 6-21(b).

To generate a ramp, charge the capacitor with constant *current*, not constant voltage.

The circuit in Fig. 6-22 will drive a constant current into the capacitor. This causes the output voltage (which is also the voltage across the capacitor) to vary linearly with time. The current charging the capacitor is provided by the source E, through R. Since the inverting input is at virtual ground,

$$I = \frac{E}{R}$$

Ideally, none of I will flow into the op amp. All will go to charge the capacitor.

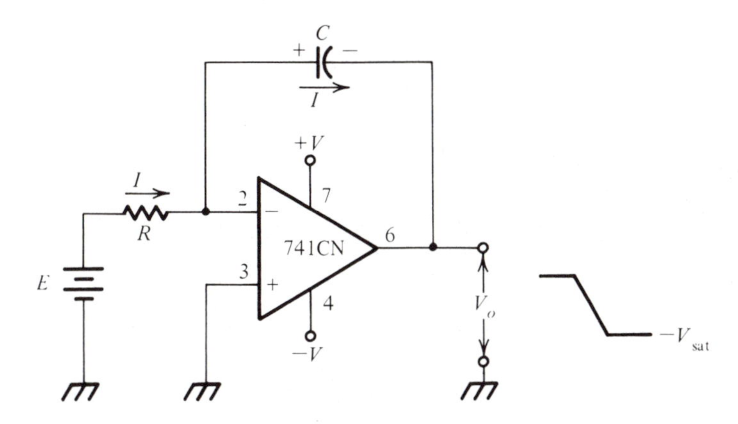

Figure 6-22 Charging a capacitor with constant current

When E is applied, the voltage across the capacitor, and therefore the output voltage, will ramp linearly downward. This will continue at a *constant rate* until the output reaches $-V_{SAT}$.

The *rate* at which this ramp drops is the important parameter.

$$V = \frac{I}{C} t$$

$$\text{Rate} = \frac{V}{t}$$

$$\text{Rate} = \frac{I}{C} \tag{6-13b}$$

$$I = \frac{E}{R}$$

$$\text{Rate} = \frac{E}{RC} \tag{6-14}$$

The rate or slope of the ramp is directly proportional to the charging voltage and inversely proportional to the time constant, RC.

The RC network, as modified in Fig. 6-22, allows you to produce a linear ramp. To build a ramp generator, you only need to add a two-level, *noninverting* comparator. The circuit in Fig. 6-1 is a two-level inverting comparator. To convert it to a noninverting comparator you ground the inverting input, and apply the signal to R2. This is illustrated in Fig. 6-23.

A positive potential at e_i will drive the noninverting input high, sending the output to $+V_{SAT}$. The output will stay at $+V_{SAT}$ even when e_i falls and then goes

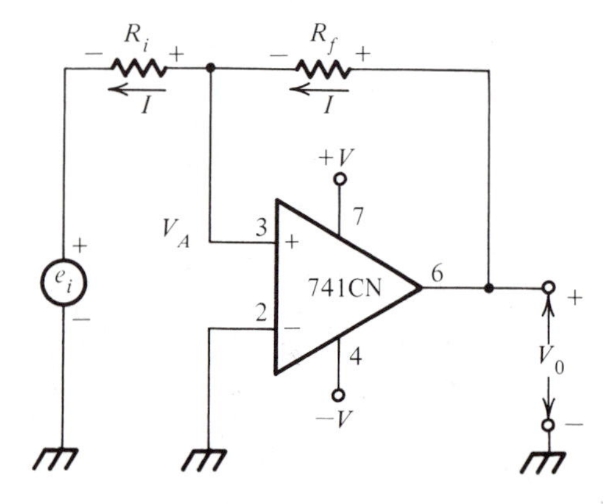

Figure 6-23 Two level noninverting comparator

negative. This is because R_f and R_i form a voltage divider between e_i and $+V_{SAT}$. The large, positive $+V_{SAT}$ keeps the noninverting input positive even for small *negative* values of e_i. The output will not switch to $-V_{SAT}$ until e_i has gone negative enough to force the noninverting input to just below ground. The input voltage necessary to do that is the lower trigger point. You can derive that voltage as follows:

The current in the outer loop, I, is

$$I = \frac{V_{SAT} - e_i}{R_i + R_f}$$

$$V_A = e_i + IR_i$$

$$V_A = e_i + R_i\left(\frac{V_{SAT} - e_i}{R_i + R_f}\right)$$

$$V_A = \frac{e_i R_i + e_i R_f + V_{SAT} R_i - e_i R_i}{R_i + R_f}$$

$$V_A(R_i + R_f) = e_i R_f + V_{SAT} R_i$$

The output will switch when V_A goes to ground. Here e_i is the lower trigger point.

$$0 = V_{LT} R_f + V_{SAT} R_i$$

$$V_{LT} = \frac{-R_i}{R_f} V_{SAT} \tag{6-15}$$

When the input falls below this lower trigger, the output goes to $-V_{SAT}$. This large negative on the end of the R_f holds the noninverting input of the op amp negative, keeping the output at $-V_{SAT}$. Even when e_i goes a little positive, $-V_{SAT}$ through the voltage divider R_f and R_i keeps the noninverting input of the op amp negative. This, in turn, forces the output to stay at $-V_{SAT}$. When e_i goes positive enough to drive the noninverting op amp input above ground, the output switches to $+V_{SAT}$. That is the upper trigger point and can be derived just as the lower trigger point was.

$$V_{UT} = \frac{R_i}{R_f} V_{SAT} \tag{6-16}$$

The input/output relationship of the noninverting two-level op amp comparator is illustrated in Fig. 6-24.

The linear charge circuit of Fig. 6-22 can be combined with the two-level

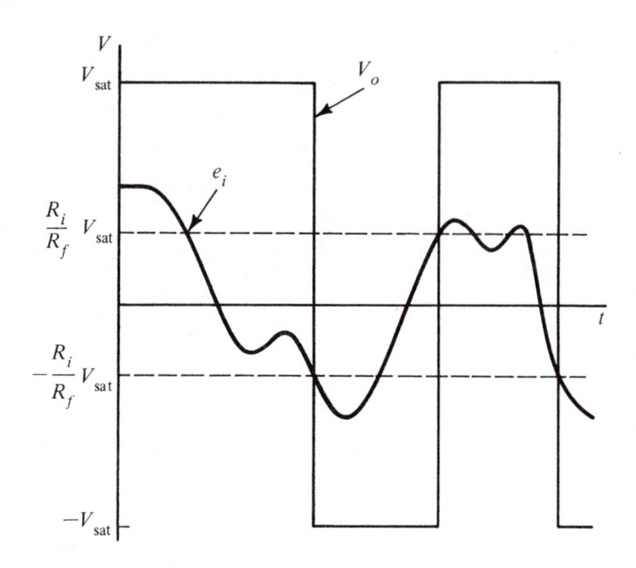

Figure 6-24 Input-output relationship of two level noninverting comparator

noninverting comparator of Fig. 6-23 to produce a triangle wave generator. Its schematic is shown in Fig. 6-25. The output of the linear charge circuit (U1) is a ramp. This ramp is the input to the comparator (U2). The square wave output of the comparator provides the charge voltage for the linear charge circuit.

Assume that the output of U2 is initially low. Current from the output of U1 flows through its feedback capacitor and through R. The capacitor charges up at

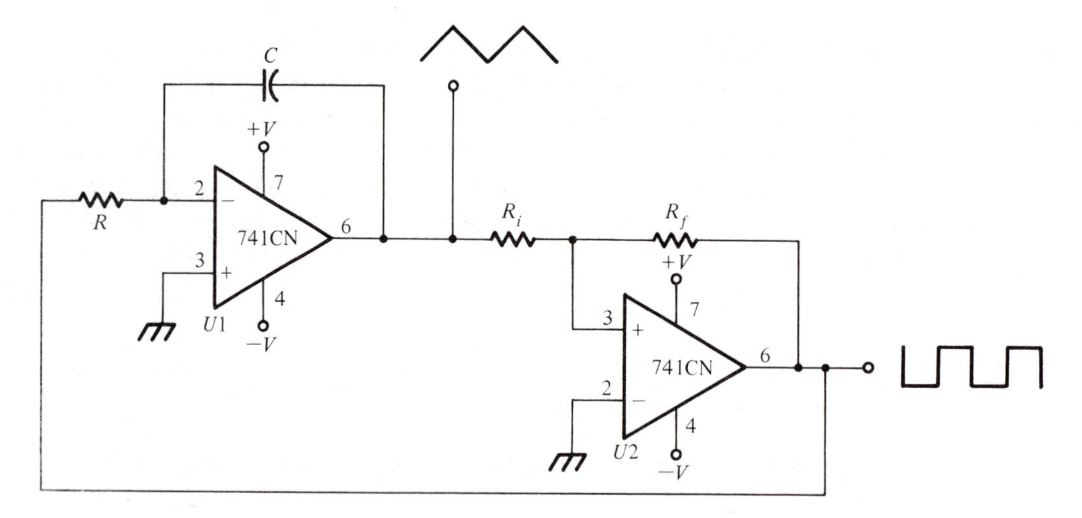

Figure 6-25 Op amp triangle wave generator

a linear rate, as set by equation (6-14). The output voltage of U1 ramps up. When it reaches the upper trigger voltage of the comparator [equation (6-16)], the comparator's output switches to $+V_{SAT}$.

With $+V_{SAT}$ as the charge voltage for the linear charge circuit, the current through the capacitor reverses. It now flows from the output of U2, through R, through C, and into U1. The capacitor charges down at the same rate as it charged up. The charge on the capacitor, and the output of U1 go to zero and then negative. When the output of U1 goes negative enough to cross the lower trigger point of the comparator, the comparator's output switches to $-V_{SAT}$. This reverses the charge direction of U1, and the cycle repeats. The result is a square wave out of the comparator U2, and a triangle at the output of U1.

The rate of rise and fall of the triangle wave, and the trigger points of the comparator set the amplitude and frequency. The rate of change of the capacitor voltage (triangle wave) is given by equation (6-17), where V_{SAT} is the charging voltage.

$$\text{Rate} = \frac{V_{SAT}}{RC} \tag{6-17}$$

The triangle charges between V_{LT} and V_{UT}, so the distance the triangle waves have to travel is

$$\text{Distance} = V_{UT} - V_{LT} = 2\frac{R_i}{R_f} V_{SAT}$$

The time it takes to travel a given distance is

$$\text{Time} = \frac{\text{distance}}{\text{rate}}$$

So for the triangle wave,

$$t = \text{pulse width} = \frac{2\dfrac{R_i}{R_f} V_{SAT}}{\dfrac{V_{SAT}}{RC}}$$

$$t = \frac{(2R_i V_{SAT})RC}{R_f V_{SAT}}$$

$$t = \frac{2R_i}{R_f} RC$$

But this is only the time for one half-cycle. The full period is

$$T = 2t$$

and the frequency is

$$f = \frac{1}{2t}$$

$$f = \frac{R_f}{4R_i RC} \tag{6-18}$$

Varying *any* of the components will alter the frequency.

The amplitude of the square wave is determined by the saturation voltage of U2. You can limit the amplitude with zener diodes, just as you did with the square wave generator described in section 6-1.1.

The peak-to-peak amplitude of the triangle wave is determined by the trigger levels of the comparator. The triangle wave must go from the lower trigger point to the upper trigger point (or vice versa) to cause the comparator to switch.

$$V_{\text{peak-peak}} = V_{UT} - V_{LT}$$

$$V_{\text{peak-peak}} = 2\frac{R_i}{R_f} V_{SAT} \tag{6-19}$$

The amplitude of the square wave and triangle wave are directly proportional to the output level (either V_{SAT} or V_{zener}) of U2. In addition, R_i and R_f, which set the comparator trigger points, help set the triangle amplitude, but (along with R and C) also alter the frequency.

Example 6-4

For the circuit in Fig. 6-25 let:

$$R = 10\ \text{k}\Omega$$

$$C = 0.1\ \mu\text{F}$$

$$R_i = 100\ \text{k}\Omega$$

$$R_f = 820\ \text{k}\Omega$$

$$\pm V_{\text{supply}} = 12\ \text{V}$$

a. Calculate the amplitude of the triangle wave and the square wave.

b. Calculate the signal's frequency.

Solution

a. Square wave amplitude

$$V_{p\text{-}p} = +V_{SAT} - (-V_{SAT}) \cong 10\text{ V} - (-10\text{ V})$$

$$V_{p\text{-}p} \cong 20\text{ V}_{p\text{-}p}$$

Triangle wave amplitude

$$V_{p\text{-}p} = 2\frac{R_i}{R_f}V_{SAT}$$

$$V_{p\text{-}p} = \frac{2(100\text{ k}\Omega)(10\text{ V})}{820\text{ k}\Omega}$$

$$V_{p\text{-}p} = 2.44\text{ V p-p}$$

b.

$$f = \frac{R_f}{4R_i RC}$$

$$f = \frac{820\text{ k}\Omega}{4(100\text{ k}\Omega)(10\text{ k}\Omega)(0.1\ \mu\text{F})}$$

$$f = 2.05\text{ kHz}$$

Example 6-5

For the circuit in Example 6-4,

a. What can you do to double the amplitude of the triangle wave?

b. What effect will this change have on the frequency?

c. What must be done to compensate for the amplitude variation to keep the frequency constant?

Solution

a.

$$V_{peak\text{-}peak} = \frac{2R_i}{R_f}V_{SAT}$$

To double the amplitude, you must alter the ratio of R_i/R_f. Lowering R_f (by half) is better than doubling R_i, since this would decrease the effects of offset current. But beware of making R_f so small that the op amp is forced into short circuit current limiting. Change $R_f = 410\text{ k}\Omega$.

b.
$$f = \frac{R_f}{4R_i RC}$$

Cutting R_f in half should also cut the frequency in half. This should seem intuitively correct, since doubling the amplitude means the capacitor must charge twice as far to reach a trip point. At a fixed rate this should take twice as long.

c. To compensate for this increased distance, you must double the charge rate. This is easily done by cutting either R or C in half [equation (6-17)]. Again do not lower R to the point of forcing the driving op amp into current limiting.

6-2.2 Norton Amp Triangle Wave Generator

You saw in the previous section that a triangle wave can be generated by a ramp circuit (capacitor in the negative feedback loop) and a comparator. This can be done with Norton amps as well as with op amps.

The circuit in Fig. 6-26 will charge and discharge the capacitor with constant current. This produces linear rising and falling output voltages. The current into the noninverting input terminal is determined by

$$I_+ = \frac{e_i}{R_i}$$

Because of the current mirror, an equal current will be pulled into the inverting input from the external circuit. This current must come through R_B and through C.

$$I_- = I_+ = I_{R_B} + I_C$$

or

$$I_C = I_+ - I_{R_B}$$

$$I_C = \frac{e_i}{R_i} - \frac{(+V)}{R_B}$$

So the current which charges the capacitor depends on R_B, R_i, the input voltage, and the supply voltage. If

$$e_i = +V$$

$$I_C = \frac{+V}{R_i} - \frac{(+V)}{R_B} = \frac{+V(R_B - R_i)}{R_B R_i}$$

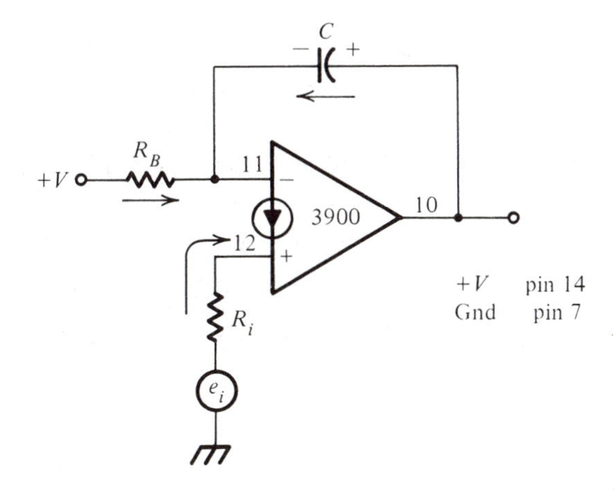

Figure 6-26 Charging a capacitor linearly with Norton amp

$$\text{Charge rate} = \frac{I}{C} = \frac{+V(R_B - R_i)}{CR_B R_i}$$

and the output voltage ramps.

The current, and therefore the charge rate, depend on the *difference* in R_B and R_i. For positive charge (as shown in Fig. 6-26), R_B must be larger than R_i.

If

$$e_i = 0$$

$$I_C = -\frac{(+V)}{R_B}$$

$$\text{Rate} = -\frac{(+V)}{R_B C}$$

With $e_i = 0$ there is no current into the noninverting input, so there can be no current into the inverting input. Current flowing through R_B, then, must flow left to right through the capacitor. Since this is opposed to the capacitor current shown in Fig. 6-26, I_C is negative. The output current ramps down. Since the Norton is powered from +V and ground, the output (and therefore the charge on the capacitor) will never go negative. So an electrolytic capacitor may be used, connected with the (+) to the output.

The special case where

$$e_i = +V \text{ or } 0$$

and

$$R_i = 1/2\ R_B$$

provides a symmetric charge and discharge rate. When

$$e_i = +V$$

$$\text{Charge rate} = \frac{+V(R_B - R_i)}{R_B R_i C} = \frac{+V(R_B - \frac{1}{2}R_B)}{R_B(\frac{1}{2}R_B)C}$$

$$\text{Charge rate} = \frac{+V}{R_B C} \tag{6-20}$$

When

$$e_i = 0$$

$$\text{Discharge rate} = -\frac{(+V)}{R_B C} \tag{6-21}$$

The two-level Norton amp comparator can be used to produce e_i (either +V or 0). The input to the comparator can come from the output of the linear charge circuit of Fig. 6-26. This composite schematic is shown in Fig. 6-27. Initially, with no voltage on the capacitor, the output of U1A is low, forcing the output of the comparator U1B to +V (actually +V − 0.8 V). This causes e_i to the capacitor charge circuit to +V. The capacitor charges up at the rate of

$$\frac{+V(R_{B1} - R_{i1})}{R_{B1} R_{i1} C}$$

or, if

$$R_{i1} = \tfrac{1}{2}\ R_{B1}$$

$$\text{Charge rate} = \frac{+V}{R_{B1} C}$$

The charge on the capacitor rises, as does the output voltage. When the output of U1A reaches

$$V_{UT} = \frac{R_{i2}(R_{B2} + R_f)}{R_{B2} R_f}(+V)$$

the comparator (U1B) switches, and outputs 0 V.

With e_i at 0 V, there is no current into the noninverting input of U1A. Current from R_{B1} now flows into C, reversing its charge.

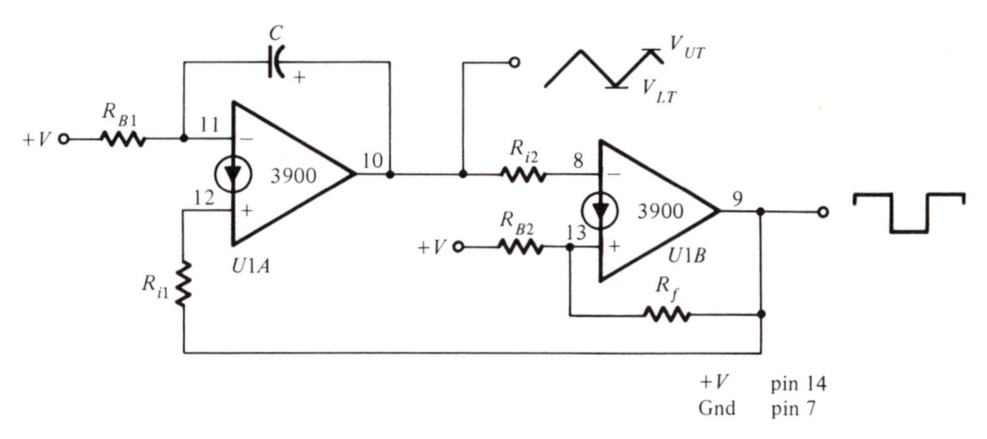

Figure 6-27 Norton amp triangle wave generator

$$\text{Discharge rate} = -\frac{(+V)}{R_{B1}C}$$

Charge circuit U1A will continue to drive current (left to right) into C, discharging it, as the output voltage falls. When the output of U1A falls below

$$V_{LT} = \frac{R_{i2}}{R_{B2}}(+V)$$

the comparator switches its output to +V. The capacitor begins to charge as the cycle repeats itself.

The output of the comparator is a square wave going from 0 V to +V (actually from 0.2V to +V − 0.8 V). The output of U1A is the triangle wave. It goes from V_{LT} to V_{UT}.

$$\text{Triangle amplitude} = V_{UT} - V_{LT}$$

$$= \frac{R_{i2}(R_{B2} + R_f)(+V)}{R_{B2}R_f} - \frac{R_{i2}(+V)}{R_{B2}}$$

$$= \frac{R_{i2}R_{B2}(+V) + R_{i2}R_f(+V) - R_{i2}R_f(+V)}{R_{B2}R_f}$$

$$\text{Triangle amplitude} = \frac{R_{i2}}{R_f}(+V) \tag{6-22}$$

The time it takes the triangle wave to go from V_{UT} to V_{LT} (or vice versa) is

$$\text{Time} = \frac{\text{distance}}{\text{rate}} = \frac{\dfrac{R_{i2}(+V)}{R_f}}{\dfrac{(+V)}{R_{B1}C}}$$

$$t = \frac{R_{i2}R_{B1}C}{R_f}$$

But this is just half the period, if a symmetric wave is assumed ($R_{i1} = \frac{1}{2}R_{B1}$)

$$T = \frac{2R_{i2}R_{B1}C}{R_f}$$

$$f = \frac{R_f}{2R_{i2}R_{B1}C} \tag{6-23}$$

Consequently, you can alter the amplitude by changing either R_{i2} or R_f [equation (6-22)]. Since this changes the distance the capacitor must charge, this will also change the frequency. The frequency can be adjusted without affecting the amplitude by changing either R_{B1} or C.

There are several precautions you must observe. First, the frequency calculation assumed a symmetric wave (charge time and rate equal to the discharge time and rate). To do this you must keep $R_{i1} = \frac{1}{2}R_{B1}$.

If you plan to vary the frequency with a potentiometer, you must use two (R_{B1} and R_{i1}) ganged so that they will track, always keeping $R_{i1} = \frac{1}{2}R_{B1}$. Upper limits on R_{i2} and R_{B2}, R_f, R_{i1}, and R_{B1} are established by the need of about 10 μA of input current in order to ensure good matching of input transistor characteristics. (See the section on the Norton amp square wave generator.) Resistors R_{i2}, R_f, and R_{i1} must be large enough (usually tens of kilohms) to not load down the Norton amps. This will limit high frequency operation.

The output of U1A can go no lower than 0.2 V and no higher than +V − 0.8 V. So you must set

$$V_{UT} < +V - 0.8\text{ V}$$

$$V_{LT} > 0.2\text{ V}$$

Otherwise, the capacitor will fully charge (or discharge) without causing the comparator to switch.

Example 6-6

Design a Norton triangle wave generator to meet the following:

$$f = 2\text{ kHz}$$

$$V_{\text{p-p}}\text{ triangle wave} = 2.5\text{ V}$$

$$+\text{V} = 9\text{ V}$$

(These are the same specifications you calculated for the op amp triangle wave generator in Example 6-4.)

Solution

To set the amplitude,

$$V_{\text{p-p}} = \frac{\text{R}_{i2}}{\text{R}_\text{f}}(+\text{V})$$

Pick

$$\text{R}_{i2} = 330\text{ k}\Omega$$

(This was chosen to give other resistors values which were close to standard value resistors and to keep the other resistors small enough to allow adequate bias currents.)

$$\text{R}_\text{f} = \frac{\text{R}_{i2}(+V)}{V_{\text{p-p}}} = \frac{330\text{ k}\Omega \times 9\text{ V}}{2.5\text{ V}}$$

$$\text{R}_\text{f} = 1.2\text{ M}\Omega$$

To set the frequency,

$$f = \frac{\text{R}_\text{f}}{2\text{R}_{i2}\,\text{R}_{\text{B}1}\,\text{C}}$$

or

$$\text{R}_{\text{B}1} = \frac{\text{R}_\text{f}}{2\text{R}_{i2}\,f\text{C}}$$

Pick C = 0.1 μF

$$\text{R}_{\text{B}1} = \frac{1.2\text{ M}\Omega}{2 \times 330\text{ k}\Omega \times 2\text{ kHz} \times 0.1\ \mu\text{F}}$$

$$\text{R}_{\text{B}1} = 9.1\text{ k}\Omega$$

$$\text{R}_{i1} = \tfrac{1}{2}\text{R}_{\text{B}1}$$

so select $R_{i1} = 4.2\ k\Omega$ with a small (500 Ω) series potentiometer to allow you to fine-tune the duty cycle.

The only component remaining to select is R_{B2}. It does not appear in the amplitude or frequency calculation. However, it helps set V_{LT} and V_{UT}.

$$V_{LT} = \frac{R_{i2}}{R_{B2}}(+V)$$

Since no V_{LT} was specified, you can pick any value V_{LT} and solve for R_{B2}. V_{LT} must be chosen, though, so that

$$(V_{LT} + V_{p\text{-}p}) = V_{UT} < (+V - 0.8\ V)$$

Pick $V_{LT} = 3$ V.

$$R_{B2} = \frac{R_{i2}(+V)}{V_{LT}} = \frac{330\ k\Omega \times 9\ V}{3\ V}$$

$$R_{B2} = 990\ k\Omega$$

Pick $R_{B2} = 1\ M\Omega$.

6-2.3 Sawtooth Generator

You may find it necessary to provide a relatively slow linear ramp with a rapid drop (or rise in the case of a negative ramp) at its end. This is a sawtooth wave. Also, in applications such as time base generators and power control circuits, the sawtooth must be triggered by (or be synchronized with) some control signal. The circuits of this section will allow you to meet these requirements.

The circuit in Fig. 6-28 gives you the ability to control ramp generation with an external signal. An NPN bipolar junction transistor (BJT) has been placed around the charging capacitor. Notice that the emitter of the transistor is tied to the inverting input of the op amp, which is at virtual ground. Resistor R_B limits the base current to protect the transistor. However, R_o should be kept relatively small to assure that the transistor can be driven into saturation.

With a zero or negative control input voltage, Q1 is off. The capacitor charges up (from the op amp's output, through the capacitor, through R_i and to $-V$). The charge rate is given by equation (6-14).

$$\text{Rate} = \frac{-V}{R_i \times C}$$

Figure 6-28 BJT controlled ramp generator

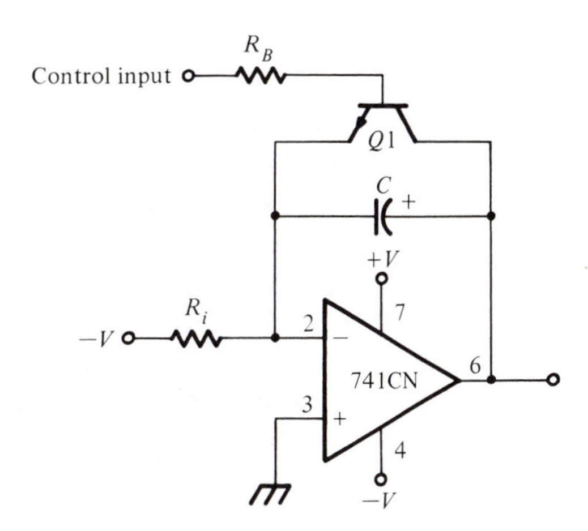

If you do not change the control voltage, the capacitor will eventually charge up, and hold the output at $+V_{SAT}$.

However, when you apply a positive control input voltage, Q1 is turned *on*. If this voltage is large enough to force Q1 into saturation, you have effectively placed a short circuit around the capacitor. The capacitor rapidly discharges. The output voltage falls to zero (actually ~0.2 V) and stays there as long as your positive control voltage keeps Q1 saturated. An example of the waveforms you can expect is given in Fig. 6-29.

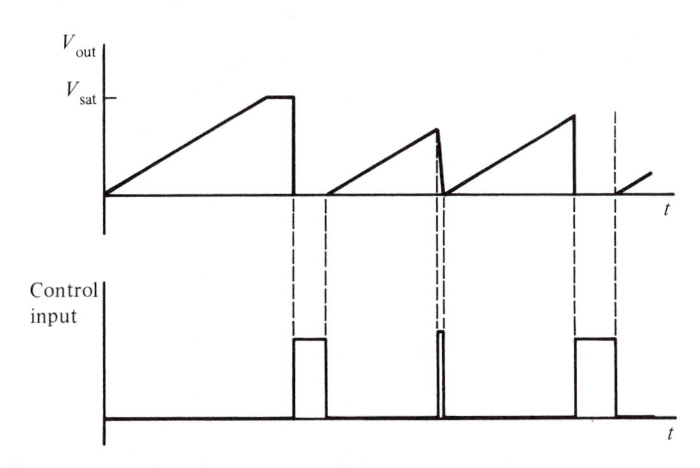

Figure 6-29 Control input signal's effect on the ramp generator's output

If you want to produce and control negative going ramps, several minor changes must be made to the circuit in Fig. 6-28. First, the charging voltage attached to R_i must be changed to $+V$. This reverses the direction of the charging current. Reversing the charge polarity also requires you to reverse the capacitor if it is electrolytic. The emitter of the transistor must be connected to virtual ground (the op amp's inverting input). To allow the capacitor to discharge left to right, you must replace the NPN transistor with a PNP transistor. Finally, a zero or positive base voltage (control input) keeps a PNP transistor off, while a negative control input is necessary to turn the transistor on. This is opposite to Figs. 6-28 and 6-29. Consequently, you must invert the control signal, making sure that it goes negative to turn the transistor on.

A programmable unijunction transistor (PUT) can be used to replace the BJT across the capacitor in Fig. 6-28. The PUT is a thyristor with an anode, a cathode, and an anode gate. Normally, the PUT is off, presenting a large resistance between anode and cathode. However, when the voltage on the anode (with respect to a grounded cathode) exceeds the voltage on the gate by approximately 0.7 volt, the PUT goes on. Like other thyristors, once fired, the anode-to-cathode resistance falls to about 10 Ω. The gate loses control. The PUT will remain on until anode current falls below the holding current. The PUT then goes off. The anode voltage must be taken above the gate voltage to fire the PUT again.

A PUT controlled sawtooth generator is shown in Fig. 6-30. When power is first applied, the PUT is off. The capacitor begins to charge up and the output rises. The rate is determined by equation (6-14). This continues until the output voltage (which is also the anode voltage of the PUT) is ~0.7 volt above the control input (the gate voltage). The PUT goes on. The capacitor is shorted out. It immediately discharges through the PUT. The output voltage (voltage across the capacitor) falls. When the current through the PUT falls below its holding current, it goes off and the cycle repeats. When the PUT turns off, approximately 1 V is usually left on the capacitor. The output waveform is shown in Fig. 6-31.

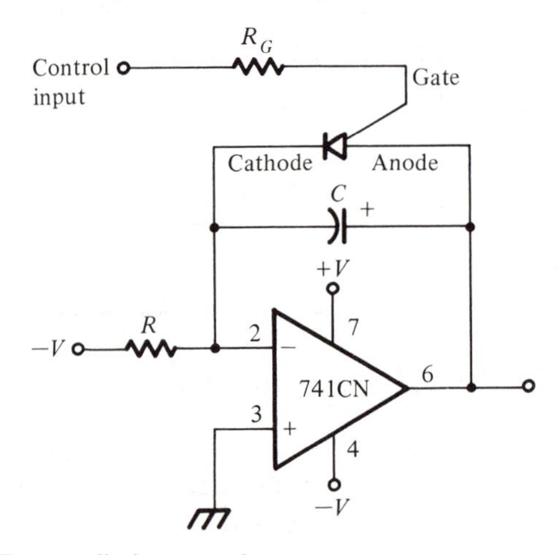

Figure 6-30 PUT controlled sawtooth generator

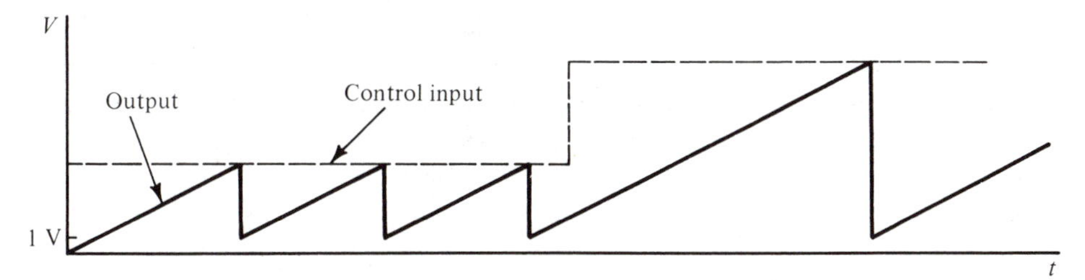

Figure 6-31 Output from a PUT controlled sawtooth generator

The period of the PUT sawtooth generator depends on the charge rate (V/RC) and on the control voltage. This is evident in Fig. 6-31.

$$\text{Time} = \frac{\text{distance}}{\text{rate}}$$

$$t = \frac{(V_{\text{control}} + 0.7\text{ V}) - 1\text{ V}}{\dfrac{V}{RC}}$$

$$t \cong \frac{V_{\text{control}}RC}{V} \tag{6-24}$$

$$f \cong \frac{V}{V_{\text{control}}RC}$$

The PUT controlled sawtooth generator can be used as a voltage-to-frequency converter.

You must be aware of two precautions. First, the cathode of the PUT must be tied to ground (or virtual ground), and current flows only from anode to cathode. Consequently, you cannot use a PUT to control a negative ramp generator. The polarities are all wrong. Second, to turn off, the current through the PUT must fall below its holding current, I_H (specified by the manufacturer). When the PUT is on, not only does the capacitor discharge current flow through the PUT, but also a current equal to that used to charge the capacitor flows through the PUT. This current flows through R to $-V$.

$$I = \frac{-V}{R} < I_H$$

must be below I_H. Otherwise, once the PUT goes on, it will be held on by this charge current, even when the capacitor has fully discharged. You can lower this charge current by increasing R or decreasing the negative voltage. However, both of these affect the charge rate and therefore the frequency. [See equation (6-24).] You will have to balance changes in either R or $-V$ with appropriate changes in C if the frequency is not to be affected.

6-3 SINE WAVE GENERATORS

The circuits you have seen so far in this chapter are often used in timing and control applications. However, the test and evaluation of linear circuits require a sine wave input.

Production of this low distortion, audio frequency sine wave is not quite as simple as square wave and triangle wave generation. In this section you will learn what is required to generate a low distortion sine wave and will see several ways to meet these requirements with both op amps and Norton amps.

6-3.1 Requirements for Sinusoidal Oscillations

There are four elements necessary for a circuit to produce a low distortion sine wave. They are a frequency determining device, an amplifier, positive (regenerative) feedback, and adaptive negative feedback. Any circuit you build with all four of these elements may generate a sine wave (even if the circuit was not supposed to oscillate at all). Conversely, remove any one of the elements, and you lose the sine wave.

The frequency determining device must select one frequency. This becomes the operating frequency of the oscillator. All other frequencies (those not selected) rapidly die out. For audio frequency oscillators, RC networks are normally used. Often they take the form of either passive bandpass or passive notch filters. Two of these are illustrated in Fig. 6-32. For the component relationships shown, the center frequency (f_0) of both the Wien-bridge bandpass and the twin T notch is

$$f_0 = \frac{1}{2\pi RC} \tag{6-25}$$

To recover the energy lost in the frequency determining device, an amplifier is necessary. Either an op amp or a Norton amp will work. However, if you choose a Norton amp, you must remember two things. First, a Norton amp is current driven, so series resistors must be included at the inputs to convert the voltage signals to the current signals needed. Second, since the output of the Norton amp is biased at $\frac{1}{2}(+V)$, the sine wave must be RC coupled when used as either an output or as a feedback signal.

To produce sustained oscillations, the output of the amplifier must be fed

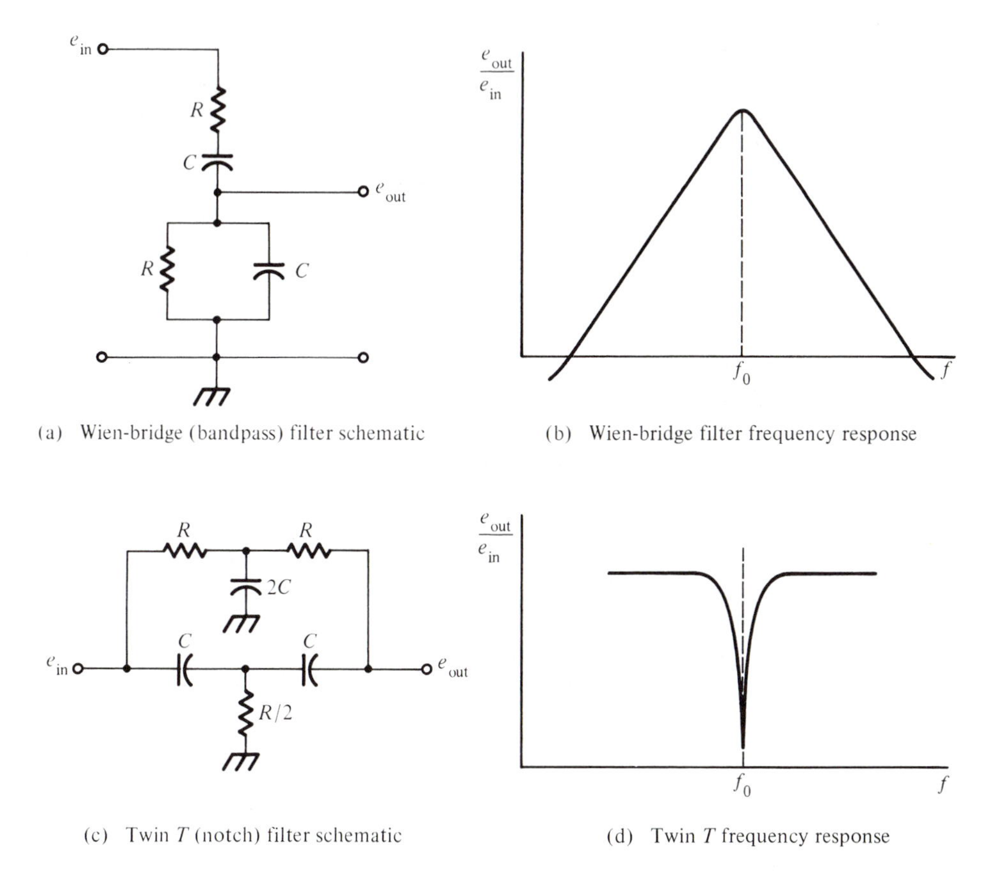

(a) Wien-bridge (bandpass) filter schematic

(b) Wien-bridge filter frequency response

(c) Twin T (notch) filter schematic

(d) Twin T frequency response

Figure 6-32 RC frequency determining devices

back to its input, through the frequency determining device. In addition, the signal fed back must end up in phase with the amplifier's input. Only in this way does the feedback add to, and therefore sustains, the amplifier's input signal. Any phase difference at all between the feedback signal and the amplifier's input will cause the output to decrease. The oscillations would die. This feedback of a portion of the output so that it is in phase with the amplifier's input is positive feedback. It is the third requirement for oscillations.

Figure 6-33 illustrates two possible ways of getting positive feedback. The noninverting amplifier of Fig. 6-33(a) requires a frequency determining device which does not shift the phase of the selected frequency signal. The inverting amplifier of Fig. 6-33(b) must have a frequency determining device which inverts (180 degree shift) the signal.

The last requirement for a low distortion sine wave oscillator is *adaptive* negative feedback. The amplifier (Fig. 6-33) increases its input by a factor of A. The frequency determining device reduces the output signal by β as it is fed back. If the attenuation (β) just balances the amplification (A), you will get a stable sine wave. However, if the frequency determining device attenuates the signal more than the amplifier increases it, the output gets smaller each cycle and eventually dies. On the other hand, if the amplifier increases the signal more than the frequency determining device attenuates it, the signal will get larger and larger.

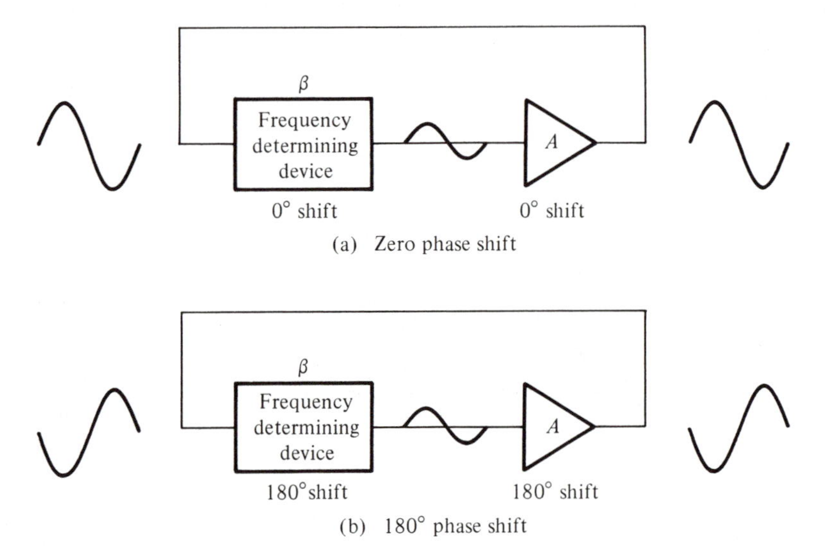

Figure 6-33 Positive feedback

As the amplitude approaches saturation, the signal distorts and eventually becomes a square wave. These facts are summarized in equations (6-26a, b, and c).

Amplitude decreases	$\beta A < 1$	(6-26a)
Stable amplitude	$\beta A = 1$	(6-26b)
Amplitude increase	$\beta A > 1$	(6-26c)

To produce a stable, low-distortion sine wave, the loop gain (βA) must be *exactly* 1; less and the oscillations die; more and distortion occurs. This is referred to as the Barkhausen Criterion.

You have seen in previous chapters that precise stable gain (A) can be obtained by using an op amp (or Norton amp) with high open loop gain (A_{OL}) and applying negative feedback. Properly done, the resulting gain is stable and controlled by the negative feedback. You then can use an op amp or Norton amp with negative feedback to produce a precise stable gain A coupled with a stable frequency determining device (β); equation (6-26b) is satisfied.

However, when power is first applied, the input (and therefore the output) is small. If you had carefully set $\beta A = 1$, this signal would probably never become large enough to be usable. Or, if a load reduces the output of your oscillator (where you had set $\beta A = 1$), the input falls, causing the output to fall more. Both of these problems can be solved with *adaptive* negative feedback. When the output is low, negative feedback becomes small. The gain A is large. Loop gain $\beta A > 1$ and the output signal amplitude increases. As the amplitude increases, the adapative negative feedback lowers the amplifier's gain (A). At a preset level, $\beta A = 1$ and stable oscillations occur. Should the output try to increase, the amplifier's gain is reduced, forcing $\beta A < 1$, and the output falls. For stable, low distortion sine wave generation, *adaptive* negative feedback (i.e., automatic level control) is necessary.

6-3.2 Wien-bridge Oscillator

One of the more popular, simple audio frequency oscillators is based around the bandpass filter. Its schematic is shown in Fig. 6-32(a). Analysis of this network is easier if you split the circuit into two parts. This is shown in Fig. 6-34(a) and (b).

The impedance of the series element is called Z1. For R1 = R2 = R and C1 = C2 = C,

$$\text{Z1} = \text{R} - \frac{j}{\omega \text{C}} \tag{6-27}$$

The impedance of the parallel element is Z2.

$$\text{Z2} = \frac{1}{\dfrac{1}{\text{R}} + j\omega \text{C}}$$

$$\text{Z2} = \frac{\text{R}}{1 + j\omega \text{RC}} \tag{6-28}$$

The output from the bandpass filter comes from the voltage divider law.

$$v_o = \frac{\text{Z2}}{\text{Z1} + \text{Z2}} e_i$$

$$\frac{v_o}{e_i} = \frac{\text{Z2}}{\text{Z1} + \text{Z2}} \tag{6-29}$$

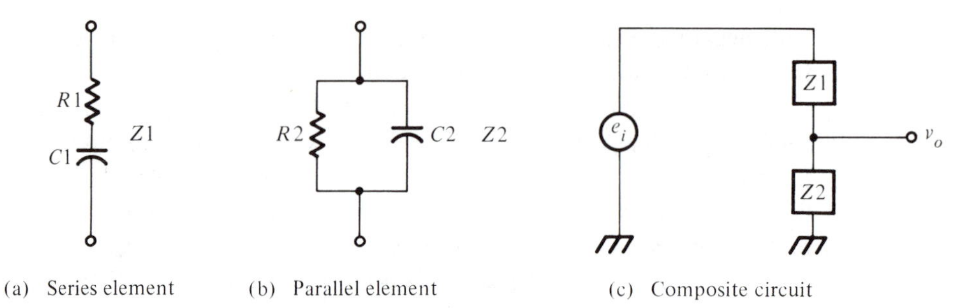

Figure 6-34 Wien-bridge analysis

Substituting equations (6-27) and (6-28) into equation (6-29),

$$\frac{v_o}{e_i} = \frac{\dfrac{R}{1 + j\omega RC}}{\left(R - \dfrac{j}{\omega C}\right) + \left(\dfrac{R}{1 + j\omega RC}\right)}$$

After a bit of algebraic manipulation, to gather all of the imaginary terms together,

$$\frac{v_o}{e_i} = \frac{R}{3R + j\left(\omega R^2 C - \dfrac{1}{\omega C}\right)} \tag{6-30}$$

The input to this network (e_i) is the output from the oscillator's amplifier. The output from the filter (v_o) forms the input to the amplifier. The factor β in the loop gain equation

$$\beta A = 1$$

is

$$\beta = \frac{v_o}{e_i}$$

For $\beta A = 1$, β must be real. To make this happen, the imaginary term in equation (6-30) must equal zero.

$$\omega R^2 C - \frac{1}{\omega C} = 0$$

or

$$\omega = \frac{1}{RC}$$

$$f_o = \frac{1}{2\pi RC} \tag{6-31a}$$

For unequal resistors (R1 $\neq$ R2) and unequal capacitors (C1 $\neq$ C2), equation (6-31a) becomes

$$f_o = \frac{1}{2\pi\sqrt{R1R2C1C2}} \tag{6-31b}$$

At this frequency, and this frequency only, there will be no phase shift across the bandpass filter. Only at this frequency is there positive (regenerative) feedback. Only at this frequency is β real, so that equation (6-26b) can be realized. At f_o, the imaginary part of equation (6-30) is zero. So at f_o, equation (6-30) becomes

$$\frac{v_o}{e_i} = \beta = \frac{R}{3R} = \frac{1}{3}$$

The Wien-bridge attenuates the operating frequency signal by 1/3. For sustained oscillations, the amplifier must have a gain of precisely 3. This is shown in Fig. 6-35. You can build the $A = 3$ amplifier with an op amp. This is shown in Fig. 6-36. However, should R_f be adjusted just a bit low, $A < 3$, oscillations will die out, or will fail to start when power is first applied. On the other hand, setting R_f a little high (so that $A > 3$) will ensure that oscillations will begin. But since the overall loop gain is greater than one, the amplifier increases the signal more than the network attenuates it. The signal amplitude will grow and distort as it approaches the saturation levels of the op amp.

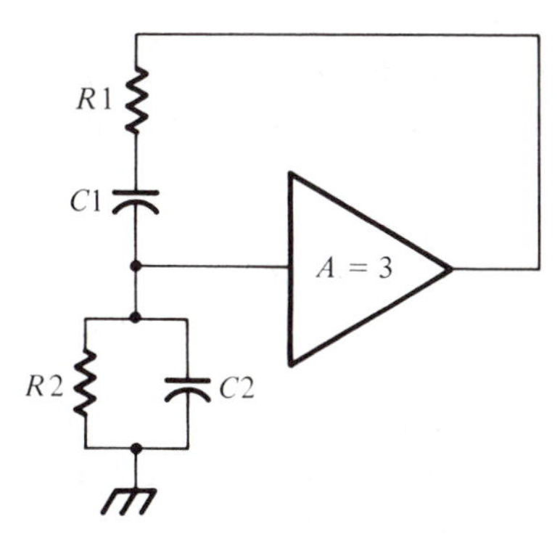

Figure 6-35 Theoretical Wien-bridge oscillator

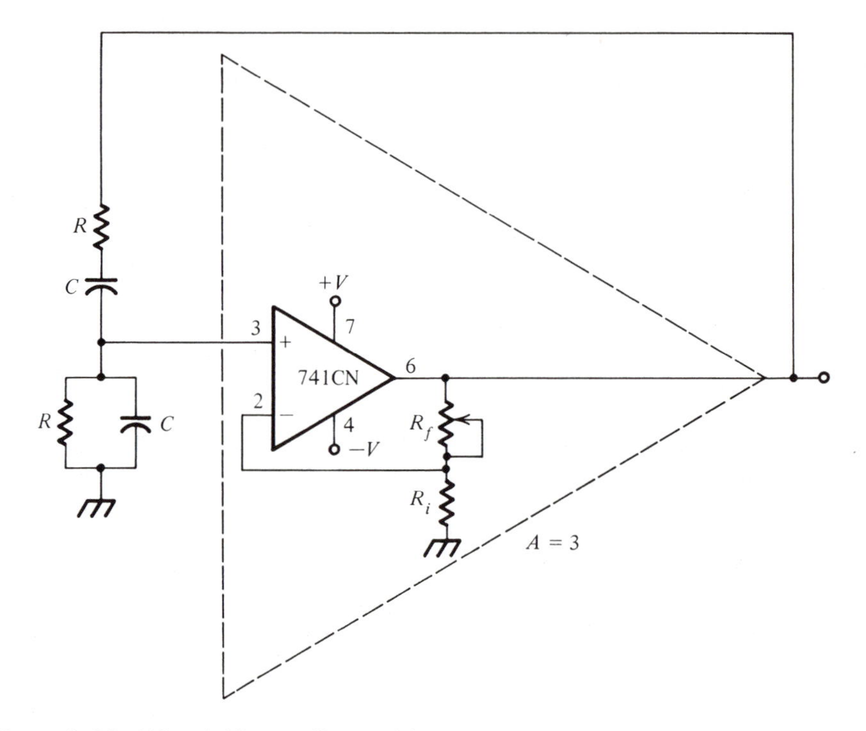

Figure 6-36 Wien-bridge oscillator with an op amp

Adaptive negative feedback is the solution. A Wien-bridge oscillator with adaptive negative feedback is shown in Fig. 6-37. The op amp is configured as a noninverting amplifier. Remember that the gain is set by $A = 1 + (R_f/R_i)$, where R3 and the part of the potentiometer above the wiper is R_f and the portion of the potentiometer below the wiper is R_i. Without the *signal* diodes around R3, you adjust R4 until the gain (A) of the amplifier is high enough to ensure that $\beta A > 1$. The circuit will oscillate at or near f_o [equation (6-31)]. However, obtaining a stable, undistorted sine wave would be difficult, since A must be adjusted to make βA *precisely* equal to 1.

The two signal diodes solve this problem automatically. Initially R4 is adjusted to give a gain such that $\beta A > 1$. The output signal increases in amplitude until the peak voltage across R3 approaches ± 0.5 volt. The diodes begin turning on (one on the positive half-cycle, the other on the negative). They lower the resistance in parallel with R3, effectively reducing R_f. This lowers the gain of the amplifier (provides more negative feedback), which in turn lowers the output amplitude. Should the signal amplitude fall, the diodes would begin to turn off, raising the value of R_f, increasing the gain (lowering negative feedback), and increasing the output signal.

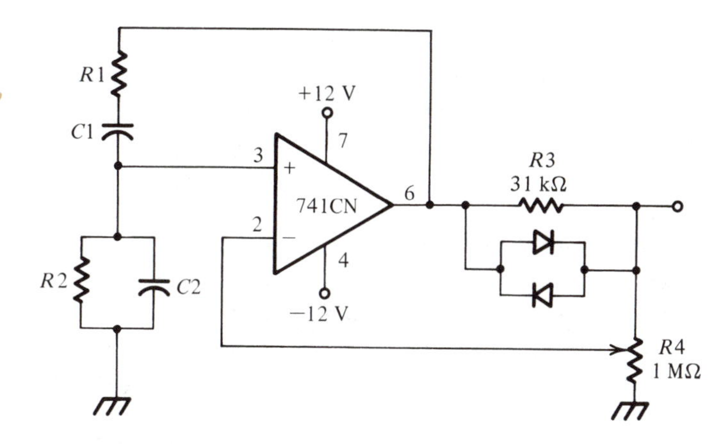

Figure 6-37 Practical Wien-bridge oscillator with adaptive negative feedback

The end result of these two signal diodes in the negative feedback path is a stable, low distortion sine wave. The circuit samples its own output and adjusts its gain accordingly. An additional benefit of this particular adaptive feedback technique is that you can adjust the amplitude of the output from about 1 V p-p to within a few volts of the supplies. The size of the output signal, when the diodes have ± 0.5 volt across them, depends on the voltage divider relationship of R3 and precisely where on R4 the wiper is. Careful adjustment of R4 not only ensures oscillations, but gives you an output amplitude control.

A second adaptive feedback technique for the Wien-bridge oscillator is shown in Fig. 6-38. The parallel signal diodes of Fig. 6-37 have been replaced with back-to-back zener diodes. When the signal is small, the zeners stay off. Potentiometer R4 is set to give an amplifier gain of $A > 3$, so that the signal amplitude grows. When the signal is large enough to *begin* turning the zener

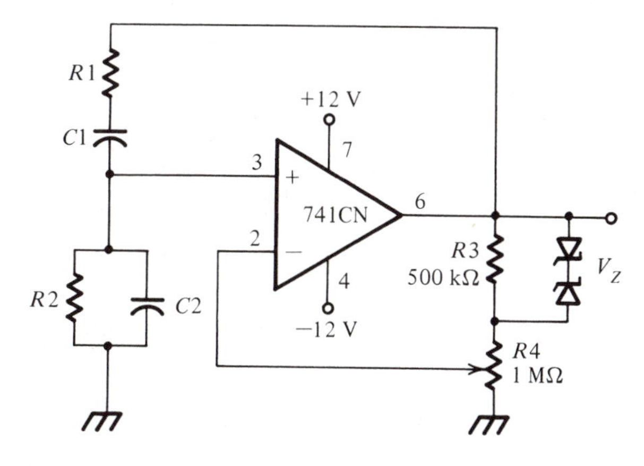

Figure 6-38 Wien-bridge oscillator with zener diode adaptive feedback

diodes on (one goes on; the other breaks down), their resistance falls, lowering the gain and stabilizing the signal amplitude. The minimum output amplitude you can get is set by the zener diode breakdown voltage.

$$V_{min}(\text{peak}) \simeq V_z \tag{6-32}$$

Adjusting R4 for a smaller output will kill oscillations. When you adjust R4 for a larger output, the signal amplitude will increase, but it also begins to distort. Minimum distortion occurs at *minimum* amplitude.

A better adaptive feedback technique is shown in Fig. 6-39. The lower resistor in the negative feedback path (R4) is paralleled by the channel resistance of the N-channel JFET, Q1. Zener diode CR2 senses the amplitude of the output. Diode CR1 is a half-wave rectifier, while C3 and R5 form a simple capacitive filter. When the output is small, CR2 is off. There is no charge on the capacitor, so Q1 has very little reverse bias (gate to channel). The channel resistance of Q1 is low. This lowers the effect of R4. Resistor R4 and Q1 develop only a small negative feedback. This allows the output amplitude to grow. When the negative peak output signal is large enough to break down CR2 and turn on CR1, capacitor C3 charges. This negative input increases the channel resistance of Q1. The increased resistance develops more negative feedback voltage, and the amplifier's gain is automatically lowered. If you properly select fixed values of R3 and R4 to

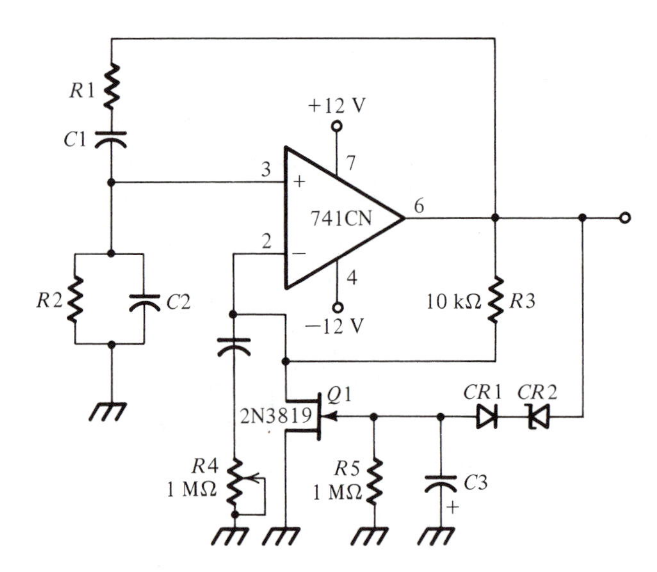

Figure 6-39 Wien-bridge oscillator with JFET controlled adaptive feedback (*Courtesy of National Semiconductor*)

match Q1, the output will be a low distortion sine wave whose amplitude is determined by the breakdown voltage of the zener diode.

You can also build a Wien-bridge oscillator with a Norton amp. When using the Norton amp, you must remember two things. First, the Norton must be driven with *current*, not voltage, so convert all input voltages to currents with series resistance. Second, since the Norton amp operates from a single supply, you must bias the output to half of the supply voltage.

To produce a Norton amp Wien-bridge oscillator, you must take several steps. First is proper bias. This is shown in Fig. 6-40(a). Adjustment of R_B alters the bias of the amplifier's output. The second requirement for oscillator is a frequency determining device. This is provided by the bandpass filter, as shown in Fig. 6-40(b). The feedback voltage (v_{fb}) is taken across R2 and C2. To produce positive feedback (the third oscillation requirement), you must convert v_{fb} to a current, and feed it into the Norton amp's noninverting input. This is shown in Fig. 6-40(c). You must be careful to keep the feedback resistor large compared to the parallel impedance of R2 and C2. Otherwise, the filter will be loaded down, producing a shift in operating frequency.

$$R3 \gg R2 \; // \; Z_{C2}$$

The final requirement for low distortion oscillation is adaptive negative feedback. This is shown in Fig. 6-40(d). Resistors R4 and R5 are capacitively coupled from the output to the inverting input to provide negative feedback current. The diodes are off for small output signals, providing low negative feedback and therefore a large gain. At larger output signal voltages, the diodes are turned on (one on the positive half-cycle, the other on the negative). They short out a part of the negative feedback resistance; increasing the negative feedback current lowers the gain and, therefore, the output signal amplitude. The capacitor C3 blocks the bias voltage from its branch. This prevents R4 and R5 from altering the bias. It also prevents the *bias* voltage from turning the diodes on or off. Since C3 forms a RC coupler, you should select

$$f_c = \frac{1}{2\pi R5C3} \ll f_o \qquad f_o = \text{oscillation frequency}$$

Usually, selecting

$$C1 = C2 = C3$$

will meet this requirement. You should select the series combination of R4 and R5 to be small as compared to the 330 kΩ bias feedback resistor. This removes any effect it may have on the AC operation.

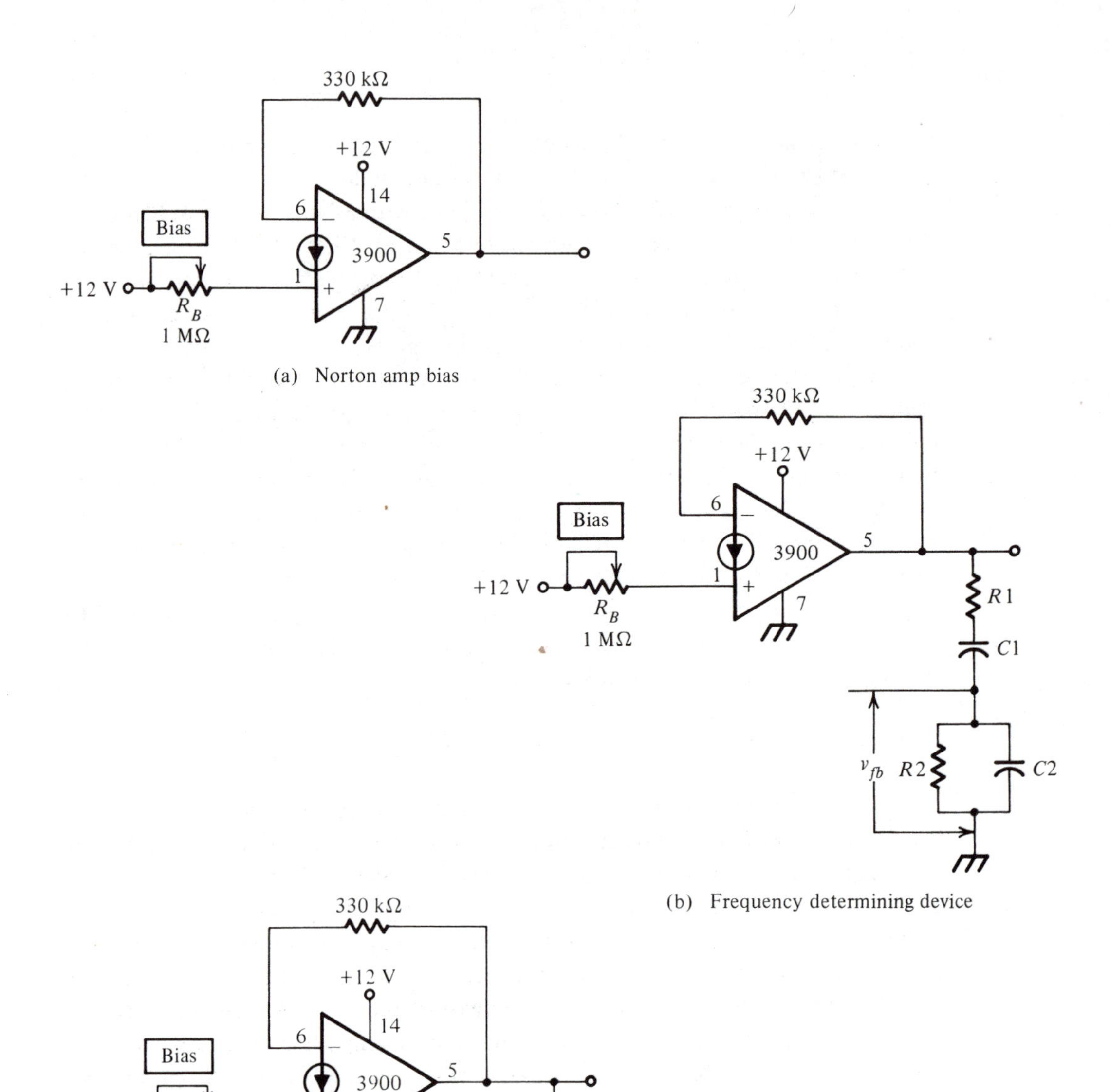

(a) Norton amp bias

(b) Frequency determining device

(c) Positive feedback

Figure 6-40 Norton amp Wien-bridge oscillator

Figure 6-40 *continued*

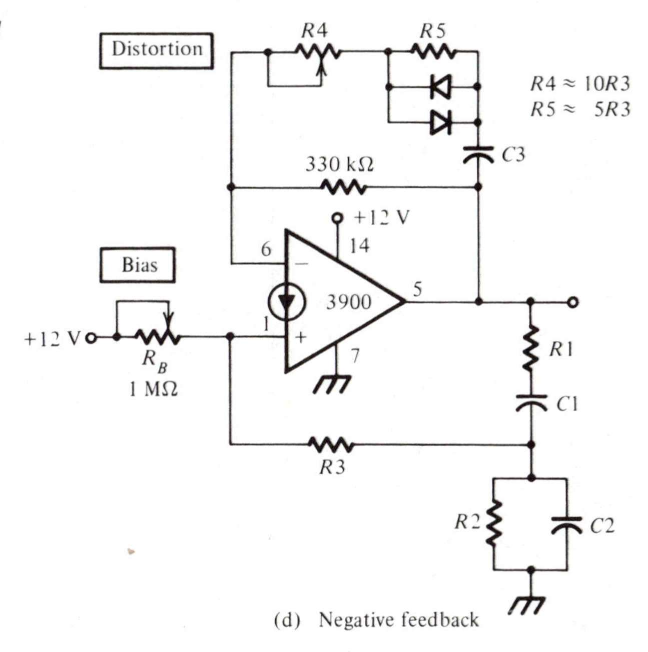

(d) Negative feedback

In summary, for the Wien-bridge oscillator, even though many configurations are possible,

> **The frequency determining device is a bandpass filter in the positive feedback path of an amplifier. Low distortion sine waves are produced by setting the loop gain to 1 with adaptive negative feedback.**

6-3.3 Twin T Oscillator

The Wien-bridge oscillator uses a *bandpass* filter in the *positive* feedback path. You can also make an audio frequency oscillator by placing a *notch* filter in the *negative* feedback path. The frequency response of a notch filter is shown in Fig. 6-41. Signals of all frequencies are passed without attenuation, except for a narrow band around the operating frequency, f_o. Signals with frequencies near f_o will be severely attenuated.

There are several passive (RC) networks which form notch filters. One of these is the twin T filter shown in Fig. 6-42. If you select the resistors and capacitors with the relationship specified in Fig. 6-42, then

$$f_o = \frac{1}{2\pi RC} \tag{6-33}$$

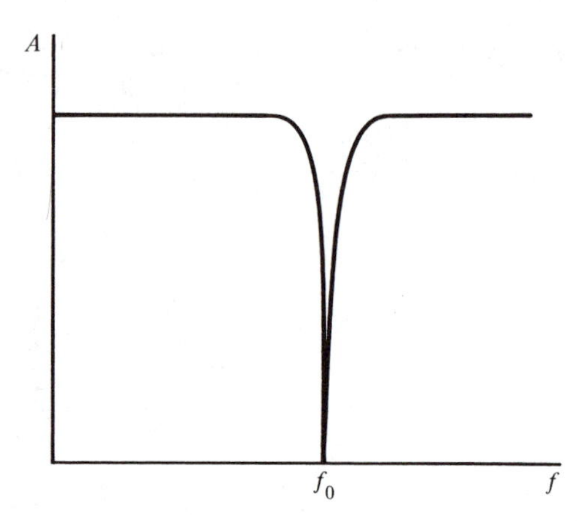

Figure 6-41 Frequency response of a notch filter

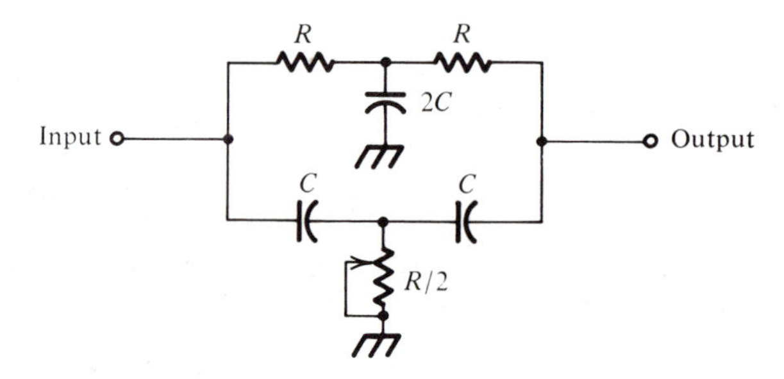

Figure 6-42 Twin T notch filter

The derivation of equation (6-33) follows the same basic procedures as was used in the derivation for the Wien-bridge [equation (6-31a)]. However, Laplace transforms must be used extensively, and the algebraic manipulation is rather lengthy, so the derivation will not be included here. Varying the R/2 potentiometer allows you to easily alter f_o.

An oscillator using the twin T filter is shown in Fig. 6-43. The op amp provides the amplification. The voltage divider, R1 and R2, provides positive feedback for signals of *all* frequencies. The notch filter in the *negative* feedback path passes signals of all frequencies, *except* the desired oscillation frequency, so all other frequencies receive large negative feedback and consequently very low gain. Only the desired oscillation frequency signal receives positive feedback with little or no counterbalancing negative feedback. For that frequency alone does the amplifier produce a significant gain. All other frequency signals die out. Only the signal at the notch frequency receives positive feedback without negative feedback, resulting in high gain and therefore sustained oscillations.

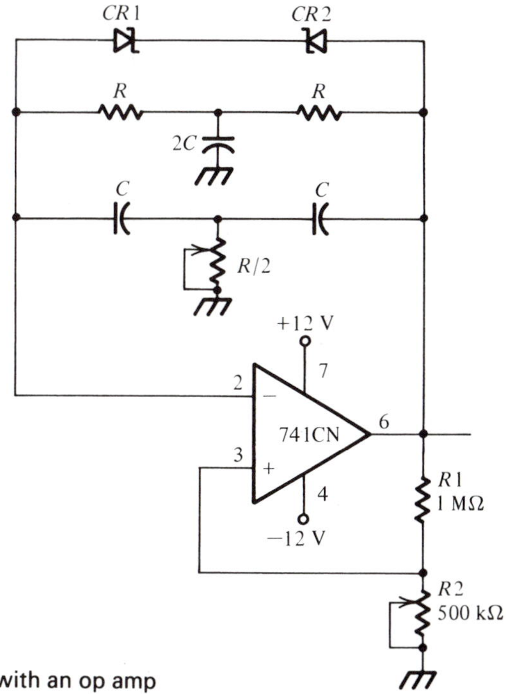

Figure 6-43 Twin T oscillator built with an op amp

The zener diodes CR1 and CR2 provide the adaptive negative feedback. At small output amplitudes, the zeners are off and very little negative feedback is provided at the oscillation frequency. This results in a large amplifier gain. As the output peak signal increases to a point that the zeners begin to turn on, their resistance drops, producing more negative feedback (even at the oscillation frequency) and lowering the amplifier's gain. As with the Wien-bridge oscillator, the minimum sustainable output voltage is approximately

$$V_{\text{min}}(\text{peak}) \approx V_z$$

Increasing R2 will increase the amount of positive feedback. The output will increase, increasing the size of the negative feedback as compensation. Although you can control the amplitude this way, increases in amplitude also yield an increase in the amount of distortion. As with the Wien-bridge oscillator, minimum distortion appears at minimum output amplitude.

You can also build a twin-T oscillator with a Norton amp. As with the Wien-bridge oscillator, you must be sure to provide bias and convert feedback voltages to currents. A schematic for the Norton amp twin-T oscillator is given in Fig. 6-44. The component values in the twin-T filter are selected to obtain the

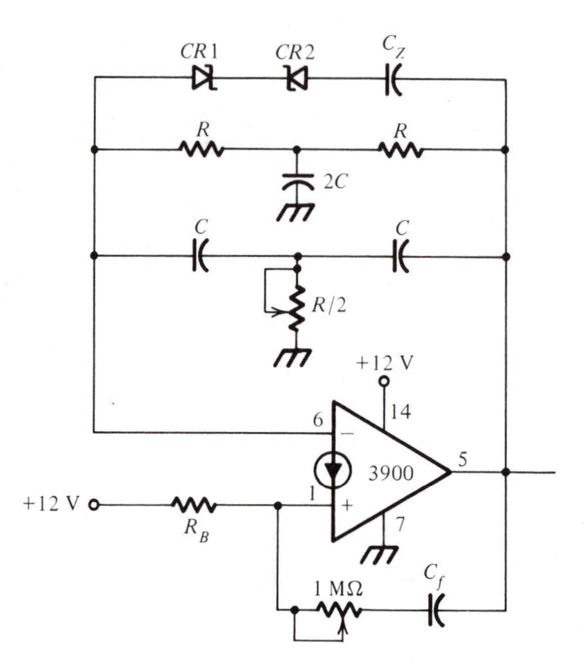

Figure 6-44 Norton amp twin-T oscillator

desired oscillation frequency. The R/2 potentiometer allows you to adjust that frequency. Bias is set with R_B. The two resistors in the filter provide the DC feedback, so there is no need for a separate bias feedback resistor as there was for the Wien-bridge oscillator in Fig. 6-40. Selecting

$$R_B = 4R = 2(R + R)$$

will set the output DC at approximately half of the supply voltage. The precise size of C_Z and C_f is best test selected. However, values in the 100's pF to 1000's pF work well for oscillation frequencies in the kHz. The potentiometer in the positive feedback loop allows you to adjust the amplifier's gain to obtain sustained oscillations, and set the signal's amplitude and distortion (to some degree). The zener diodes provide adaptive negative feedback, as you have seen in other oscillators.

Consequently, for the twin-T oscillator,

A notch filter is the frequency determining device and is placed in the negative feedback line. Adaptive feedback is provided by zeners paralleling the filter. Adjustable positive feedback must also be provided.

You must consider several points when comparing Wien-bridge and twin-T oscillators. The twin-T oscillator frequency can be adjusted with a single resistor. For the Wien-bridge, both resistors should be adjusted simultaneously to tune the frequency. The twin-T filter is more selective than the Wien-bridge, making the twin-T a more stable oscillator, less sensitive to loading. Also, for the twin-T oscillator, the precise level of feedback is not as critical as it is for the Wien-bridge oscillator, making the twin-T oscillator easier to set up. On the other hand, the Wien-bridge oscillator is simpler, requiring fewer components.

As you have seen, waveform generators can be built from either op amps or Norton amps. For sine wave generators, the output of the Norton amp should be biased to half the supply. This requires additional biasing components and coupling capacitors to block the DC from the signal feedback paths. Also, the higher output impedance of the Norton amp makes Norton amp oscillators more susceptible to loading than op amp oscillators. However, Norton amps are considerably less expensive than op amps and work easily from a single power supply.

6-4 FUNCTION GENERATORS

The circuits you have studied so far will generate either a triangle wave, or a square wave, or a sine wave. A function generator will produce all three shapes simultaneously. In addition, it is quite easy to vary the frequency from 0.01 Hz to well over 100 kHz without significant distortion of any of the signals. Often, frequency can be controlled with a DC voltage. This diversity and flexibility often make the function generator the most popular of waveform generators.

6-4.1 Multi-Op Amp Function Generator

You saw in Fig. 6-25 how to use two op amps to produce a triangle wave and a square wave simultaneously. If you are not familiar with that circuit's operation, review section 6-2.1 before going on.

The circuit in Fig. 6-25 will produce both the triangle wave and the square wave. To convert this to a function generator, you must add a sine shaper. This is a circuit which rounds the tips of the triangle wave, producing a fairly low-distortion (<1 percent) sine wave from the triangle wave. It is done with an array of resistors and diodes.

The schematic for the multi-op amp function generator is given in Fig. 6-45. Op amp U1 produces the triangle wave and U2 produces the square wave, just as was done in Fig. 6-25. The **Freq** potentiometer should be large (100 kΩ to 1 MΩ), since the capacitors in the negative feedback path cannot be electrolytic. Moving **S1** selects a different capacitor and therefore a different frequency range. The **Freq** potentiometer allows you to adjust the frequency continuously across a given

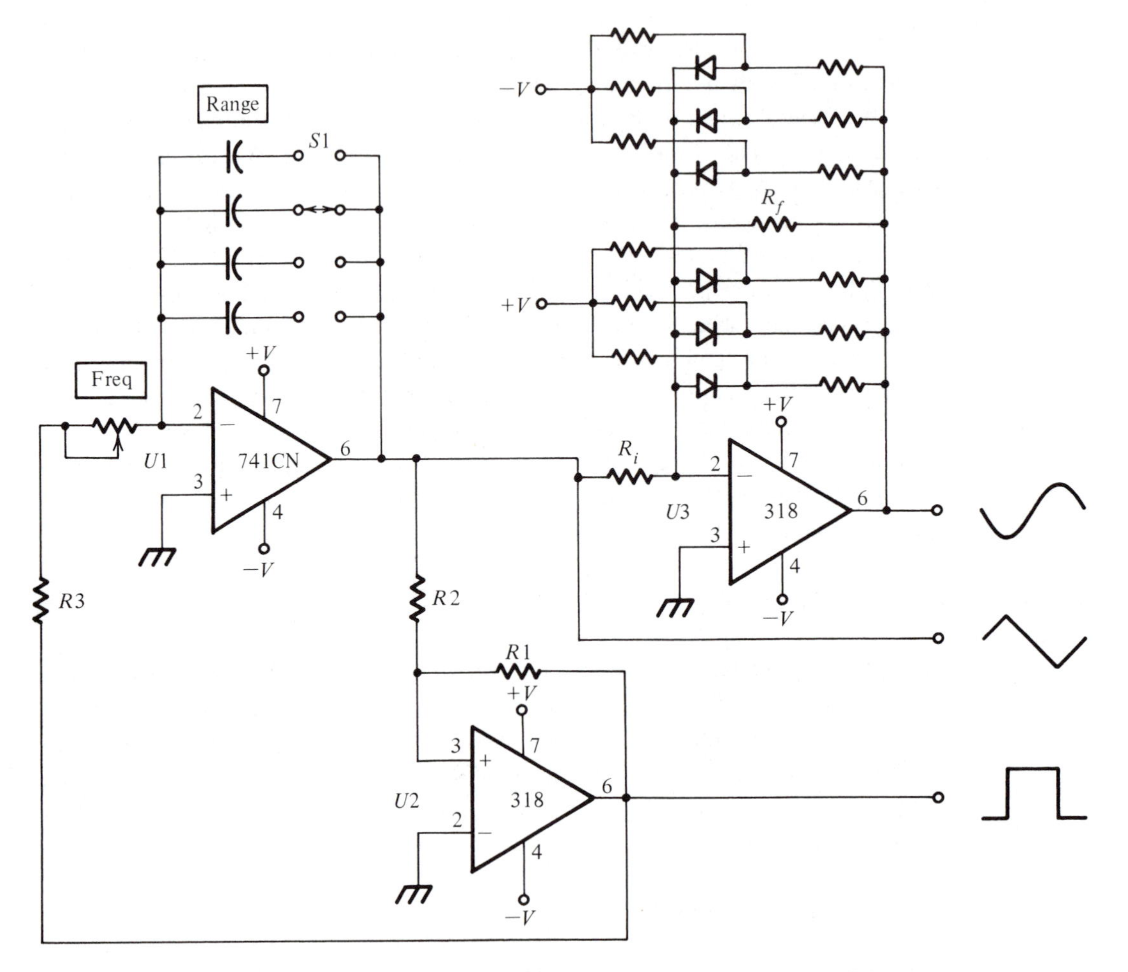

Figure 6-45 Multi-Op amp function generator

range. You may want to place a resistor in series with the frequency potentiometer to keep from shorting out the potentiometer, thereby shorting the output of U2 to ground.

The two-level comparator U2 is built with a LM318 high speed op amp. It has a slew rate of typically 50 volts/μs, which produces a square wave with much faster rise and fall times than from the less expensive 741C. Be sure to use an R3 of at least 1 kΩ. Selection criteria for R1, R2, the **Range** capacitors, and the **Freq** potentiometer are covered in section 6-2.1.

The circuit around U3 is a nonlinear, sine shaper amplifier. Its operation, analysis and design are covered in detail in Chapter 8. It is used to shape the triangle wave into a sine wave. When the triangle wave's amplitude is low, all of the diodes in U3's feedback path are off. The amplifier's gain is $-R_f/R_i$. This is shown as line segment A in Fig. 6-46.

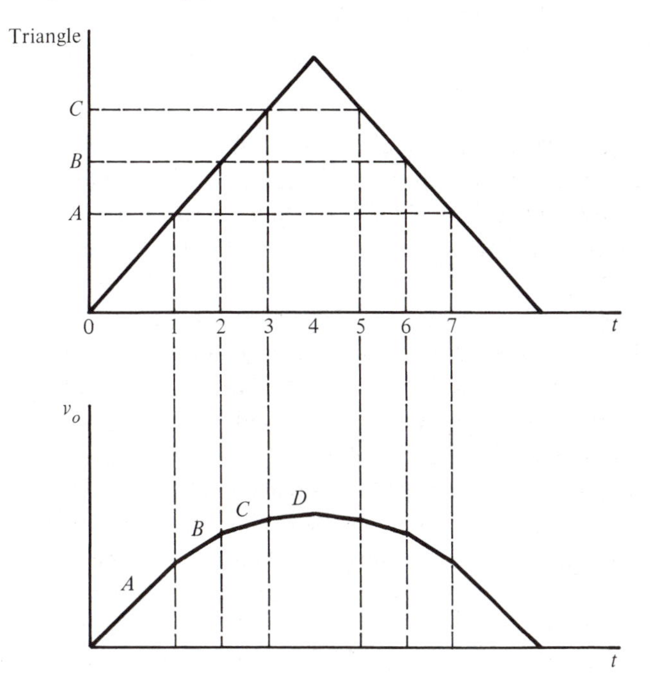

Figure 6-46 Operation of a sine shaper

When the output reaches amplitude A at time 1, a diode in the feedback path forward biases. This places a resistor in parallel with R_f, lowering the total feedback resistance. A drop in feedback resistance lowers the gain of the amp. Consequently, after time 1, the output rises at a slower rate (segment B).

When the output reaches amplitude B at time 2, a second diode in the feedback path is turned on. This further reduces the feedback resistance, reducing the amplifier's gain and the slope of the output wave once more. This process continues for line segments C and D.

The sine shaper produces a sine wave by rounding off the tips of the triangle wave. The distortion of the resulting sine wave may be several percent, which is much worse than the sine waves produced by the RC oscillators you have studied. However, this is the only *practical* way to produce sine waves whose frequencies vary from very low (0.01 Hz) to very high (over 10 MHz). No significant changes need to be made to the sine shaper as you adjust either the **Freq** or **Range** controls.

To produce an instrument quality function generator, you must also add a function switch, and an output amplifier with amplitude and DC offset controls. Such a circuit is shown in Fig. 2-18. You may also want to replace the 741C with a LM318 to improve the amplifier's speed.

6-4.2 IC Function Generator

There are several monolithic integrated circuits available that allow you to build a function generator with the addition of external resistors and capacitors. The 8038 is widely available and quite flexible, while still being easy to use. Its functional diagram is given in Fig. 6-47(a) and the connection diagram in Fig. 6-47(b).

As with the multi-op amp function generator of the previous section, the heart of the 8038 is the linear charge and discharge of a capacitor. This produces the triangle wave from which the square wave and sine wave are derived. Two current sources (one sources current, the other sinks current) alternately charge and discharge an *external* capacitor. When current source 1 drives the triangle voltage above the positive reference level, comparator 1 flips the flip-flop. This in turn switches in current source 2. Current source 2 discharges the capacitor at the same rate as it was charged. When the triangle wave falls below the negative reference level, the other comparator flops the flip-flop. Current source 2 is switched out and current source 1 begins charging up the capacitor again. The voltage across the external capacitor is buffered and becomes the triangle output. It also drives the sine shaper. The output from the flip-flop is buffered and becomes the square wave output.

The square wave output is open collector. This means that during the negative half-cycle, there is a short (0.2 V) between that output and the chip's $-V_{cc}$ pin. But during the positive half-cycle the output is *open.* To obtain an output *voltage* you must connect a pull-up resistor between the square wave output pin and some positive supply. A 10 kΩ to 15 kΩ resistor works well. Although this may appear to be an inconvenience, you can pull up the output to some voltage other than that from which you are running the 8038. In this way you can make the square wave compatible with any digital logic family. With a pull-up resistor the peak-to-peak amplitude of the square wave is approximately 90 percent of the difference in potential between the pull-up supply and $-V_{cc}$.

The triangle wave has a linearity of typically 0.1 percent or better and an amplitude of approximately 30 percent of the difference in potential between $+V_{cc}$ and $-V_{cc}$. The sine wave can typically be adjusted to a distortion of less than 1 percent with an amplitude of approximately 20 percent of the chip's supply difference. This distortion, however, may vary rather widely with variation in frequency.

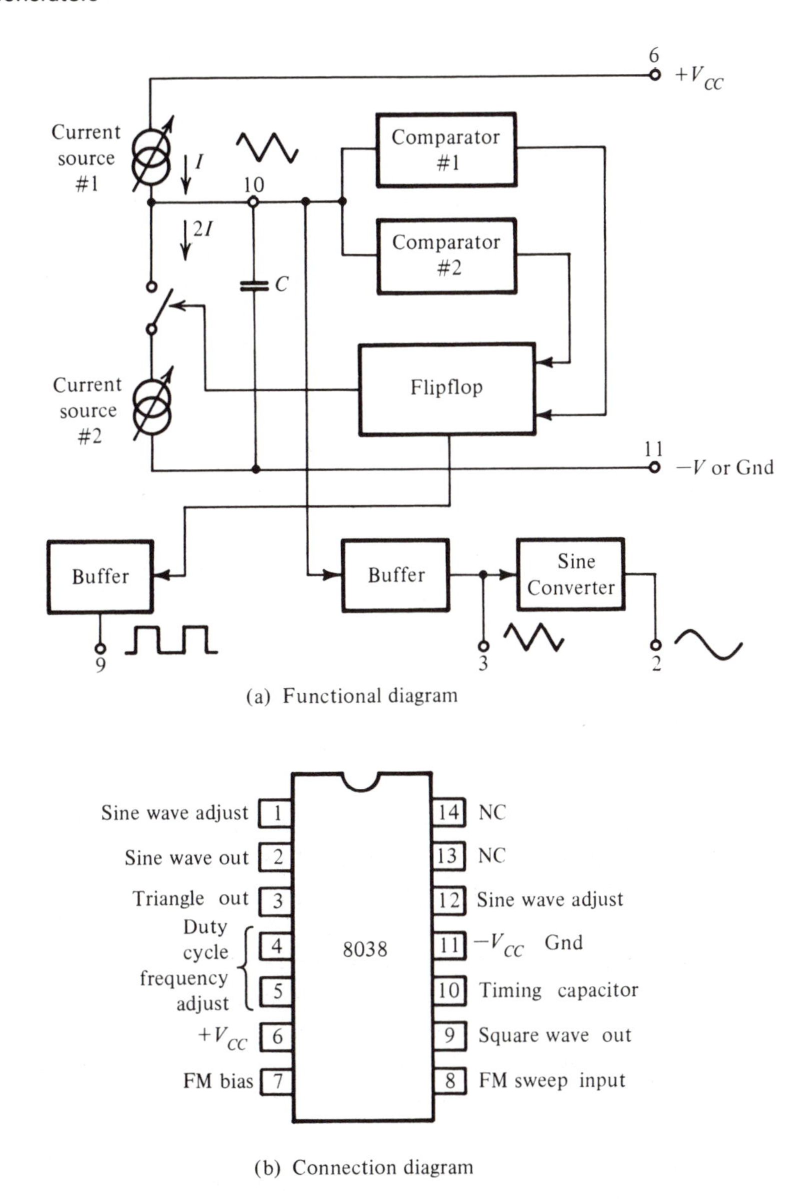

Figure 6-47 8038 Function Generator IC (*Courtesy of Intersil*)

The recommended external connections are shown in Fig. 6-48. Either a single power supply voltage (pin 11 ground) or $\pm V$ split supplies may be used. If a split supply is used, the output signals ride evenly above and below ground. If you use a single supply, the outputs will automatically self-bias to place the signals riding at half $+V$. The external timing capacitor, which is charged and

Figure 6-48 8038 basic schematic (*Courtesy of Intersil*)

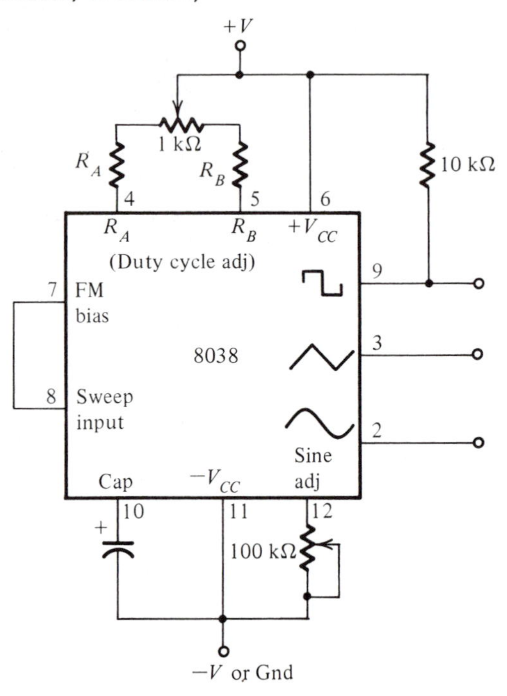

discharged by the current sources, may be electrolytic, but be sure you connect it properly. The internal sine shapers can be slightly adjusted by varying the 100 kΩ potentiometer. If this is not important, it can be replaced with an 82 kΩ fixed resistor.

The frequency is determined by the rate at which the capacitor charges and discharges. This, in turn, is inversely proportional to the size of the time capacitor, and directly proportional to the value of the current generators [equation 6-13(b)]. The value of current source 1 is set by the resistance between pin 4 (R_A) and $+V$. Current source 2 is set by the resistance between pin 5 (R_B) and $+V$.

$$f = \frac{1}{\frac{5}{3} R_A C\left(1 + \frac{R_B}{2R_A - R_B}\right)} \tag{6-34}$$

For 50 percent duty cycle, which is necessary for minimum sine distortion,

$$R_A = R_B = R$$

$$f = \frac{0.3}{RC} \tag{6-35}$$

However, there is some slight nonsymmetry between the two current sources. Even if you have R_A exactly equal to R_B, the duty cycle may not be precisely 50 percent. The 1 kΩ potentiometer allows you to adjust the duty cycle to 50 percent. [For calculations, assume that the potentiometer is centered, so that $R = R_A + 500\ \Omega$. Then use equation (6-35).] Resistors R_A and R_B should be selected to keep the charging currents between 10 μA and 1 mA.

$$I_c = \frac{+V}{5R_A} \quad \text{or} \quad I_c = \frac{+V}{5R_B}$$

The voltage applied to pin 8, the sweep input, also affects the size of the internal current sources, and, therefore, the frequency. Pin 7, FM bias, provides the proper voltage to ensure that equation (6-35) holds. However, you may provide the sweep input voltage from an external source. If you do, equation (6-35) becomes

$$f = \frac{3(+V - V_{\text{pin 8}})}{2RCV_{\text{total}}} \tag{6-36}$$

where $R_A = R_B = R$. V_{total} = total difference in potential between the $+V_{cc}$ and $-V_{cc}$ pins of the 8038.

There are several things that you should notice about this relationship. First, the frequency is no longer independent of the supply voltage, so you not only must take this into account, but should also provide a regulated power supply. Second, the frequency varies as the difference in potential between $+V$ and $V_{\text{pin 8}}$. Lowering the sweep voltage *increases* the frequency. The sweep voltage must be kept between

$$+V > V_{\text{pin 8}} > 2/3(V_{\text{total}}) + (-V_{cc}) + 2V \tag{6-37}$$

where $-V_{cc}$ is the potential at the $-V_{cc}$ pin with respect to ground.

Varying the frequency by changing the sweep input has many applications. Driving the proper level and polarity ramp into the sweep input will cause the output frequency to vary linearly over the entire audio range. This swept frequency oscillator is quite useful in testing circuit frequency response. Driving the sweep input with a digital-to-analog converter would allow you to digitally control (from a computer) the function generator's frequency.

Simpler, though, than these examples is single potentiometer variation of the frequency. Replacing R_A and R_B with a single variable resistor allows the duty cycle to vary as the resistor is tuned. Trying to gang R_A and R_B together also results in some duty cycle variation. A better technique is illustrated in Fig. 6-49.

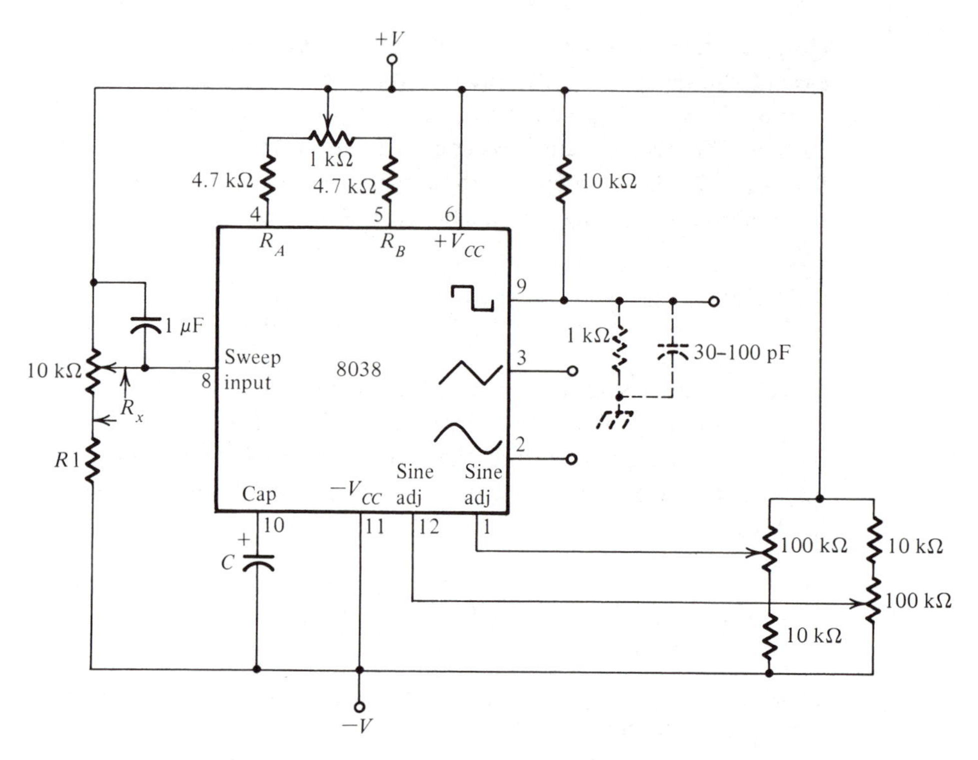

Figure 6-49 8038 improved function generator

The duty cycle of the square wave is set by the 1 kΩ potentiometer. To alter the frequency, you vary the 10 kΩ potentiometer, changing the voltage on pin 8 and therefore the output frequency. Resistor R1 is selected to set the upper adjustment limits for a given range [see equations (6-36) and (6-37)].

To minimize sine wave distortion, you may need to add several other components. These are also illustrated in Fig. 6-49. The single 100 kΩ potentiometer at pin 12 has been replaced with a network. One potentiometer will adjust the top of the sine wave; the other will adjust the bottom. The rapidly rising and falling edges of the square wave may couple high frequency noise onto the triangle and sine waves. This appears as a spike at the two tips. If this happens, you can absorb some of this extra energy with a 1 kΩ resistor to ground, a small capacitor to ground, or both at the *square* wave output.

Example 6-7

For the circuit in Fig. 6-49, with $\pm V = \pm 12$ V, calculate the value of R1 and C necessary to set the upper frequency at

a. 200 Hz **b.** 2 kHz **c.** 20 kHz

Solution

$$V_{total} = 24 \text{ V}$$

$$V_{pin\ 8} = \left(\frac{2}{3}\right)(24 \text{ V}) - 12 \text{ V} + 2 \text{ V}$$

$$V_{pin\ 8} = 6 \text{ V with respect to ground}$$

$$V_{10\ k\Omega} = 12 \text{ V} - 6 \text{ V} = 6 \text{ V}$$

$$\text{I} = \frac{6 \text{ V}}{10 \text{ k}\Omega} = .6 \text{ mA}$$

$$\text{R1} = \frac{6 \text{ V} - (-12 \text{ V})}{.6 \text{ mA}} = 30 \text{ k}\Omega$$

$$\text{f} = \frac{3(+V - V_{pin\ 8})}{2\text{RC}V_{total}}$$

$$\text{C} = \frac{3(+V - V_{pin\ 8})}{2\text{R}V_{total}\text{f}_{max}}$$

$$\text{C} = \frac{3(12 \text{ V} - 6 \text{ V})}{2(5.2 \text{ k}\Omega)(24 \text{ V})(\text{f}_{max})}$$

$$\text{f}_{max} = 200 \text{ Hz} \qquad \text{C} = .36\ \mu\text{F}$$

$$\text{f}_{max} = 2 \text{ k}\Omega \qquad \text{C} = .036\ \mu\text{F}$$

$$\text{f}_{max} = 20 \text{ k}\Omega \qquad \text{C} = .0036\ \mu\text{F}$$

Example 6-8

For the circuit designed in Example 6-7, calculate the voltage on pin 8 necessary to produce an output frequency of 100 Hz with $C = 0.47\ \mu F$.

Solution

$$\text{f} = \frac{3(+\text{V} - \text{V}_{pin\ 8})}{2\text{RCV}_{total}}$$

$$+\text{V} - \text{V}_{pin\ 8} = \frac{2}{3}\text{RCfV}_{total}$$

$$\text{V}_{pin\ 8} = +\text{V} - \frac{2}{3}\text{RCfV}_{total}$$

$$V_{pin\ 8} = +12\ \text{V} - \frac{2}{3}(5.2\ \text{k}\Omega)(0.47\ \mu\text{F})(100\ \text{Hz})(24\ \text{V})$$

$$V_{pin\ 8} = 8.09\ \text{V}$$

SUMMARY

In this chapter you have studied square wave, triangle wave, sine wave, and function generators. These circuits have been built with op amps, Norton amps, and special purpose integrated circuits.

All low frequency square wave generators rely upon the charge of a capacitor through a resistor to establish the timing. For the op amp and Norton amp versions, the RC charge circuit is driven by the output of a two-level inverting comparator, while the input of the comparator senses the voltage across the capacitor. When the charge reaches one of these levels, the output of the comparator switches to the other level. This switch also reverses the charge direction of the capacitor. When the capacitor charges to the opposite level, the comparator switches again. The process repeats itself. The output amplitude is set either by the saturation limits of the amplifier, or by the output clamping zener diodes. Frequency is set by the trigger levels of the comparator and the size of the RC timing components.

Several special purpose ICs can also produce rectangular waveshapes. The 555 timer senses the charge and discharge of an external RC network with two internal comparators. Internal control circuits either short out or allow the charge of the external timing capacitor between $1/3V$ and $2/3V$. The 555 can also be configured as a triggered timer (monostable).

The 2240 is a 555 timer and an 8 bit binary counter in a single package. Extremely long timing intervals can be established. Because the counter outputs are open-collector, programmable oscillators and timers, bit pattern generators and waveform synthesizers can be built.

Ramps must be generated by charging a capacitor with a constant current. This is easily done by placing the capacitor in the negative feedback loop of an op amp or Norton amp. The charge rate is directly proportional to the applied voltage while being inversely proportional to the input resistor and capacitance. By driving a two-level noninverting comparator (either op amp or Norton amp) with the ramp generator and using the output of the comparator to drive the ramp, a triangle wave can be produced. The amplitude is determined by the trip points of the comparator. Frequency depends on this amplitude, the RC charge components, and the output level of the comparator. One side benefit is the square wave generated by the comparator.

Sawtooth waveforms can be produced by placing a switch around the capacitor of the ramp generator. Using a bipolar transistor for this switch allows you to synchronize the sawtooth with some external signal. A programmable unijunction transistor (PUT) around the capacitor will terminate the ramp when it reaches an externally set voltage.

To produce a low distortion audio sine wave you need four elements. The frequency determining device can be either a bandpass filter (Wien-bridge) or a notch filter (twin T). It selects the oscillation frequency. An amplifier is necessary to return energy to the system loss in the frequency determining device. Both op amps and Norton amps may be used. Positive (in phase) feedback is necessary to sustain the oscillations. Adaptive negative feedback is necessary to automatically adjust the loop gain to precisely one. Larger loop gain causes distortion. Less will allow oscillations to die out. This adaptive negative feedback can be provided by shorting part of the negative feedback with signal diodes, zener diodes, or a field effect transistor.

Function generators produce a ramp by charging a capacitor with constant current. A two-level comparator monitors the ramp level, and switches the charge directions (charging current), when the ramp crosses one of the levels. This results in a triangle wave from the capacitor and a square wave from the comparator. A sine wave can be produced by rounding the tips of the triangle wave. As the triangle wave increases, diodes switch resistors in parallel with the feedback resistance of an inverting amplifier. This lowers the gain (slope) of the output from the amplifier, rounding the signal into a sine wave.

The function generator provides the major electronic test signals over a very wide frequency range. The circuitry is so simple that only two op amps or a single special purpose monolithic IC is needed. The 8038 chip also allows you to adjust the frequency with a DC voltage.

Amplifiers with RC networks placed in their feedback paths can be made to oscillate, generating a desired waveform, if given enough gain. When you lower the gain of the amplifier and drive the frequency determining network with an external signal, you have converted the oscillator to an active filter.

The active filters of Chapter 7 allow you to precisely control the response of a circuit to variations in frequency. Signal processing circuits in industrial, measurement, audio, and control equipment make extensive use of active filters. You will see the characteristics, terminology and mathematics of these filters in Chapter 7. Analysis and design for several low pass, high pass, band pass, notch, and state variable filters will be covered.

PROBLEMS

6-1 For the circuit in Fig. 6-1, calculate the comparison levels (trigger points) if $\pm V = \pm 12$ volts and

a. R1 = 10 kΩ, R2 = 4.7 kΩ

b. R1 = 10 kΩ, R2 = 9 kΩ

6-2 Design a two-level comparator like that in Fig. 6-1 with a 2 volt difference between the positive and negative comparison level when operating $\pm V = \pm 12$ volts.

6-3 For the circuit in Fig. 6-4, with R = 10 kΩ, C = 0.1 μF, R1 = 4.7 kΩ, R2 = 9 kΩ, $\pm V = \pm 12$ V, $R_{SC} = 2.2$ kΩ, CR1 = CR2 = 5.2 V; Calculate the frequency and the peak-to-peak amplitude of the output.

6-4 Design a square wave generator to produce a TTL compatible output ($V_{low} \sim 0$ V, $V_{high} \sim 5$ V). Use a LM318 to provide fast rise times. Limit the output current from the op amp to less than 10 mA. Provide two switch selectable frequencies: one at 0.3 Hz, the other at 10 kHz.

6-5 Calculate the upper and the lower trigger voltages for the comparator of Fig. 6-5 if $+V = 9$ V, $R_B = 330$ kΩ, $R_i = 100$ kΩ, $R_f = 330$ kΩ.

6-6 Design a two-level Norton amp comparator with a two volt difference between the upper trigger point and the lower trigger point when operating for a 12 volt supply. (See Problem 6-2.)

6-7 Calculate the output amplitude and frequency for the circuit in Fig. 6-6 given that $+V = 9$ V, $R_i = 100$ kΩ, $R_B = 330$ kΩ, $R_f = 330$ kΩ, R = 4.7 kΩ, and C = 0.1 μF. (See Problem 6-3.)

6-8 Repeat Problem 6-4, using a LM 3900 Norton amp. (Note: The LM 3900 will work from a single +5 volt supply, outputting a minimum of +0.2 volt and a maximum of 4.2 volts.)

6-9 Calculate the time high, the time low, and the frequency of the output from the circuit in Fig. 6-8, given that $R_A = 470$ Ω, $R_B = 330$ Ω, and C = 0.1 μF.

6-10 For the circuit in Fig. 6-50, explain operation for

a. enable = 2.5 V

b. enable = 0 V

6-11 Repeat Problem 6-4, using the 555 timer.

6-12 Design a monostable multivibrator with trigger pulse shaping which will drive a LED on for 0.5 second each time it is pulsed. (This is called a pulse stretcher. It will allow you to see a narrow, or nonperiodic, pulse.)

6-13 For the circuit in Fig. 6-14, calculate the *frequency* at pin 1, pin 5, and pin 8 for R = 10 kΩ, C = 0.1 μF.

6-14 For the circuit in Fig. 6-16, draw the output wave shape for the following pins connected to the output:

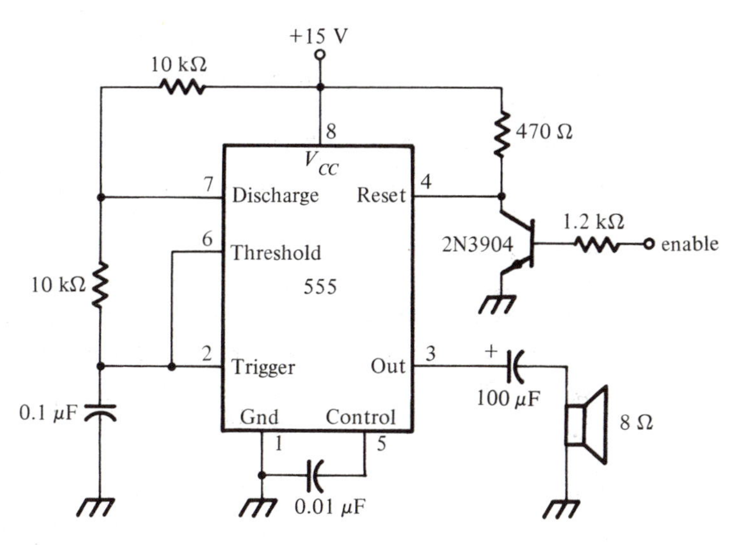

Figure 6-50 Schematic for problem 6-10

a. Pin 1, pin 4
b. Pin 2, pin 4, pin 5
c. Pin 1, pin 5, pin 7
d. Pin 2, pin 8

6-15 Repeat Problem 6-14 for the circuit in Fig. 6-18, with R1 = 10 kΩ, C1 = 0.1 μF.

6-16 Using a 2240 and a toggle flip-flop, design a programmable oscillator which will output a maximum frequency symmetric square wave of 20 kHz.
a. Which *one* pin is connected to the output to produce that frequency?
b. For the frequencies below, indicate the closest frequency your oscillator can produce and which pins must be connected to the output to produce that frequency.
200 Hz, 500 Hz, 1 kHz, 8 kHz, 12 kHz

6-17 **a.** Design a programmable timer with a resolution of 5 ms using a 2240.
b. How must the output pins be connected to the output to produce a 1 second delay?
c. Explain why this method of producing 1 second delay is better than using a 555 monostable.

6-18 For the circuit in Fig. 6-22, calculate the rate (or slope) of the output, given that E = 12 V, R = 10 kΩ, and C = 0.1 μF.

6-19 For Problem 6-18, could you use a 741C if E = 12 V, R = 470 Ω, and C = 0.001 μF? Explain.

6-20 Repeat Problem 6-1 for the circuit in Fig. 6-23. (R_f = R1; R_i = R2).

6-21 Repeat Problem 6-2 for the circuit in Fig. 6-23.

6-22 For the circuit in Fig. 6-25, calculate the items listed below, given that $\pm V = \pm 12$ V, $R_f = 100$ kΩ, $R_i = 22$ kΩ, R = 8.2 kΩ, C = 0.1 μF.

a. Frequency of the output.

b. Rate of rise of the triangle wave (slope).

c. Triangle wave amplitude.

d. Square wave amplitude.

6-23 **a.** Design a triangle wave/square wave generator (Fig. 6-25) with a triangle peak-to-peak amplitude of 10 volts. Provide two switch-selectable frequencies: one at 0.3 Hz, the other at 10 kHz. Use a LM 318 for the comparator to provide fast rise times.

b. Design a clipper circuit for the output of the square wave to make it TTL compatible ($V_{low} \simeq 0$ V, $V_{high} \simeq 5$ V) without altering the operation of the triangle generator. (See Problems 6-4, 6-8, 6-11.)

6-24 For the circuit in Fig. 6-27, given that C = 0.1 μF, $R_{B1} = 1$ MΩ, $R_{i2} =$ 100 kΩ, $R_{B2} = 470$ kΩ, $R_f = 220$ kΩ, $+V = +12$V calculate the output frequency, the triangle amplitude and the square wave amplitude for

a. $R_{i1} = 470$ kΩ

b. $R_{i1} = 220$ kΩ

c. $R_{i1} = 720$ kΩ

6-25 Repeat Problem 6-23a for the circuit in Fig. 6-27.

6-26 For the circuit in Fig. 6-28, draw the output and accurately label the voltage and time axis if $-V = -12$ V, $R_i = 220$ kΩ, C = 0.1 μF, and the control input is positive pulses occurring at every zero crossing of the line voltage (i.e., one pulse every 8.3 ms).

6-27 **a.** Using the circuit of Fig. 6-30, design a circuit whose output frequency will be 8 kHz when the control voltage is 4 volts, $\pm V = \pm 12$ V.

b. What will the frequency of your circuit be when the control voltage rises to 8 volts?

6-28 The circuit in Problem 6-27 is a voltage-to-frequency converter. However, an increase in the control voltage produces a *decrease* in the frequency. Design a signal conditioner using an op amp to drive the control input of Fig. 6-30 so that 4 volts into the signal conditioner will produce an output frequency of 4 kHz and 8 volts into the signal conditioner will produce 8 kHz out of the sawtooth generator.

6-29 Design a Wien-bridge oscillator, using an op amp to oscillate at 1 kHz with an adjustable output amplitude.

6-30 Repeat Problem 6-29, using a Norton amp.

6-31 Design a twin-T oscillator, using an op amp to oscillate at 1 kHz with an amplitude of approximately 4 volts peak.

6-32 Repeat Problem 6-31, using a Norton amp.

6-33 Repeat Problem 6-4, using a 8038 function generator IC. Careful! The 8038 requires at least 10 volts between $+V_{cc}$ and $-V_{cc}$. Also, when the 8038's square wave goes low, it is providing a short to $-V_{cc}$.

6-34 For the circuit in Fig. 6-48, do all required calculations to complete Table 6-2. $R_A = R_B = 4.7\ k\Omega$, 1 kΩ potentiometer exactly centered.

TABLE 6-2. 8038 performance, Problem 6-34

V_{cc} *(Volts)*	*C* (μF)	*Frequency Hertz*	*Amplitudes* (Volts p-p)		
			Square	*Triangle*	*Sine*
±10	0.001				
±10	0.1				
±10	50				

6-35 For the circuit in Fig. 6-49, with $\pm V_{cc} = \pm 15$ volts, calculate the required value of R1 and C to produce the following maximum frequencies (potentiometer adjusted to give $R_x = 0$):

a. 200 Hz
b. 5 kHz
c. 20 kHz

6-36 For the circuit in Fig. 6-49, with $\pm V_{cc} = \pm 12$ volts, $C = 0.1\ \mu F$, and $R1 = 51\ k\Omega$, calculate the sweep input voltage and the output frequency for the following potentiometer settings:

a. $R_x = 0$
b. $R_x = 1.25\ k\Omega$
c. $R_x = 2.5\ k\Omega$
d. $R_x = 5\ k\Omega$
e. $R_x = 7.5\ k\Omega$
f. $R_x = 8.75\ k\Omega$

6-37 Equation (6-36) indicates that frequency changes as a function of voltage applied to pin 8 *with respect to* $+V$. This means that increasing the voltage at pin 8 with respect to ground lowers the output frequency. Design a signal conditioner to drive pin 8, so that 0 volt into the signal conditioner will cause a minimum output frequency and five volts into the signal conditioner will cause a maximum frequency (for the set R_A, R_B, and C).

CHAPTER 7

Active Filters

Electric filters are used in practically all circuits which require the separation of signals according to their frequencies. Applications include (but are certainly not limited to) noise rejection and signal separation in industrial and measurement circuits, feedback of phase and amplitude control in servo loops, smoothing of digitally generated analog (D-A) signals, audio signal shaping and sound enhancement, channel separation, and signal enhancement in communication circuits.

Such filters can be built from passive RLC components, electromechanical devices, crystals, or with resistors, capacitors, and op amps (active filters). Active filters are applicable over a wide frequency range, are inexpensive, and offer high input impedance, low output impedance, gain, and a wide variety of responses.

In this chapter you will be introduced to the characteristics, terminology, and mathematics of active filters. The analysis, design, and mathematics behind several types of low and high pass filters, wide and narrow bandpass filters, the notch filter, and the state variable filter are presented.

OBJECTIVES

After studying this chapter, you should be able to do the following:

1. Describe in detail a frequency response plot.
2. Convert between ratio and decibel gain.
3. Define the critical frequency, and locate it on a frequency response plot.
4. Use a frequency response plot with a normalized frequency axis.
5. Briefly state the purpose of Laplace transforms.
6. Analyze a simple circuit to determine its transfer function, gain versus frequency response, and phase versus frequency response.
7. Draw the ideal and practical frequency response plots for a low pass filter, high pass filter, bandpass filter, and notch filter.

8. Identify the stop and pass bands of a given filter.
9. Describe the rolloff rate, dB/decade, dB/octave, and their relationship to filter order.
10. Given the response curve for a bandpass or notch filter, determine the center frequency, low frequency cutoff, high frequency cutoff, bandwidth, Q, and notch depth.
11. Describe one application for the low pass, high pass, bandpass, and notch filters.
12. List four disadvantages of passive filters and explain how active filters overcome them.
13. List four disadvantages of active filters.
14. Derive the transfer function and gain and phase equations for a first order filter and a second order low pass and high pass active filter.
15. Given filter specifications (low pass or high pass of any order), determine type, order, and component values necessary to build the filter using a Sallen-Key unity gain, equal component, or state variable configuration.
16. Analyze a given unity gain, equal component, or state variable active filter to determine critical frequency, response shape, rolloff rate, and pass band gain.
17. List specific advantages, disadvantages, limitations, and precautions to be observed when designing or building low pass, high pass, bandpass, or state variable active filters.
18. Derive the transfer function for the single op amp bandpass and notch filters, defining all appropriate parameters.
19. Given bandpass or notch filter specifications, select the proper type (single op amp, cascaded, or low pass-high pass) and calculate required component values.
20. Analyze a given bandpass or notch filter circuit to determine center frequency, bandwidth, Q, low and high frequency cutoffs, gain, and rolloff rate.
21. Derive the transfer function for each output of a unity gain state variable active filter and define all appropriate parameters.
22. Discuss state variable applications, including cascading, output summing, and the AF100 universal active filter IC.

7-1 INTRODUCTION TO FILTERING

Filters are specified, analyzed, and designed somewhat differently than are the circuits discussed in the previous chapters. Performance is specified in terms of the frequency response, how the gain and phase shift change with frequency. Analysis and design often use Laplace transforms, and the circuit's transfer function. *Angular frequency, critical frequency, bands, ripple, rolloff rate, center frequency,* and *Q* are among the terms unique to filters. In this section frequency response plots, basic transform math, and terminology common to most filters are presented. Later sections of the chapter apply these fundamentals to specific filter circuits.

7-1.1 Frequency Response

The gain and phase shift of a filter changes as the frequency changes. Indeed, this is the purpose of a filter. Performance is often described with a graph of gain (and phase) versus frequency.

You have already seen such a frequency response plot. Figure 4-13 gave the frequency response of an op amp. The vertical axis is the gain, and the horizontal axis is the frequency. Notice in that graph that the frequency axis gives equal distance for each *decade* of frequency. The distance between 1 Hz and 10 Hz is the same as the distance between 100 kHz and 1 MHz. On a linear graph, the distance between 100 kHz and 1 MHz should be 100,000 larger. However, this would make it impossible to plot any large range of frequencies on a single graph. So the horizontal axis on frequency response plots is normally scaled logarithmically.

A log scaled horizontal axis is shown in Fig. 7-1. In Fig. 7-1(a), each decade (factor of ten) increase or decrease moves you the same distance along the scale. There are two other points you should notice. First, the starting point is not at zero. If you went farther left, each increment would lower the scale by 10 (to 0.1, then 0.01, then 0.001, etc.), but you would never read zero. So start your log scales at the lowest frequency decade of interest, not at zero. Second, the divisions between decades are not uniform (linear). This is easily seen in Fig. 7-1(b), an expansion of a single decade. Five does not fall halfway between 1 and 10; $3\frac{1}{3}$ does. Be very careful about this when interpolating between major scale divisions.

Gain, plotted on the vertical axis, is normally expressed in decibels.

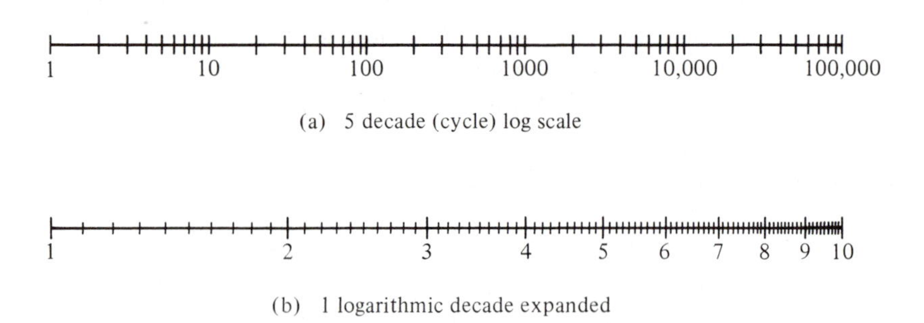

Figure 7-1 Logarithmic scale

$$dB = 10 \log_{10} \frac{\text{power out}}{\text{power in}} \tag{7-1}$$

$$dB = 10 \log_{10} \frac{v_O^2/R_L}{v_{in}^2/R_{in}}$$

$$dB = 10 \log_{10} \left(\frac{v_O}{v_{in}}\right)^2 \left(\frac{R_{in}}{R_L}\right)$$

If it is assumed that

$$R_{in} = R_L$$

$$dB = 10 \log_{10} \left(\frac{v_O}{v_{in}}\right)^2$$

$$dB = 20 \log_{10} \frac{v_O}{v_{in}} \tag{7-2}$$

For equation (7-2) to be valid, the load impedance must equal the filter's input impedance. This is seldom the case. However, equation (7-2) is normally used anyway.

You can readily convert ratio gain (v_O/v_{in}) to dB gain with most scientific calculators. Notice that log base 10 is used, not the natural log (1n or log base *e*). It is handy to know several dB and ratio gain points. These are listed in Table 7-1. Decreasing the ratio gain by 10 subtracts 20 dB; increasing by 10 adds 20 dB. Doubling the gain adds 6 dB; halving it subtracts 6 dB. Cutting the gain by $\sqrt{2}$ gives -3 dB. A ratio gain of 1 ($v_{out} = v_{in}$) is a gain of zero dB.

TABLE 7-1. Ratio and dB gain comparison

$\frac{v_O}{v_{in}}$	dB	$\frac{v_O}{v_{in}}$	dB
1000.	60	0.707	−3
100.	40	0.5	−6
10.	20	0.1	−20
2.	6	0.01	−40
1.	0	0.001	−60

A typical low pass filter frequency response is given in Fig. 7-2. The vertical axis is scaled in decibels and the horizontal axis logarithmically (two decades or cycles in the figure). By plotting dB versus log frequency, you are actually plot-

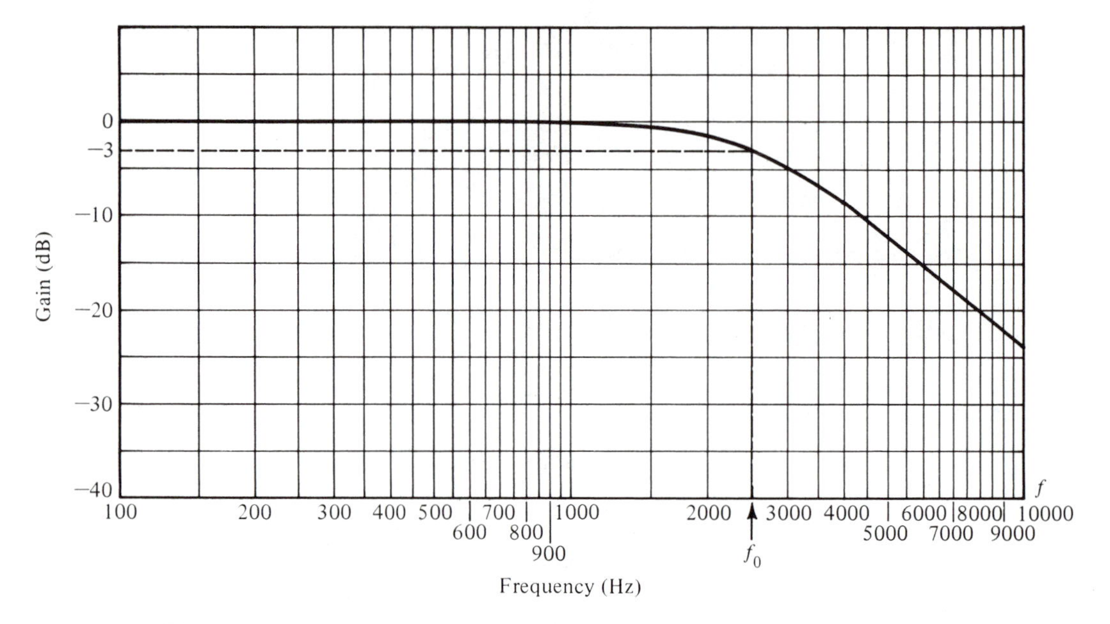

Figure 7-2 Typical low pass frequency response curve

ting log gain versus log frequency, so any linear relationship will be displayed as a straight line. A parameter of major importance is the critical frequency, f_o. At the critical frequency the *power* delivered to the load has been cut in half. What has happened to the *voltage* gain at the critical frequency?

$$P = \frac{v^2}{R}$$

or

$$v = \sqrt{PR}$$

At the critical frequency,

$$v_{f_o} = \sqrt{\frac{P}{2} R}$$

$$v_{f_o} = \sqrt{\frac{PR}{2}}$$

$$v_{f_o} = \frac{\sqrt{PR}}{\sqrt{2}}$$

$$v_{f_o} = \frac{1}{\sqrt{2}}\sqrt{PR}$$

$$v_{f_o} = 0.707\sqrt{PR}$$

$$v_{f_o} = 0.707 v_{\text{low frequency}} \tag{7-3}$$

Consequently, at the critical frequency, the power has been cut in half, and the voltage (and voltage gain) have fallen by 0.707. Refer to Table 7-1. A decrease in gain of 0.707 is the same as a drop of −3 dB. The critical frequency in Fig. 7-2 is 2.5 kHz.

You may see frequency expressed as angular velocity in radians per second (ω). There are 2π radians in a circle, just as there are 360 degrees. The conversion between radians per second (ω) and Hertz (f) is

$$\omega = 2\pi f \tag{7-4}$$

The units convert:

$$\omega = 2\pi \frac{\text{radians}}{\text{cycle}} \times f \frac{\text{cycles}}{\text{seconds}}$$

$$\omega = 2\pi f \frac{\text{radians}}{\text{second}}$$

Each filter produces its own frequency response. These responses can be much more easily grouped, identified, analyzed, or circuits designed if the frequency axis is normalized. This is done by dividing all points on the horizontal axis by the filter's critical frequency. Consequently, the critical frequency of all normalized frequency response plots is

$$\frac{f_o}{f_o} = 1$$

The plot of Fig. 7-2 has been normalized and replotted in Fig. 7-3. To denormalize or scale a frequency response curve, you simply multiply all frequencies by the desired critical frequency.

$$f_{\text{actual}} = f_{\text{normalized}} \times f_o \tag{7-5}$$

As you have seen, the gain of a filter changes with frequency, and this is plotted on a frequency response graph. Often overlooked, however, is the fact that the phase relationship between the input and output voltage of a filter also

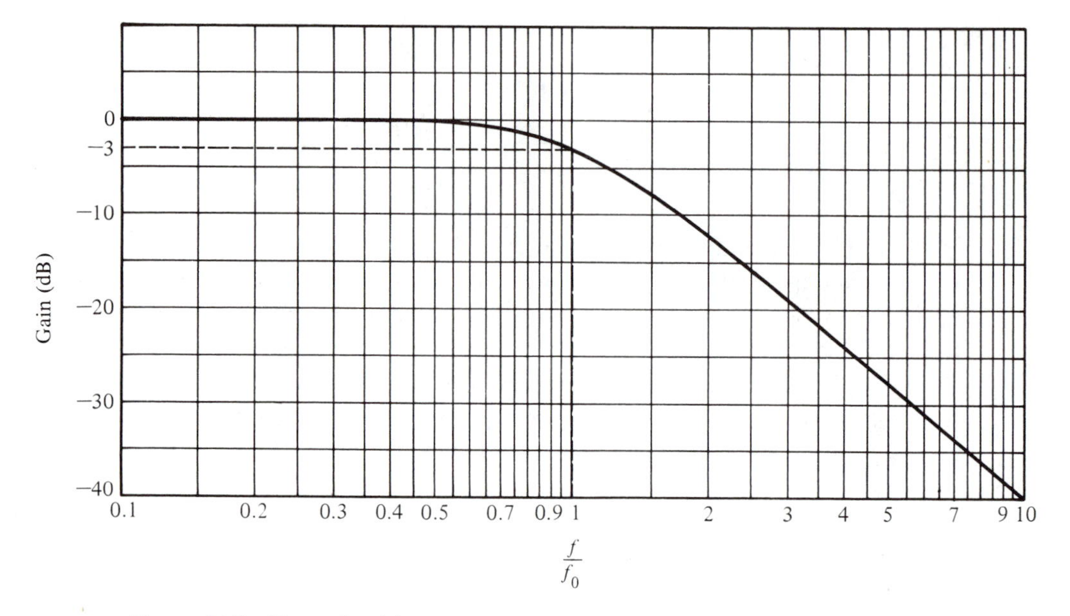

Figure 7-3 Normalized frequency response curve

changes with frequency. A typical plot of phase shift versus frequency is given in Fig. 7-4. At the critical frequency (1 on the normalized graph), there will be a *multiple* of 45 degrees phase shift, depending on the order filter. In fact, often you can locate the critical frequency of a filter more precisely and quickly by measuring the phase shift than you can be looking for a 3 dB drop in gain. The combination of the gain variation and phase shift as a function of frequency forms a complete frequency response plot (Fig. 7-5). Not only is plotting both graphs on the same axis convenient, but it will also allow you to determine if the filter (or any system) you anticipate building will be stable (i.e., will not oscillate).

Analysis of circuits containing several reactive components is most easily handled by using Laplace transforms. It is far beyond the scope of this text to present a rigorous coverage of Laplace transform circuit analysis. However, a very simplified touch will be useful.

When the Laplace transform of an equation (or a circuit) is used, differential and integral terms are replaced by s and $1/s$. You can then manipulate and simplify the equation, using the rules of algebra rather than those of differential equations.

As far as active filters are concerned, the primary term is

$$i_c = C\frac{dv}{dt}$$

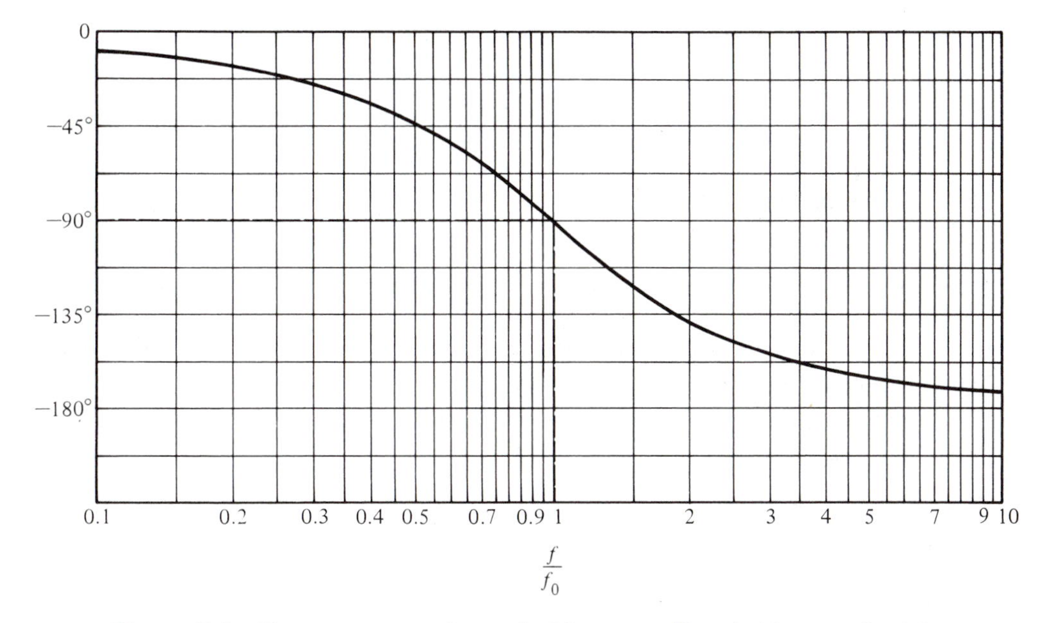

Figure 7-4 Phase response of a typical low pass filter (with normalized frequency axes)

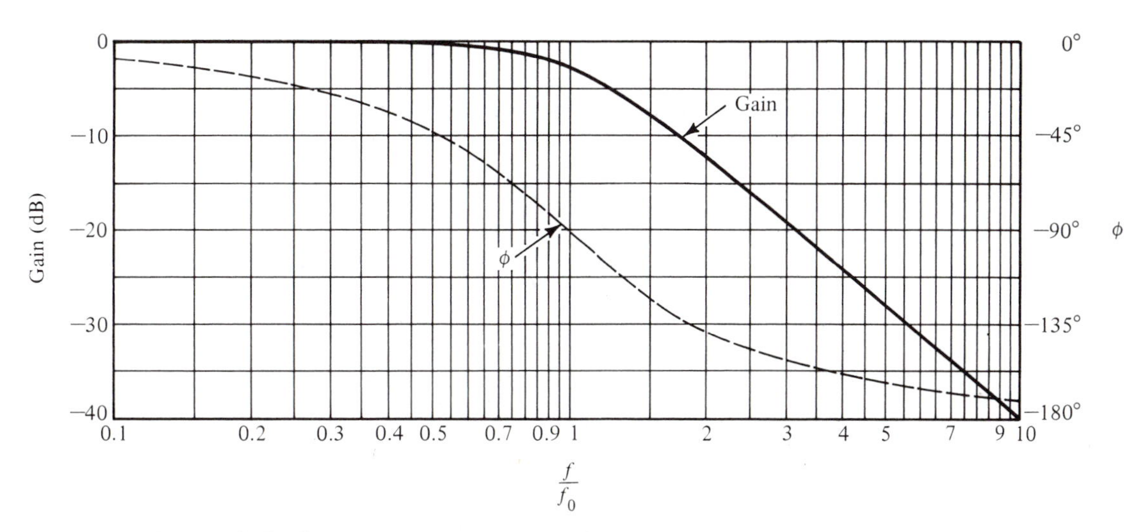

Figure 7-5 Composite gain and phase frequency response

To take the Laplace transform of this, just replace the derivative with s:

$$I = CsV$$

The Laplace impedance, then, of a capacitor is

$$Z = \frac{V}{I} = \frac{1}{Cs} \tag{7-6}$$

and the Laplace admittance is

$$Y = \frac{I}{V} = Cs \tag{7-7}$$

The term s contains both the amplitude and phase information about an equation's (or a circuit's) response.

To obtain frequency response information from a Laplace equation, make the substitution

$$s = j\omega \tag{7-8}$$

where j is the imaginary number $\sqrt{-1}$ and $\omega = 2\pi f$.

Example 7-1

Verify that equation (7-6) gives the correct impedance when converted back to the frequency domain.

Solution To convert from the Laplace domain (equation containing s) to the frequency domain (equation containing f or ω), apply equation (7-8).

$$Z = \frac{1}{Cs}$$

$$Z = \frac{1}{j\omega C}$$

$$Z = \frac{1}{j\,2\pi f C}$$

$$Z = -\frac{j}{2\pi f C}$$

$$Z = -jX_c$$

This is the impedance of a capacitor, as you learned from AC circuits.

Using Laplace functions, you can determine the frequency response (both gain and phase) of a circuit. You must first determine the circuit's transfer function. The transfer function is just the gain expressed in Laplace terms.

Example 7-2

1. Determine the transfer function for the circuit in Fig. 7-6(a).
2. Calculate and plot the frequency response (both gain and phase) for $R = 500\ \text{k}\Omega$, $C = 1\ \mu\text{F}$.

Solution

1. To determine the transfer function, you substitute the Laplace function of each variable in Fig. 7-6(a).

$$v_{in} \rightarrow V_{in}, \qquad v_{out} \rightarrow V_{out}, \qquad R \rightarrow R$$

$$Z_c \rightarrow \frac{1}{sC}$$

The transformed circuit is shown in Fig. 7-6(b). Applying the voltage divider law, you obtain

$$v_{out} = \frac{Z_c}{R + Z_c} v_{in}$$

$$V_{out} = \frac{\frac{1}{sC}}{R + \frac{1}{sC}} V_{in}$$

$$\frac{V_{out}}{V_{in}} = \frac{\frac{1}{sC}}{R + \frac{1}{sC}}$$

$$\frac{V_{out}}{V_{in}} = \frac{1}{RsC + 1}$$

$$\frac{V_{out}}{V_{in}} = \frac{1}{RCs + 1} \tag{7-9}$$

Equation (7-9) is the transfer function of the circuit.

2. To determine the frequency response, substitute

$s = j\omega$ into equation (7-9).

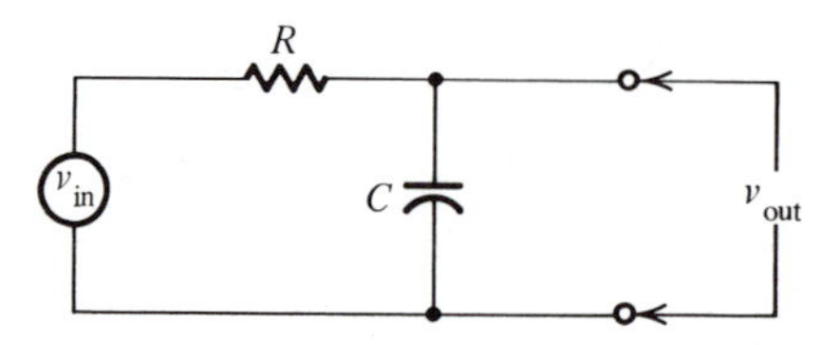

(a) Frequency domain (regular) schematic

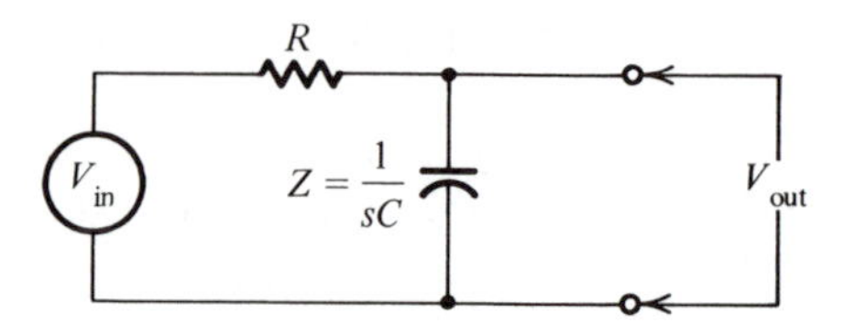

(b) Laplace transformed circuit

Figure 7-6 Circuits for Example 7-2

$$\frac{V_{out}}{V_{in}} = \frac{1}{j\omega RC + 1}$$

$$\frac{V_{out}}{V_{in}} = \frac{1}{1 + j\omega RC}$$

You now have an equation with a real and an imaginary part. To separate these parts, you must multiply both numerator and denominator by the complex conjugate of the denominator.

$$\frac{V_{out}}{V_{in}} = \frac{1}{1 + j\omega RC} \times \frac{1 - j\omega RC}{1 - j\omega RC}$$

$$\frac{V_{out}}{V_{in}} = \frac{1 - j\omega RC}{1 - j^2\omega^2 R^2 C^2}$$

$$\frac{V_{out}}{V_{in}} = \frac{1 - j\omega RC}{1 + \omega^2 R^2 C^2}$$

$$\frac{V_{out}}{V_{in}} = \frac{1}{1 + \omega^2 R^2 C^2} - j\frac{\omega RC}{1 + \omega^2 R^2 C^2}$$

$$\frac{V_{out}}{V_{in}} = \text{real} + j\ \text{imaginary}$$

$$\text{Real} = \frac{1}{1 + \omega^2 R^2 C^2}$$

$$\text{Imaginary} = -\frac{\omega RC}{1 + \omega^2 R^2 C^2}$$

$$|G| = \text{magnitude} = \sqrt{\text{real}^2 + \text{imaginary}^2}$$

$$|G| = \sqrt{\frac{1}{(1 + \omega^2R^2C^2)^2} + \frac{\omega^2R^2C^2}{(1 + \omega^2R^2C^2)^2}}$$

$$|G| = \frac{\sqrt{1 + \omega^2R^2C^2}}{1 + \omega^2R^2C^2}$$

$$|G| = \frac{1}{\sqrt{1 + \omega^2R^2C^2}} \tag{7-10}$$

$$\phi = \text{phase angle} = \arctan \frac{\text{imaginary}}{\text{real}}$$

$$\phi = \arctan \left[\frac{\dfrac{-\omega RC}{1 + \omega^2R^2C^2}}{\dfrac{1}{1 + \omega^2R^2C^2}} \right]$$

$$\phi = -\arctan \omega RC \tag{7-11}$$

Equation (7-10) gives gain versus frequency and equation (7-11) gives phase shift versus frequency. A tabulation of $|G|$, dB gain, and ϕ for $R = 500$ kΩ and $C = 1$ μF is given in Table 7-2.

TABLE 7-2. Values obtained by substituting various ω, R = 500 kΩ, C = 1 μF into equations (7-10) and (7-11) for Example 7-2.

ω	$\|G\|$	dB = 20 log $\|G\|$	ϕ
0.1	0.999	−0.010	−2.86°
0.6	0.958	−0.374	−16.7°
1	0.894	−0.969	−26.6°
2	0.707	−3.00	−45.0°
3	0.555	−5.12	−56.3°
4	0.447	−6.99	−63.4°
5	0.371	−8.60	−68.2°
6	0.316	−10.0	−71.6°
8	0.243	−12.3	−76.0°
10	0.196	−14.1	−78.7°
30	0.067	−23.5	−86.2°
100	0.020	−34.0	−88.9°

The frequency response plots of equations (7-10) and (7-11) are given in Figure 7-7.

Theoretically, the technique of Example 7-2 can be used to determine the frequency response of any network. However, each *RC* combination in the circuit increases the order of the denominator by one. (Two *RC* pairs cause $s^2 + as + b$; three *RC* pairs cause $s^3 + bs^2 + Cs + d$, etc.). Fortunately, the mathematics involved in breaking down and solving these higher-order equations has already been worked out for the popular, more useful circuits.

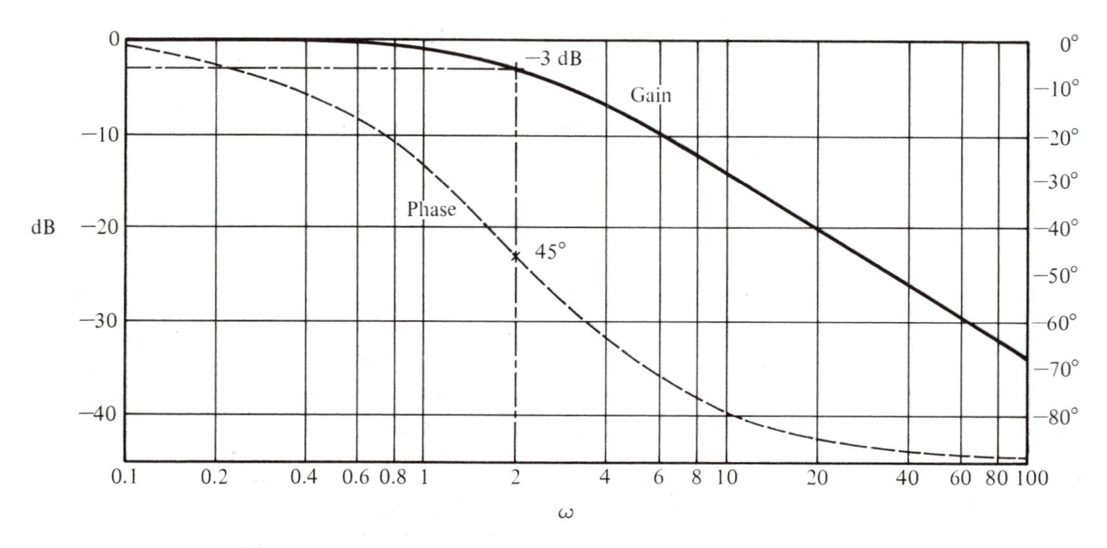

Figure 7-7 Frequency response plot for Example 7-2

7-1.2 Characteristics and Terminology

The frequency responses you have seen so far are for low pass filters. The purpose of a low pass filter is to pass low frequency signals while stopping high frequency signals. Ideally, a low pass filter would have a frequency response as shown in Fig. 7-8. All frequencies below the critical frequency, f_o, would be uniformly passed. Any frequency above f_o would be completely stopped.

Of course, such a filter cannot be built. The response of a practical low pass filter can be divided into two bands, as shown in Fig. 7-9. The critical frequency, f_o, forms the boundary between the pass band and the stop band. For some filters gain varies up and down (or ripples) in the pass band, in the stop band, or both. The amount of pass band ripple allowable (and, to a lesser degree, stop band ripple) is an important parameter to keep in mind when you are designing a filter.

Figure 7-8 Ideal low pass filter response

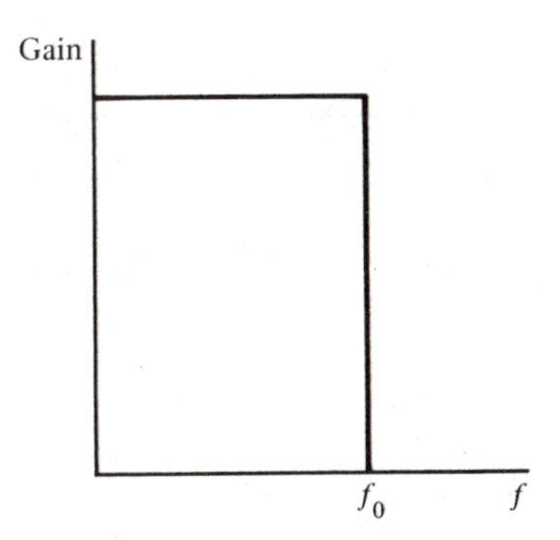

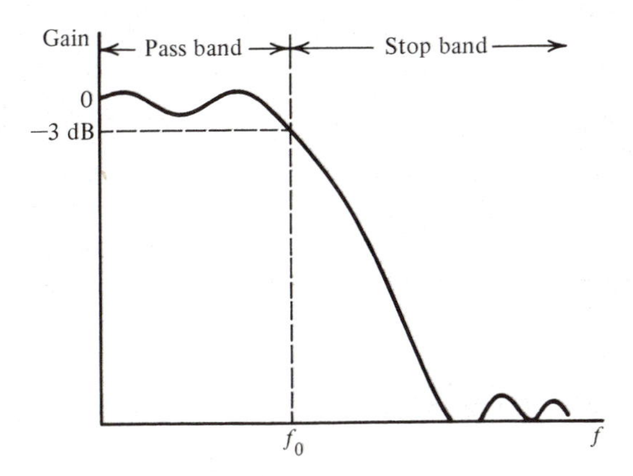

Figure 7-9 Practical low pass filter

How rapidly the gain falls off as the stop band is entered is called the rolloff. The first five rolloff rates are illustrated in Fig. 7-10. The rolloff rate is determined by the order of the filter (1, 2, 3, ...). Each increase in order increases the rolloff by 20 dB/decade. In turn, as was mentioned in the Laplace transform section, the order of the filter (and its transfer function denominator) equals the number of resistor/capacitor pairs in the circuit. So increasing the circuit complexity by adding an RC pair increases the circuit's order, and the difficulty of the mathematics, but also increases the rolloff rate by 20 dB/decade.

You will hear rolloff specified in dB/decade and dB/octave. A decade increase in frequency means that the frequency has gone up by a factor of 10. An octave increase means that the frequency has doubled. Decibels per octave ratings are used primarily with audio/music applications. Table 7-3 correlates filter order (number of RC pairs and transfer function denominator order) with the rolloff rate in dB/decade and dB/octave.

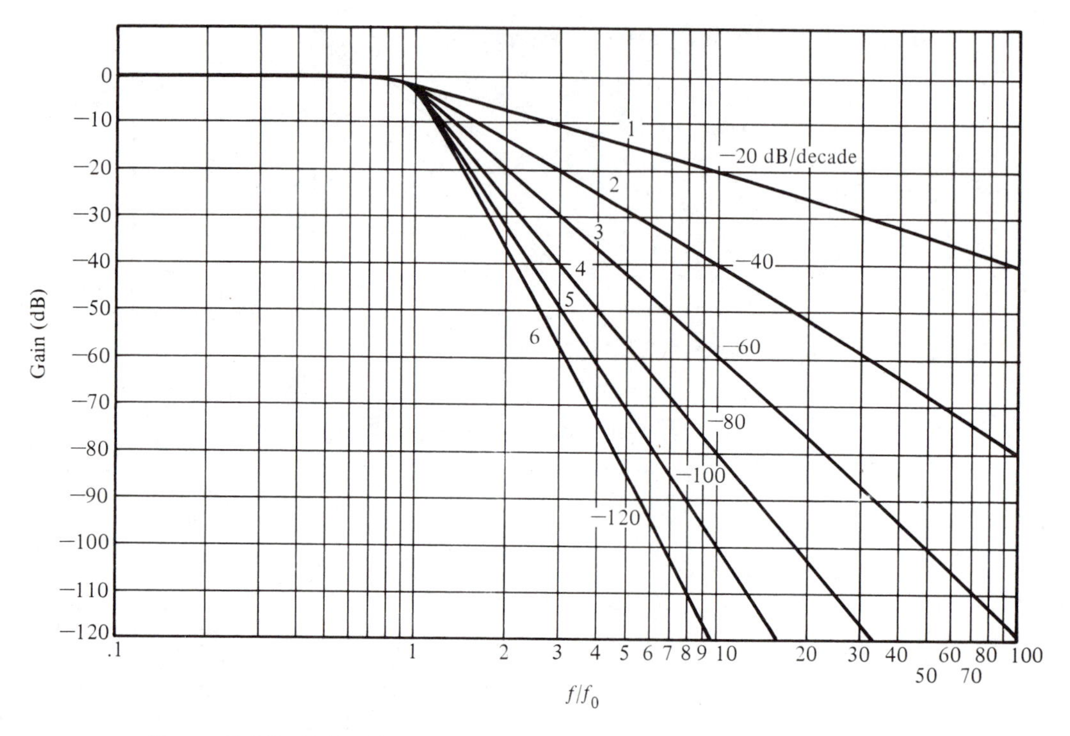

Figure 7-10 Roll-off rate comparison

TABLE 7-3. Rolloff rate correlation

ORDER	*dB/DECADE*	*dB/OCTAVE*
1	20	6
2	40	12
3	60	18
4	80	24
5	100	30

The opposite of the low pass filter is the high pass filter. A high pass response is illustrated in Fig. 7-11. Low frequency signals and DC (which is 0 Hz) are blocked, while high frequency signals are passed. Specification, analysis, and design of high pass filters are closely analogous to those you have already seen for low pass filters. The critical frequency is the frequency at which the gain is 3 dB below the pass band gain. There is a multiple of 45 degrees phase shift at the critical frequency. There may be ripple in the pass band and/or the stop band. The transfer function and frequency response plots are obtained just as was done

Figure 7-11 High pass filter response

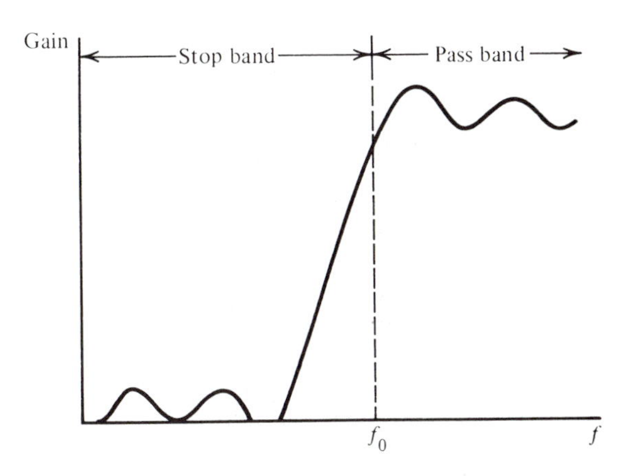

in Example 7-2 for the low pass filter. The order of the denominator is set by the number of RC pairs, which in turn establishes the rolloff rate.

You have already used a simple high pass filter. The RC couplers of Chapter 5 (Figs. 5-10 and 5-11) are used to block the DC bias from previous stages or the signal generator, while passing the signal whose frequency is of interest. This RC coupler is the complement of the circuit analyzed in Example 7-2. With a single RC pair, the RC coupler will have a 20 dB/decade rolloff and 45 degrees phase shift at the critical frequency.

The bandpass filter passes only those signals within a given band. Signals above and below that band are blocked. Figure 7-12(a) is the response of an ideal bandpass filter, while Fig. 7-12(b) is a more realistic response curve. Since the response rises, peaks out, and then falls, there are three frequencies of interest. The center frequency is f_c. Depending on component configuration and values, there may be considerable gain at the center frequency (even for filters with no built-in amplifier). The cutoff frequencies (f_l and f_h) occur where the gain has fallen by 3 dB (below the center frequency gain).

Instead of specifying rolloff rate, bandwidth and Q are given. Bandwidth is the distance between the low frequency and high frequency cutoffs.

$$\Delta f = f_h - f_l \tag{7-12a}$$

or

$$B = \omega_h - \omega_l \tag{7-12b}$$

The Q is the ratio of center frequency to bandwidth

$$Q = \frac{f_c}{\Delta f} = \frac{\omega_c}{B} \tag{7-13}$$

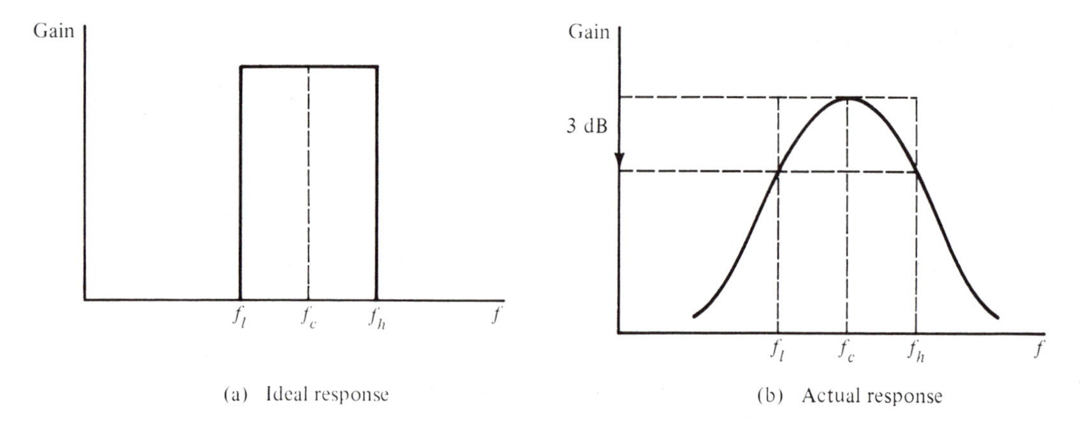

Figure 7-12 Bandpass filter response

This gives a measure of how sharp or narrow the bandpass filter is. The higher the Q, the more selective the filter.

Simple bandpass filters can be made with two RC pairs. This makes their transfer function second order. However, since the two networks are opposing each other, the rolloff rate will not be 40 dB/decade. In fact, the slope may vary significantly. By properly cascading several such filters and staggering their tuning, you can produce a response with steep edges and a flat top.

Bandpass filters are used in audio, communications, and instrumentation circuits. Equalizers and speech filters are audio bandpass filters. Station tuning in radio and television uses bandpass filters. Spectrum analyzers measure a circuit's frequency response with a bandpass filter.

The notch filter is the complement of the bandpass filter. It rejects those signals in a given band of frequencies and passes all others. The response of a notch filter is illustrated in Fig. 7-13. Like the bandpass filter, center frequency, low frequency cutoff, and high frequency cutoff are specified. Both bandwidth and Q are defined for the notch as they were for the bandpass filter. The other specification needed is the notch depth. This indicates how severely a signal at center frequency will be rejected.

Notch filters are used to reject unwanted signals at a known frequency. One major application of the notch filter is in sensitive instruments in an environment of high electrical (line) noise. This noise will be at 60 Hz and will vary in frequency only slightly. By passing all input signals through a 60 Hz notch filter, the electrical line noise is removed. Distortion analyzers also use a notch filter to notch out the fundamental of the signal being analyzed. The remaining signal is the harmonic distortion, which is then measured and displayed.

Figure 7-13 Notch filter response

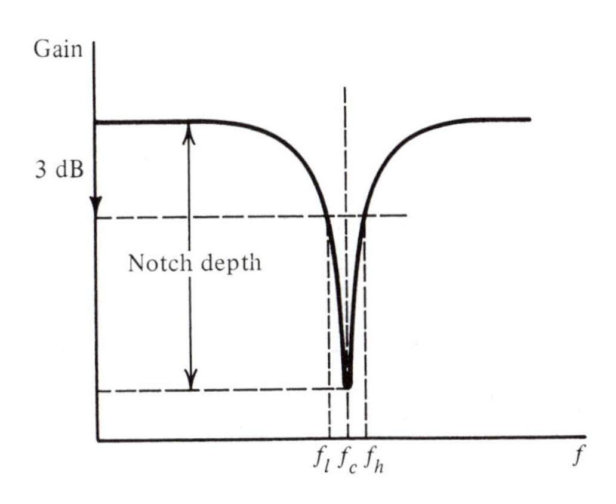

7-1.3 Active Versus Passive Filters

The descriptions of sections 7-1.1 and 7-1.2 apply (more or less) to filters in general, independently of how they are built. The simplest approach to building a filter is with passive components (resistors, capacitors, and inductors). In the radio frequency range this works quite well. However, as the frequency comes down, inductors begin to have problems. Audio frequency inductors are physically large, heavy, and therefore expensive. To increase inductance (for low frequency applications), more turns of wire must be used, which adds to the series resistance, degrading the inductor's performance.

Input and output impedance of passive filters (especially below *rf*) are both a problem. The input impedance is low, which loads down the source, and it varies with frequency. The output impedance is often relatively high, which limits the load impedance that the passive filter can drive. There is no isolation between the load impedance and the passive filter. This means that the load must be considered as a component of the filter and must be taken into consideration when one is determining filter response or design. Any change in load impedance may significantly alter one or more of the filter's response characteristics.

Active filters incorporate an amplifier with resistor/capacitor networks to overcome these problems of passive filters. Originally built with vacuum tubes and then transistors, active filters now normally are centered around op amps. By enclosing a capacitor in a feedback loop, the inductor (with all of its low frequency problems) can be eliminated. Properly configured, input impedance can be increased. The load is driven from the output of the op amp, giving an output impedance as low as a few ohms. Not only does this improve load drive capability, but the load is now isolated from the frequency determining network. Variation in load will have no effect on the active filter's characteristics.

The amplifier allows you to specify and easily adjust passband gain, passband ripple, cutoff frequency, and *initial* rolloff. Because of the high input impedance of the op amp, large value resistors can be used, allowing you to reduce the value (size and cost) of the capacitors. By selecting a quad op amp IC, steep rolloffs can be built in very little space and for very little money.

Active filters also have their limitations. High frequency response is limited by the gain bandwidth and slew rate of the op amps. High frequency op amps are more expensive, making passive filters a more economical choice for *rf* applications. Active filters require a power supply. For op amps this may be two supplies. Variation in these power supplies' output voltage shows up, to some degree, in the signal output from the active filter. In multistage applications, the common power supply provides a bus for high frequency signals. Feedback along the power supply lines can cause oscillations unless decoupling techniques are rigorously applied. Active devices, and therefore active filters, are much more susceptible to radio frequency interference and ionization than are passive RLC filters. Practical considerations limit the Q of active bandpass and notch filters to less than 50. For circuits requiring very selective (narrow) filtering, a crystal filter is best.

7-2 LOW PASS FILTER

There is a wide variety of ways to implement active filters of different orders and types. In this section you will begin by considering the first order filter, a small extension of the circuit analyzed in Example 7-2. The general second order filter comes next, which introduces the concepts of damping and types of response. Two techniques for designing specific types of second order low pass filters follow. Finally, you will see how to cascade first and second order filters to produce filters of whatever order is necessary to achieve the rolloff rate you need.

7-2.1 First Order Low Pass Active Filters

A first order filter consists of a single RC network and has a rolloff rate of 20 dB/decade. At the critical frequency the phase shifts 45 degrees. The denominator of the transfer function has s raised only to the first power (i.e., no s^2, s^3, etc.). This description should sound familiar. It matches the response of the simple passive filter of Example 7-2.

To convert this passive filter to an active filter, you simply connect its output to a noninverting amplifier. If you want a gain of 1 (0 dB), connect the passive filter to a voltage follower. These are shown in Fig. 7-14. For these filters the op amp's main purpose is to drive the load (providing a low output impedance, and

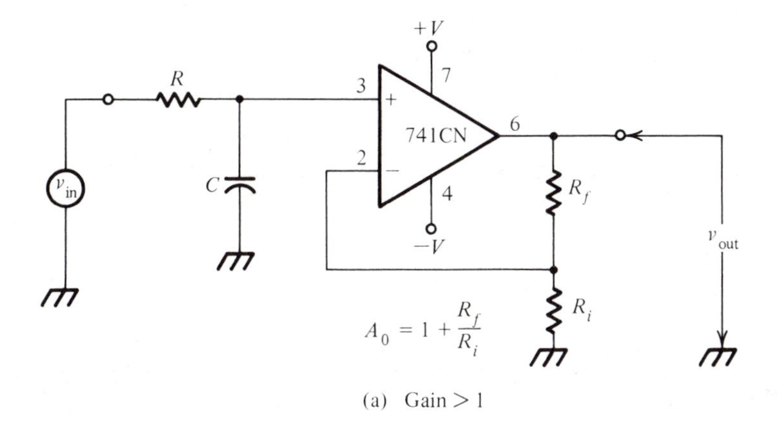

(a) Gain > 1

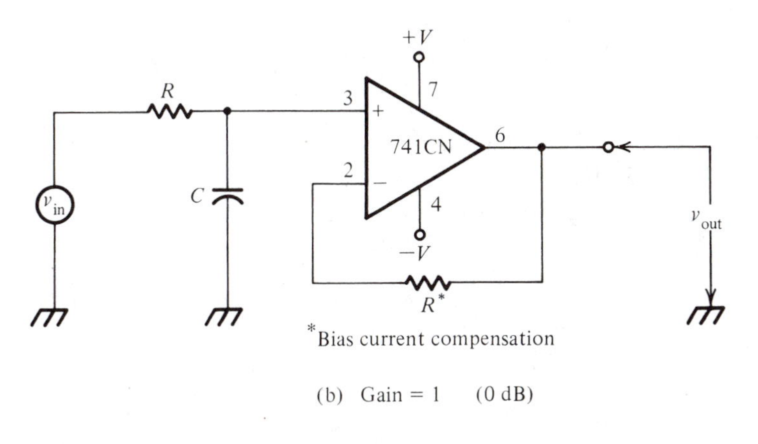

(b) Gain = 1 (0 dB)

Figure 7-14 First order low pass active filter

minimizing the effect of the load impedance on the filter's characteristics), and to provide gain if you want it.

Consequently, the only change in the transfer function for the circuits of Fig. 7-14 from that of Example 7-2, equation (7-9), is the gain A_0.

$$A_0 = 1 + \frac{R_f}{R_i}$$

$$\frac{V_{out}}{V_{in}} = \frac{A_0}{RCs + 1}$$

Letting $\omega_0 = \frac{1}{RC}$, you obtain

$$\frac{V_{out}}{V_{in}} = \frac{A_0}{\frac{s}{\omega_0} + 1}$$

or

$$\frac{V_{out}}{V_{in}} = \frac{A_0 \omega_0}{s + \omega_0} \tag{7-14}$$

Equation (7-14) is the standard form for the transfer function of a first order system.

The gain equation (7-10) need only be changed by A_0.

$$|G| = \frac{A_0}{\sqrt{1 + \omega^2 R^2 C^2}} \tag{7-15}$$

The critical frequency is that frequency at which $|G|$ has dropped to $0.707A_0$ (or $A_0 - 3$ dB).
Substituting into equation (7-15) at f_o, you obtain

$$|G| = 0.707A_0 = \frac{A_0}{\sqrt{1 + \omega_0^2 R^2 C^2}}$$

$$0.707 = \frac{1}{\sqrt{1 + \omega_0^2 R^2 C^2}}$$

$$0.5 = \frac{1}{1 + \omega_0^2 R^2 C^2}$$

$$1 + \omega_0^2 R^2 C^2 = \frac{1}{0.5} = 2$$

$$\omega_0^2 R^2 C^2 = 1$$

$$\omega_0^2 = \frac{1}{R^2 C^2}$$

$$\omega_0 = \frac{1}{RC} \tag{7-16a}$$

$$\omega = 2\pi f$$

$$2\pi f_0 = \frac{1}{RC}$$

$$f_0 = \frac{1}{2\pi RC} \tag{7-16b}$$

At the critical frequency, the phase shift is

$$\phi_o = -\arctan \omega_0 RC$$

$$\phi_o = -\arctan \left(\frac{1}{RC}\right) RC$$

$$\phi_o = -\arctan 1$$

$$\phi_o = -45 \text{ deg}$$

The frequency response plot of the first order low pass active filter is the same as for the circuit in Example 7-2 (Fig. 7-7).

In selecting components to build this filter there are several small points to keep in mind. The op amp, along with R_f and R_i, sets the gain. Its gain bandwidth and slew rate must allow it to operate well beyond the critical frequency ($f_0 \times 10$). Otherwise, the filter response may be partially caused by the op amp's frequency response (which is not entirely predictable). The capacitor should be nonelectrolytic if a bipolar input signal is expected. The filter resistor R strongly affects the input impedance as well as f_o. Keep it high to minimize loading the source. However, the larger R is, the larger are the nonideal effects of offset current at the output. So R should be 100 kΩ or less, unless a precision op amp is to be used. To minimize the effects of bias currents on the output, remember to select R_f and R_i so that their parallel resistance equals R.

$$R = R_f // R_i$$

7-2.2 Second order Active Filter Model

Improved and considerably variable performance is obtained by going to a second order (rather than first order) active filter. The second order filter consists of two RC pairs and has a rolloff rate of 40 dB/decade. At the critical frequency, the phase is shifted 90 degrees. The denominator of the transfer function is a quadratic in s (i.e. $s^2 + as + b$).

The schematic of a general second order active filter is shown in Fig. 7-15. Impedances Z1, Z2, Z3, Z4 may be either resistors or capacitors. This general form is being used so that the results will be applicable to low pass, or high pass filters.

The output v_0 is determined by the signal at the op amp's noninverting input times the amplifier's gain:

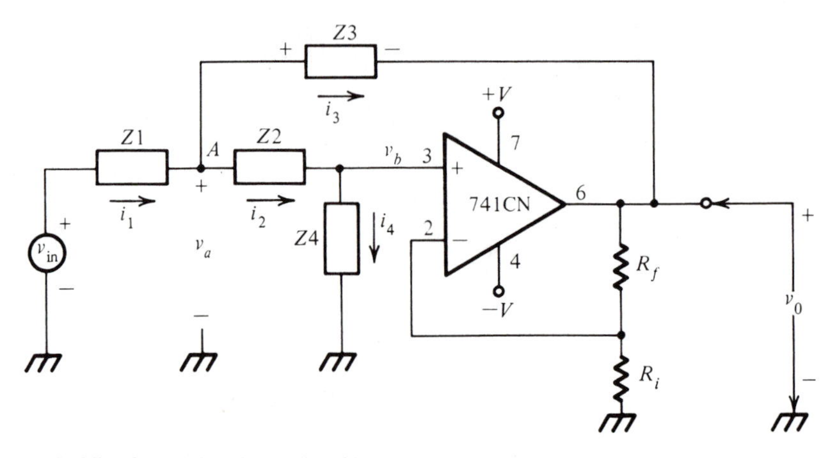

Figure 7-15 Second order active filter model

$$v_0 = A_0 v_b$$

where
$$A_0 = 1 + \frac{R_f}{R_i}$$

$$v_b = \frac{v_0}{A_0} \tag{7-17}$$

Assuming no current into the op amp, you have

$$\mathbf{i}_4 = \frac{v_b}{Z4} \tag{7-18}$$

and

$$\mathbf{i}_2 = \mathbf{i}_4 = \frac{v_b}{Z4}$$

At the node *A*

$$v_a = \mathbf{i}_4(Z2 + Z4) \tag{7-19}$$

Substituting (7-18) into (7-19), you obtain

$$v_a = \frac{v_b}{Z4}(Z2 + Z4) \tag{7-20}$$

Current into the filter, $\mathbf{i}_1$, is the difference in potential across Z1 divided by Z1

$$\mathbf{i}_2 = \frac{v_{in} - v_a}{Z1} = \frac{v_{in}}{Z1} - \frac{v_a}{Z1} \tag{7-21}$$

Substituting equation (7-20) into equation (7-21) gives

$$\mathbf{i}_1 = \frac{v_{in}}{Z1} - \frac{v_b(Z2 + Z4)}{Z1Z4} \tag{7-22}$$

The current through the feedback impedance, $\mathbf{i}_3$, can be calculated by summing the currents at node *A*.

$$\mathbf{i}_3 = \mathbf{i}_1 - \mathbf{i}_2 \tag{7-23}$$

Substituting equations (7-22) and (7-18) into (7-23), you obtain

$$\mathbf{i}_3 = \frac{v_{in}}{Z1} - \frac{v_b(Z2 + Z4)}{Z1Z4} - \frac{v_b}{Z4} \tag{7-24}$$

Summing the loop from node *A*, the feedback loop, and the output, you have

$$v_a - \mathbf{i}_3 Z3 - v_o = 0$$

or

$$v_o = v_a - \mathbf{i}_3 Z3 \tag{7-25}$$

Substituting equations (7-20) and (7-24) into (7-25) gives

$$v_O = \frac{v_b}{Z4}(Z2 + Z4) - \left[\frac{v_{in}}{Z1} - \frac{v_b(Z2 + Z4)}{Z1Z4} - \frac{v_b}{Z4}\right]Z3 \tag{7-26}$$

Substituting equation (7-17) into (7-26), you have

$$v_O = \frac{v_O}{A_O Z4}(Z2 + Z4) - \frac{v_{in} Z3}{Z1} + \frac{v_O(Z2 + Z4)Z3}{A_O Z1Z4} + \frac{v_O Z3}{A_O Z4} \tag{7-27}$$

Equation (7-27) is an expression containing v_O, v_{in}, and circuit components. The circuit analysis is complete. To obtain the transfer equation you must manipulate equation (7-27) to group and separate terms, isolating v_O/v_{in} on the left side of the equation.

When this is done,

$$\frac{v_0}{v_{\text{in}}} = \frac{A_0\, Z3Z4}{Z1Z2 + Z2Z3 + Z3Z4 + Z1Z3 + Z1Z4(1 - A_0)} \tag{7-28a}$$

If you prefer an expression containing admittances,

$$\frac{v_0}{v_{\text{in}}} = \frac{A_0\, Y1Y2}{Y3Y4 + Y1Y4 + Y1Y2 + Y2Y4 + Y2Y3(1 - A_0)} \tag{7-28b}$$

7-2.3 Second Order Low Pass Active Filter Characteristics

To convert Fig. 7-15 into a second order low pass active filter, the resistors (of the two RC pairs) must be in series with the main signal path, and the capacitors must be tied to ground (or to the output which is only a few ohms from ground). This is shown in Fig. 7-16.

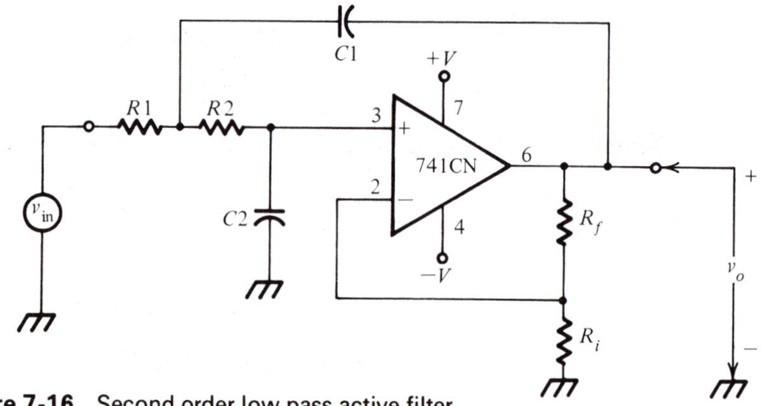

Figure 7-16 Second order low pass active filter

Compare Fig. 7-16 with Fig. 7-15. Equation (7-28a) applies to the low pass active filter of Fig. 7-16 if

$$Z1 = \text{R1}$$

$$Z2 = \text{R2}$$

$$Z3 = \frac{1}{\text{C1}s}$$

$$Z4 = \frac{1}{\text{C2}s}$$

Making these substitutions into equation (7-28a), you have

$$\frac{v_0}{v_{in}} = \frac{\dfrac{A_0}{C1s \cdot C2s}}{R1R2 + \dfrac{R2}{C1s} + \dfrac{1}{C1s \cdot C2s} + \dfrac{R1}{C1s} + \dfrac{R1}{C2s}(1 - A_0)}$$

Finding a common denominator for the denominator, you obtain

$$\frac{v_0}{v_{in}} = \frac{\dfrac{A_0}{C1C2s^2}}{\dfrac{R1R2C1C2s^2 + R2C2s + 1 + R1C2s + R1C1s(1 - A_0)}{C1C2s^2}}$$

$$\frac{v_0}{v_{in}} = \frac{A_0}{R1R2C1C2s^2 + R2C2s + 1 + R1C2s + R1C1s(1 - A_0)}$$

Grouping terms by power of s gives

$$\frac{v_0}{v_{in}} = \frac{A_0}{R1R2C1C2s^2 + [R2C2 + R1C2 + R1C1(1 - A_0)]s + 1}$$

The quadratic in the denominator is more easily solved if the coefficient of s^2 is 1. Dividing numerator and denominator by R1R2C1C2, you obtain

$$\frac{v_0}{v_{in}} = \frac{A_0/(R1R2C1C2)}{s^2 + \left[\dfrac{R2C2 + R1C2 + R1C1(1 - A_0)}{R1R2C1C2}\right]s + \dfrac{1}{R1R2C1C2}} \tag{7-29}$$

Second order physical systems have been studied extensively for many years. Mechanical, hydraulic, and chemical as well as electrical second order systems behave similarly. The transfer function for one group of such second order systems is

$$\frac{A_0 \omega_0^2}{s^2 + \alpha\omega_0 s + \omega_0^2} \tag{7-30}$$

where A_0 is the gain.
ω_0 is the critical frequency in radians/second.
α is the damping coefficient.

Compare the general second order transfer function (7-30) to that derived for the second order low pass active filter [equation (7-29)].

$$A_0 = 1 + \frac{R_f}{R_i}$$

$$\omega_0^2 = \frac{1}{R1R2C1C2}$$

or

$$\omega_0 = \frac{1}{\sqrt{R1R2C1C2}} \tag{7-31a}$$

$$\alpha = \frac{R2C2 + R1C2 + R1C1(1 - A_0)}{\sqrt{R1R2C1C2}} \tag{7-31b}$$

Since it is often more convenient to work with a normalized frequency axis, if

$$\omega_0 = 1 \quad \text{(normalized)}$$

the transfer function becomes

$$\frac{v_0}{v_{in}} = \frac{A_0}{s^2 + \alpha s + 1}$$

$$\alpha = R2C2 + R1C2 + R1C1(1 - A_0)$$

To determine the gain and phase relationships, substitute $s = j\omega$ into the transfer function.

$$\frac{v_0}{v_{in}} = \frac{A_0}{(j\omega)^2 + \alpha j\omega + 1}$$

$$\frac{v_0}{v_{in}} = \frac{A_0}{-\omega^2 + j\alpha\omega + 1}$$

$$\frac{v_0}{v_{in}} = \frac{A_0}{(1 - \omega^2) + j\alpha\omega}$$

To obtain two terms, one real and one imaginary, you must multiply numerator and denominator by the complex conjugate of the denominator.

$$\frac{v_0}{v_{in}} = \frac{A_0}{(1-\omega^2)+j\alpha\omega} \times \frac{(1-\omega^2)-j\alpha\omega}{(1-\omega^2)-j\alpha\omega}$$

$$\frac{v_0}{v_{in}} = \frac{A_0[(1-\omega^2)-j\alpha\omega]}{(1-\omega^2)^2+\alpha^2\omega^2}$$

$$\frac{v_0}{v_{in}} = \frac{A_0(1-\omega^2)}{(1-\omega^2)^2+\alpha^2\omega^2} - j\frac{A_0\,\alpha\omega}{(1-\omega^2)^2+\alpha^2\omega^2}$$

$$\text{Real} = \frac{A_0(1-\omega^2)}{(1-\omega^2)^2+\alpha^2\omega^2} \qquad \text{Imaginary} = -\frac{A_0\,\alpha\omega}{(1-\omega^2)^2+\alpha^2\omega^2}$$

$$|G| = \sqrt{\text{real}^2 + \text{imaginary}^2}$$

$$|G| = \sqrt{\frac{A_0^2(1-\omega^2)^2}{[(1-\omega^2)^2+\alpha^2\omega^2]^2} + \frac{A_0^2\,\alpha^2\omega^2}{[(1-\omega^2)^2+\alpha^2\omega^2]^2}}$$

$$|G| = \frac{A_0\sqrt{[(1-\omega^2)^2+\alpha^2\omega^2]}}{(1-\omega^2)^2+\alpha^2\omega^2} \tag{7-32a}$$

$$|G| = \frac{A_0}{\sqrt{(1-\omega^2)^2+\alpha^2\omega^2}}$$

The phase relationship is

$$\phi = \arctan\frac{\text{imaginary}}{\text{real}}$$

$$\phi = -\arctan\frac{\alpha\omega}{(1-\omega^2)} \tag{7-33a}$$

For the special case where the amplifier is a voltage follower, $A_0 = 1$,

$$|G| = \frac{1}{\sqrt{(1-\omega^2)^2+\alpha^2\omega^2}} \tag{7-32b}$$

$$\phi = -\arctan\frac{\alpha\omega}{1-\omega^2} \tag{7-33b}$$

The equations for the first order filter [(7-10) and (7-11)] show that the normalized response is only a function of frequency. However, equations (7-32) and (7-33) indicate that the normalized *second order* filter depends on both the frequency and α, the damping coefficient. This term describes how stable the filter is. A heavily damped filter ($\alpha > 1.7$) is very stable. Its transient response (response to pulses) is excellent. However, it begins to roll off very early in the pass band, producing significant attenuation to signals near the top of the pass band. As the damping coefficient is reduced, filter transient response produces more and more overshoot and ringing. High pass band response becomes better. However with $\alpha < 1.4$, ripple begins to appear in the gain at the top end of the pass band. Should the damping coefficient be reduced too far, the filter may become unstable and begin to oscillate.

The frequency response of unity gain second order low pass filters is plotted in Fig. 7-17 for various damping coefficients.

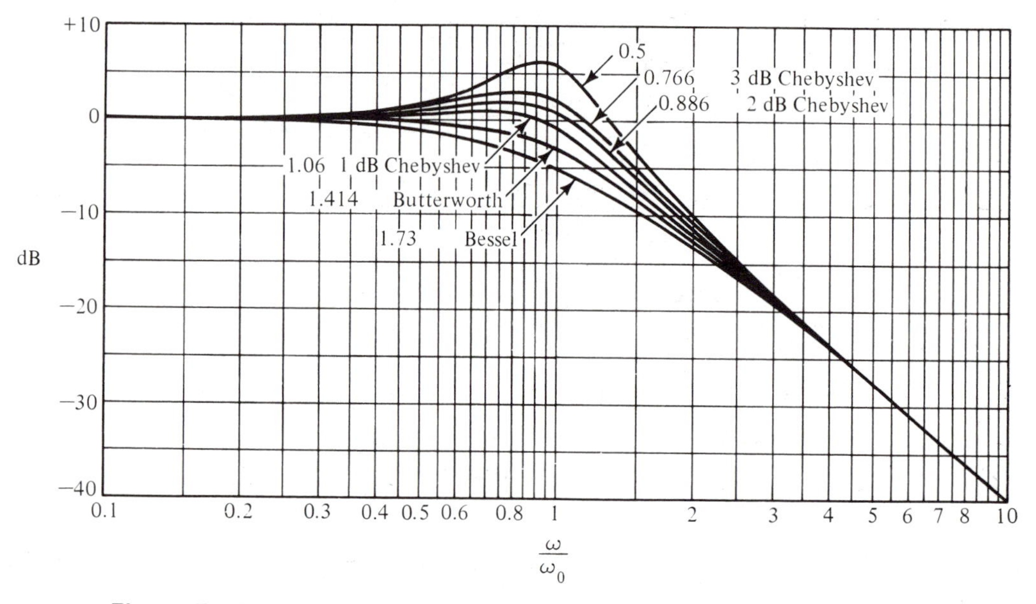

Figure 7-17 Second order low pass unity gain active filter responses for different damping coefficients

The *flattest* pass band response occurs with a damping coefficient of 1.414. This is a Butterworth filter. It will allow some overshoot when a pulse is applied, but gives reasonably fast *initial* rolloff. Audio filters are very often Butterworth.

The Bessel filter has a damping coefficient of 1.73. It is heavily damped. This gives good pulse response and the best possible time delay for sinusoidal signals. However, noticeable rolloff begins at $0.3\omega_0$, causing significant attenuation of the upper end of the *pass* band.

The Chebyshev filters are more lightly damped than the Butterworth. This causes the gain to increase with frequency at the upper end of the pass band. This ripple amplitude is inversely proportional to the damping coefficient. A 3 dB amplitude boost (peak) is usually considered the largest usable ripple. Allowing pass band ripple with the Chebyshev filters provides much faster *initial* rolloff than the Bessel or Butterworth filters. However, decreasing the damping coefficient increases overshoot and ringing, giving Chebyshev filters the poorest transient (pulse) response.

The damping coefficient determines the shape of the frequency response plots *near the critical frequency*. The Butterworth filter offers optimum flatness. Increasing the damping coefficient (Bessel) causes significant pass band attenuation and slow *initial* rolloff, but assures good transient and time delay response. Decreasing the damping coefficient (Chebyshev) allows much faster *initial* rolloff, but introduces pass band ripple and poor transient response. You must realize, however, that the damping coefficient affects the response only *near* the critical frequency. At two octaves above or below the critical frequency ($f < 0.25f_0$; $f > 4f_0$) the gain responses for each type of filter are virtually identical.

Figure 7-18 shows the phase responses of the unity gain second order low pass filter for different damping coefficients. Each curve begins at 0 degrees phase shift, passes through -90 degrees phase shift at the critical frequency, then asymptotes toward -180 degrees. The heavier the damping (higher damping coefficient), the flatter the response.

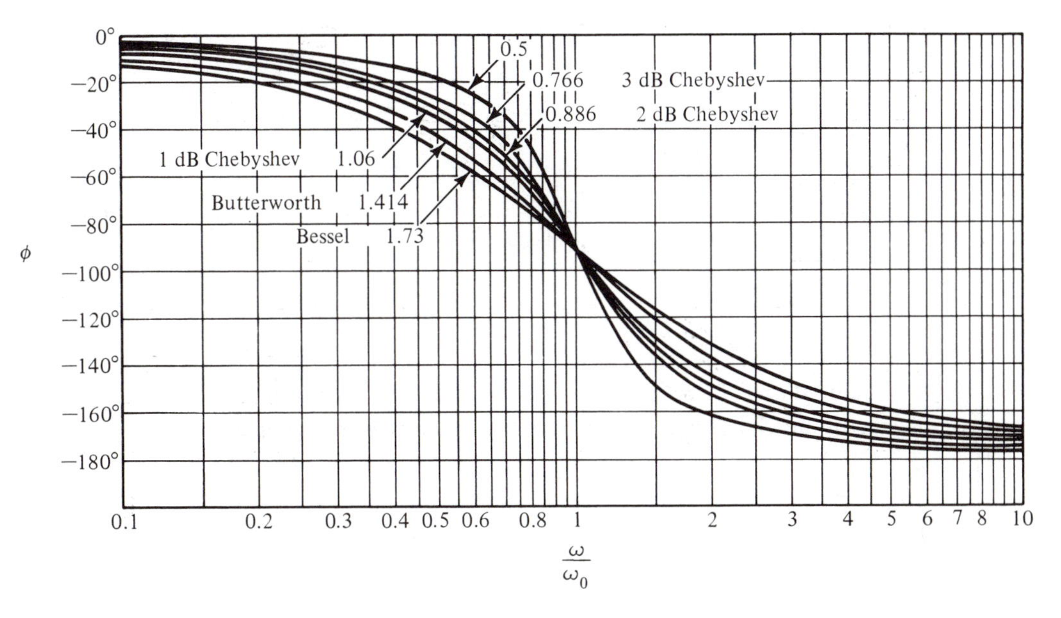

Figure 7-18 Phase shift of second order low pass unity gain active filter for different damping coefficients

7-2.4 Sallen-Key Unity Gain Filter

When designing or analyzing a second order filter, you are concerned with the critical frequency, shape (damping), and possibly pass band gain. But you must deal with R1, R2, C1, C2, R_f and R_i. Changing any one will affect all three parameters. Many different combinations of components will produce the same performance. One of the simplest ways to deal with this tangle of interaction is the Sallen-Key unity gain filter.

The amplifier gain is defined as 1. This means that equations (7-32b) and (7-33b) apply. Second, the two frequency determining resistors are set equal. So for the Sallen-Key unity gain low pass active filter,

$$A_0 = 1$$

$$R1 = R2 = R$$

The parameters [equations (7-31a) and (7-31b)] then become

$$\omega_0 = \frac{1}{\sqrt{R^2 C1C2}} = \frac{1}{R\sqrt{C1C2}}$$

or

$$f_0 = \frac{1}{2\pi R\sqrt{C1C2}} \tag{7-34a}$$

$$\alpha = \frac{2RC2}{R\sqrt{C1C2}}$$

$$\alpha = 2\sqrt{\frac{C2}{C1}} \tag{7-34b}$$

Equations (7-34a) and (7-34b) allow you to analyze a given Sallen-Key unity gain circuit. To design such a circuit, they must be rearranged.

$$R = \frac{1}{2\pi f_0\sqrt{C1C2}} \tag{7-35a}$$

$$C2 = \frac{C1\alpha^2}{4} \tag{7-35b}$$

Example 7-3

Design a filter using the Sallen-Key unity gain low pass active filter to meet the following specifications:

a. Rolloff rate: 40 dB/decade
b. Critical frequency: 2 kHz
c. Pass band as flat as possible
d. Gain of 5 at DC

Solution

A rolloff rate of 40 dB/decade requires a second order filter. The flattest pass band comes from a Butterworth type filter.

$$\alpha = 1.414$$

Pick C1 = 0.1 μF (something convenient).

$$C2 = \frac{C1\alpha^2}{4} = \frac{0.1\ \mu F\ (1.414)^2}{4}$$

$$C2 = 0.05\ \mu F$$

$$R = \frac{1}{2\pi f_0 \sqrt{C1C2}}$$

$$R = \frac{1}{2\pi(2\ kHz)\sqrt{0.1\ \mu F \times 0.05\ \mu F}}$$

$$R = 1.13\ k\Omega$$

To obtain a DC gain of 5, you must add an amplifier with a gain of 5 following the filter, since this filter must have unity gain.

This straightforward technique produces correct results for Butterworth filters. However, look carefully at the critical frequencies of the curves in Fig. 7-17. For filters with pass band ripple (Chebyshev), the peak frequency, not the -3dB frequency, is normally specified. The actual frequency (peak frequency for Chebyshev, -3 dB for Butterworth and Bessel) depends on the damping coefficient. Looking at equation (7-32b), this should not surprise you. Equations (7-34a) and (7-35a) should be corrected.

$$f_\alpha = \frac{k_{1p}}{2\pi R\sqrt{C1C2}} \qquad (7\text{-}36)$$

$$R = \frac{k_{1p}}{2\pi f_\alpha \sqrt{C1C2}}$$

where k_{1p} is the damping coefficient correction factor from Table 7-4 and f_α is the -3dB frequency for $\alpha \geq 1.414$ and the peak's frequency for $\alpha < 1.414$.

TABLE 7-4. Second order filter parameters

Filter Type	*Damping* (α)	*Low Pass Frequency Correction Factor* (k_{lp})
Bessel	1.732	0.786*
Butterworth	1.414	1*
0.5 dB Chebyshev	1.158	0.574†
1 dB Chebyshev	1.054	0.673†
2 dB Chebyshev	0.886	0.779†
3 dB Chebyshev	0.766	0.841†

* 3 dB frequency
† Peak's frequency

Example 7-4

Determine the response shape (type of filter) and frequency of the filter designed in Example 7-3 if C2 were changed to 0.033 μF.

Solution

$$\alpha = 2\sqrt{\frac{C2}{C1}}$$

$$\alpha = 2\sqrt{\frac{0.033\ \mu F}{0.1\ \mu F}}$$

$$\alpha = 1.15$$

From Table 7-4

$$k_{lp} = 0.574$$

$$f_\alpha = \frac{k_{lp}}{2\pi R\sqrt{C_1 C_2}}$$

$$f_\alpha = \frac{0.574}{2\pi(1.1\ k\Omega)\sqrt{0.1\ \mu F \times 0.033\ \mu F}}$$

$$f_\alpha = 1.446\ kHz$$

Looking at Fig. 7-17, you can see that this filter is more heavily damped than the 1 dB Chebyshev, but less than the Butterworth, so there should be a small ($\sim$0.5 dB peak) pass band ripple.

The Sallen-Key unity gain low pass active filter appears easy to use. However, it has several problems. The damping (and therefore the response shape and to a lesser degree the critical frequency) depends on the *ratio* of two capacitors. Obtaining the precise standard value capacitor you need is as much chance as skill. Fine tuning of the damping coefficient is not practical. Small damping coefficients require a very wide ratio between capacitors, forcing you to use both very small and very large capacitors. Finally, any attempt to adjust the damping coefficient (by changing either capacitor) affects the critical frequency doubly; once through k, and also because both capacitors are part of the equation for f_α.

7-2.5 Sallen-Key Equal Component Filter

By allowing the amplifier to have a precise, adjustable *gain* many of the disadvantages encountered with the Sallen-Key unity gain filter can be eliminated. The Sallen-Key equal component filter requires

$$\mathrm{R1} = \mathrm{R2} = \mathrm{R}$$

and

$$\mathrm{C1} = \mathrm{C2} = \mathrm{C}$$

for the filter of Fig. 7-16.

This means that equation (7-31a)

$$\omega_0 = \frac{1}{\sqrt{\mathrm{R1R2C1C2}}}$$

becomes

$$\omega_0 = \frac{1}{\mathrm{RC}} \tag{7-37a}$$

or

$$f_0 = \frac{1}{2\pi \mathrm{RC}} \tag{7-37b}$$

Equation (7-31b)

$$\alpha = \frac{\mathrm{R2C2} + \mathrm{R1C2} + \mathrm{R1C1}(1 - A_0)}{\sqrt{\mathrm{R1R2C1C2}}}$$

becomes

$$\alpha = 3 - A_0 \tag{7-38}$$

The critical frequency is primarily determined by RC, and the shape (damping) is controlled *only* by the amplifier *gain*. This means that damping can now be tuned with a single potentiometer. Capacitors may be much more easily selected from stock values.

The type of filter you want (shape) governs not only the damping coefficient, but also the gain. Lower damping causes higher ripple in the pass band (as it did for the unity gain filter), but it also causes the overall pass band gain to go up. Figure 7-19 is a gain plot for various damping coefficients for the equal component filters. Compare it closely with Fig. 7-17. Figure 7-20 is the phase plot for the equal component filters. Compare it carefully with Fig. 7-18. Phase has a sharper change near 90 degrees for any given damping on the equal component filter.

Example 7-5

Design a filter using the Sallen-Key equal component low pass active filter to meet the following specifications:

a. Rolloff rate: 40 dB/decade
b. Critical frequency: 2 kHz
c. Pass band as flat as possible
d. Gain of 5 at DC

This is a repeat of Example 7-3 using the Sallen-Key equal component filter.

Solution

As with Example 7-3, a second order Butterworth filter is required.

$$f_\alpha = \frac{k_{lp}}{2\pi RC}$$

$$R = \frac{k_{lp}}{2\pi f_\alpha C}$$

Pick C = 0.01 μF.

$$k_{lp} = 1 \text{ for the Butterworth filter}$$

$$R = \frac{1}{2\pi(2\text{ kHz})(0.01\ \mu\text{F})} = 7.96\text{ k}\Omega$$

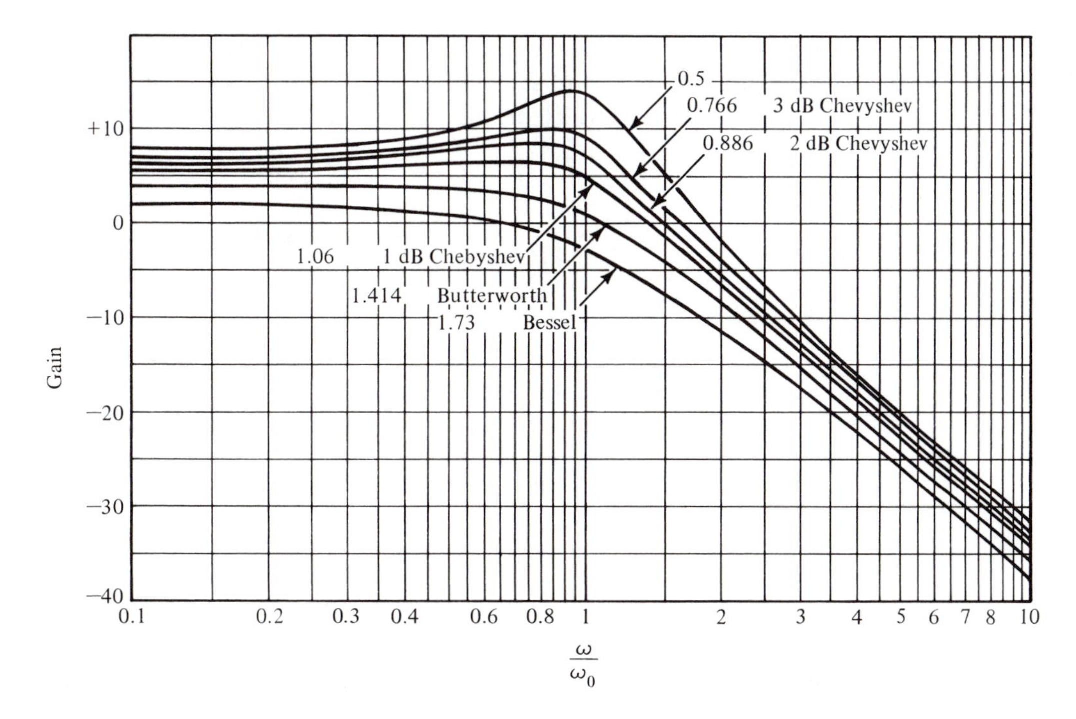

Figure 7-19 Frequency response of the Sallen Key equal component second order low pass filter

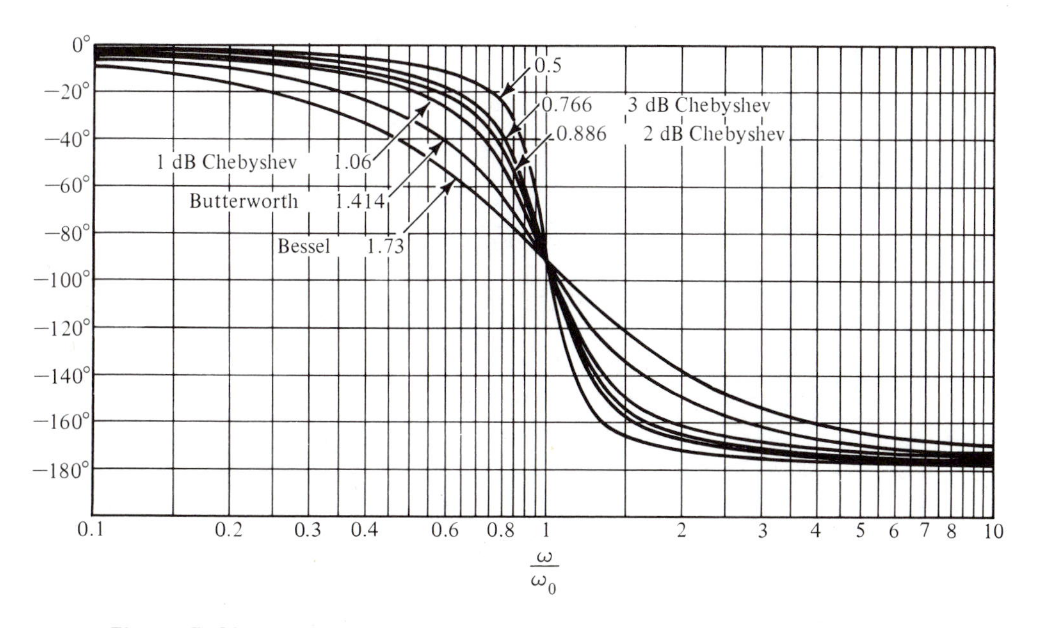

Figure 7-20 Phase shifts of the Sallen Key equal component second order filter

Since this is not close to standard resistor, pick another standard capacitor and repeat the calculation until you have selected a standard capacitor and standard resistor (in the 10 kΩ range) which produces $f_\alpha = 2$ kHz.

Pick C = 0.0068 μF

$$R = \frac{1}{2\pi(2\text{ kHz})(0.0068\ \mu\text{F})} = 11.7\text{ k}\Omega$$

Pick R = 12 kΩ

$$\alpha = 1.414 \text{ for a Butterworth filter}$$

$$A_0 = 3 - \alpha = 1.586$$

$$A_0 = 1 + \frac{R_f}{R_i} = 1.586$$

$$\frac{R_f}{R_i} = 0.586 \quad \text{or} \quad R_f = 0.586R_i$$

However, to minimize the effect of offset current (which may be quite important in a *low* pass filter), the resistance at the inverting terminal must equal the resistance at the noninverting terminal.

$$2R = \frac{R_f \times R_i}{R_f + R_i}$$

Substituting for R_f, you have

$$2R = \frac{0.586R_i^2}{1.586R_i}$$

$$24\text{ k}\Omega = .369\ R_i$$

$$R_i = 65.0\text{ k}\Omega$$

Pick $R_i = 68$ kΩ.

$$R_f = 0.586R_i = 0.586\ (68\text{ k}\Omega)$$

$$R_f = 39.8\text{ k}\Omega$$

Pick $R_f = 39$ kΩ with 1k Ω in series.

To obtain an overall gain of five you must add an amplifier after the filter. It should have a gain of

$$A_{\text{amp}} = \frac{5}{A_0} = \frac{5}{1.586} = 3.15$$

Like the unity gain filter, the damping coefficient has some effect on the critical frequency of the equal components filter. This relationship is tabulated in Table 7-4.

Example 7-6

For the circuit in Fig. 7-16, calculate the critical frequency, the gain in the pass band, and the response shape for

$$\text{R1} = \text{R2} = 22\ \text{k}\Omega$$

$$\text{C1} = \text{C2} = 0.47\ \mu\text{F}$$

$$\text{R}_f = 91\ \text{k}\Omega$$

$$\text{R}_i = 82\ \text{k}\Omega$$

Solution

$$A_0 = 1 + \frac{\text{R}_f}{\text{R}_i}$$

$$A_0 = 1 + \frac{91\ \text{k}\Omega}{82\ \text{k}\Omega} = 2.11$$

The pass band gain is 2.11 or

$$\text{dB} = 20 \log_{10} 2.11 = 6.48\ \text{dB}$$

$$\alpha = 3 - A_0$$

$$\alpha = 3 - 2.11 = 0.89$$

Comparing this to the curves in Fig. 7-20 indicates that this is a 2 dB Chebyshev.

$$f_\alpha = \frac{k_{\text{lp}}}{2\pi \text{RC}}$$

where $k_{\text{lp}} = 0.779$ from Table 7-4

$$f = \frac{0.779}{2\pi(22\ \text{k}\Omega)(0.47\ \mu\text{F})}$$

$$f_\alpha = 12\ \text{Hz}$$

When you are using the Sallen-Key equal component filter there are several points to remember. First, resistors of equal size and capacitors of equal size must be used in the frequency determining network. This in turn allows the damping (and, therefore, the response shape) to be set by the amplifier gain alone. This gain must be kept below 3. With a gain greater than 3 the filter becomes unstable and oscillates. (This should make sense, since the circuit has all three requirements for oscillations: a frequency determining device, an amplifier, and positive feedback).

To minimize the effects of offset current the resistance at the + input (2R) should equal the resistance at the − input ($R_i // R_f$).

$$2R = \frac{R_f R_i}{R_f + R_i}$$

Although the equal component filter requires two more resistors than the unity gain filter, it is much easier to obtain standard components for, and is much easier to tune (both f_α and α). An additional benefit is that an identical high pass equal component filter can be made simply by interchanging the resistors and capacitors in the frequency determining network.

7-2.6 Higher Order Filters

The first order and second order filters, though relatively easy to build, may not provide an adequate rolloff rate. The only way to improve the rolloff rate is to increase the order of the filter. This was illustrated in Fig. 7-10. Each increase in order produces a 20 dB/decade increase in rolloff rate.

Example 7-7

It is necessary to build a filter with a critical frequency of 3 kHz. At 10 kHz the gain must be −40 dB below the pass band gain. Determine the order of filter necessary.

Solution

Figure 7-21a is a repeat of Fig. 7-10. However, the frequency axis has been denormalized (scaled) by multiplying by 3 kHz, the given critical frequency. The intersection of the 10 kHz, −40 dB projection is just above the fourth order filter curve. The fourth order filter will slightly exceed the specifications. The third order filter provides only −32 dB attenuation at 10 kHz, which is inadequate.

Higher order filters can be built by cascading the proper number of first and second order filters (sections). This technique results in a transfer function

$$\frac{A_0}{(s^2 + \alpha_1 s + \omega_1^2)(s^2 + \alpha_2 s + \omega_2^2)(s + \omega_3)\ \ldots}$$

second-order section — another second-order section — first-order section

Each term in the denominator has its own damping coefficient and critical frequency. To obtain a given well-defined frequency response (Bessel, Butterworth, Chebyshev), the transfer function *as a whole* must be solved and the appropriate coefficients must be determined. You will not get a fourth order 1 kHz Bessel filter by cascading two second order 1 kHz Bessel filters. Figure 7-22 illustrates the effects of cascading a first order filter with a second order filter without correcting for damping coefficient and critical frequency.

The mathematics used to solve the higher order polynomials and derive the coefficients necessary to produce the required frequency response is beyond the scope of this book. The results for third order through sixth order filters are presented in Table 7-5.

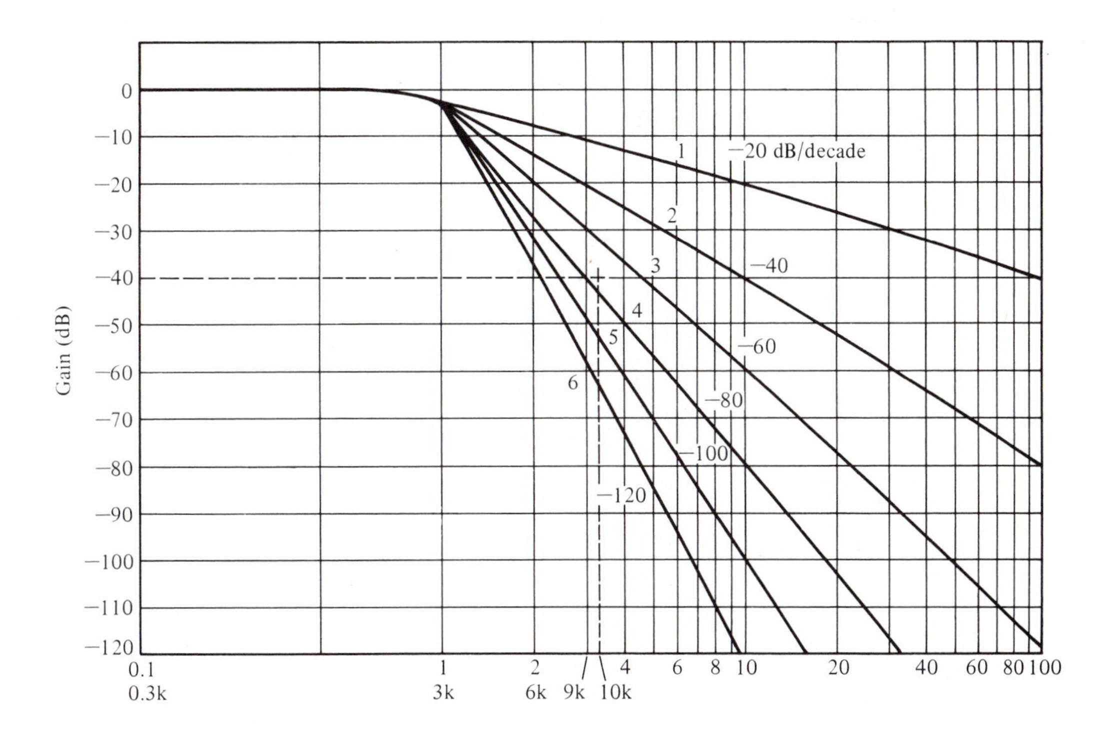

Figure 7-21a Roll-off rate comparison for Example 7-7

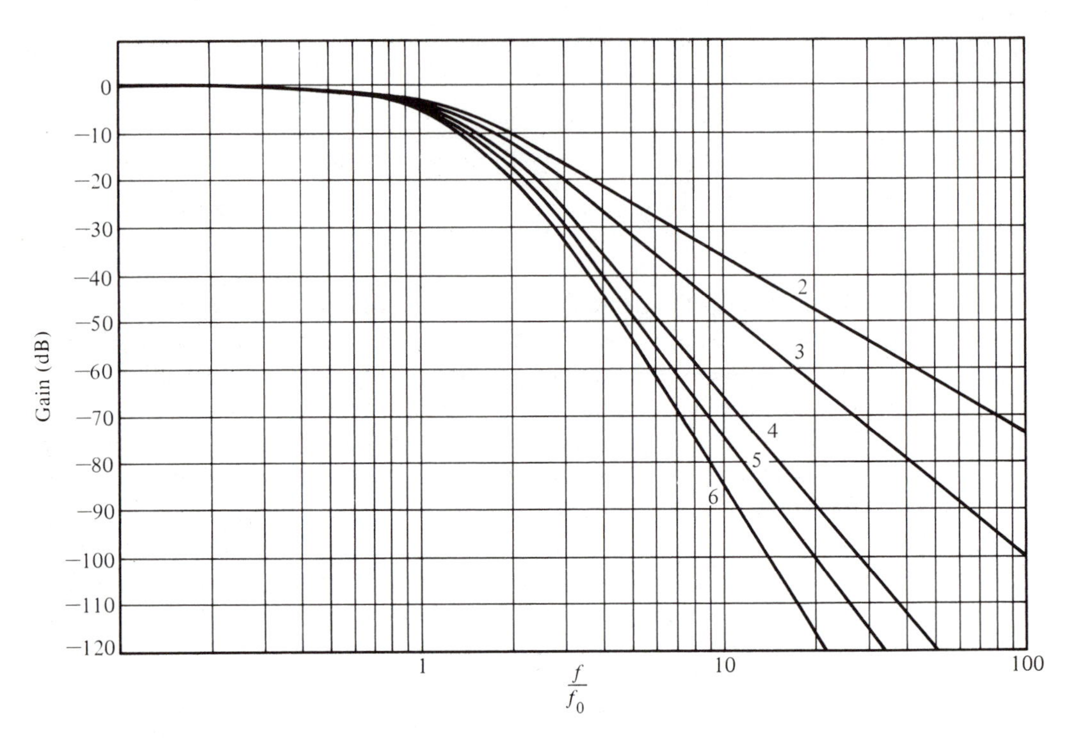

Figure 7-21b Bessel higher order response

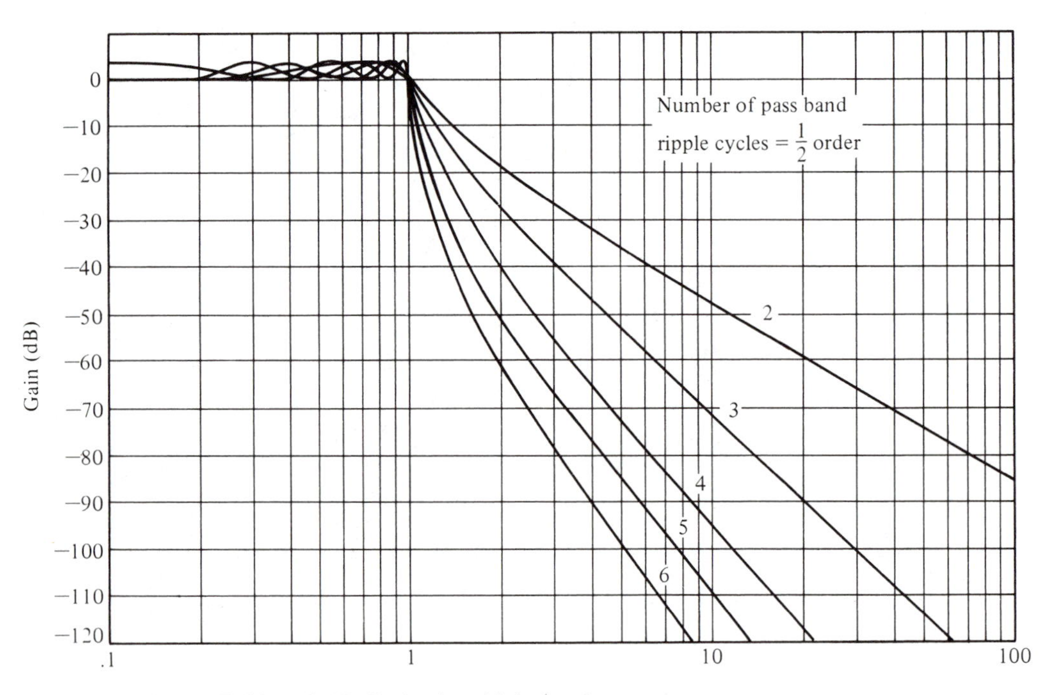

Figure 7-21c 3 dB Chebyshev higher order response

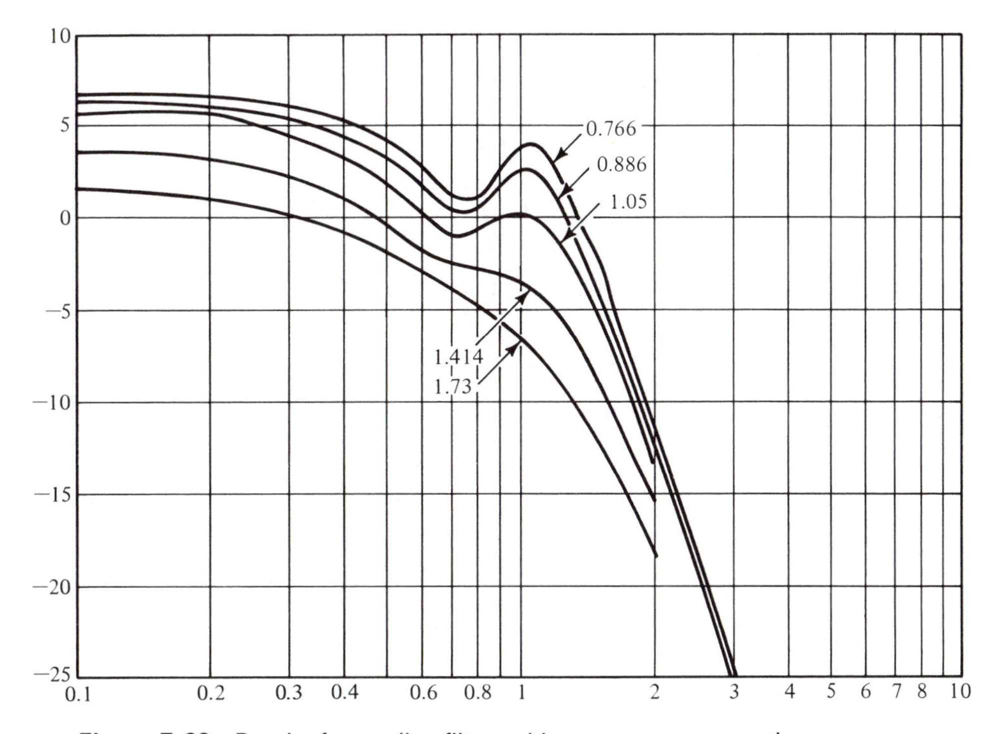

Figure 7-22 Result of cascading filters without parameter correction

Example 7-8

Design a 2 dB peak Chebyshev filter to meet the rolloff requirements of Example 7-7.

Solution

Example 7-7 indicated that a fourth order filter is needed. You can build this by cascading two second order filters. For a 2 dB Chebyshev, from Table 7-5, the first stage must have a damping coefficient of

$$\alpha_1 = 1.076$$

so

$$A_1 = 3 - \alpha = 1.924$$

For that section, the frequency is set by

$$f_\alpha = \frac{k_{lp}}{2\pi RC}$$

$$k_{lp} = 0.648$$

Pick C1 = 0.0033 μF

$$R1 = \frac{k_{1p}}{2\pi f_\alpha C1}$$

$$R1 = \frac{0.648}{2\pi(3\ \text{kHz})(0.0033\ \mu\text{F})} = 10.4\ \text{k}\Omega$$

$$R1 = 10\ \text{k}\Omega$$

$$A_1 = 1 + \frac{R_{f_1}}{R_{i_1}} = 1.924$$

$$\frac{R_{f_1}}{R_{i_1}} = 0.924 \quad \text{or} \quad R_{f_1} = 0.924\ R_{i_1}$$

But

$$2R1 = \frac{R_{f_1} \times R_{i_1}}{R_{f_1} + R_{i_1}}$$

$$20\ \text{k}\Omega = \frac{(0.924R_{i_1})R_{i_1}}{1.924R_{i_1}}$$

$$20\ \text{k}\Omega = 0.480R_{i_1}$$

$$R_{i_1} = 41.6\ \text{k}\Omega \quad (39\ \text{k}\Omega \text{ in series with } 2.7\ \text{k}\Omega)$$

$$R_{f_1} = 0.924R_{i_1} = 0.924\ (41.6\ \text{k}\Omega)$$

$$R_{f_1} = 38.5\ \text{k}\Omega \quad (39\ \text{k}\Omega)$$

The second section is handled the same way, by using

$$\alpha_2 = 0.218$$

$$A_2 = 3 - \alpha_2 = 2.782$$

and

$$k_{21p} = 0.986$$

giving

$$C2 = 0.0033\ \mu\text{F}$$

$$R2 = 16\ \text{k}\Omega$$

$$R_{i_1} \cong 50\ \text{k}\Omega$$

$$R_{f_2} \cong 89\ \text{k}\Omega$$

Neither section, *alone*, will exhibit a 2 dB Chebyshev response, or will have

TABLE 7-5. Higher order filter damping and low pass frequency correction factors*

Filter Order	Sect. Order		Bessel	Butter-worth	1 dB Che.	2 dB Che.	3 dB Che.
	1	α	—	—	—	—	—
		k_{lp}	1	1	1	1	1
3	2	α	1.447	1	0.496	0.392	0.326
		k_{lp}	0.977	0.707	0.937	0.961	0.973
	2	α	1.916	1.848	1.275	1.076	0.929
		k_{lp}	0.684	0.719	0.433	0.648	0.754
4	2	α	1.241	0.765	0.281	0.218	0.179
		k_{lp}	0.479	0.841	0.980	0.986	0.993
	1	α	—	—	—	—	—
		k_{lp}	1	1	1	1	1
5	2	α	1.775	1.618	0.715	0.563	0.468
		k_{lp}	0.761	0.859	0.863	0.917	0.943
	2	α	1.091	0.618	0.180	0.138	0.113
		k_{lp}	0.636	0.899	0.992	0.995	0.997
	2	α	1.960	1.939	1.314	1.109	0.958
		k_{lp}	0.662	0.676	0.368	0.620	0.735
6	2	α	1.636	1.414	0.455	0.352	0.289
		k_{lp}	0.847	1.000	0.947	0.968	0.979
	2	α	0.977	0.518	0.125	0.096	0.078
		k_{lp}	0.723	0.981	0.996	0.998	1

* Jerald G. Graeme, Gene E. Tobey, Lawrence P. Huelsman, *Operational Amplifiers Design and Applications* (New York, N.Y.: McGraw Hill Book Co., 1971), pp 320–25.

the correct cutoff frequency. However, together the proper response is produced. The overall pass band filter gain is the product of the gain of each stage:

$$A_0 = A_1 \times A_2$$

or

$$A_{0\,\mathrm{dB}} = A_{1\,\mathrm{dB}} + A_{2\,\mathrm{dB}}$$

$$A_0 = 1.924 \times 2.782$$

$$A_0 = 5.353$$

This is the low frequency (DC) gain of the entire filter.

When cascading filter sections to produce higher order filters, be sure to use the correct damping and frequency correction factor, as illustrated in Table 7-5. Use the lowest order filter which will meet the given specifications. Damping coefficient (and therefore filter stability) becomes quite small as you increase the order. Use a Butterworth filter if possible. Go to the Bessel if you need better transient response. But this will give poorer *initial* rolloff. Use the Chebyshev filter if *initial* rolloff of the Butterworth is not adequate. However, transient response and pass band flatness will suffer.

7-3 HIGH PASS ACTIVE FILTER

The complement of the low pass filter is the high pass active filter. It is formed by exchanging the place of resistors and capacitors in the frequency determining section of the filter. This is shown in Fig. 7-23. Compare it carefully with Fig. 7-16, the schematic of the second order low pass active filter.

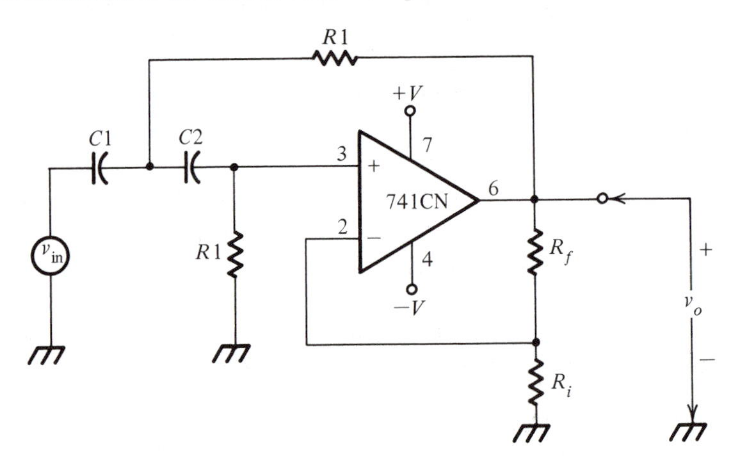

Figure 7-23 Second order high pass active filter

The general second order active filter is shown in Fig. 7-15. Its transfer function was derived in section 7-2.2. The result is

$$\frac{v_0}{v_{in}} = \frac{A_0 Z3Z4}{Z1Z2 + Z2Z3 + Z3Z4 + Z1Z3 + Z1Z4(1 - A_0)} \quad \text{(7-28a)}$$

For the high pass filter of Fig. 7-23,

$$Z1 = \frac{1}{C1s} \qquad Z2 = \frac{1}{C2s}$$

$$Z3 = R1 \qquad Z4 = R2$$

Substituting these into equation (7-28a) gives

$$\frac{v_0}{v_{in}} = \frac{A_0 R1R2}{\dfrac{1}{C1C2s^2} + \dfrac{R1}{C2s} + R1R2 + \dfrac{R1}{C1s} + \dfrac{R2}{C1s}(1 - A_0)}$$

$$\frac{v_0}{v_{in}} = \frac{A_0 R1R2C1C2s^2}{1 + R1C1s + R1R2C1C2s^2 + R1C2s + R2C2s(1 - A_0)}$$

$$\frac{v_0}{v_{in}} = \frac{A_0 R1R2C1C2s^2}{(R1R2C1C2)s^2 + [R1C1 + R1C2 + R2C2(1 - A_0)]s + 1}$$

$$\frac{v_0}{v_{in}} = \frac{A_0 s^2}{s^2 + \dfrac{[R1C1 + R1C2 + R2C2(1 - A_0)]s}{R1R2C1C2} + \dfrac{1}{R1R2C1C2}} \tag{7-39}$$

Second order physical systems have been studied extensively for many years. Mechanical, hydraulic, and chemical, as well as electrical second order systems behave similarly. The transfer function for one group of such second order systems is

$$\frac{A_0 s^2}{s^2 + \alpha\omega_0 s + \omega_0^2} \tag{7-40}$$

where A_0 is the gain.
ω_0 is the critical frequency.
α is the damping coefficient.

Compare the general second order transfer function (7-39) with equation (7-40).

$$A_0 = 1 + \frac{R_f}{R_i} \tag{7-41}$$

$$\omega_0^2 = \frac{1}{R1R2C1C2}$$

$$\omega_0 = \frac{1}{\sqrt{R1R2C1C2}} \tag{7-42}$$

$$\alpha = \frac{R1C1 + R1C2 + R2C2(1 - A_0)}{\sqrt{R1R2C1C2}} \tag{7-43}$$

Compare the critical frequency of the low pass filter equation (7-31a) with the critical frequency of the high pass filter 7-42. They are the same! Look carefully at the damping coefficients of the low pass filter (7-31b) and the high pass filter (7-43). They are *not* the same. But the equations are very complementary.

To normalize the frequency axis

$$\omega_0 = 1$$

The transfer function becomes

$$\frac{v_0}{v_{in}} = \frac{A_0 s^2}{s^2 + \alpha s + 1}$$

$$\alpha = R1C1 + R1C2 + R2C2(1 - A_0)$$

To determine the gain and phase relationships substitute $s = j\omega$ into the transfer function

$$\frac{v_0}{v_{in}} = \frac{-A_0 \omega^2}{-\omega^2 + j\alpha\omega + 1}$$

$$\frac{v_0}{v_{in}} = \frac{(-A_0 \omega^2)[(1-\omega^2) - j\alpha\omega]}{[(1-\omega^2) + j\alpha\omega][(1-\omega^2) - j\alpha\omega]}$$

$$\frac{v_0}{v_{in}} = \frac{-A_0 \omega^2(1-\omega^2) + jA_0 \omega^3 \alpha}{(1-\omega^2)^2 + \alpha^2\omega^2}$$

$$\text{Real} = \frac{-A_0 \omega^2(1-\omega^2)}{(1-\omega^2)^2 + \alpha^2\omega^2}$$

$$\text{Imaginary} = \frac{A_0 \alpha\omega^3}{(1-\omega^2)^2 + \alpha^2\omega^2}$$

The gain is

$$|G| = \sqrt{\text{real}^2 + \text{imaginary}^2}$$

$$|G| = \frac{\sqrt{A_0^2 \omega^4(1-\omega^2)^2 + A_0^2 \alpha^2\omega^6}}{(1-\omega^2)^2 + \alpha^2\omega^2}$$

$$|G| = \frac{A_0 \omega^2}{\sqrt{(1-\omega^2)^2 + \alpha^2\omega^2}} \tag{7-44}$$

Compare this to the gain for the low pass active filter, equation (7-32b). The denominators are identical. However, for the high pass filter, the gain is *directly* proportional to the *square* of the frequency. An increase in frequency causes an increase in gain. This is plotted in Fig. 7-24. Carefully compare it to Fig. 7-19, the gain responses for the low pass filters.

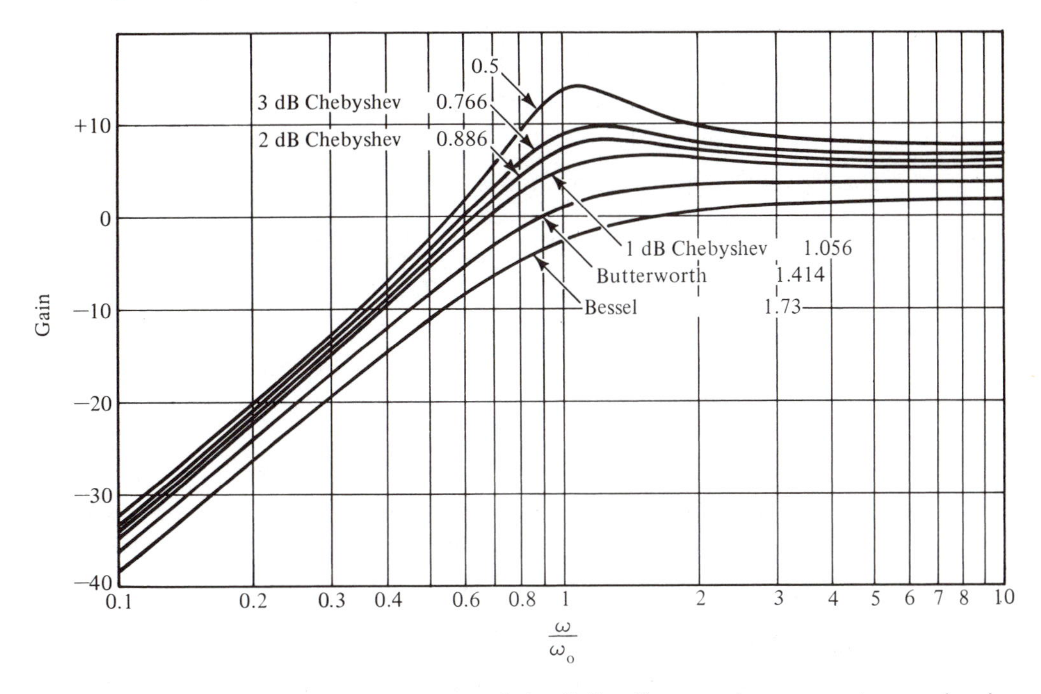

Figure 7-24 Frequency response of the Sallen Key equal component second order high pass filter

The phase shift is

$$\phi = \arctan \frac{\text{imaginary}}{\text{real}}$$

$$\phi = -\arctan \frac{\alpha\omega}{1 - \omega^2}$$

This is the same as equation (7-33b), the phase response of the low pass filter. So the phase shift for the high pass and low pass filters are the same, plotted in Fig. 7-18.

The Sallen-Key equal component filter configuration has several advantages over the unity gain circuit. These were outlined in section 7-2.4. They are true for the high pass as well as for the low pass filter. To build a Sallen-Key equal component high pass filter, set

$$R1 = R2 = R$$

$$C1 = C2 = C$$

$$A_0 = 1 + \frac{R_f}{R_i} = 3 - \alpha$$

The high pass frequency correction is

$$f = \frac{k_{hp}}{2\pi RC}$$

where

$$k_{hp} = \frac{1}{k_{1p}}$$

You can verify this by comparing the peaks or -3 dB points of the low pass filter (Fig. 7-19 and Table 7-4) with the peaks or -3 dB points of the high pass filter (Fig. 7-24).

Example 7-9

For a high pass filter whose components are listed below, calculate the peak or -3dB frequency, filter type, pass band gain, and high frequency response.

$$R = 10\ k\Omega \qquad C = 0.1\ \mu F \qquad R_f = 5.8\ k\Omega$$

$$R_i = 10\ k\Omega$$

Solution

$$A_0 = 1 + \frac{R_f}{R_i} = 1 + \frac{5.8\ k\Omega}{10\ k\Omega}$$

$$A_0 = 1.58 \text{ pass band gain}$$

$$\alpha = 3 - A_0 = 3 - 1.58$$

$$\alpha = 1.42$$

This is a Butterworth filter.

$$k_{hp} = 1 \text{ for a Butterworth filter}$$

$$f_\alpha = \frac{1}{2\pi RC}$$

$$f_\alpha = \frac{1}{2\pi(10\ \text{k}\Omega)(0.1\ \mu\text{F})}$$

$$f_\alpha = 159\ \text{Hz}$$

This is the -3 dB frequency.

The gain bandwidth product for a 741 op amp is 1 MHz. With a pass band gain of 1.58 (4 dB), the filter's gain will fall off at high frequencies. In fact,

$$f_{\text{high}} = \frac{\text{GBW}}{A_0} = \frac{1\ \text{MHz}}{1.58}$$

$$f_{\text{high}} = 633\ \text{kHz}$$

The gain will have fallen off by 3 dB, to 1 dB or 1.117. Above this frequency the gain *decreases* at 20 dB/decade because of the internal compensation of the op amp.

In addition, the slew rate requirement must be met.

$$f_{\max} = \frac{\text{SR}}{2\pi E_{\text{PK}}}$$

The higher order response comparison for the high pass filters is given in Fig. 7-25.

Example 7-10

Design a filter to remove line noise and drift from a transmitted pulse train produced by a voltage-to-frequency converter. The minimum frequency is 500 Hz and must be passed with an amplitude of 2.5 volts or more. The maximum frequency is 10 kHz with a rise and fall time of 1 μs. Line noise (60 hertz) must be reduced by 40 dB. The transmitted signal is 2.8 volts at the output of the voltage-to-frequency converter.

Solution

Since you want to remove drift (DC) and 60 Hz while passing higher frequencies, you will need a high pass filter. To obtain optimum response for the *pulses* (best transient response), a Bessel type filter should be used.

Where should the critical frequency be selected? It would be convenient to select 500 Hz, since this is the minimum signal frequency to be passed.

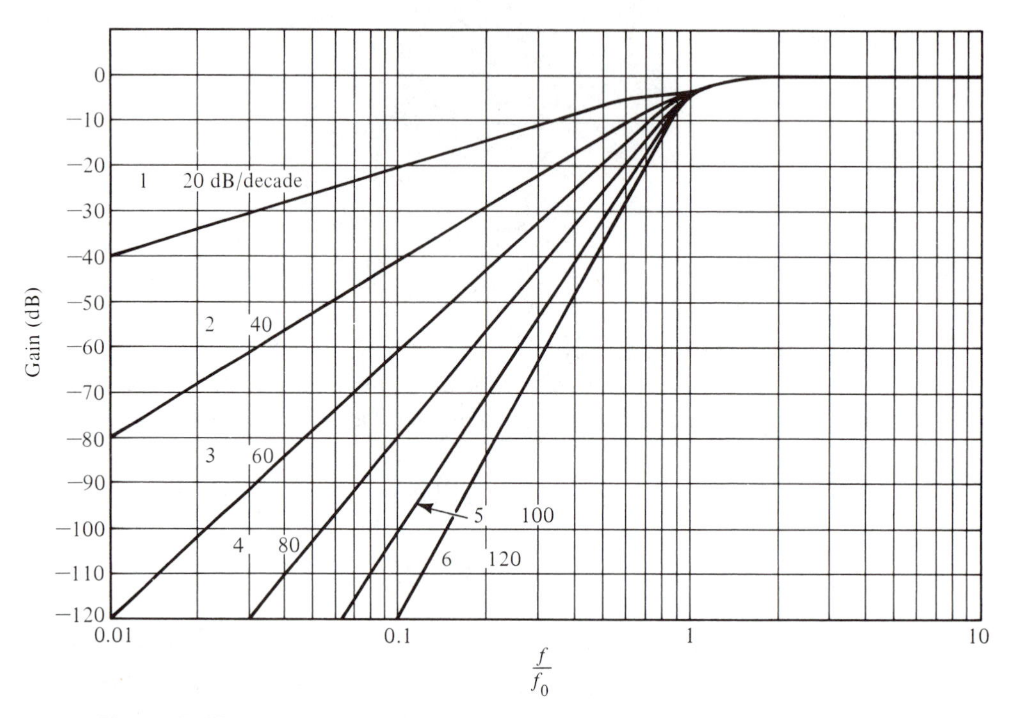

Figure 7-25a Butterworth high pass, higher order responses

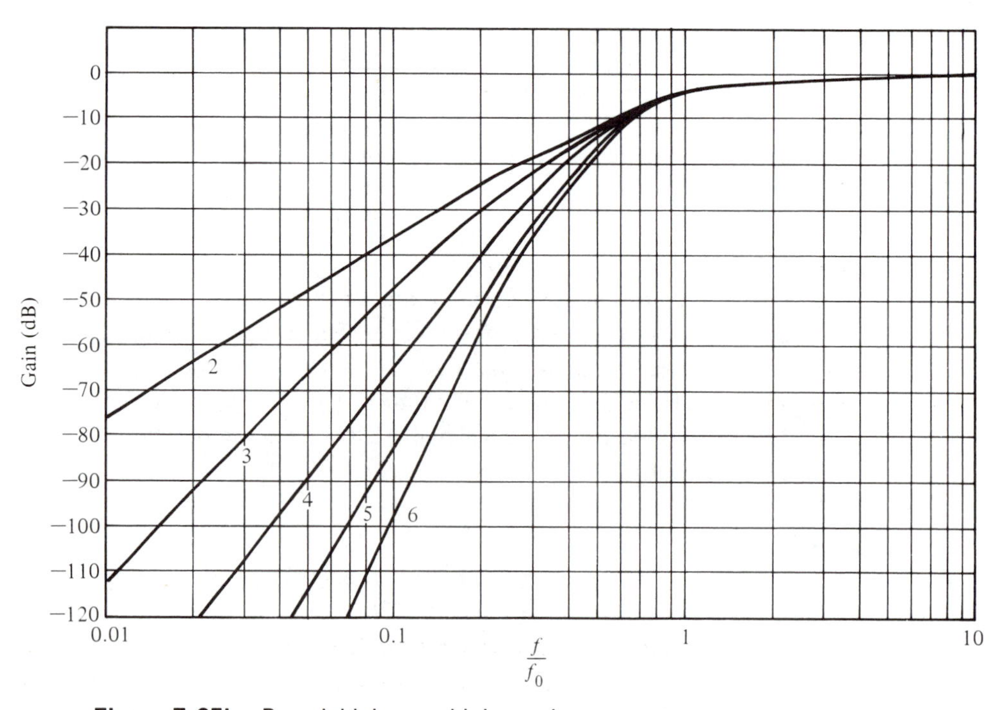

Figure 7-25b Bessel high pass, higher order responses

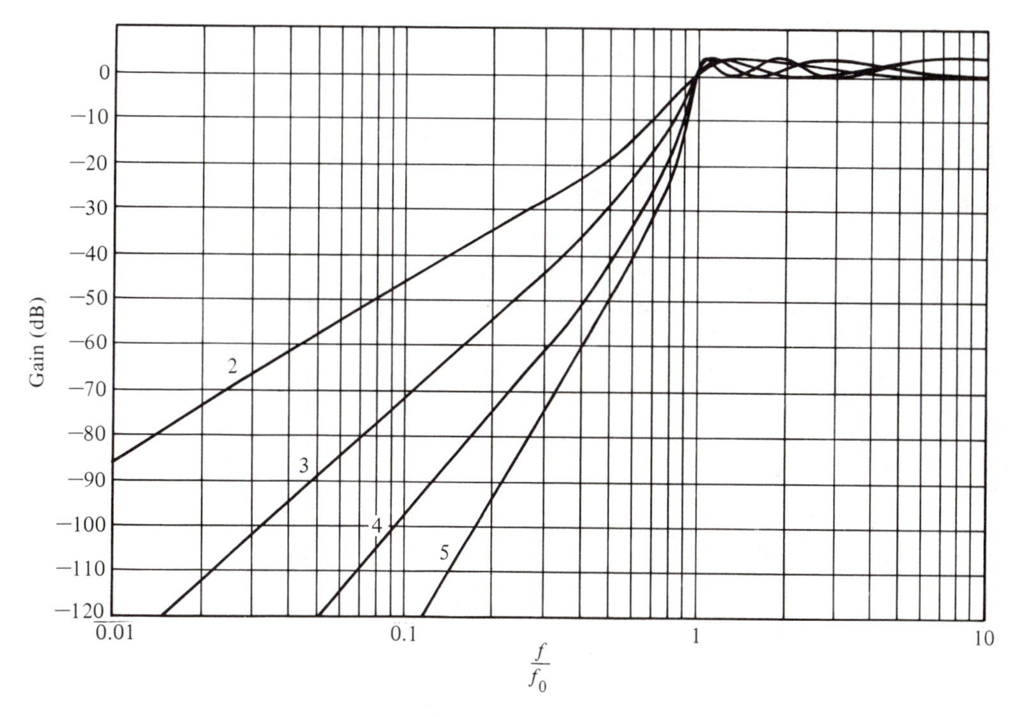

Figure 7-25c 3 dB Chebyshev high pass, higher order responses

Line frequency (60 hertz) signals must be reduced by at least 40 dB. The higher order high pass filters responses are shown in Figs. 7-25(a), (b), (c). The horizontal axis of the Bessel curves must be denormalized, and the proper order must be selected. A third order filter is necessary. This can be built by cascading a first order filter with a second order. Refer to Table 7-5.

For the first order filter, α has no significance, and

$$f_\alpha = \frac{1}{2\pi RC}$$

Pick $C = 0.068\ \mu F$

$$R = \frac{1}{2\pi \times 500\ \text{Hz} \times 0.068\ \mu F} = 4.7\ \text{k}\Omega$$

The second order section has an α of

$$\alpha = 1.447$$

$$A_0 = 3 - \alpha = 1.553$$

$$A_0 = 1 + \frac{R_f}{R_i} = 1.553$$

$$R_f = 0.553R_i$$

Since DC is being blocked, drift and offset of the active filter op amps are not as important as for the low pass filter. It probably is not worth forcing

$$R_{+\text{ input}} = R_{-\text{ input}}$$

So just pick $R_i = 6.2\ \text{k}\Omega$.

$$R_f = 3.4\ \text{k}\Omega$$

$$f_\alpha = \frac{k_{\text{hp}}}{2\pi RC}$$

$$k_{\text{hp}} = \frac{1}{k_{\text{lp}}} = \frac{1}{0.977} = 1.024$$

$$f_\alpha = \frac{1.024}{2\pi RC}$$

Pick $C = 0.1\ \mu F$

$$R = \frac{1.024}{2\pi(500\ \text{Hz})(0.1\ \mu\text{F})}$$

$$R = 3.3\ \text{k}\Omega$$

The pass band gain for a higher order filter is the product of the gain of each stage (or the sum of the dB gain). The pass band gain of a first order filter is 1 (or 0 dB). For the second order stage, $A_0 = 1.553$.

$$A_{\text{filter}} = 1 \times 1.553 = 1.553$$

So the 2.8 volt input will be

$$V_{\text{pass band}} = 2.8\ \text{V} \times 1.553 = 4.35\ \text{V}$$

in the pass band, which is within TTL levels.

At 500 Hz, that signal will have fallen by 0.707.

$$V_{(\text{at } f_x)} = 4.35\text{ V} \times 0.707 = 3.08\text{ V}$$

which exceeds the minimum frequency amplitude specification.

The final thing to consider is which op amp to use. It must have a gain bandwidth of

$$\text{GBW} = A_0 \times f_{(\text{max})} \times 10 \quad \text{(for 10 percent gain error)}$$

$$\text{GBW} = 1.553 \times 10\text{ kHz} \times 10$$

$$\text{GBW} = 155\text{ kHz}$$

The 741C will certainly exceed this requirement.

The rise time is the time required for the output to go from 10 percent to 90 percent of its amplitude. That is, it is the time for the output to slew up by 80 percent of the amplitude.

$$\text{SR} = \frac{\Delta V}{\Delta t} = \frac{80\% \times 4.35\text{ V}}{1\ \mu\text{s}}$$

$$\text{SR} = 3.48\text{ V}/\mu\text{s}$$

The 741C has adequate gain bandwidth, but its slew rate is too slow to meet the rise and fall time specifications. You must use a faster op amp, such as the 318.

7-4 BANDPASS FILTER

The bandpass filter was introduced in section 7-1.2. Its frequency responses, both ideal and actual, are given in Fig. 7-12. Please refer back to these.

The parameters of importance in a bandpass filter are the high and low frequency cutoffs (f_h and f_l), the bandwidth (Δf or B), the center frequency (f_c or ω_c), center frequency gain (A_0), and the selectivity or Q. Several of these parameters were introduced in section 7-1.2. The bandwidth is the distance between the upper frequency and lower frequency -3 dB points.

$$\Delta f = f_h - f_l \tag{7-12a}$$

In radians/second,

$$B = \omega_h - \omega_l \tag{7-12b}$$

The center frequency is at the *geometric* mean of these cutoff frequencies.

$$f_c = \sqrt{f_h f_l} \tag{7-45a}$$

or

$$\omega_c = \sqrt{\omega_h \omega_l} \tag{7-45b}$$

On a *log* plot this puts f_c halfway between f_h and f_l. This is *not* true on a linear plot. You must be careful. With a low frequency cutoff of 1 kHz and a high frequency cutoff of 3 kHz, the center frequency is at 1732 Hz, not 2 kHz.

The selectivity Q is the ratio of the center frequency to the bandwidth.

$$Q = \frac{f_c}{\Delta f} \tag{7-13}$$

or

$$Q = \frac{\omega_c}{B}$$

As such, it gives a measure of the relative narrowness of the filter. The higher the Q, the narrower the bandwidth, and therefore the more selective the filter.

Conversely, knowing the circuit's Q and center frequency, you can calculate the upper and lower cutoff frequencies.

$$f_l = f_c\left[\sqrt{\frac{1}{4Q^2} + 1} - \frac{1}{2Q}\right] \tag{7-45c}$$

$$f_h = f_c\left[\sqrt{\frac{1}{4Q^2} + 1} + \frac{1}{2Q}\right] \tag{7-45d}$$

The schematic of a single op amp band pass filter is given in Fig. 7-26. Capacitor C1 and resistor R2 produce a high pass effect while resistor R1 and capacitor C2 are low pass. Resistor R3 lowers the filter *gain requirement* to a manageable level.

With this simple filter, stable, predictable operation occurs for

$$1 < Q < 10 \text{ to } 20$$

For lower or higher values of Q, multi-op amp circuits are required.

7-4.1 Single Op Amp Bandpass Filters

The transfer function of a second order bandpass filter is

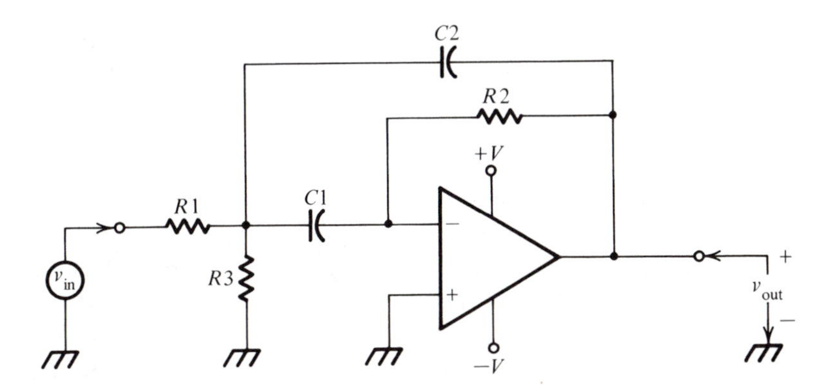

Figure 7-26 Single op amp bandpass filter

$$\frac{v_o}{v_{in}} = \frac{A_0 \alpha \omega_0 s}{s^2 + \alpha \omega_0 s + \omega_0^2} \tag{7-46}$$

where

α = damping coefficient.

ω_0 = center frequency.

A_0 = gain at the center.

It has the same denominator as the second order low pass and the second order high pass filters. The numerator of the bandpass filter contains the first power of s. The numerator of the low pass filter's transfer function had no s term. The numerator of the high pass filter had s^2.

To obtain a plot of gain versus frequency, you substitute $j\omega = s$ in equation (7-46), just as was done for the low pass and high pass filters. Then multiply the numerator and denominator by the complex conjugate of the denominator and simplify. This allows you to separate the real and the imaginary parts of the equation. The gain, then, is

$$\text{Gain} = \sqrt{\text{real}^2 + \text{imaginary}^2}$$

or

$$\text{Gain} = \sqrt{\frac{A_0^2 \alpha^2 \omega_0^2 \omega^2}{\omega^4 + \omega_0^2(\alpha^2 - 2)\omega^2 + \omega_0^4}}$$

The Q, defined in terms of center frequency and bandwidth, is determined by the circuit's damping.

$$Q = \frac{1}{\alpha}$$

Figure 7-27 is a plot of the frequency response (gain) of the circuit in Fig. 7-26. The horizontal axis and vertical axis have been normalized for convenience. To convert to a particular center frequency f_c, multiply the horizontal axis by f_c. Add A_0 (dB) to the vertical axis to scale it for a given A_0 (dB).

There are two points you should notice. The higher the Q, the sharper (and more selective) the filter. This corresponds to lowering the damping. However, this difference appears only between approximately $0.5 f_c$ to $2 f_c$. Below $0.5 f_c$ and above $2 f_c$ the filters all rolloff at 20 dB/decade independent of the Q. This is limited by the two RC pairs in the circuit. One pair causes the 20 dB/decade rolloff at low frequencies. The other governs high frequencies. To obtain sharper rolloff away from the center frequency, you must use a multi-op amp circuit.

The circuit in Fig. 7-26 can be analyzed mathematically to determine how component values affect each of these parameters. Figure 7-28 is the analysis schematic of the circuit. The capacitors have been set equal to each other, and the

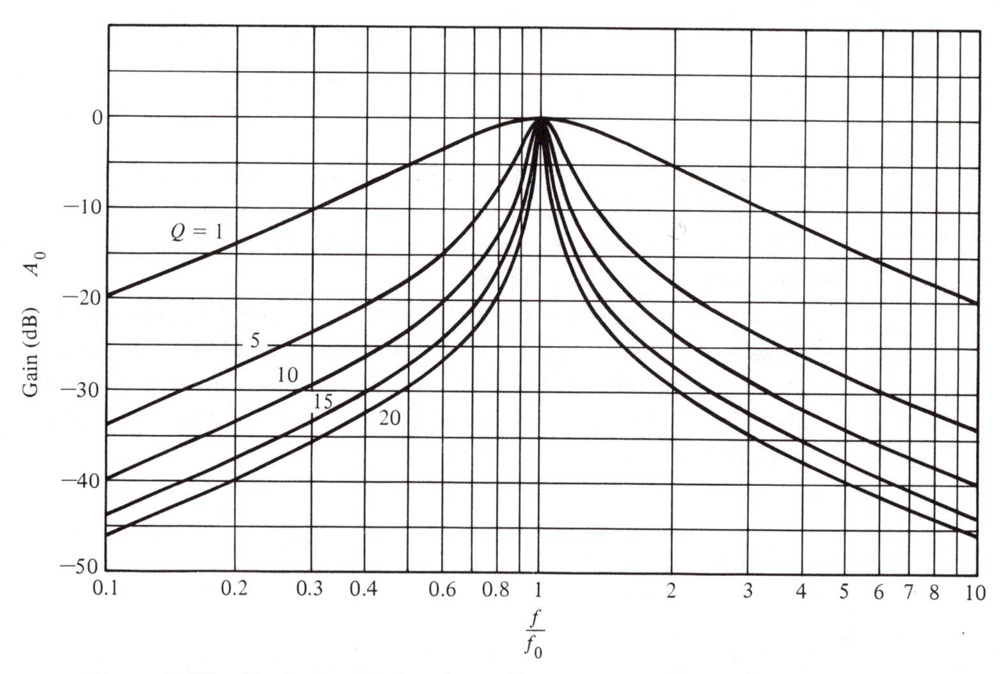

Figure 7-27 Single op amp bandpass filter response (Gain and frequency both normalized)

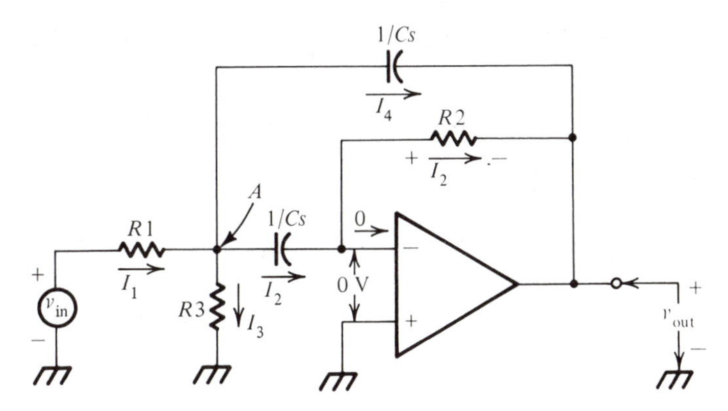

Figure 7-28 Analysis of single op amp bandpass filter

Laplace transform of each circuit element is used. As with the analysis of other op amp circuits, because of high open loop gain and negative feedback, there is no difference in potential between the op amp's inverting and noninverting inputs. Also, no significant signal current flows into either op amp input.

Therefore, I2 flows through both the capacitor and R2. The left end of R2 is at virtual ground. This means that all of v_{out} is dropped across R2 by I2.

$$\mathbf{I}_2 = -\frac{v_{out}}{\mathrm{R2}} \tag{7-47}$$

The voltage at node A with respect to ground is the voltage dropped across the capacitor by I_2.

$$v_A = \frac{\mathbf{I}_2}{\mathrm{C}s}$$

$$v_A = -\frac{v_{out}}{\mathrm{R2C}s} \tag{7-48}$$

The current through R3 is produced by v_A.

$$\mathbf{I}_3 = \frac{v_A}{\mathrm{R3}}$$

$$\mathbf{I}_3 = -\frac{v_{out}}{\mathrm{R2R3C}s} \tag{7-49}$$

The current through the upper feedback capacitor, I4, is

$$I_4 = -\frac{v_A - v_{out}}{\dfrac{1}{Cs}} = -(v_A - v_{out})(Cs)$$

After some algebraic manipulations, this becomes

$$I_4 = -\frac{v_{out}(R2Cs + 1)}{R2} \tag{7-50}$$

The input current I_1 is

$$I_1 = \frac{v_{in} - v_A}{R1}$$

$$I_1 = \frac{v_{in}}{R1} + \frac{v_{out}}{R1R2Cs} \tag{7-51}$$

The current entering node A (I_1) must equal the current leaving node A $(I_2$, I_3, and $I_4)$.

$$I_1 = I_2 + I_3 + I_4 \tag{7-52}$$

Substituting equations (7-47), (7-49), (7-50), and (7-51) into equation (7-52), you obtain

$$\frac{v_{in}}{R1} + \frac{v_{out}}{R1R2Cs} = -\frac{v_{out}}{R2} - \frac{v_{out}}{R2R3Cs} - \frac{v_{out}(R2Cs + 1)}{R2}$$

With diligence and some luck, this can be simplified to

$$\frac{v_{out}}{v_{in}} = \frac{-R2R3Cs}{(R1R2R3C^2)s^2 + (2R1R3C)s + R1 + R3}$$

or

$$\frac{v_{out}}{v_{in}} = \frac{\left(\dfrac{1}{-R1C}\right)s}{s^2 + \dfrac{2}{R2C}s + \dfrac{R1 + R3}{R1R2R3C^2}} \tag{7-53}$$

Compare equation (7-53) to equation (7-46).

$$\omega_0^2 = \frac{\text{R1} + \text{R3}}{\text{R1R2R3C}^2}$$

or

$$\omega_0 = \sqrt{\frac{\text{R1} + \text{R3}}{\text{R1R2R3C}^2}} \tag{7-54a}$$

$$f_c = \frac{1}{2\pi}\sqrt{\frac{\text{R1} + \text{R3}}{\text{R1R2R3C}^2}} \tag{7-54b}$$

All components combine to determine the center frequency.

The coefficients of s in the denominator of the two equations are

$$\alpha\omega_0 = \frac{2}{\text{R2C}} \tag{7-55}$$

The coefficients of s in the numerators of the two equations are

$$A_0\alpha\omega_0 = -\frac{1}{\text{R1C}} \tag{7-56}$$

Dividing equation (7-56) by equation (7-55) gives

$$A_0 = -\frac{\text{R2}}{2\text{R1}} \tag{7-57}$$

At the center frequency, the gain is set by the ratio of the feedback resistor to the input resistor, similar to the inverting amplifier.

The coefficient of the second term in the denominator of equation (7-46) is $\alpha\omega_0$. This must equal the coefficient of the second term in the denominator of equation (7-53).

$$\alpha\omega_0 = \frac{2}{\text{R2C}}$$

or

$$\alpha = \frac{2}{\text{R2}\omega_0\text{C}}$$

It turns out, however, that the Q defined by equation (7-13)

$$Q = \frac{\omega_0}{B} \tag{7-13}$$

is the reciprocal of the damping coefficient α.

$$Q = \frac{1}{\alpha}$$

or

$$Q = \frac{\text{R}2\omega_0 \text{C}}{2}$$

This simply says that the higher the Q, the lower the damping coefficient and the more peaked the response.

$$Q = \frac{\omega_0}{B} = \frac{\text{R}2\omega_0 \text{C}}{2}$$

or

$$B = \frac{2}{\text{R2C}}$$

$$\Delta f = \frac{2}{2\pi \text{R2C}}$$

The bandwidth (in radians) is set *solely* by R2 and C. Tuning of R1 and R3 will not alter the bandwidth.

The equations above allow you to analyze a given single op amp active filter to determine center frequency, gain (at ω_0), Q, bandwidth, and low frequency and high frequency cutoff.

For design, the same group of equations must be manipulated a bit. This is illustrated in Example 7-11.

Example 7-11

Design a bandpass active filter with a single op amp to meet the following specifications:

$$f_l = 3 \text{ kHz}$$

$$f_h = 3.5 \text{ kHz}$$

$$A_0 = -5$$

Solution

The center frequency

$$f_c = \sqrt{f_l \cdot f_h}$$

$$f_c = \sqrt{3 \text{ kHz} \cdot 3.5 \text{ kHz}}$$

$$f_c = 3240 \text{ Hz}$$

The bandwidth is

$$\Delta f = f_h - f_l$$

$$\Delta f = 3.5 \text{ kHz} - 3 \text{ kHz}$$

$$\Delta f = 500 \text{ Hz}$$

$$Q = \frac{f_c}{\Delta f}$$

$$Q = \frac{3240 \text{ Hz}}{500 \text{ Hz}}$$

$$Q = 6.5$$

This is within the limits set for a single op amp bandpass filter.

$$1 < Q < 10 \text{ to } 20$$

To calculate component values, first pick a convenient size capacitor. Let

$$\text{C} = 0.027\ \mu\text{F}$$

Next R2, which alone sets the bandwidth, must be determined.

$$\text{R2} = \frac{2}{2\pi\, \Delta f \text{C}} \tag{7-60}$$

This is obtained from equation (7-59b).

$$R2 = \frac{2}{2\ \pi\ 500\ \text{Hz} \cdot 0.027\ \mu\text{F}}$$

$$R2 = 23.6\ \text{k}\Omega$$

Set $$R2 = 24\ \text{k}\Omega$$

The gain is set by both R1 and R2.

$$A_0 = -\frac{R2}{2R1} \tag{7-57}$$

or

$$R1 = -\frac{R2}{2A_0}$$

$$R1 = -\frac{24\ \text{k}\Omega}{2(-5)}$$

$$R1 = 2.4\ \text{k}\Omega$$

Finally, the value of R3 must be calculated to set the center frequency. From equation (7-54b)

$$f_c = \frac{1}{2\pi}\sqrt{\frac{R1 + R3}{R1R2R3C^2}}$$

You must obtain an equation for R3. After some algebraic manipulation, the result is

$$R3 = \frac{R1}{4\pi^2 R1R2 f_c^2 C^2 - 1} \tag{7-58}$$

$$R3 = \frac{2.4\ \text{k}\Omega}{4\pi^2(2.4\ \text{k}\Omega)(24\ \text{k}\Omega)(3240\ \text{Hz})^2(0.027\ \mu\text{F})^2 - 1}$$

$$R3 = 146\ \Omega$$

To fine tune the filter, lower R1, R2, and R3 to the next smaller standard value. Then add a small series potentiometer to each. The *order* in which you adjust these potentiometers is important. Since bandwidth is only affected by R2 [see equation (7-59)], tune R2 first to obtain the desired bandwidth. Gain is affected by R2 and R1 [equation (7-57)]. Without changing R2 (since the bandwidth is set where you want it), adjust R1 to get the desired center frequency gain. Finally, the center frequency is affected by all three resistors [equation (7-54)]. Without changing R1 or R2, lastly adjust R3 to obtain the desired center frequency.

Only center frequency is affected by R3, so you can move the center frequency about (by changing R3) without altering gain or bandwidth. If you need to adjust the gain (with R1) you must also retune the center frequency. Changing the bandwidth with R2 will require you to readjust both gain and center frequency.

There are two component limitations to this single op amp bandpass filter. First, look at equation (7-58). To get a real value for R3, the denominator may not be zero or negative. That is

$$4\pi^2 \mathrm{R1R2} f_c^2 \mathrm{C}^2 - 1 > 0$$

or

$$f_c > \frac{1}{2\pi\sqrt{\mathrm{R1R2}}\,\mathrm{C}} \tag{7-59a}$$

This same condition can be expressed in terms of the filter's specifications.

$$2Q^2 > A_0 \tag{7-59b}$$

Otherwise, the filter cannot be built by using the schematic of Fig. 7-26. This condition is not really all that restrictive. If you need more gain [equation (7-59b) is violated], set the filter gain to a lower, workable level and add an amplifier after the filter.

The second component limitation is the op amp's gain bandwidth. At the center frequency the op amp's open loop gain must be at least $20Q^2$ to ensure less than 10 percent gain error. In terms of the op amp's gain bandwidth,

$$\mathrm{GBW} \geq 20Q^2 \times f_c \tag{7-61}$$

7-4.2 Multistage Bandpass Filters

The single op amp bandpass filter works reasonably well for moderate values of Q if 20 dB/decade rolloff is adequate away from the center frequency. However, if you need wide band operation ($Q < 1$), very narrow band operation ($Q > 20$), or faster rolloff, you must cascade several filters.

Cascading two single op amp bandpass filters together increases the rolloff rate from 20 dB/decade to 40 dB/decade. A third stage sets the rolloff rate to 60 dB/decade. However, as with cascading low pass filters or high pass filters, the overall resultant damping coefficient and, therefore, Q change.

For a multistage bandpass filter to produce maximum sharpness, set the center frequency of all stages equal to the desired overall center frequency.

$$f_c = f_{c1} = f_{c2} = f_{c3}$$

Set the Q of each stage equal to the desired overall Q corrected by the Q correction factor.

$$Q_{\text{corrected}}\ 2\ \text{stage} = (Q_{\text{desired}})(0.644) \tag{7-62a}$$

$$Q_{\text{corrected}}\ 3\ \text{stage} = (Q_{\text{desired}})(0.510) \tag{7-62b}$$

$$Q_{\text{corrected}}\ \text{n stages} = Q_{\text{desired}}\sqrt{2^{1/n} - 1} \tag{7-62c}$$

Cascading single op amp bandpass filters will allow you to obtain reasonable results for

$$Q_{\text{desired}} < 50$$

Higher Qs are better obtained from more sophisticated circuits. The overall cascaded filter gain is the product of the gain of each stage (or the sum of the dB gains of each stage).

$$A_{0\ \text{overall}} = A_{01} \times A_{02} \times A_{03} \times \cdots \times A_{0n}$$

Example 7-12

Design a circuit to meet the following specifications:

$$A_0 = -9$$

$$Q = 30$$

$$f_c = 60\ \text{Hz}$$

60 dB/decade rolloff rate

Solution

To obtain a 60 dB/decade rolloff rate, you must cascade three single op amp bandpass filters.

$$f_c = 60 \text{ Hz} = f_{c1} = f_{c2} = f_{c3}$$
$$Q_1 = Q_2 = Q_3 = (0.51)(Q_{\text{desired}})$$
$$Q_1 = Q_2 = Q_3 = (0.51)(30) = 15.3$$
$$A_0 = A_{01} \times A_{02} \times A_{03}$$

Letting $A_{01} = A_{02} = A_{03}$, you obtain

$$A_0 = A_{01}^3$$
$$A_{01} = \sqrt[3]{A_{01}}$$
$$A_{01} = \sqrt[3]{-9} = -2.08$$

Now you simply design a single op amp bandpass filter with

$$f_c = 60 \text{ Hz}$$
$$A_{01} = 2.08$$
$$Q_1 = 15.3$$

Follow the procedure you saw in Example 7-11. Then cascade three of these identical stages to produce the overall filter.

You can alter the shape of two cascaded filter stages by keeping the Q of each stage equal, but by slightly staggering the center frequencies. To produce a symmetrically shaped response, one filter's frequency should be shifted up by the same proportion that the other filter is shifted down. This is easily done by

$$f_{c\,\text{low}} = mf_c$$
$$f_{c\,\text{high}} = f_c/m$$
$$m = \text{stagger factor}$$

With $m = 1$ a *maximum peak* frequency response is produced, as previously discussed. As m falls, the top of the pass band spreads out. However, the initial and final rolloff rates are not affected. For

$$m = 0.946 \quad \text{(2 stages)}$$

the pass band is at *maximum flatness.* Decreasing the stagger factor more will cause the pass band to have ripple.

To stagger-tune a three stage filter, keep one stage tuned at f_c, and shift the center frequency of the other two.

$$f_{c\,\text{low}} = mf_c$$

$$f_{c\,\text{mid}} = f_c$$

$$f_{c\,\text{high}} = f_c/m$$

For maximum flatness in a three stage filter, set

$$m = 0.877 \quad \text{(3 stages)}$$

Also, when you are stagger-tuning a three stage filter, best shape is obtained if

$$Q_{\text{low}} = Q_{\text{high}}$$

$$Q_{\text{mid}} = \tfrac{1}{2}Q_{\text{low}}$$

The Q of the mid-frequency stage should be set to half of the other stages' Q.

A wide band bandpass filter requires a low Q. If you were to build this with a single op amp bandpass filter, the rolloff rate would be only 20 dB/decade right up to the center frequency (see Fig. 7-27, $Q = 1$). This will not produce a selective filter.

However, in many audio and speech processing applications low Q filters (wide band) are needed with flat tops and sharp edges. You can build such a filter by cascading a Butterworth low pass filter with a Butterworth high pass filter. The effect is shown in Fig. 7-29.

The high frequency cutoff of the bandpass filter is the critical frequency of the low pass filter. Conversely, the low frequency cutoff is the critical frequency of the high pass filter. A fourth order low pass cascaded with a fourth order high pass filter has been shown. You should use the lowest order filters capable of providing adequate performance. The gain has been normalized. The actual center frequency gain will be the product of the pass band gains of the two cascaded filters.

By keeping the critical frequencies of the two cascaded filters well separated, there will be very little interaction between the damping coefficients. This means that you need not apply damping coefficient correction factors (Tables 7-4 and 7-5) as was necessary when cascading low pass filter sections (or high pass filter sections) together. However, this restricts this technique of bandpass filter construction to wide band (low Q) applications only.

7-5 NOTCH FILTER

The bandpass filter enhanced one set of frequencies while rejecting all others. The notch filter does just the opposite. It rejects a band of frequencies, while passing

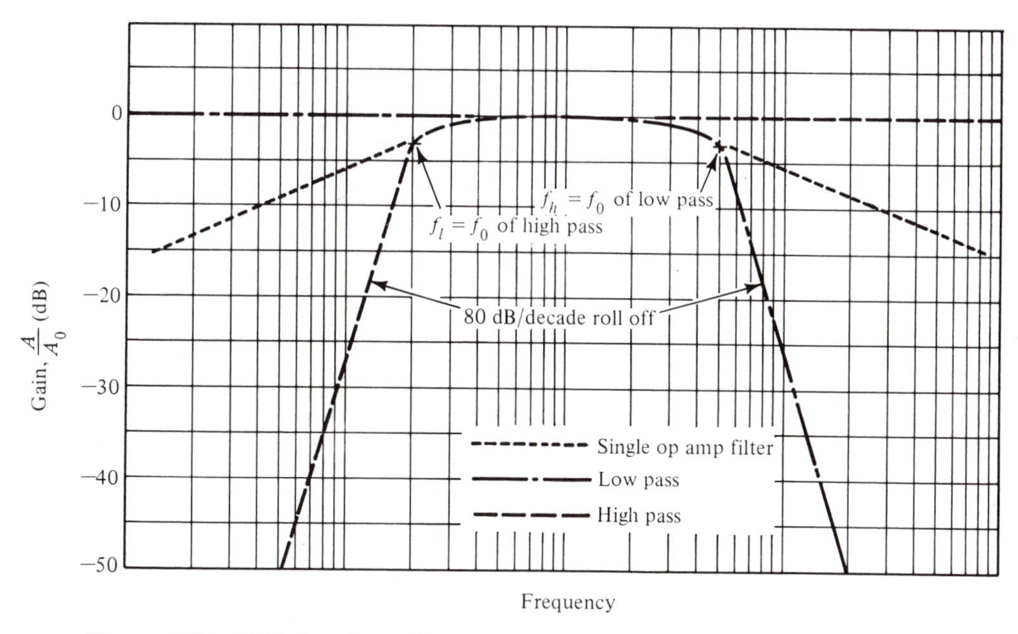

Figure 7-29 Wide bandpass filter response of 4th order low pass filter cascaded with 4th order high pass filter

all others. This becomes quite handy, for recovering signals which are buried in line (60 Hz) noise. Also, harmonic distortion is easily measured when you notch out the fundamental. The remaining signal is the sub- and superharmonics which are a direct indication of harmonic distortion.

There are several different circuits which produce notch filters. However, perhaps the most straightforward technique is to subtract the bandpass filter output from its input. The result is a notch. This is illustrated in block form in Fig. 7-30. The center frequency, f_c, is the largest out of the bandpass filter, subtracting the most from the input signal. This yields a very small output at f_c. Away from the center frequency, the bandpass filter has only a small output. This subtracts only a small amount from the input signal, giving a large output away from the center (or notch) frequency.

The bandpass filters you have seen all have an inverted output. That means that their gain or transfer is already negative. Consequently, when building a circuit to implement Fig. 7-30, use a summer rather than a subtractor. Also, the bandpass filter has a gain of A_0, so the center frequency will be $-A_0 \times v_{in}$. To completely subtract this out, the input of the summer must be precisely $A_0 v_{in}$. A gain of A_0 must be added between the input signal and the summer. Refer to Fig. 7-31.

Figure 7-30 Notch filter block diagram

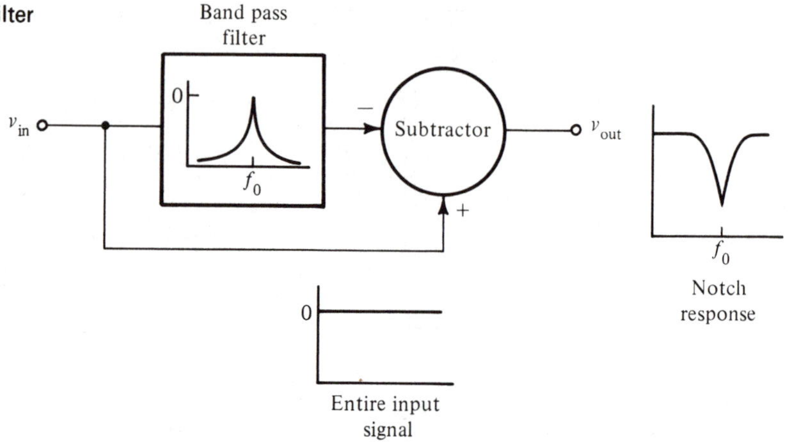

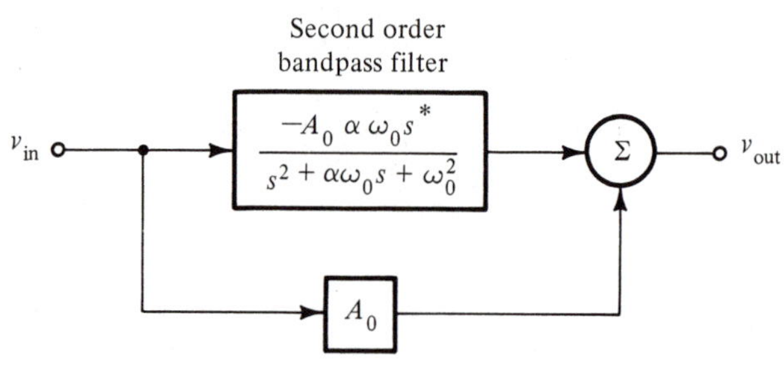

Figure 7-31 Practical notch filter block diagram

The output of the circuit is the sum of the outputs from the two blocks.

$$v_{out} = A_0\, v_{in} + \left(\frac{-A_0\,\alpha\omega_0\, s v_{in}}{s^2 + \alpha\omega_0\, s + \omega_0^2}\right)$$

$$\frac{v_{out}}{v_{in}} = A_0 - \frac{A_0\,\alpha\omega_0\, s}{s^2 + \alpha\omega_0\, s + \omega_0^2}$$

$$\frac{v_0}{v_{in}} = A_0\left(1 - \frac{\alpha\omega_0\, s}{s^2 + \alpha\omega_0\, s + \omega_0^2}\right)$$

$$\frac{v_0}{v_{in}} = A_0\,\frac{s^2 + \alpha\omega_0\, s + \omega_0^2 - \alpha\omega_0\, s}{s^2 + \alpha\omega_0\, s + \omega_0^2}$$

$$\frac{v_0}{v_{in}} = \frac{A_0(s^2 + \omega_0^2)}{s^2 + \alpha\omega_0\, s + \omega_0^2} \tag{7-63}$$

This is the transfer function for a second order notch filter. The terms are defined precisely as they are for the bandpass filter of equation (7-46).

As with the other filters, to obtain the gain equation, substitute $s = j\omega$ into equation (7-63). Manipulate the result to separate the real and imaginary parts. Then apply

$$\text{Gain} = \sqrt{\text{real}^2 + \text{imaginary}^2}$$

The result is

$$|G| = \frac{A_0\,|\omega_0^2 - \omega^2|}{\sqrt{\omega^4 + (\alpha^2 - 2)\omega_0^2\omega^2 + \omega_0^4}} \tag{7-64}$$

This is plotted in Fig. 7-32, for different values of Q. The gain and frequency have both been normalized.

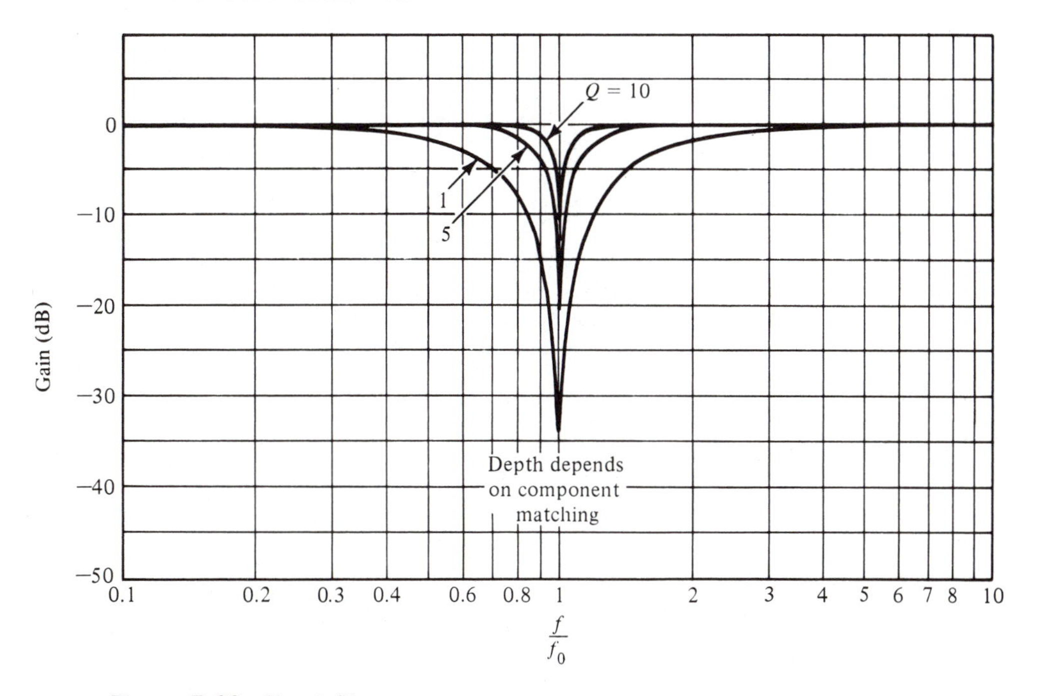

Figure 7-32 Notch filter response

To build a notch filter you build a bandpass filter with the desired Q and bandwidth. Then, connect its output into a summer. The other input of the summer is connected back to the input of the bandpass filter. Finally, the gains must be adjusted. The schematic of a second order notch filter is shown in Fig.

7-33. The bandpass filter is R1, R2, R3, and C. The summer consists of R4, R5, and R6. Build and tune the bandpass filter as you saw in the previous section. To select values for the summer, you must make the gain at center frequency for the bandpass filter through the summer equal the input's gain through the summer. That is,

$$\frac{\mathrm{R2}}{\mathrm{2R1}} \times \frac{\mathrm{R5}}{\mathrm{R4}} = \frac{\mathrm{R5}}{\mathrm{R6}} \tag{7-65a}$$

or if

$$\mathrm{R5} = \mathrm{R4}$$

$$\frac{\mathrm{R2}}{\mathrm{2R1}} = \frac{\mathrm{R5}}{\mathrm{R6}} \tag{7-65b}$$

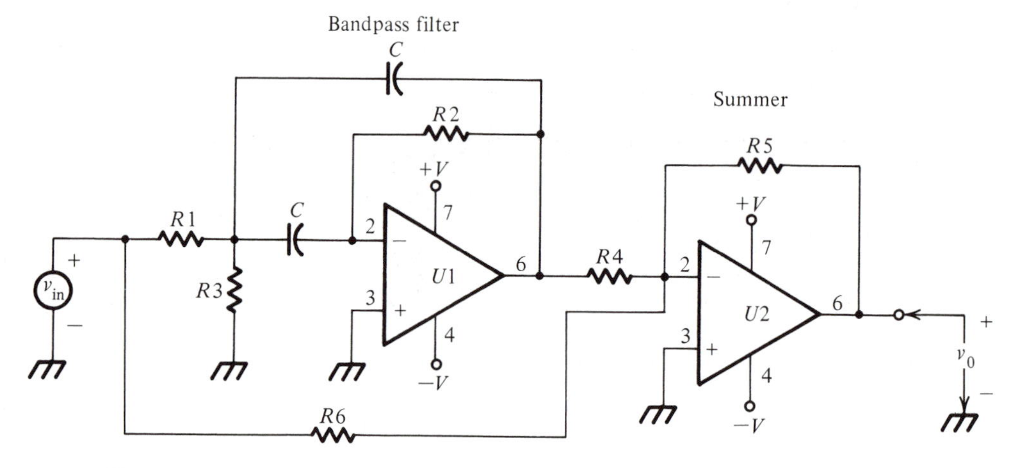

Figure 7-33 Notch filter schematic

Example 7-13

Design a notch filter to meet the following specifications:

$$f_l = 3 \text{ kHz}$$

$$f_h = 3.5 \text{ kHz}$$

$$A_0(\text{outside of the notch}) = 5$$

Solution

This is the same problem as Example 7-11. So use the bandpass filter response determined in that example.

$$C = 0.027\ \mu F$$

$$R1 = 2.4\ k\Omega$$

$$R2 = 24\ k\Omega$$

$$R3 = 150\ \Omega$$

Applying equation (7-65b), you obtain

$$R6 = \frac{2R1\ R5}{R2}$$

Picking $R4 = R5 = 12\ k\Omega$, you have

$$R6 = \frac{2 \times 2.4\ k\Omega \times 12\ k\Omega}{24\ k\Omega}$$

$$R6 = 2.4\ k\Omega$$

Select R6 = 2.2 kΩ resistor with a series 500 Ω potentiometer. This will allow you to tune the gain of the input. The more closely the two gains match, the deeper the notch. So adjust R1, R2, and R3 as described in the bandpass filter section to get the pass band gain, bandwidth, and center frequency. Then correct and adjust R6 for maximum notch depth.

The bandpass filter section used in the notch filter is not restricted to a single op amp. You may use any of the multi-op amp (narrow and wide band types) to obtain the notch characteristics needed.

7-6 STATE VARIABLE FILTER

The filters you have seen so far are relatively simple single op amp circuits, or several single op amp circuits cascaded. The state variable filter, however, uses three or four op amps and two feedback paths. Though a bit more complicated, the state variable configuration offers several features not available with the simpler Sallen-Key filters. First, all three filter types (low pass, bandpass, and high pass) are available *simultaneously*. By properly summing these outputs some very interesting responses can be made. Bandpass filters with high Q can be built. The damping and/or critical frequency could be *electronically* tuned.

Figure 7-34 is the schematic of a three op amp, unity gain state variable filter. Op amps U2 and U3 are integrators. Op amp U1 sums the input with the low pass output and a portion of the bandpass output. The circuit is actually a small analog computer, designed to solve the differential equation (transfer function) for each filter type. For proper operation,

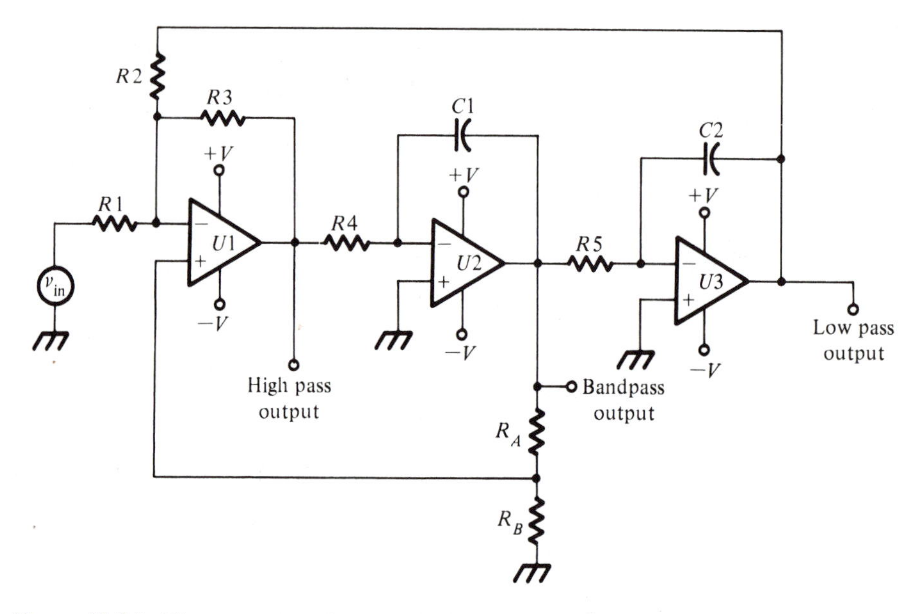

Figure 7-34 Three op amp state variable filter

$$R1 = R2 = R3 = R$$

$$R4 = R5 = R_f$$

$$C1 = C2 = C$$

You will see from the following derivations that the critical frequencies of each of the three filters are equal.

$$f_0 = \frac{1}{2\pi R_f C} \tag{7-66}$$

The damping is set by R_A and R_B. This determines the type of low pass and high pass response (Bessel, Butterworth, or Chebyshev).

$$\alpha = 3\left[\frac{R_B}{R_A + R_B}\right] \tag{7-67}$$

It also sets the Q and gain of the bandpass filter

$$Q = 1/\alpha$$

$$A_0 = Q \tag{7-68}$$

The detailed analysis of the state variable filter must be done in several steps. First, to determine the effect U2 has on the output of U1, notice that the inverting input pin of U2 is at virtual ground. So the current flowing through R4 is

$$\mathbf{I}_4 = \frac{V_{\mathrm{HP}}}{\mathrm{R4}}$$

Since none of this current flows into the op amp, it must all go to charge (and discharge) C1. In the Laplace domain,

$$V_c = \frac{I}{Cs}$$

so

$$V_{c1} = \frac{I_4}{Cs}$$

$$V_{\mathrm{BP}} = -\frac{I_4}{Cs}$$

$$V_{\mathrm{BP}} = -\frac{\frac{V_{\mathrm{HP}}}{\mathrm{R4}}}{Cs}$$

$$V_{\mathrm{BP}} = -\frac{V_{\mathrm{HP}}}{\mathrm{R4}Cs}$$

or

$$V_{\mathrm{BP}} = -\frac{V_{\mathrm{HP}}}{\mathrm{R}_f Cs} \tag{7-69}$$

Applying the same technique to U3 yields

$$V_{\mathrm{LP}} = -\frac{V_{\mathrm{BP}}}{\mathrm{R}_f Cs} \tag{7-70}$$

Op amp U1 is a summer with three inputs. To determine its output (V_{HP}) you can use superposition. The superposition circuits are shown in Fig. 7-35. When considering either $\mathrm{V_{in}}$ or $\mathrm{V_{LP}}$, the other inputs are all grounded, leaving a simple inverter with unity gain. The bandpass output V_{BP} is voltage divided by a ratio of

$$\rho = \frac{\mathrm{R_B}}{\mathrm{R_A} + \mathrm{R_B}}$$

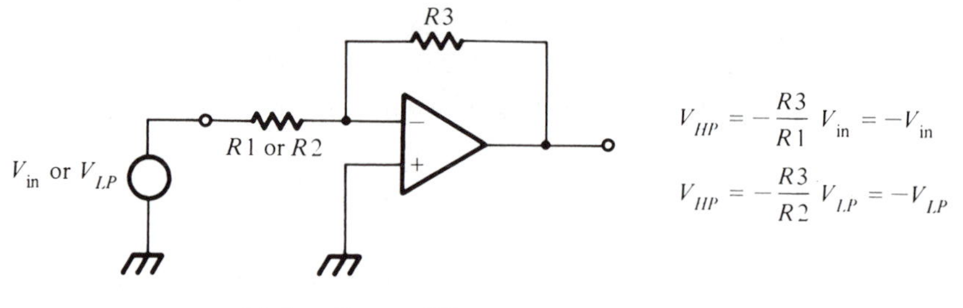

(a) Inverting amplifier

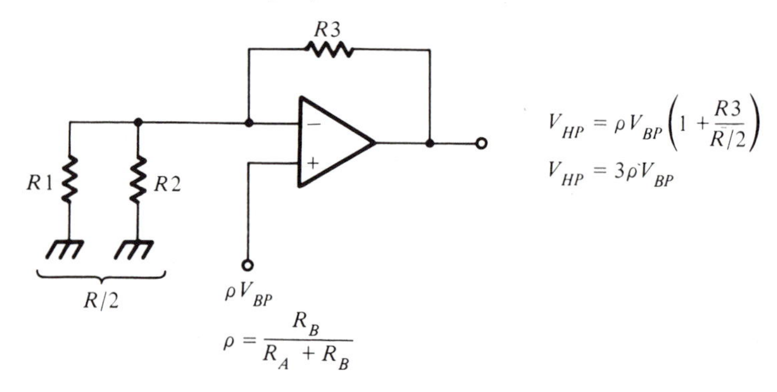

(b) Noninverting amplifier

Figure 7-35 Summer superposition

and then applied to U1 as a noninverting amp. Since R1 = R2 = R and since they are in parallel, the gain of U1 (as a noninverting amp for the signal ρV_{BP}) is

$$1 + \frac{R3}{R/2}$$

But since $R_3 = R$, that gain is 3.

By superposition, the output is the sum of the effects of each input.

$$V_{HP} = -V_{in} - V_{LP} + 3\rho V_{BP} \tag{7-71}$$

You now have three simultaneous equations which describe the behavior of the state variable filter

$$V_{BP} = -\frac{V_{HP}}{R_f Cs} \tag{7-69}$$

$$V_{LP} = -\frac{V_{BP}}{R_f Cs} \tag{7-70}$$

$$V_{HP} = -V_{in} - V_{LP} + 3\rho V_{BP} \tag{7-71}$$

For the high pass output, substitute equation (7-70) into equation (7-71).

$$V_{HP} = -V_{in} + \frac{V_{BP}}{R_f Cs} + 3\rho V_{BP} \tag{7-72}$$

Now substitute equation (7-69) for V_{BP} into (7-72).

$$V_{HP} = -V_{in} + \frac{-\dfrac{V_{HP}}{R_f Cs}}{R_f Cs} + 3\rho\left[-\frac{V_{HP}}{R_f Cs}\right]$$

After a bit of algebraic manipulation,

$$V_{HP}\left(\frac{R_f^2 C^2 s^2 + 3\rho R_f Cs + 1}{R_f^2 C^2 s^2}\right) = -V_{in}$$

or

$$V_{HP} = -\frac{s^2 V_{in}}{s^2 + \dfrac{3\rho}{R_f C}s + \dfrac{1}{R_f^2 C^2}} \tag{7-73}$$

$$\frac{V_{HP}}{V_{in}} = \frac{-s^2}{s^2 + \dfrac{3\rho}{R_f C}s + \dfrac{1}{R_f^2 C^2}} \tag{7-74}$$

Compare equation (7-74) with equation (7-40) (the standard form of a high pass filter's transfer function).

$$\frac{A_0 s^2}{s^2 + \alpha\omega_0 s + \omega_0^2} \tag{7-40}$$

For the high pass filter part of the state variable filter,

$$A_0 = -1$$

$$\omega_0 = \frac{1}{R_f C}$$

$$\alpha\omega_0 = \frac{3\rho}{R_f C}$$

$$\alpha = 3\rho = \frac{3R_B}{R_A + R_B}$$

For the low pass output,

$$V_{LP} = -\frac{V_{BP}}{R_f Cs} \tag{7-70}$$

Substituting equation (7-69) into equation (7-70), you obtain

$$V_{LP} = \frac{\dfrac{V_{HP}}{R_f Cs}}{R_f Cs}$$

(7-75)

$$V_{LP} = \frac{V_{HP}}{R_f^2 C^2 s^2} \tag{7-75}$$

Substituting equation (7-73) into (7-75) gives

$$V_{LP} = \frac{\dfrac{-s^2 V_{in}}{s^2 + \dfrac{3\rho}{R_f C}s + \dfrac{1}{R_f^2 C^2}}}{R_f^2 C^2 s^2}$$

$$V_{LP} = \frac{-s^2 V_{in}}{s^2 + \dfrac{3\rho}{R_f C}s + \dfrac{1}{R_f^2 C^2}} \times \frac{1}{R_f^2 C^2 s^2}$$

$$\frac{V_{LP}}{V_{in}} = \frac{-\dfrac{1}{R_f^2 C^2}}{s^2 + \dfrac{3\rho}{R_f C}s + \dfrac{1}{R_f^2 C^2}} \tag{7-76}$$

Compare this with the standard form for a low pass filter transfer function, equation (7-30).

$$\frac{A_0 \omega_0^2}{s^2 + \alpha\omega_0 s + \omega_0^2} \tag{7-30}$$

Just like the high pass filter, the low pass filter gives

$$A_0 = -1$$

$$\omega_0 = \frac{1}{R_f C}$$

$$\alpha = \frac{3R_B}{R_A + R_B}$$

Finally, the bandpass response is obtained from equation (7-69).

$$V_{BP} = -\frac{V_{HP}}{R_f Cs} \tag{7-69}$$

Substituting equation (7-73) into (7-69) gives

$$V_{BP} = \frac{\dfrac{-s^2 V_{in}}{s^2 + \dfrac{3p}{R_f C}s + \dfrac{1}{R_f^2 C^2}}}{R_f Cs}$$

$$V_{BP} = \frac{-s^2 V_{in}}{s^2 + \dfrac{3\rho}{R_f C}s + \dfrac{1}{R_f^2 C^2}} \times \frac{1}{R_f Cs}$$

$$\frac{V_{BP}}{V_{in}} = \frac{\left[-\dfrac{1}{R_f C}\right]s}{s^2 + \dfrac{3\rho}{R_f C}s + \dfrac{1}{R_f^2 C^2}} \tag{7-77}$$

Compare this with the standard bandpass transfer function, equation (7-46).

$$\frac{A_0 \alpha \omega_0 s}{s^2 + \alpha \omega_0 s + \omega_0^2} \tag{7-46}$$

Again,

$$\omega_0 = \frac{1}{R_f C}$$

$$\alpha = \frac{3R_B}{R_A + R_B}$$

Equating the numerators of equations (7-77) and (7-46), you obtain

$$-A_0 \alpha \omega_0 = -\frac{1}{R_f C}$$

$$A_0 = \frac{1}{R_f C \alpha \omega_0}$$

$$A_0 = \frac{1}{R_f C} \times \frac{1}{\omega_0} \times \frac{1}{\alpha}$$

$$A_0 = \frac{1}{R_f C} \times R_f C \times Q$$

$$A_0 = Q$$

The state variable filter produces the standard second order low pass, bandpass, and high pass responses. The critical frequencies of each are equal, and the damping is set by the feedback from the bandpass output. For all three outputs this damping has precisely the same effect (at the same numerical values) as it did for the single op amp filters. For low pass and high pass, the damping coefficient of 1.414 produces a Butterworth response. Damping of 1.73 gives a Bessel response, and $\alpha = 0.766$ causes 3 dB peaks (Chebyshev). Also, the -3 dB or peak frequencies of the low pass sections are shifted (from the critical frequency) according to the correction factors in Table 7-4. The high pass -3 dB or peak frequencies are similarly shifted by the high pass correction factor

$$k_{hp} = \frac{1}{k_{lp}}$$

For the bandpass section, changing the damping coefficient inversely alters the Q and gain (at critical frequency).

$$Q = \frac{1}{\alpha}$$

$$A_0 = Q$$

But the critical frequency is set by R_f and C. It is not altered by changes in the damping coefficient. This means that changes in damping only (and directly) affect the bandwidth. So tuning the band pass filter is very convenient. Resistors R_f adjust the center frequency (only). Resistors R_A and R_B adjust the bandwidth (only).

At this point, it is critical that you realize that *optimum* performance from *all* three outputs cannot be obtained simultaneously. If you want maximum flatness in the pass bands of the low pass and high pass outputs, you must select a Butterworth response with $\alpha = 1.414$. But a damping coefficient of 1.414 gives a Q of

$$Q = \frac{1}{\alpha}$$

$$Q = 0.707$$

$$A_0 = 0.707$$

The bandpass filter will not be very selective and will attenuate even the center frequency by 30 percent.

On the other hand, if you select $Q = 20$ to obtain reasonable selectivity and center frequency gain, the low and high pass outputs will have a damping coefficient of only 0.05. This will cause a pass band peak of over 25 dB. You can *either* optimize the bandpass output or the low pass and high pass outputs.

The frequency responses for four different values of damping are shown in Fig. 7-36 through 7-39. Bessel response is shown in Fig. 7-36. Gain and Q are low. Notice that all three responses coincide at f_0. This is also true for the Butterworth response of Fig. 7-37, though Q and gain are up. Decreasing the damping further causes the curves to begin to peak as in Fig. 7-38. Higher Q can

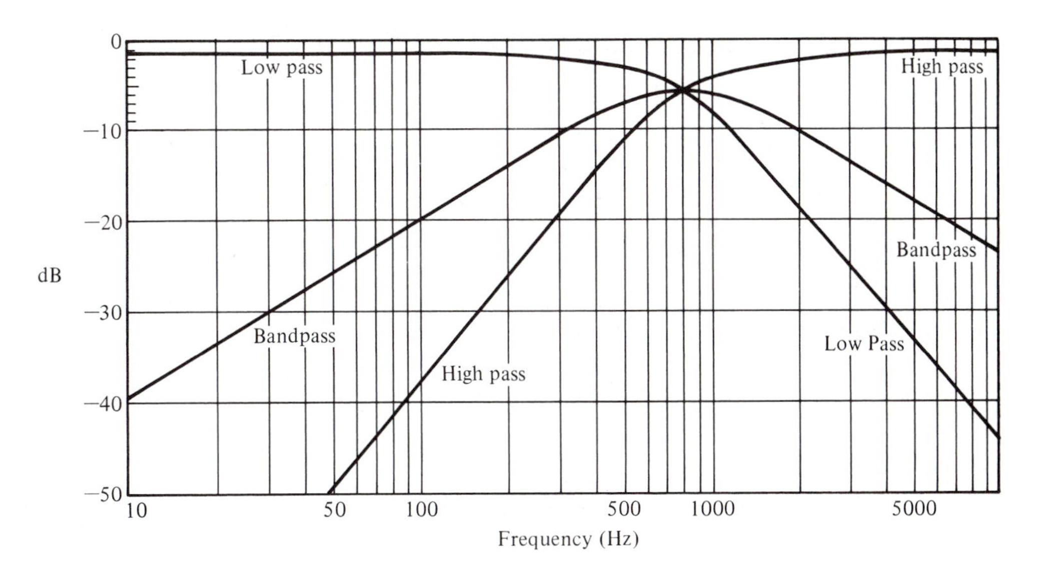

Figure 7-36 State variable Bessel response

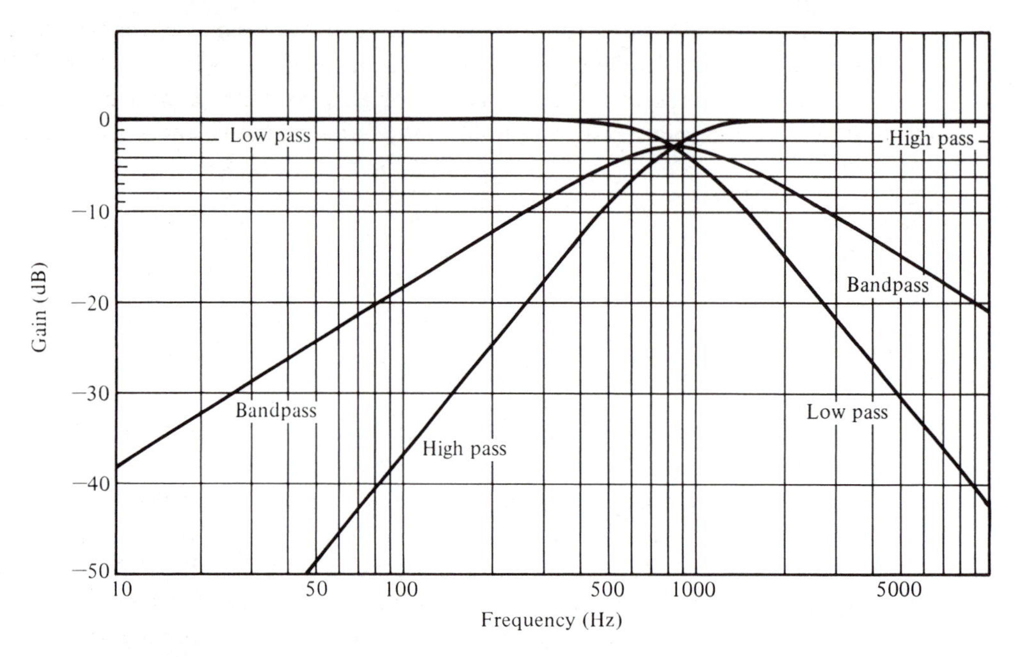

Figure 7-37 State variable Butterworth response

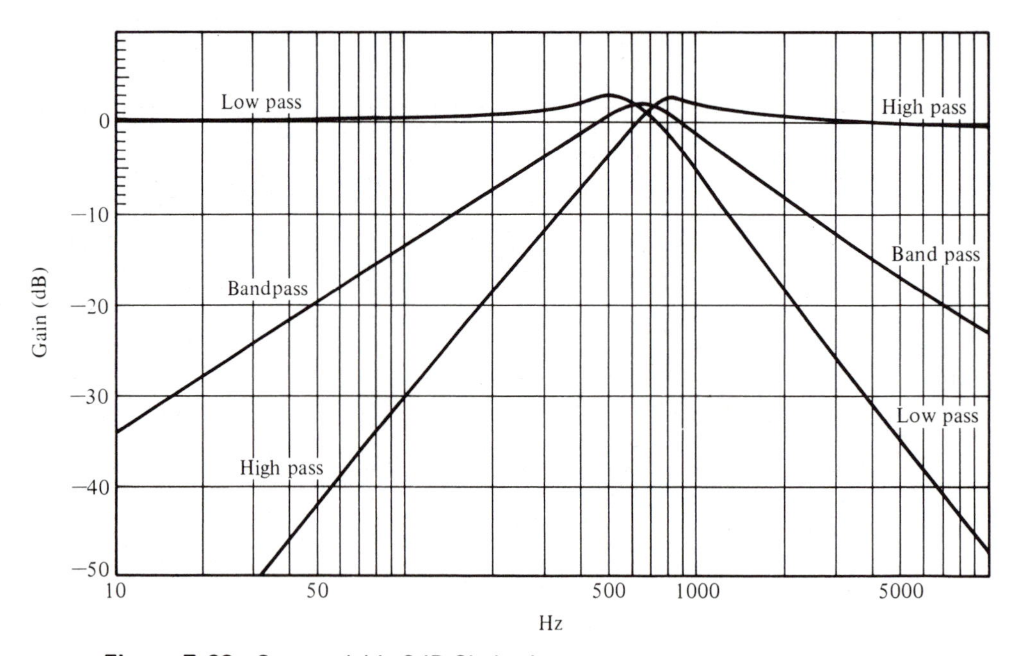

Figure 7-38 State variable 3dB Chebyshev response

be obtained only by lowering the damping coefficient further. A filter with $Q \approx 8$ is shown in Fig. 7-39. Notice the large peaks in the low pass and high pass bands as well as for the band pass output.

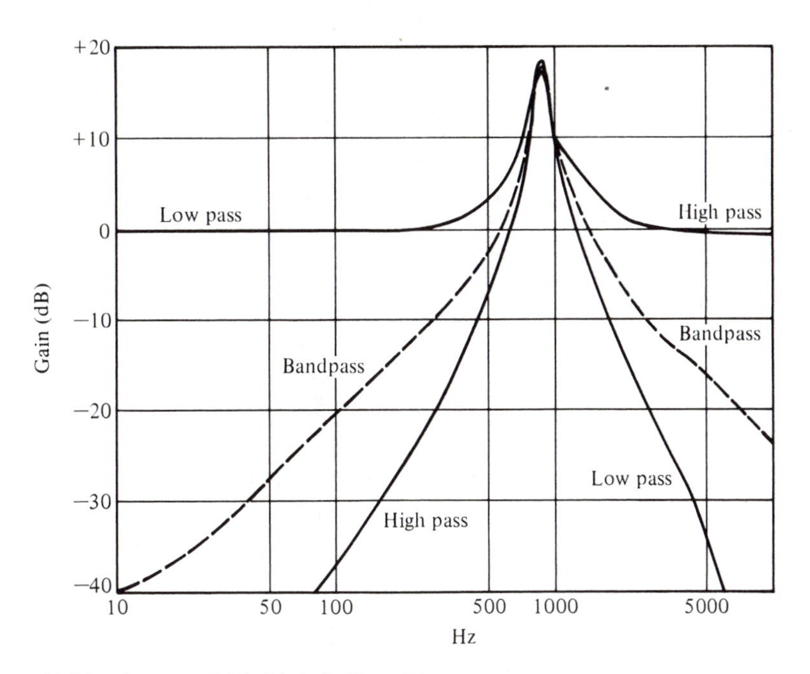

Figure 7-39 State variable high Q ($Q \approx 8$) response

Example 7-14

Given a state variable filter with the following component values (see Fig. 7-34):

$$R1 = 10\ k\Omega,\ R2 = 10\ k\Omega,\ R3 = 10\ k\Omega,\ R4 = 4.7\ k\Omega,\ R5 = 4.7\ k\Omega,$$

$$R_A = 2.2\ k\Omega,\ R_B = 1.2\ k\Omega,\ C1 = 0.01\ \mu F,\ C2 = 0.01\ \mu F$$

determine filter response. Calculate the following:

a. Type of response (Bessel, Butterworth, Chebyshev).
b. Low pass and high pass −3 dB or peak frequencies.
c. Bandpass center frequency.
d. Bandpass bandwidth.
e. Bandpass gain at center frequency.

Solution

a. The type of response is determined by the damping coefficient

$$\alpha = \frac{3R_B}{R_A + R_B} \tag{7-67}$$

$$\alpha = \frac{3(1.2\ k\Omega)}{2.2\ k\Omega + 1.2\ k\Omega} = 1.059$$

Comparing this with the damping coefficients of Table 7-4, you can see that this is a 1 dB peak Chebyshev.

b. The critical frequency is

$$f_0 = \frac{1}{2\pi R_f C} \tag{7-66}$$

$$f_0 = \frac{1}{2\pi(4.7\ k\Omega)(0.01\ \mu F)} = 3.386\ kHz$$

The low pass peak frequency must be corrected by the factor from Table 7-4.

$$f_{\alpha\ \text{low pass}} = f_0 \times k_{lp}$$

$$f_{\alpha\ \text{low pass}} = 3.386\ kHz \times 0.673 = 2.279\ kHz$$

For the high pass output

$$f_{\alpha\ \text{high pass}} = f_0 \times k_{hp}$$

$$k_{hp} = 1/k_{lp}$$

$$f_{\alpha\ \text{high pass}} = f_0 \times 1/k_{lp}$$

$$f_{\alpha\ \text{high pass}} = 3.386\ kHz \times 1/0.673 = 5.031\ kHz$$

Remember, these are the frequencies of the 1 dB peaks. The actual -3 dB point is not very far shifted from f_0. Look again at Fig. 7-17.

c. The bandpass center frequency is f_0.

$$f_c = f_{BP\ \text{center}} = 3.386\ kHz$$

d. The bandpass bandwidth comes from Q.

$$Q = 1/\alpha$$

$$Q = 1/1.059 = 0.9443$$

$$Q = \frac{f_c}{\Delta f} \tag{7-13}$$

$$\Delta f = \frac{f_c}{Q}$$

$$\Delta f = \frac{3.386 \text{ kHz}}{0.9443} = 3.586 \text{ kHz}$$

e. The bandpass gain at center frequency is

$$A_0 = Q$$

$$A_0 = 0.9443$$

If you wish to adjust the pass band gain to something other than 1, the circuit in Fig. 7-40 can be used. The bandpass feedback, which sets the damping, has been fed through an inverter and then summed at the inverting terminal of the summer. The damping coefficient for this configuration is

$$\alpha = \frac{R_B}{R_A}$$

Also, the input signal's resistor has been lowered, in comparison to the other resistors to the summer. This causes a gain of

$$A_{\text{pass band}} = \frac{R}{R_i}$$

for the low pass and high pass outputs. The bandpass center frequency gain is

$$A_0 = \frac{R}{R_i} Q$$

With the circuit of Fig. 7-40, you can set the Q by adjusting R_B, then set the pass band gain to some other value by adjusting R_i. Be careful, however. To increase gain you must lower R_i. This is the filter's input impedance. Lowering it may load the source, dropping the input (and therefore output) signals.

The rolloff rate of the high pass and low pass outputs is 40 dB/decade. For the bandpass it is 20 dB/decade. This is because each output is second order in nature. To obtain steeper rolloffs you must have higher order transfer functions, which means multiple stages.

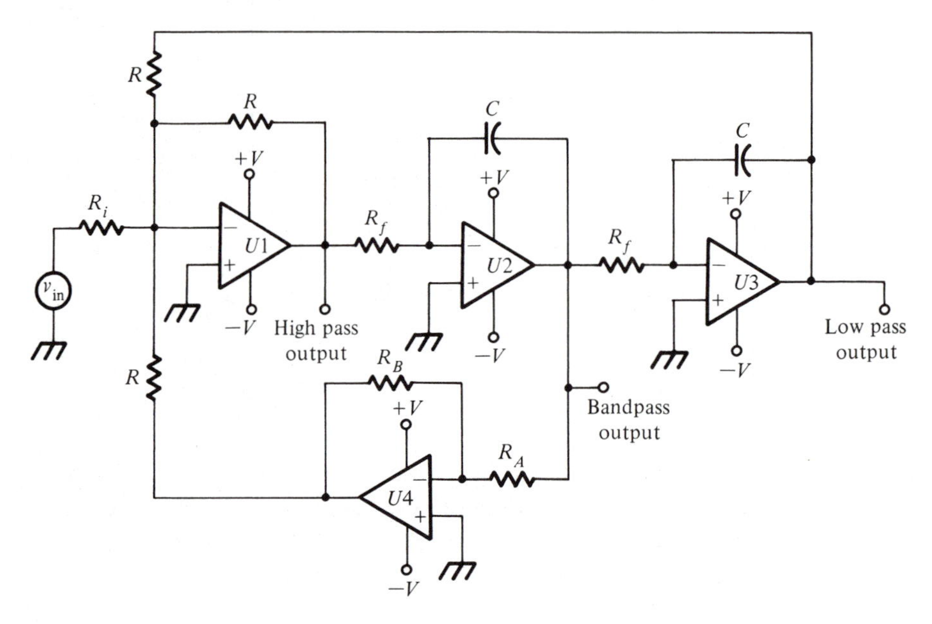

Figure 7-40 Adjustable gain state variable filter

Cascading state variable filters is only a bit more complex than cascading single op amp filters. The difference appears in the *three* outputs of the state variable filter. One possible solution is shown in Fig. 7-41. Cascading three state variable filters gives a rolloff rate of 120 dB/decade (sixth order). To produce a 120 dB/decade low pass filter, the low pass output of one stage must drive the input of the next. To switch to a 60 dB/decade bandpass, the bandpass outputs must drive the next stage's input. The same is true for a 120 dB/decade high pass filter. So, the switches shown could be a single three-pole, three-position switch (3P3T).

As with cascading single op amp filters, the damping coefficient and frequency correction factor are different for each stage. Those values listed in Table 7-5 for the single op amp filters also hold true for the cascaded state variable filters. Since each state variable stage is itself second order, only order 4 and order 6 are of interest.

Also remember that

$$k_{hp} = 1/k_{lp}$$

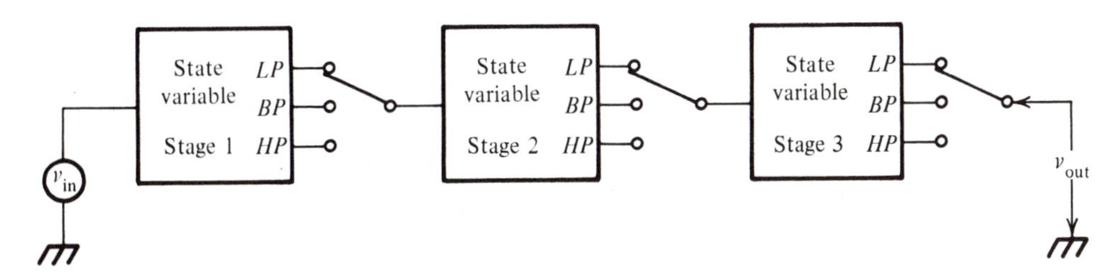

Figure 7-41 Cascaded state variable filters

If you switch from low pass to high pass, you will have to adjust f_0 of each stage if you need to keep the same -3 dB or peak frequency.

As with the low pass and high pass outputs, the characteristics of the cascaded state variable bandpass outputs parallel the characteristics of cascading single op amp bandpass filters.

If you choose to set the switches in Fig. 7-41 to different positions (i.e., not all LP or all HP or all BP), rather unusual results may occur. However, the effect of properly adjusted critical frequencies and damping may be just the overall filter's response needed for a particular application. Experiment. Another way to produce unusual (though possibly useful) frequency response is to sum together (using an op amp summer) the low pass, high pass, and even bandpass outputs. By adjusting the gain of each summer input, the overall filter response can be shaped.

When one is designing a circuit for production, the total parts count determines the final production cost. Building the circuit with five components is cheaper to mass produce than building it with thirteen components, even if the five components are much more expensive than the thirteen.

The AF 100, universal filter from National Semiconductor is just such a component. It is a single integrated circuit with all op amps, capacitors, and most resistors already internally configured as a state variable filter. Its schematic (and pin diagram) is given in Fig. 7-42. To build a state variable filter you add four resistors. That is all. This is shown in Fig. 7-43. Resistors R_{f1} and R_{f2} (which should be equal) set the critical frequency. Resistor R_Q sets the damping (and Q). The input resistor, R_{in}, determines gain. The undedicated op amp can be used to sum the other outputs, to provide additional gain, or to build a single op amp filter to cascade with the state variable filter.

Look carefully at the component values within the filter in Fig. 7-43. The summer's feedback resistor is only 10 kΩ, while the low pass feedback resistor and bandpass feedback resistor are 100 kΩ. This causes a change in the transfer functions and resulting parameter equations. For the AF 100

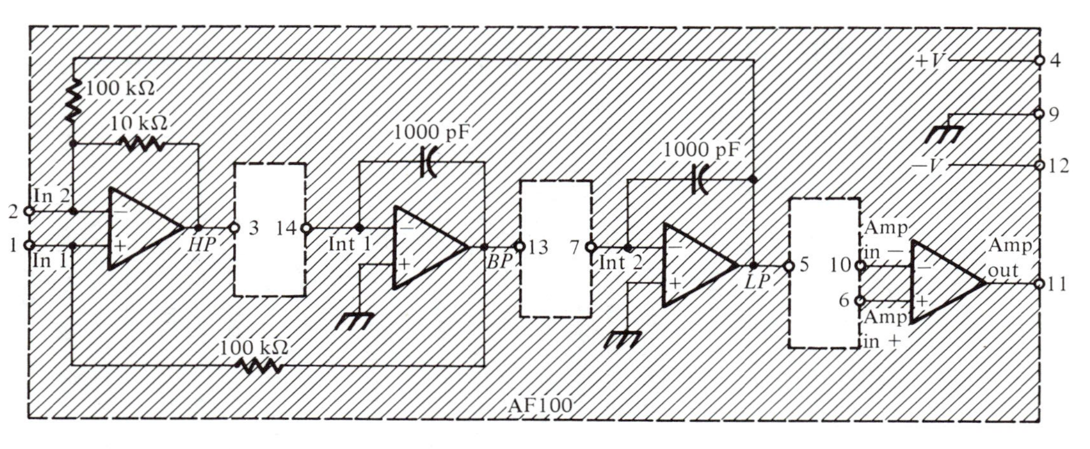

Figure 7-42 AF 100 universal active filter (state variable) IC

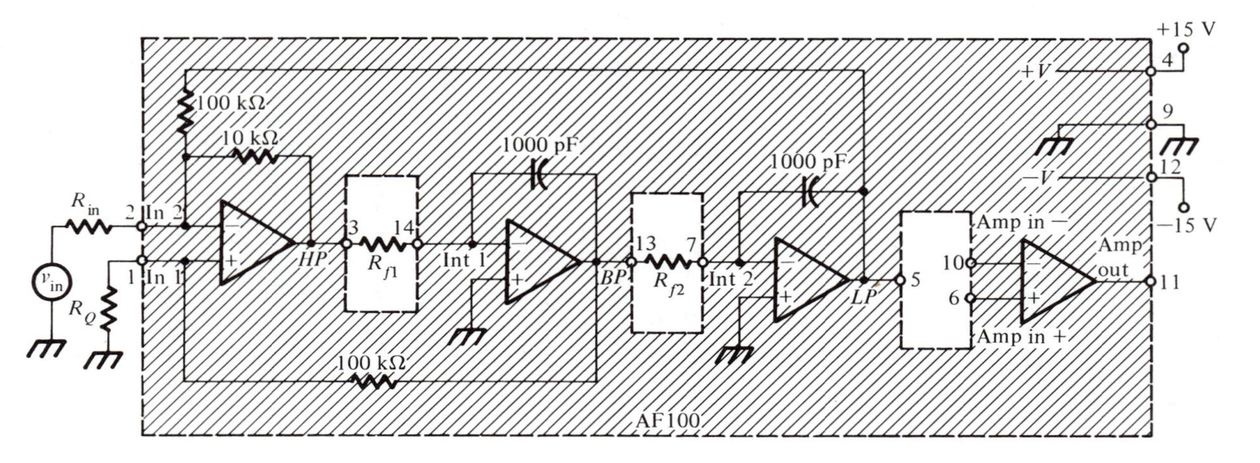

Figure 7-43 State variable filter built with an AF 100

$$\omega_0 = \frac{\sqrt{0.1}}{R_f C} = \frac{0.316}{R_f C}$$

or

$$f_0 = \frac{0.316}{2\pi R_f C}$$

The low pass gain (in its pass band) is ten times greater than the high pass gain (in its pass band).

$$A_{LP} = -\frac{10^5}{R_{in}}$$

$$A_{HP} = -\frac{10^4}{R_{in}}$$

Of course, the bandpass center frequency gain is also affected.

$$A_0 = \frac{\dfrac{10^4}{R_{in}}\left[1 + \dfrac{10^5}{R_Q}\right]}{1.1 + \dfrac{10^4}{R_{in}}}$$

Finally, the damping (and therefore Q) is altered.

$$\alpha = \frac{1.1 + \dfrac{10^4}{R_{in}}}{(0.316)\left[1 + \dfrac{10^5}{R_Q}\right]}$$

This can be manipulated to give

$$R_Q = \frac{10^5}{\left[\dfrac{1}{0.316\alpha}\right]\left[1.1 + \dfrac{10^4}{R_{in}}\right] - 1}$$

SUMMARY

Filter behavior is usually described in a frequency response plot. Gain (and often phase) is plotted on the vertical axis, while frequency is plotted logarithmically on the horizontal axis. Gain is normally expressed in decibels, where

$$\text{dB} = 20 \log \frac{v_o}{v_{in}}$$

if it is assumed that the circuit's input and load impedance are equal. The critical frequency occurs where the gain has dropped by -3 dB below the pass band gain. This corresponds to a 30 percent decrease in amplitude and a 50 percent reduction in power to the load. Often the axis of filter response plots is normalized. This makes the critical frequency 1 and the pass band gain 0 dB. To denormalize, multiply the horizontal axis by f_0 and add the pass band dB gain to the vertical axis.

The critical frequency serves as the dividing line between the stop band and the pass band. Low pass filters pass low frequencies and stop high frequency signals. High pass filters do just the opposite. The bandpass filter passes only a certain range of frequencies. All lower and higher frequency signals are stopped. The notch filter stops the band of frequencies, while passing higher and lower frequencies.

Laplace transforms allow you to easily analyze the response of circuits containing differential or integral terms. Capacitors' impedances are replaced by 1/Cs and inductors' impedances with Ls. You then analyze the circuit algebraically obtaining a "gain" in terms of s. This "gain" is the transfer function and contains the full gain and phase response information. To "decode" the transfer function substitute $j\omega = s$, separate real and imaginary parts and then solve

$$\text{Gain} = \sqrt{\text{real}^2 + \text{imaginary}^2}$$

$$\phi = \arctan \frac{\text{imaginary}}{\text{real}}$$

The order of the denominator of the transfer function determines the rolloff rate in the stop band. Increasing the denominator one order (from s to s^2 or s^2 to s^3) increases the rolloff rate by 20 dB/decade (from 20 dB/decade to 40 dB/decade). Twenty dB/decade is the same as 6 dB/octave.

For bandpass and notch filters, f_0 lies at the *geometric* mean of the low frequency and high frequency cutoffs. This is called the center frequency. Bandwidth is the distance between the low and high frequency cutoffs. Q is the ratio of center frequency to bandwidth. It is a measure of how selective the filter is.

Passive filters usually have low input impedance and high output impedance, and are easily loaded and frequency shifted by the load. There is poor isolation between load and source. To obtain second order or higher response you must use inductors. At audio frequencies or below these inductors are expensive, bulky, and have rather poor electrical characteristics.

Active filters, especially those built with op amps, have none of these problems. However, they require power supplies, and may be more expensive than simple passive filters. They are limited by the op amp's amplitude and high frequency responses. Parasitic feedback can cause oscillations in active filters. They are also subject to the same environmental hazards and limitations (heat, radiation, etc.) as any other semiconductor.

Several transfer functions were derived in the chapter. The most important ones are listed below.

First order low pass

$$\frac{A_0 \omega_0}{s + \omega_0} \tag{7-14}$$

General second order filter

$$\frac{A_0 Z3Z4}{Z1Z2 + Z2Z3 + Z3Z4 + Z1Z3 + Z1Z4(1 - A_0)} \tag{7-28a}$$

Second order low pass

$$\frac{A_0 \omega_0^2}{s^2 + \alpha\omega_0 s + \omega_0^2} \tag{7-30}$$

Second order high pass

$$\frac{A_0 s^2}{s^2 + \alpha\omega_0 s + \omega_0^2} \tag{7-40}$$

Second order bandpass

$$\frac{A_0 \alpha\omega_0 s}{s^2 + \alpha\omega_0 s + \omega_0^2} \tag{7-46}$$

Second order notch

$$\frac{A_0(s^2 + \omega_0^2)}{s^2 + \alpha\omega_0 s + \omega_0^2} \tag{7-63}$$

A_0 = pass band gain

ω_0 = critical frequency (radians/sec)

α = damping coefficients

The critical frequency is determined by input and feedback resistors and capacitors. Damping is determined by the amplifier's gain. Highly damped filters (Bessel) give good transient response, but are slow with early initial rolloff. The Butterworth response gives optimum flatness in the pass band. Less damping allows pass band ripple and poor transient response, but gives fast initial rolloff (Chebyshev). For Sallen-Key unity gain filters, the pass band gain is set to 1, and damping is determined by the frequency setting capacitors. The Sallen-Key equal component filters require pass band gain and damping to be directly related. This allows damping and critical frequency to be determined independently. However, -3 dB or peak frequency is affected by the damping, so you must use appropriate correction factors.

To obtain higher order response, several single op amp filters can be cascaded. The orders add. However, each stage's damping and frequency correction factors are often radically different from those of a single op amp filter and from the other stages of the filter.

Wide-band bandpass filters are best obtained by cascading a low pass with a high pass filter. For $1 < Q < 20$ you can build a single op amp filter.

$$Q = \frac{1}{\alpha}$$

For this filter there are several parameters which are interdependent. Be sure to follow the recommended design and tuning sequence. One op amp bandpass filter will give an ultimate rolloff rate of 20 dB/decade. Higher rolloff rates can be obtained by cascading. As with other cascaded filters, adjustments in α (or Q) and overall gain are required.

The notch filter is obtained by subtracting the output of a bandpass filter from its input signal. All parameters are then similar to those of the bandpass filter. Notch depth is set by the closeness of matching between the gains of the bandpass filter and the summer.

The state variable filter is actually a small scale analog computer which solves all three transfer functions simultaneously. This allows simultaneous high pass, bandpass, and low pass second order outputs. Critical frequency, damping, and gain may be set independently in some versions. By summing and/or cascading various outputs complex responses can be produced. The AF 100 IC allows you to build a state variable filter with as few as five components (one IC, two frequency determining resistors, one gain setting resistor, and one damping coefficient setting resistor).

PROBLEMS

7-1 For the frequency response plots in Figs. 7-36, 7-37, and 7-38, determine the following:

a. Critical frequency for the low pass filter.

b. Critical frequency for the high pass filter.

c. Center frequency for the bandpass filter.

d. f_l and f_h for the bandpass filter.

e. Bandwidth and Q for the bandpass filter.

7-2 **a.** Convert the following ratio gains to dB.

0.03, 0.125, 8, 50

b. Convert the following dB gains to ratio gains.

-22, $+18$, -65, 45

7-3 Denormalize the frequency response plot (gain only) in Fig. 7-5 to have a critical frequency of 8 kHz and a pass band gain of 2.1 dB. Accurately draw the full denormalized response.

7-4 For the circuit in Fig. 7-44:

a. Draw the Laplace transformed circuit.

b. Analyze the circuit to obtain its transfer function.

c. Substitute $j\omega = s$ into the transfer function and derive the gain equation.

d. Derive the phase equation from the transfer function.

e. For $R = 2.2$ kΩ and $C = 0.1$ μF plot the gain and the phase frequency response.

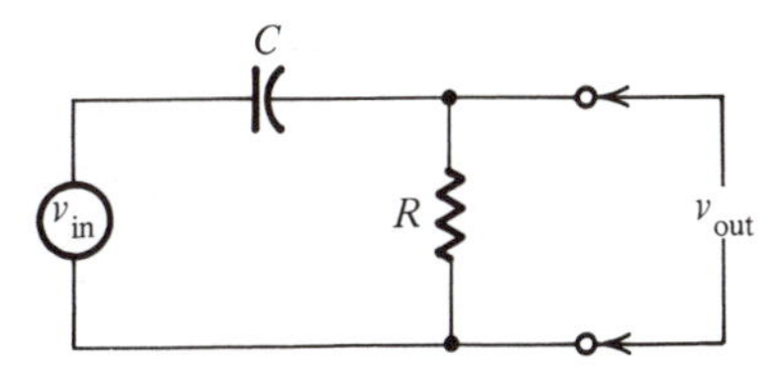

Figure 7-44 Schematic for problem 7-4

7-5 For the frequency response plots in Fig. 7-37 identify the pass band and the stop band for the

a. low pass filter

b. high pass filter

c. band pass filter

7-6 Select the order filter necessary to ensure that the output from a low pass Butterworth filter meets the following specification at $f = 2f_0$.

a. -15 dB

b. -34 dB

c. -10 dB

d. -24 dB

7-7 Express the rolloff rate for the filters selected in Problem 7-6 in dB/decade and dB/octave.

7-8 Sketch the frequency response curve of a bandpass filter with the following characteristics:

$$f_0 = 800 \text{ Hz}$$

$$Q = 12$$

$$A_0 = +20 \text{ dB (gain at } f_0)$$

7-9 Sketch the frequency response curve of a notch filter with the following characteristics:

$$f_h = 30 \text{ kHz}$$

$$f_l = 25 \text{ kHz}$$

$$\text{Notch depth} = -30 \text{ dB}$$

$$\text{Pass band gain} = 5 \text{ dB}$$

7-10 Describe two *specific* uses for
- **a.** low pass active filter.
- **b.** high pass active filter.
- **c.** bandpass active filter.
- **d.** notch active filter.

7-11 **a.** List four disadvantages of passive filters.
b. Explain how active filters overcome each.

7-12 List four disadvantages of active filters.

7-13 Repeat the procedures of Problem 7-4a-d for the circuit in Fig. 7-45.

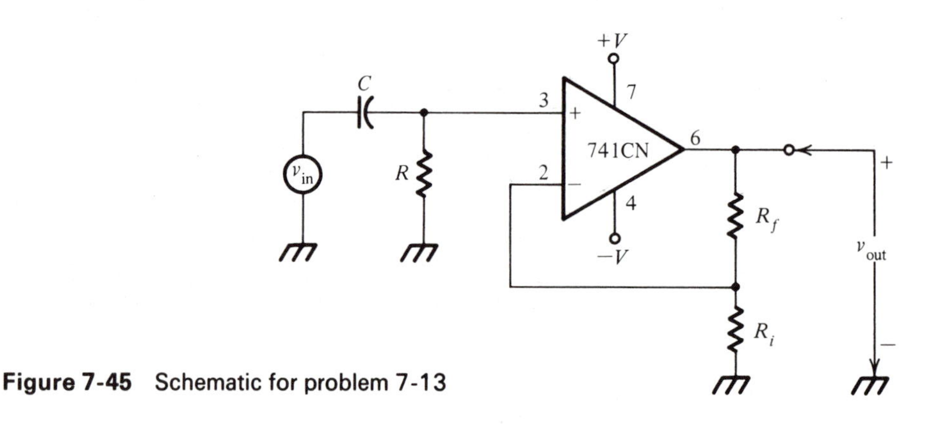

Figure 7-45 Schematic for problem 7-13

7-14 Repeat the procedures of Problem 7-4, parts a and b for the circuit in Fig. 7-46.

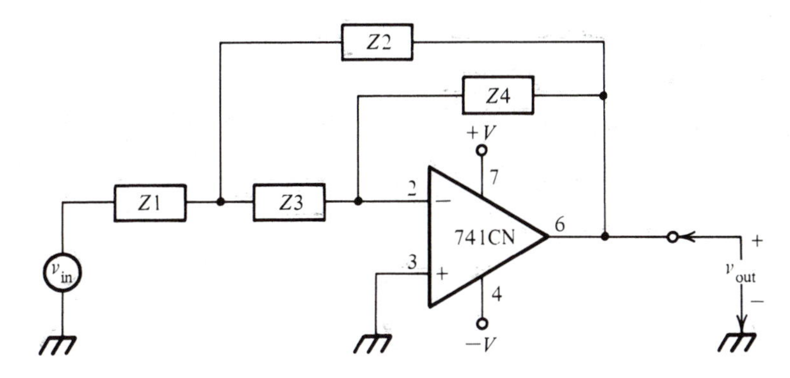

Figure 7-46 Schematic for problem 7-14

7-15 Repeat the procedures of Problem 7-4, parts c and d if

$$Z1 = R1, \qquad Z2 = \frac{1}{sC1}, \qquad Z3 = R3, \quad \text{and} \quad Z4 = \frac{1}{sC2}$$

7-16 For the circuit in Fig. 7-16 calculate the critical frequency, f_α, pass band gain, and response type (Bessel, Butterworth, or 1, 2, or 3 dB Chebyshev) for R_i open, R_f short, R1 = 3 kΩ, R2 = 3 kΩ, C1 = 0.1 μF, C2 = 0.027 μF

7-17 Repeat Problem 7-16 for

$$R_i = 10\text{ k}\Omega,\ R_f = 2.7\text{ k}\Omega,\ R1 = 1.1\text{ k}\Omega,\ R2 = 1.1\text{ k}\Omega,$$

$$C1 = 0.022\ \mu F,\ C2 = 0.022\ \mu F$$

7-18 Design a low pass 3 dB Chebyshev active filter, using a Sallen-Key unity gain configuration. Set the peak's frequency at 5 kHz.

7-19 Repeat Problem 7-18, using a Sallen-Key equal component circuit.

7-20 For the circuit in Fig. 7-47, calculate the following:

a. Critical frequency of each stage.

b. Damping of each stage.

c. Response type.

d. Overall filter f_α.

e. Overall filter rolloff rate.

f. Overall filter pass band gain.

g. Overall filter gain at $f = 2 \times f_0$.

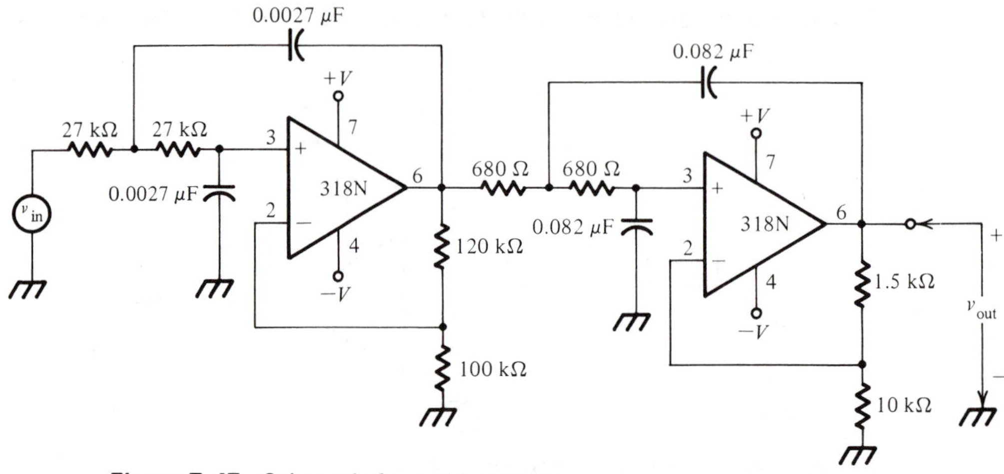

Figure 7-47 Schematic for problem 7-20

7-21 Design a low pass Bessel filter with a -3 dB frequency of 100 Hz. At 200 Hz the gain must be at least -18 dB below the pass band gain.

7-22 **a.** Repeat Problem 7-16 for the circuit in Fig. 7-23 with $R_i = 10$ kΩ, $R_f = 2.7$ kΩ, R1 = 1.1 kΩ, R2 = 1.1 kΩ, C1 = 0.022 μF, C2 = 0.022 μF

b. Compare the results with those obtained for Problem 7-17.

7-23 **a.** Design a high pass 3 dB Chebyshev active filter, using a Sallen-Key equal component circuit. Set the peak's frequency at 5 kHz.

b. Compare the results with those obtained for Problem 7-19.

7-24 **a.** Repeat Problem 7-20 for the circuit in Fig. 7-48.

b. Compare the results with those obtained for Problem 7-20.

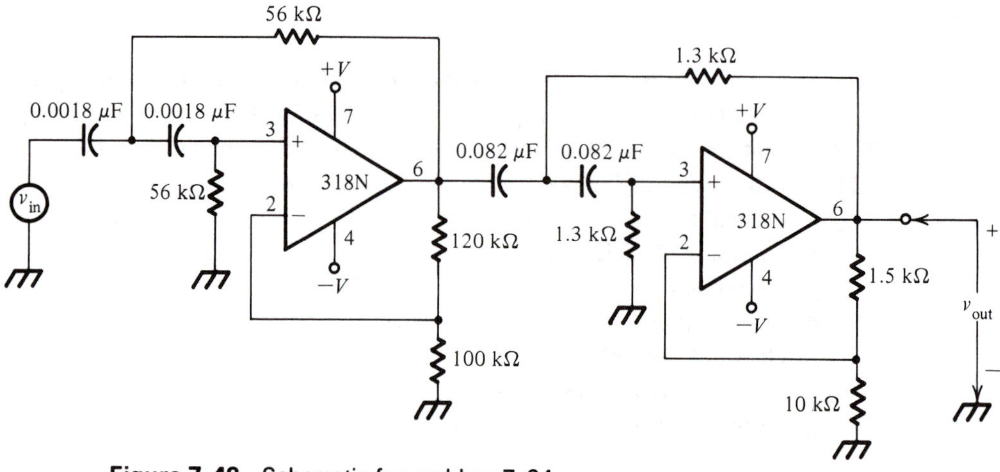

Figure 7-48 Schematic for problem 7-24

7-25 **a.** Design a high pass Bessel filter with a −3 dB frequency of 100 Hz. At 50 Hz the gain must be at least −18 dB below the pass band gain.
b. Compare the results with those obtained for Problem 7-21.
c. Calculate the gain bandwidth and slew rate requirements of the op amps.

7-26 Derive the transfer function for the circuit in Fig. 7-26 *with* R3 *removed.* Define all appropriate parameters. Compare each to those obtained for the circuit in Fig. 7-26.

7-27 Derive the transfer function for the circuit in Fig. 7-49. Define all appropriate parameters. Compare each to those obtained for the circuit in Fig. 7-26. [Hint: See Fig. 7-15 and equation (7-28a).]

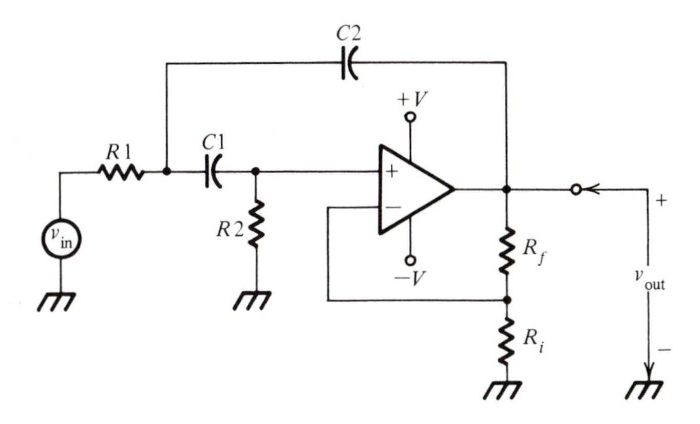

Figure 7-49 Schematic for problem 7-27

7-28 For the circuit in Fig. 7-26, calculate f_0, bandwidth, Q, f_l, f_h, and A_0 with R1 = 620 Ω, R2 = 10 kΩ, R3 = 20 Ω, C1 = 0.18 μF, C2 = 0.18 μF.

7-29 Design a bandpass filter with a center frequency of 400 Hz, $Q = 20$, $A_0 = +10$, and a 40 dB/decade rolloff rate. What are the gain bandwidth and slew rate requirements of the op amps?

7-30 Design a bandpass filter with $f_l = 500$ Hz, $f_h = 2000$ Hz, and a rolloff rate of 24 dB/octave.

7-31 Modify the circuit designed in Problem 7-29 to be a notch filter with the same specifications.

7-32 Derive the three transfer functions for the state variable filter of Fig. 7-40. Define all appropriate parameters.

7-33 **a.** Repeat Problem 7-18, using a three op amp state variable filter.
b. Sketch the high pass and bandpass outputs.

7-34 **a.** Repeat Problem 7-18, using an AF 100. Set the low pass, pass band gain to 10.
b. Sketch the high pass and bandpass outputs.

7-35 **a.** Design a bandpass filter with a center frequency of 400 Hz, $Q = 20$, using a three op amp state variable filter.

b. Sketch the high pass and low pass outputs.

7-36 Derive the three transfer functions for the AF 100 state variable filter of Fig. 7-43.

CHAPTER 8

Nonlinear Circuits

Most of the circuits you have seen so far have had a linear relationship between the inputs and outputs. In fact, considerable effort was often expended to overcome nonideal characteristics and to linearize the circuits' responses.

However, there is a class of circuits which intentionally has a highly nonlinear input to output characteristic. For these circuits a given change at the input does *not* produce a directly proportional output change. Transfer curve synthesizers, ideal rectifiers, peak detectors, logarithmic and antilog amplifiers, multipliers, and variable gain amps are all nonlinear circuits.

These nonlinear circuits find diverse applications. Industrial instrumentation, general signal processing, communications, and audio circuits all use nonlinear circuits to enhance their overall performance.

OBJECTIVES

After studying this chapter, you should be able to do the following:

1. Analyze a given transfer function synthesizer, calculating each slope and each breakpoint voltage, and drawing the transfer function.
2. Given a transfer function, design a synthesizer which will produce that transfer function.
3. Explain how enclosing a diode in an op amp's negative feedback loop eliminates the diode's offset.
4. Analyze a given half wave or full wave ideal rectifier circuit, calculating Z_{in} and v_{out}.
5. Design a half wave or full wave ideal rectifier to meet given specifications.
6. Explain why two diodes are necessary in the ideal rectifier circuit.
7. Qualitatively explain the operation of a peak detector.
8. Design a peak detector.

9. List three sources of peak detector error in the tracking mode and three sources in the hold mode.
10. Draw the schematic of a two, a three, and a four op amp log amp.
11. For each, describe circuit operation, derive the output voltage equation, and discuss relative advantages and disadvantages.
12. Draw the schematic, qualitatively describe operation, and derive the output equation for a two op amp antilog amp.
13. Explain why the feedback transistors and voltage divider must be integrated in the same wafer.
14. Illustrate proper use of a log/antilog amp functional module IC in either mode.
15. List three precautions when building and adjusting log or antilog amplifiers.
16. Describe the two different techniques of IC multiplication including the advantages and disadvantages of each.
17. Explain multiplier trimming procedures.
18. Draw the schematic of and explain the operation of each of the following applications using a multiplier IC (module):
    a. frequency doubler.
    b. true power converter.
    c. voltage controlled amplifier.
    d. balanced and conventional amplitude modulation.
    e. divider.
    f. square root converter.
    g. oscillator.
    h. automatic gain controller.
    i. true rms converter.
19. Describe the programmable op amp, indicating the techniques, effects, and applications of programming.
20. Design a circuit which will allow a programmable op amp to be electronically turned on and off. Explain two applications of such a circuit.
21. Explain the difference between a transconductance amplifier and an operational amplifier.
22. Discuss the effects of programming on an OTA.
23. Analyze and/or design the following OTA circuits:
    a. electronic parameter control.
    b. basic OTA voltage amplifier.
    c. programmable resistor.
    d. electronically variable gain amplifier.
    e. electronically tunable active filter.
    f. two quadrant multiplier.
24. Discuss OTA limitations.

8-1 DIODE FEEDBACK CIRCUITS

One of the simplest ways of producing many nonlinear circuits is by including one or more diodes in the input, or feedback loops of the op amp. This technique allows you to synthesize practically any input to output voltage relationship (transfer curve) you need. Proper diode connection also produces an ideal diode (zero turn on voltage and zero forward resistance). Adding a capacitor gives you a circuit whose output tracks the *peak* of the input voltage.

8-1.1 Transfer Curve Synthesizer

The transfer curve of a circuit is a plot of the circuit's input voltage or current (on the horizontal axis) versus its output voltage or current (on the vertical axis). For the bulk of circuits it is desirable for this transfer curve to be a straight line. This would mean that changes on the output would be directly proportional to changes on the input.

There are times, however, when a linear transfer curve is *not* desirable. If a transducer's response is itself nonlinear, running its output through a circuit with a complementary, nonlinear transfer curve would result in an overall linear response. Squaring and taking the square root of analog signals both require circuits with nonlinear transfer curves. The same is true for taking the log or antilog of analog signals. Waveshaping, from a triangle wave input to a sine wave output, is readily done with a circuit whose transfer curve is nonlinear.

Figure 8-1 is the transfer curve of a squarer. It is obviously nonlinear. However, it can be reasonably well approximated by a series of straight lines, each

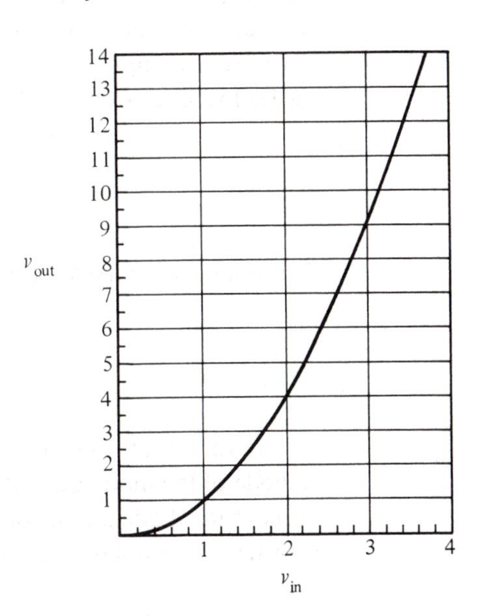

Figure 8-1 $V_{out} = V_{in}^2$ transfer curve

with a progressively steeper slope. Figure 8-2 gives the piecewise linear approximation of the squarer, using only four lines. The more line segments (pieces) you use, the closer the linear approximation is to the actual nonlinear transfer curve.

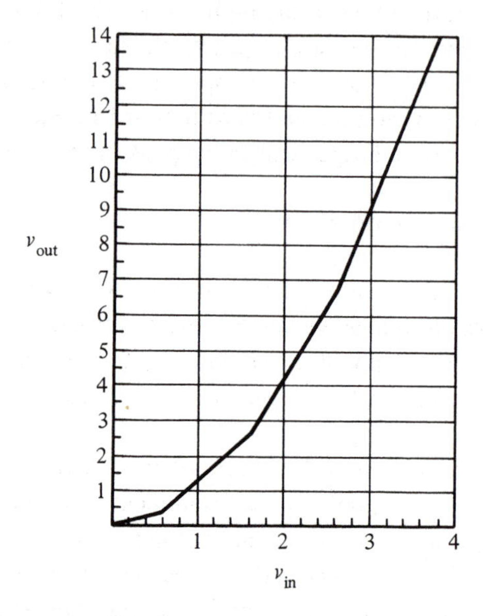

Figure 8-2 $V_{out} = V_{in}^2$ piecewise linear approximation

The circuit in Fig. 8-3(a) has a two segment, piecewise linear transfer curve, as shown in Fig. 8-3(b). For small values of negative input voltage, V_{ref} holds the cathode of the diode positive. This keeps it off. Resistor R2 sees an open and does not affect the input circuit. The result is a simple inverting amplifier with R_f as the feedback resistor and R1 as the input resistance. The gain is

$$\frac{v_{out}}{v_{in}} = \frac{-R_f}{R1}$$

This is also the slope for segment 1 of the transfer curve.

$$S1 = \frac{-R_f}{R1} \tag{8-1}$$

As the input voltage becomes more and more negative, the voltage at the cathode of the diode also falls. At $-V_{break\ 2}$ the diode goes on. Ignoring the 0.6 V offset of the diode and its on resistance, you can see that this places R2 in parallel with R1. The amplifier's gain suddenly changes to

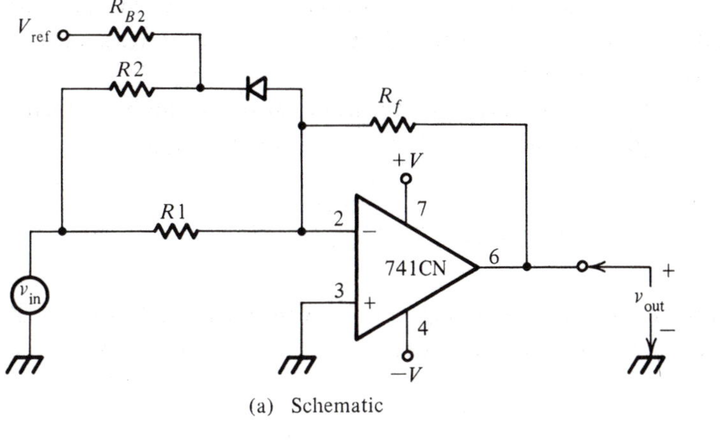

(a) Schematic

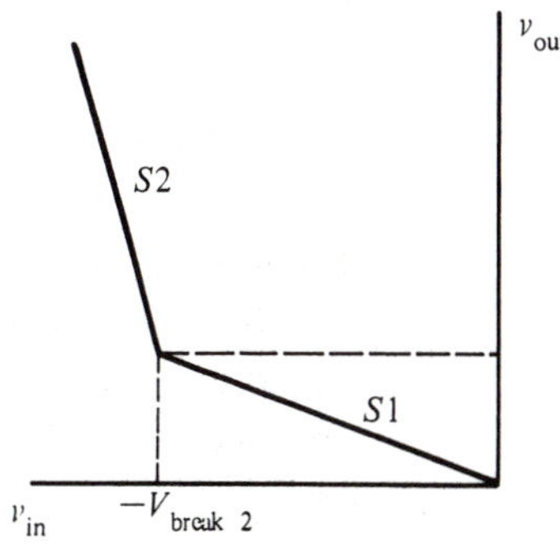

(b) Transfer curve

Figure 8-3 Two segment piecewise linear circuit

$$\frac{v_{out}}{v_{in}} = -\frac{R_f}{\dfrac{R1 \cdot R2}{R1 + R2}}$$

which is also the slope of the second line segment.

$$S2 = \frac{-R_f}{\dfrac{R1 \cdot R2}{R1 + R2}}$$

$$S2 = -R_f \times \frac{R1 + R2}{R1 \cdot R2}$$

$$S2 = -\left(\frac{R_f R1}{R1 \cdot R2} + \frac{R_f R2}{R1 \cdot R2}\right)$$

$$S2 = -\left(\frac{R_f}{R1} + \frac{R_f}{R2}\right) \tag{8-2}$$

The slope has automatically increased. Consequently, you can increase the slope of the circuit's transfer curve by causing diodes to switch in resistance to parallel the input resistor, increasing gain.

For a three segment curve, another input resistor (R3), diode and bias resistor (R_{B3}) network must be added in parallel with R1. When it switches in

$$S3 = -\left(\frac{R_f}{R1} + \frac{R_f}{R2} + \frac{R_f}{R3}\right)$$

This can be generalized for as many segments as you need. Each additional segment requires an input resistor, bias resistor, and diode network added in parallel with R1.

$$S_n = -\left(\frac{R_f}{R1} + \frac{R_f}{R2} + \frac{R_f}{R3} + \cdots + \frac{R_f}{R_n}\right) \quad (8\text{-}3)$$

The input resistors set the slopes. The break points are determined by V_{ref}, R_B, and the input resistor. Look again at figure 8-3a. The inverting input, and therefore the diode's anode, are held at virtual ground by the op amp's negative feedback. The R2, R_{B2}, diode, V_{ref} network can then be redrawn as shown in Fig. 8-4.

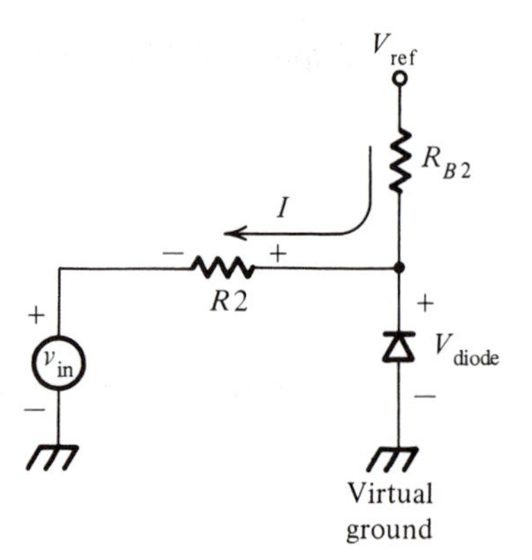

Figure 8-4 Break point analysis schematic

As long as the voltage across the diode is positive, the diode is off. When V_{diode} is driven to zero (actually −0.6 V) by v_{in}, the diode goes on and the next line segment begins.

$$I = \frac{V_{ref} - v_{in}}{R2 + R_{B2}}$$

$$V_{\text{diode}} = IR2 + v_{\text{in}}$$

$$V_{\text{diode}} = \frac{(V_{\text{ref}} - v_{\text{in}})R2}{R2 + R_{B2}} + v_{\text{in}}$$

For simplicity, assume $V_{\text{diode}} = 0$ V when it goes on.

$$0 \cong \frac{(V_{\text{ref}} - v_{\text{in}})R2}{R2 + R_{B2}} + v_{\text{in}} \tag{8-4}$$

After a bit of algebraic manipulation, equation (8-4) can be solved for the value of v_{in} which will drive V_{diode} to zero, switching it on.

$$V_{\text{break 2}} \cong -V_{\text{ref}} \frac{R2}{R_{B2}} \tag{8-5a}$$

$$R_{B2} \cong -R2 \frac{V_{\text{ref}}}{V_{\text{break 2}}} \tag{8-5b}$$

There are several observations and precautions of which you should be aware. Placing the diode network in the input circuit allows the slope of the transfer curve to *increase* at each break point. Look at equations (8-1) through (8-3). If the slope must decrease with increasing input voltage, you must use a different circuit. Second, the circuit is inherently an inverter. This puts the transfer curve in the second quadrant for negative inputs (positive outputs) and in the fourth quadrant for positive inputs (negative outputs). Adding an inverting amplifier either before or after this circuit will give you an overall noninverting circuit (first and third quadrants).

Each time a diode switches on (at each break point), the input impedance is lowered. Be careful that the circuit does not load the source at high input voltage levels.

The derivation of equations (8-1) through (8-5) assumed that

$$V_{\text{diode on}} = 0$$

$$R_{\text{diode on}} = 0$$

These approximations should cause no problems as long as you ensure that

$$V_{\text{break}} \gg 0.6 \text{ V}$$

$$R1, 2, 3 \ldots \gg R_{\text{on diode}}$$

Finally, the circuit in Fig. 8-3(a) accommodates only negative inputs. To produce a similar transfer curve for positive inputs, you must add input resistor, diode, and bias resistor networks in parallel with R1. For these positive networks, reverse the diodes and use a negative reference. A seven segment transfer curve synthesizer for both positive and negative inputs is shown in Fig. 8-5. The resulting transfer curve with appropriate parameters is shown in Fig. 8-6.

Placing the resistor-diode network in the *feedback* network produces a transfer curve whose slope *decreases* with an increase in input voltage. The schematic is given in Fig. 8-7(a) and the two segment transfer curve in Fig. 8-7(b). For proper operation, the diode must be connected to the inverting input, which provides a virtual ground.

With a small positive output (negative input), the negative reference voltage holds the diode off. The gain, and slope S1, are determined by

$$S1 = -\frac{R1}{R_i} \tag{8-6}$$

When the input becomes more negative, the output is driven more positive.

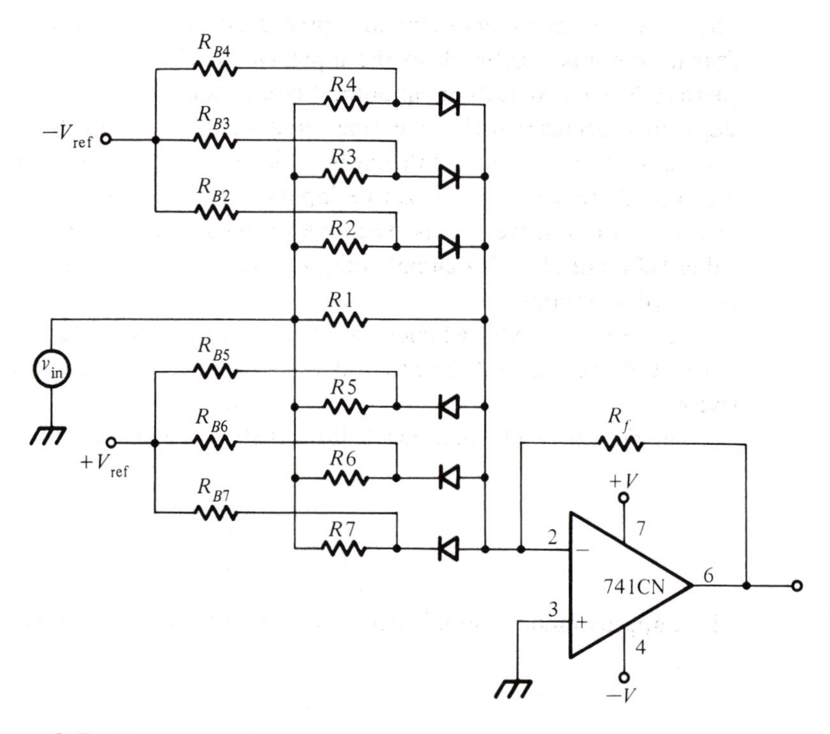

Figure 8-5 Bipolar transfer curve synthesizer

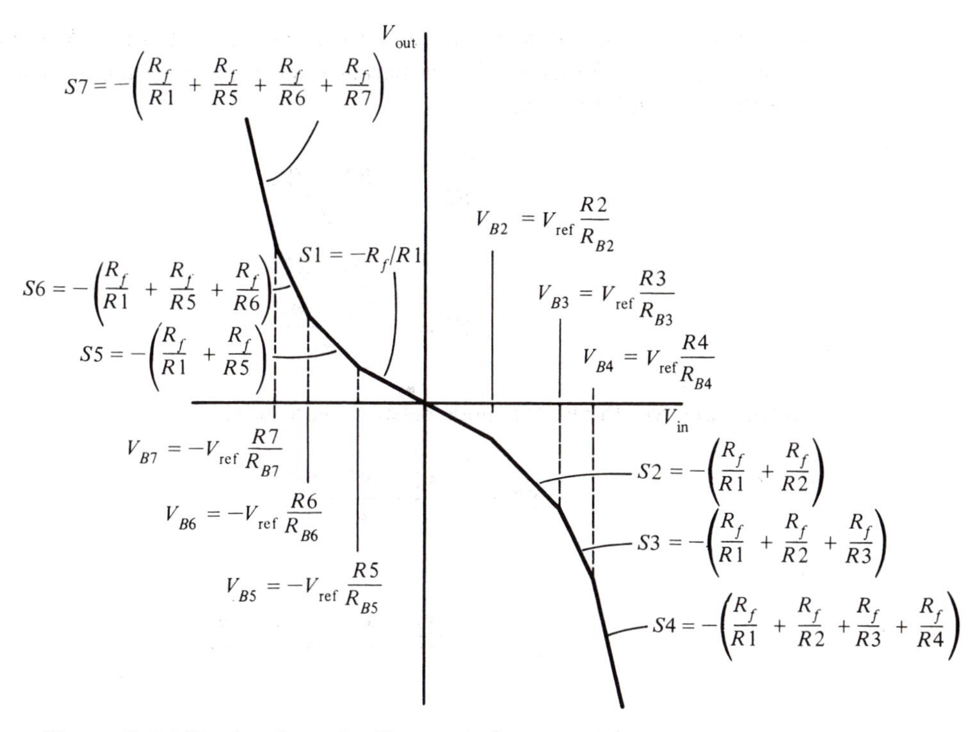

Figure 8-6 Bipolar piecewise linear transfer curve

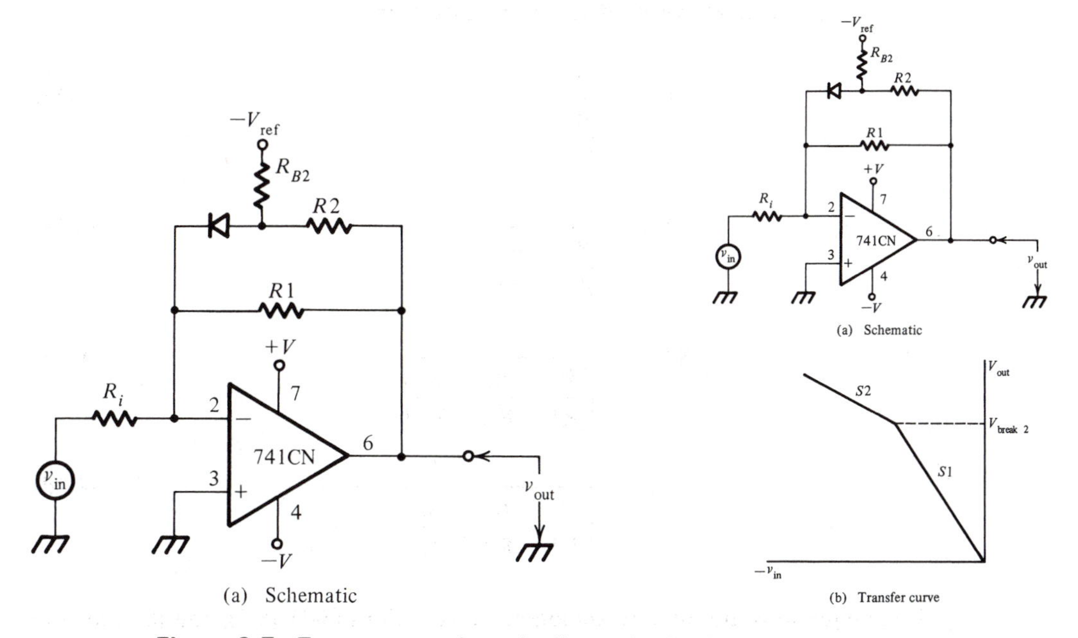

Figure 8-7 Two segment piecewise linear circuit with decreasing slope

This makes the anode of the diode more positive (or less negative). At some point, the diode goes on. Resistor R2 now parallels R1, decreasing the feedback resistance, the gain, and the slope.

$$S2 = -\frac{R1 \,//\, R2}{R_i}$$

$$S2 = -\frac{\dfrac{R1 \cdot R2}{R1 + R2}}{R_i}$$

If a third and fourth resistor-diode network were added,

$$S3 = -\frac{R1 \,//\, R2 \,//\, R3}{R_i}$$

and

$$S4 = -\frac{R1 \,//\, R2 \,//\, R3 \,//\, R4}{R_i}$$

Although straightforward, these equations are a bit difficult to manipulate for analysis or design. However, if you reorganize them,

$$S1 = -\frac{1}{\dfrac{R_i}{R1}} \tag{8-7a}$$

$$S2 = -\frac{1}{\dfrac{R_i}{R1} + \dfrac{R_i}{R2}} \tag{8-7b}$$

$$S3 = -\frac{1}{\left(\dfrac{R_i}{R1} + \dfrac{R_i}{R2} + \dfrac{R_i}{R3}\right)} \tag{8-7c}$$

$$S_n = -\frac{1}{\left(\dfrac{R_i}{R1} + \dfrac{R_i}{R2} + \dfrac{R_i}{R3} + \cdots + \dfrac{R_i}{R_n}\right)} \tag{8-7d}$$

These equations are directly analogous to equations (8-1), (8-2), and (8-3) for the circuit with the networks in the input.

The voltage at the anode of the diode in Fig. 8-7 is

$$V_{anode} = \frac{(V_{ref} - v_{out})R2}{(R2 + R_{B2})} + v_{out}$$

Consequently, for this configuration, the diode voltage is controlled by the *output* voltage, not the input. If it is assumed that the diode turns on at

$$V_{diode} \cong 0 \text{ V}$$

$$v_{out} = -V_{ref}\frac{R2}{R_{B2}}$$

is the output voltage at the break point between S1 and S2. So

$$V_{break\ 2} = -V_{ref}\frac{R2}{R_{B2}} \tag{8-8}$$

which is the same as equation (8-5a). However, equation (8-5a), for the circuit with the diode network in the *input*, refers to ($V_{break} = v_{in}$) input voltage. Equation (8-8), for the circuit with the diode network in the *output*, refers to ($V_{break} = v_{out}$) output voltage.

Example 8-1

Design a circuit which will shape a 6 volt peak input triangle wave to a 6 volt peak sine wave. Supply voltage = ±12 V.

Solution

You must first determine the transfer curve required. This is plotted in Fig. 8-8.

$$v_{out} = 6 \text{ V} \sin\theta$$

or

$$v_{out} = 6 \text{ V} \sin\left(v_{in} \times \frac{90^\circ}{6 \text{ V}}\right)$$

As the input voltage increases linearly, the output voltage changes sinusoidally.

Next, this must be converted to a piecewise linear approximation that can be implemented by the transfer curve synthesizer. After careful examination of Fig. 8-8, four segments were selected. These are plotted in Fig. 8-9, and the end point of each segment is identified.

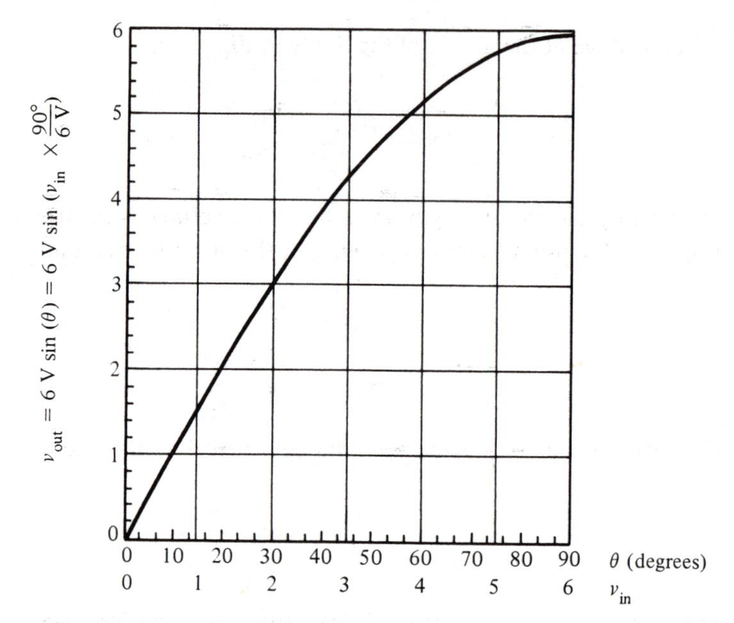

Figure 8-8 Sine shaper ideal transfer curve

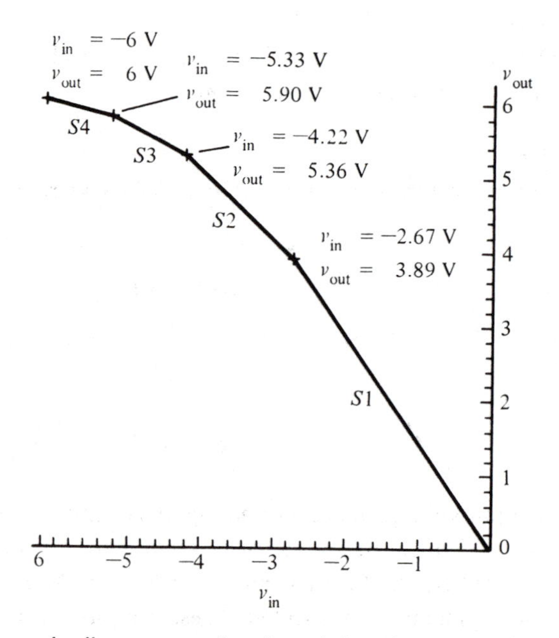

Figure 8-9 Piecewise linear approximation of sine shaper transfer curve

The slope decreases with increasing input, so the resistor-diode networks must be placed in the feedback loop. The overall circuit schematic is shown in Fig. 8-10. The slopes are

$$S1 = \frac{3.89\text{ V} - 0\text{ V}}{-2.67\text{ V} - 0\text{ V}} = -1.46 \qquad \frac{1}{S1} = -0.685$$

$$S2 = \frac{5.36\text{ V} - 3.89\text{ V}}{-4.22\text{ V} - (-2.67\text{ V})} = -0.948 \qquad \frac{1}{S2} = -1.05$$

$$S3 = \frac{5.90\text{ V} - 5.36\text{ V}}{-5.33\text{ V} - (-4.22\text{ V})} = -0.486 \qquad \frac{1}{S3} = -2.06$$

$$S4 = \frac{6\text{ V} - 5.90\text{ V}}{-6\text{ V} - (-5.33\text{ V})} = -0.149 \qquad \frac{1}{S4} = -6.71$$

$$V_{\text{break }2} = 3.89\text{ V}$$

$$V_{\text{break }3} = 5.36\text{ V}$$

$$V_{\text{break }4} = 5.90\text{ V}$$

$$\frac{1}{S1} = -\frac{R_i}{R1}$$

$$R1 = -S1 \cdot R_i = 1.46R_i \tag{8-9a}$$

$$\frac{1}{S2} = -\frac{R_i}{R1} - \frac{R_i}{R2} = \frac{1}{S1} - \frac{R_i}{R2}$$

$$-1.05 = -0.685 - \frac{R_i}{R2}$$

$$\frac{R_i}{R2} = 0.365$$

$$R2 = 2.74R_i \tag{8-9b}$$

$$\frac{1}{S3} = -\frac{R_i}{R1} - \frac{R_i}{R2} - \frac{R_i}{R3} = \frac{1}{S2} - \frac{R_i}{R3}$$

$$-2.06 = -1.05 - \frac{R_i}{R3}$$

$$\frac{R_i}{R3} = 1.01$$

$$R3 = 0.99R_i \tag{8-9c}$$

$$\frac{1}{S4} = -\frac{R_i}{R1} - \frac{R_i}{R2} - \frac{R_i}{R3} - \frac{R_i}{R4} = \frac{1}{S3} - \frac{R_i}{R4}$$

$$-6.71 = -2.06 - \frac{R_i}{R4}$$

$$\frac{R_i}{R4} = 4.65$$

$$R4 = 0.215R_i \tag{8-9d}$$

If you pick

$R_i = 10\ k\Omega$

then

$$R1 = 14.6\ k\Omega \qquad \text{pick } R1 = 15\ k\Omega$$

$$R2 = 27.4\ k\Omega \qquad \text{pick } R2 = 27\ k\Omega$$

$$R3 = 9.9\ k\Omega \qquad \text{pick } R3 = 10\ k\Omega$$

$$R4 = 2.15\ k\Omega \qquad \text{pick } R4 = 2.2\ k\Omega$$

$$V_{break\ 2} = 3.89\ V = -V_{ref}\frac{R2}{R_{B2}}$$

$$R_{B2} = -V_{ref}\frac{R2}{3.89\ V} = \frac{-(-12\ V)(27\ k\Omega)}{3.89\ V} = 83.3\ k\Omega$$

Pick $R_{B2} = 82\ k\Omega$.

$$R_{B3} = \frac{-(-12\ V)R3}{5.36\ V} = \frac{12\ V \times 10\ k\Omega}{5.36\ V} = 22.4\ k\Omega$$

Pick $R_{B3} = 22\ k\Omega$.

$$R_{B4} = \frac{-(-12\ V)R4}{5.90\ V} = \frac{12\ V \times 2.2\ k\Omega}{5.90\ V} = 4.47\ k\Omega$$

Pick $R_{B4} = 4.3\ k\Omega$

Resistors R2–R4 and R_{B2}–R_{B4} alter the slope of the positive output cycle. To cause the negative cycle to be similarly shaped, set

$$R5 = R2, \quad R6 = R3, \quad R7 = R4, \quad R_{B5} = R_{B2}, \quad R_{B6} = R_{B3},$$
$$R_{B7} = R_{B4}, \quad \text{and} \quad +V_{ref} = 12\ V$$

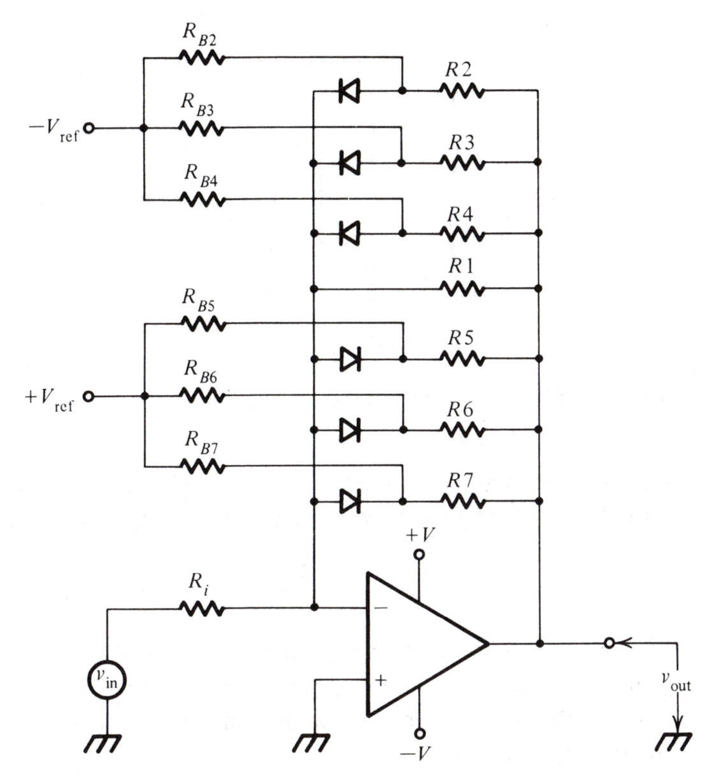

Figure 8-10 Sine shaper schematic for Example 8-1

8-1.2 Ideal Rectifier

In electrical measurements it is often necessary to convert an AC signal to an equivalent DC. This is how most meters indicate AC rms. The simplest solution is to use a diode rectifier. However, to turn on the rectifier 200 mV to 600 mV must be dropped across the diode. This portion of the signal never appears at the output. Even with a 10 V peak signal, that loss across the diode causes at least 2 percent error. Converting AC signals in the millivolt range cannot be done with a simple diode.

The circuit in Fig. 8-11 allows AC to DC conversion of small signals. It is a conventional inverting amp with a diode added to the negative feedback loop. Notice, however, that the output is taken from the diode (in the middle of the feedback loop), *not* at the op amp's output.

When the input voltage is zero, the op amp's output is zero (ignoring offsets), and the diode is off. This means that the negative feedback loop has been broken, and the op amp is operating open loop. In this open loop condition the input

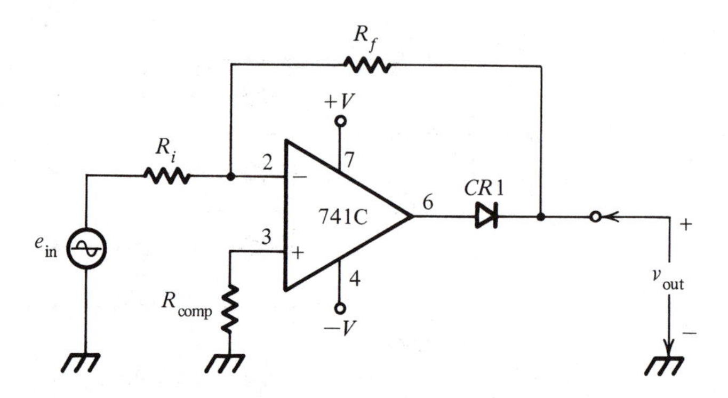

Figure 8-11 Simple ideal rectifier

signal is inverted and amplified by the op amp's open loop gain (typically 2×10^5 for the 741). A drop to $-3\ \mu V$ at the input sends the output to 0.6 V.

$$v_{out} = -(-3\ \mu V) \times 2 \times 10^5 = 0.6\ V$$

This turns the diode on. Consequently, the offset of the diode is provided by the op amp, requiring only about $3\ \mu V$ of the input signal.

Once the diode is on, the circuit functions as a regular inverting amplifier with an output of

$$v_{out} = -\frac{R_f}{R_i} e_{in}$$

This holds, as long as

$$R_f \gg R_{diode\ on}$$

Otherwise, a significant part of the signal will be dropped across the diode's forward resistance. The op amp's output will follow the circuit's output, but will be ~ 0.6 V higher.

When the input voltage goes positive, the op amp is rapidly driven into negative saturation. The diode is off. At first you may be tempted to conclude that the circuit's output must then go to zero. However, since the op amp has no negative feedback, its inverting terminal is no longer forced to virtual ground. Resistors R_i and R_f are the only things between the input and the output. They form a voltage divider with the input's source impedance and the load impedance. During that half-cycle,

$$v_{out} = \frac{R_{load}}{R_{source} + R_i + R_f + R_{load}} e_{in}$$

This may be quite different from the zero output desired.

This simple ideal rectifier has one other problem. During the positive input cycle, with CR1 off, the op amp is saturated. When the cycles switch, it may take a significant time to pull the op amp out of negative saturation and then slew to +0.6 V. This delay will significantly limit high frequency response.

Both of these problems are solved by adding a second diode and feedback resistor, as shown in Fig. 8-12. During the negative input cycle, the op amp drives CR1 on, and then functions as an inverting amplifier, as you have already seen. This produces the positive output. When the input goes positive, the op amp's output swings negative, to −0.6 V. This turns on CR2 and turns off CR1. The negative feedback loop is now provided through R_{f1} and CR2. The negative half-cycle is produced at CR2's anode just as the positive half-cycles were produced through CR1.

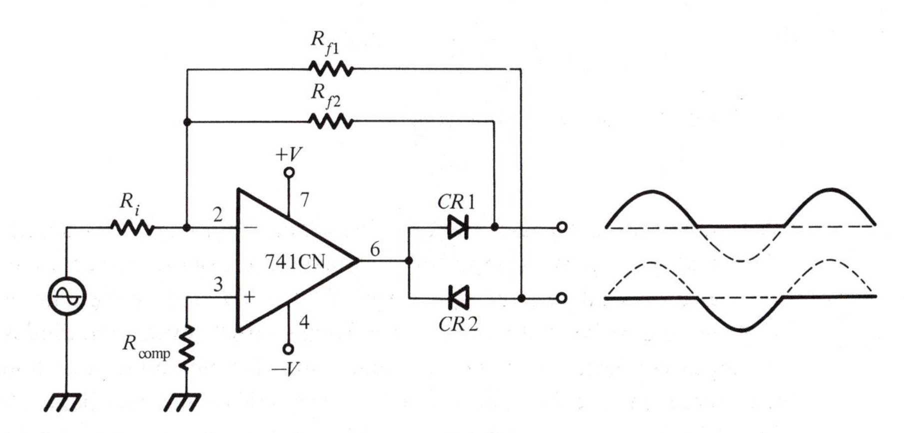

Figure 8-12 Improved ideal rectifier schematic

Since the negative feedback loop is now complete for both half-cycles, the inverting input remains at virtual ground. When CR1 is off, there is no current flow through R_{f2}, so there is no voltage across it. With the left end of R_{f2} at virtual ground, and no voltage dropped across R_{f2}, the output at CR1 must be at zero volts during positive inputs.

Also, with both cycles providing negative feedback, the op amp will never saturate. This improves speed.

You do not have to use both outputs. Use only the one which provides the polarity voltage needed. However, be sure to include both diodes and feedback resistors. Otherwise, during the off output cycle, the op amp will saturate and the output voltage will not go to zero.

The circuit in Fig. 8-12 provides both *half-wave* outputs. You can obtain a *full wave* rectified output by adding a difference amp. This is shown in Fig. 8-13.

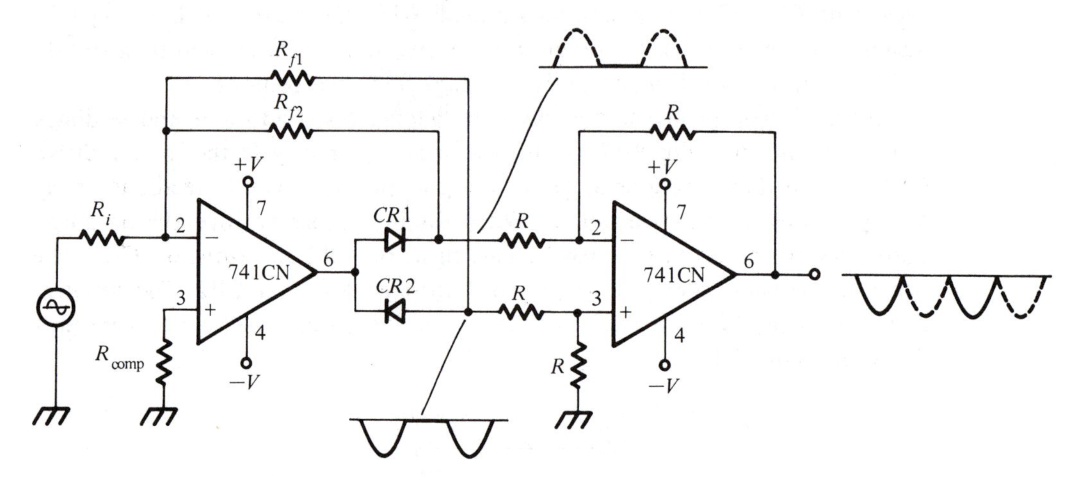

Figure 8-13 Full-wave ideal rectifier

The difference amp passes the negative signal from CR2, subtracting zero from CR1 during the positive input cycle. During the negative input cycle, CR2 provides a zero signal, and the signal from CR1 is inverted by the difference amp. If you need a positive, full wave rectified signal, either reverse the diodes or reverse the input connections to the difference amp. Do not limit your application of these circuits to ideal rectification of small signals for measurement. The circuit in Fig. 8-12 is a voltage steering network, sending the positive part of the input through CR1 and the negative part of the input through CR2. This may have many control applications. The full wave rectifier of Fig. 8-13 actually takes the absolute value of the input.

$$V_{out} = -\frac{R_f}{R_i} |e_{in}|$$

When building analog computational circuits, this could be quite handy.

8-1.3 Peak Detector

The ideal rectifiers of the previous section yield an output proportional to the average value of a repetitive waveform. There are several measurement applications where the *peak* value of a signal must be detected and then held as a DC level for later analysis. The peak of a pulse wavetrain is often of more interest than its average value. When one is performing destructive testing, it is necessary to track and hold the maximum signal. Spectral and mass spectrometer analysis also requires the use of peak detectors.

The principle of including a diode in the feedback loop to overcome its offset is used in building peak detectors. Refer to Fig. 8-14. The op amp is configured as a voltage follower. Diode CR1 is used to keep the op amp from going to negative saturation for negative inputs. This improves the circuit's response speed and lowers the reverse voltage rating needed for CR2. Diode CR2 is the ideal rectifier, enclosed in the op amp's negative feedback loop.

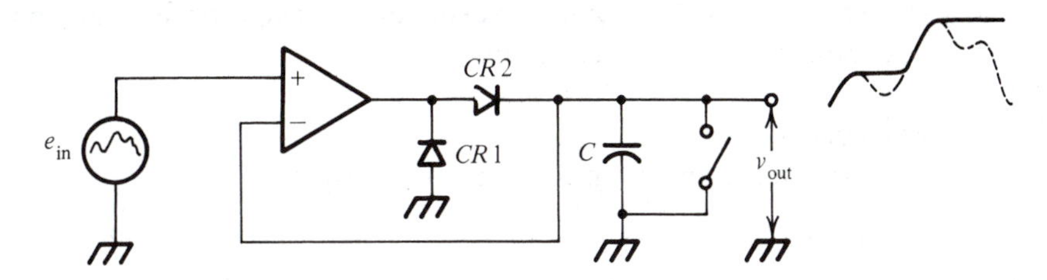

Figure 8-14 Peak detector

With no charge on the capacitor, $+3\ \mu V$ at the input will cause $+0.6$ V at the op amp's output. This turns the diode on. When the diode goes on, the op amp functions as a voltage follower. The output tries to track the input. Capacitor C charges through Z_{out} of the op amp and $R_{diode\ on}$, to the input voltage. The charge rate is limited by C, $R_{diode\ on}$, $Z_{out\ op\ amp}$, and $I_{max\ op\ amp}$. If the input voltage drops below the output voltage, the diode goes off. The capacitor then holds its charge.

There are several sources of error in the tracking mode of which you should be aware. The op amp must be able to operate stably when driving a capacitive load. Also, the circuit must not be underdamped. This means that when you apply a square wave, the circuit must not overshoot. Such overshoots would be erroneously captured as peaks. To test for proper damping, short out diode CR2 and apply a square wave input. Observe the output for overshoots. External frequency compensation capacitance may be adjusted to provide fast response to the square wave without overshoots.

Leakage is the largest source of error in the hold mode. Any current drawn by the load will discharge the capacitor, causing an error called *droop*. You can add a high impedance buffer (op amp voltage follower) between the capacitor and the load to minimize this problem. The bias current into the op amp's inverting input and into the buffer must be supplied from the capacitor during the hold mode. For precision peak detectors, carefully select the op amps to have both low bias currents and low drift.

The reverse bias leakage current through the diode and the capacitor's leakage current may cause noticeable droop over long internal holds. Finally, to discharge the capacitor and reset the circuit to detect a new peak, a switch is normally used. Electronic switches (FETs or analog switches) are convenient but cause more leakage and droop than a relay.

One tempting solution to these hold mode droop problems is simply increasing the value of the capacitor. Increasing C provides more charge and therefore lowers the effect of leakage currents. However, the larger C is, the more slowly it will accept charge during the tracking mode. The capacitor may not charge to a peak before the voltage has fallen.

Example 8-2

Explain the operation of the circuit in Fig. 8-15.

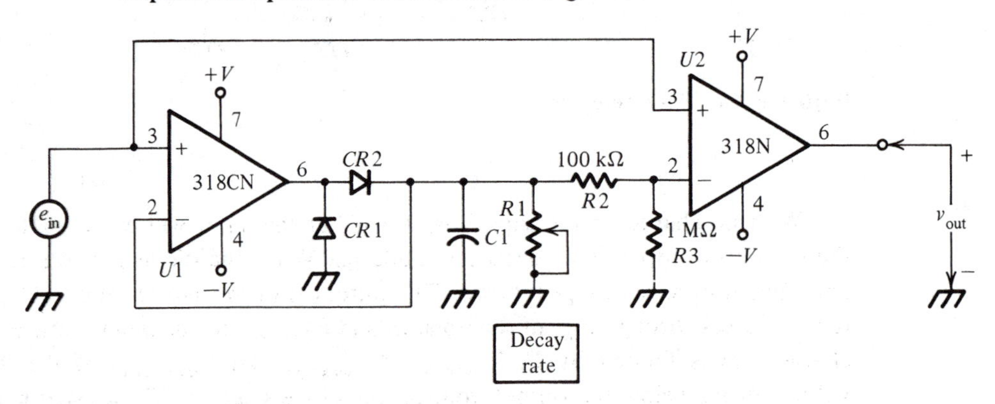

Figure 8-15 Schematic for Example 8-2

Solution

Op amp U1, CR1, CR2, and C1 form a positive peak detector. Op amp U2 is being operated without negative feedback. It is a noninverting comparator. The output from the peak detector is divided by R2 and R3, placing 0.9 V_{peak} as the reference voltage of the noninverting comparator U2.

This circuit is an adaptive comparator. It provides a square wave output for every cycle at the input. Switching occurs near the peak of the input, rather than at its zero crossing. This minimizes false switching caused by noise on the ground line (a big problem in sensitive systems).

Increases in input amplitude are automatically detected by the peak detector and the comparator's reference level is increased accordingly. The decay rate control R1 allows the capacitor to discharge slowly. It is set so that the capacitor's voltage will not approach zero over one cycle, but will fall far enough to be charged to a *lower* peak, should the input amplitude decrease.

8-2 LOGARITHMIC AND ANTILOGARITHMIC AMPLIFIERS

Logarithmic amplifiers have several areas of application. Analog computation may require ln x, log x, or sinh x. This can all be performed continuously with analog log amps. It is very convenient to have direct dB displays on digital voltmeters and spectrum analyzers. Log amps are used in these instruments. Signal dynamic range compression enhances computer signal processing as well as audio recording. Log amps can be used to compress a signal's dynamic range.

8-2.1 Log Amp

The fundamental log amp is formed by placing a transistor in the negative feedback path of an op amp. This is shown in Fig. 8-16. Because of the negative feedback, the collector is held at virtual ground. Since the base is also grounded, the transistor's voltage/current relationship is

$$I_C = \alpha I_S(e^{qV_E/kT} - 1)$$

where
$\alpha \approx 1$ and constant with I and T
I_s = emitter saturation current $\approx 10^{-13}$ A
q = one electron's charge = 1.60219×10^{-19} C
k = Boltzmann's constant 1.38062×10^{-23} $J/°K$
T = absolute temperature (in degrees K)

Assuming $\alpha = 1$, you have

$$\frac{I_c}{I_s} = e^{qV_E/kT} - 1$$

$$e^{qV_E/kT} = \frac{I_c}{I_s} + 1$$

Since

$$I_s \approx 10^{-13}\text{ A}, \quad \text{then} \quad \frac{I_c}{I_s} \gg 1$$

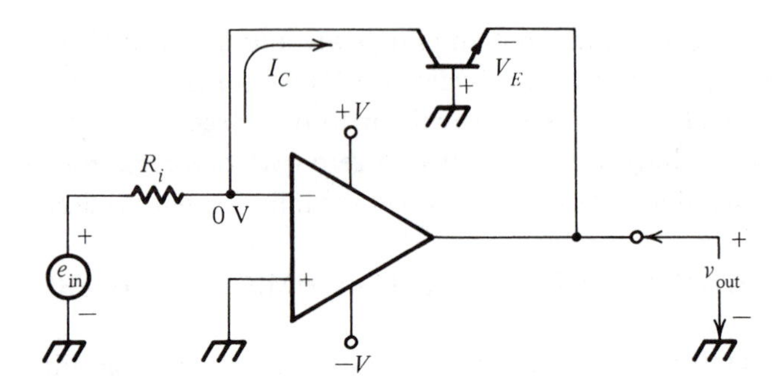

Figure 8-16 Fundamentals of log amp

Taking the natural logarithm of both sides, you obtain

$$\frac{q}{kT} V_E = \ln \frac{I_c}{I_s}$$

$$V_E = \frac{kT}{q} \ln \frac{I_c}{I_s}$$

Look at Fig. 8-16.

$$I_c = \frac{e_{in}}{R_i}$$

$$V_E = -v_{out}$$

so

$$v_{out} = -\frac{kT}{q} \ln \frac{e_{in}}{R_i I_s} \tag{8-10a}$$

or

$$v_{out} = -\frac{kT}{q} \ln \frac{e_{in}}{e_{ref}} \tag{8-10b}$$

where

$$e_{ref} = R_i I_s$$

The output of Fig. 8-16 is indeed a logarithmic function of its input. Although equation (8-10) was developed with the natural logarithm (ln) the circuit in Fig. 8-16 can be applied to $\log_{10}$ by proper scaling.

$$\log_{10} X = 0.4343 \ln X \tag{8-11}$$

The emitter saturation current I_s, though on the order of 10^{-13} A, varies significantly from one transistor to another and with temperature. Setting e_{ref}, then, is a matter of guesswork. The circuit in Fig. 8-17 allows you to eliminate this problem. The input is applied to one log amp, while a reference voltage is applied to another log amp. The two transistors are integrated close together in the same silicon wafer. This provides a close match of saturation currents and ensures good thermal tracking. For this reason,

$$I_{s1} = I_{s2} = I_s$$

The output of U1 is

$$e_1 = -\frac{kT}{q} \ln \frac{e_{in}}{R_i I_s}$$

The output of U2 is

$$e_2 = -\frac{kT}{q} \ln \frac{e_{ref}}{R_i I_s}$$

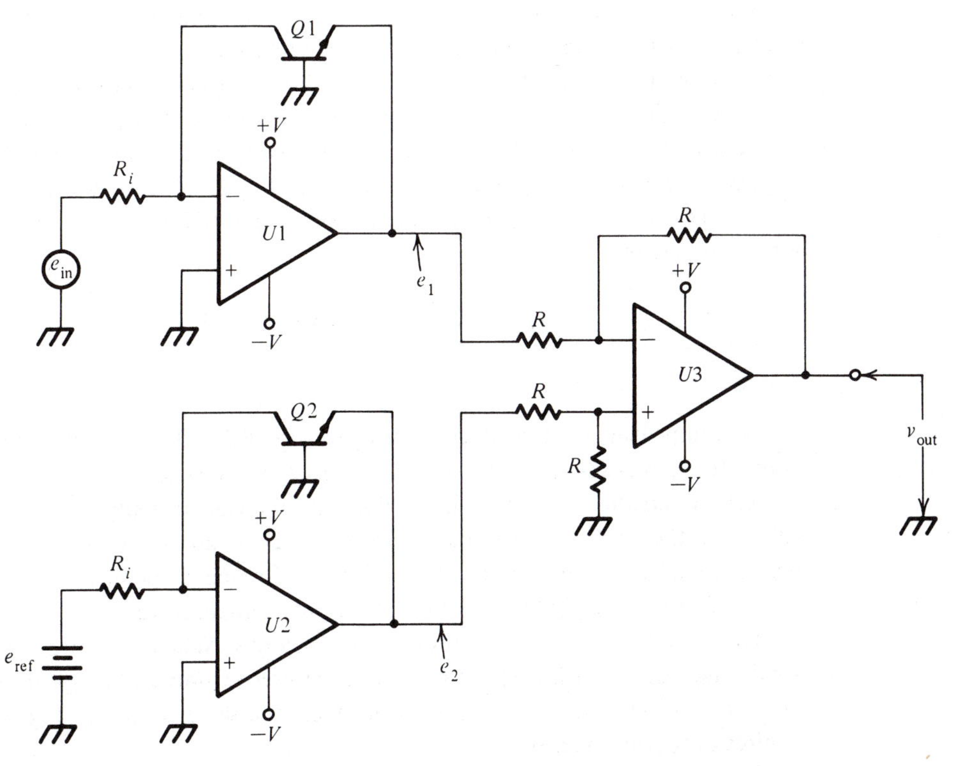

Figure 8-17 Log amp with I_s compensation (*Courtesy of Analog Devices*)

Op amp U3 subtracts these two outputs:

$$v_{out} = -\frac{kT}{q}\ln\frac{e_{ref}}{R_i I_s} - \left(-\frac{kT}{q}\ln\frac{e_{in}}{R_i I_s}\right)$$

$$v_{out} = \frac{kT}{q}\left(\ln\frac{e_{in}}{R_i I_s} - \ln\frac{e_{ref}}{R_i I_s}\right)$$

Remember, however, that subtracting logs is the same as dividing their arguments.

$$v_{out} = \frac{kT}{q}\ln\frac{\dfrac{e_{in}}{R_i I_s}}{\dfrac{e_{ref}}{R_i I_s}}$$

$$v_{out} = \frac{kT}{q}\ln\frac{e_{in}}{e_{ref}}$$

The reference level is now set with a single external voltage source. Its dependence on device and temperature has been removed.

This circuit, however, presents one further rather significant problem. The output voltage is directly proportional to temperature. Since the circuit is not intended to be a temperature sensor, this is highly undesirable. One additional stage will compensate for temperature. Refer to Fig. 8-18. A fourth stage, U4, has been added. It provides a noninverting gain of $1 + (R2/R_{TC})$, giving an overall output voltage of

$$v_{out} = \left(1 + \frac{R2}{R_{TC}}\right)\frac{kT}{q}\ln\frac{e_{in}}{e_{ref}} \qquad (8\text{-}12)$$

To compensate for the effects of temperature, R2 and R_{TC} must be integrated within the same silicon wafer as the transistors. Resistor R_{TC} has a positive temperature coefficient. When this is done, an increase in temperature of the IC will cause the second factor of equation (8-12) to go up. However, this also increases R_{TC}, lowering the first factor. Properly matched coefficients and resistor values will assure negligible variation of v_{out} with temperature.

The circuit in Fig. 8-18 requires four op amps. When building a precision circuit, high quality FET op amps should be used. Comparably speaking, these are expensive. The circuit in Fig. 8-19 produces the same output as Fig. 8-18, but requires only two op amps.

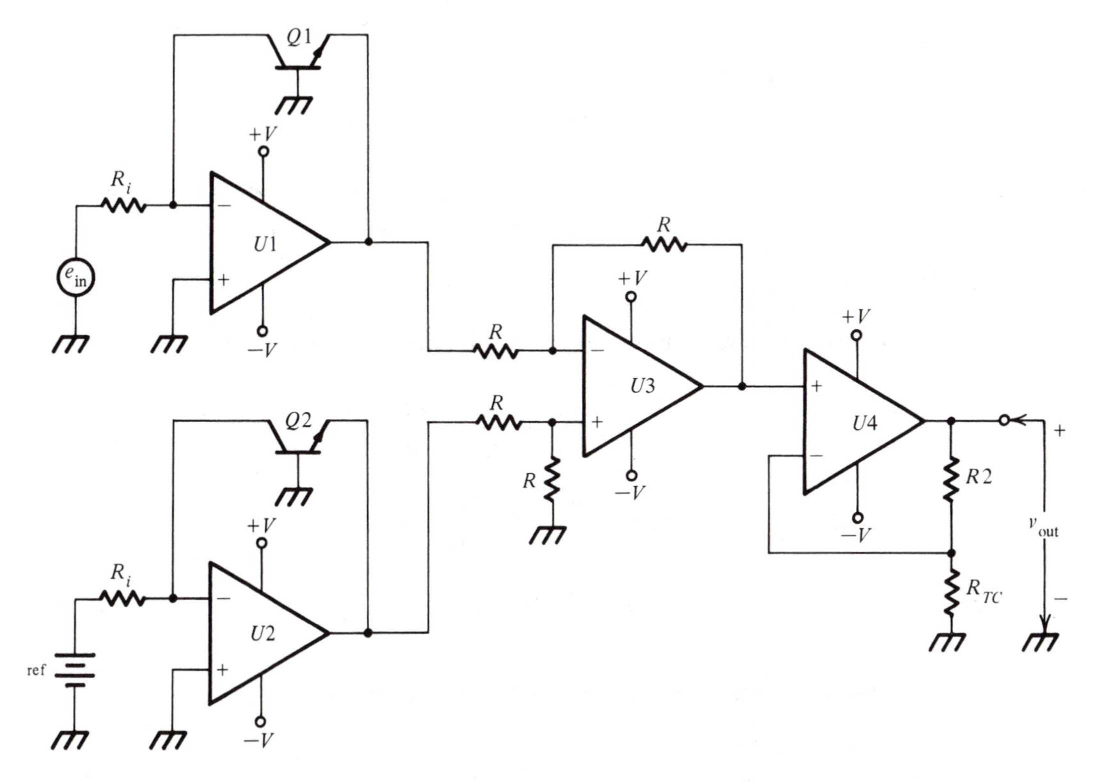

Figure 8-18 Log amp with I_S and temperature compensation (*Courtesy of Analog Devices*)

The emitter voltage of Q1 comes from equation (8-10a). This also is v_A.

$$v_A = -\frac{kT}{q} \ln \frac{e_{in}}{R_i I_s} \tag{8-13}$$

The base emitter voltage for Q2 is

$$V_{Q2B-E} = \frac{kT}{q} \ln \frac{e_{ref}}{R_i I_s} \tag{8-14}$$

The base voltage of Q2, then, is

$$V_{Q2\ Base} = V_{Q2\ emitter} + V_{Q2\ B-E}$$

But

$$V_{\text{Q2 Base}} = v_{\text{B}}$$

and

$$V_{\text{Q2 emitter}} = v_{\text{A}}$$

So

$$v_{\text{B}} = v_{\text{A}} + V_{\text{Q2 B-E}} \tag{8-15}$$

Substituting equations (8-13) and (8-14) into (8-15), you obtain

$$v_{\text{B}} = -\frac{kT}{q}\ln\frac{e_{\text{in}}}{\text{R}_i I_s} + \frac{kT}{q}\ln\frac{e_{\text{ref}}}{\text{R}_i I_s}$$

$$v_{\text{B}} = -\frac{kT}{q}\left(\ln\frac{e_{\text{in}}}{\text{R}_i I_s} - \ln\frac{e_{\text{ref}}}{\text{R}_i I_s}\right)$$

$$v_{\text{B}} = -\frac{kT}{q}\ln\frac{e_{\text{in}}}{e_{\text{ref}}} \tag{8-16}$$

This is the voltage in the center of the voltage divider.

$$v_{\text{B}} = \frac{\text{R}_{\text{TC}}}{\text{R2} + \text{R}_{\text{TC}}} v_{\text{out}} \tag{8-17}$$

Substituting equation (8-16) into (8-17), you obtain

$$-\frac{kT}{q}\ln\frac{e_{\text{in}}}{e_{\text{ref}}} = \frac{\text{R}_{\text{TC}}}{\text{R2} + \text{R}_{\text{TC}}} v_{\text{out}}$$

$$v_{\text{out}} = -\frac{\text{R2} + \text{R}_{\text{TC}}}{\text{R}_{\text{TC}}}\frac{kT}{q}\ln\frac{e_{\text{in}}}{e_{\text{ref}}}$$

$$v_{\text{out}} = -\left(1 + \frac{\text{R2}}{\text{R}_{\text{TC}}}\right)\frac{kT}{q}\ln\frac{e_{\text{in}}}{e_{\text{ref}}} \tag{8-18}$$

The circuit in Fig. 8-19 gives the same results as that in Fig. 8-18 (with an inversion) but requires only two high quality op amps.

You should carefully consider the specifications of the op amps chosen. At low levels of input current (and, therefore, input voltage), bias current and drift can cause linearity and log conformance errors. With bias currents as large as 10 percent of the input current, output errors on the order of 2.5 mV may be encountered.

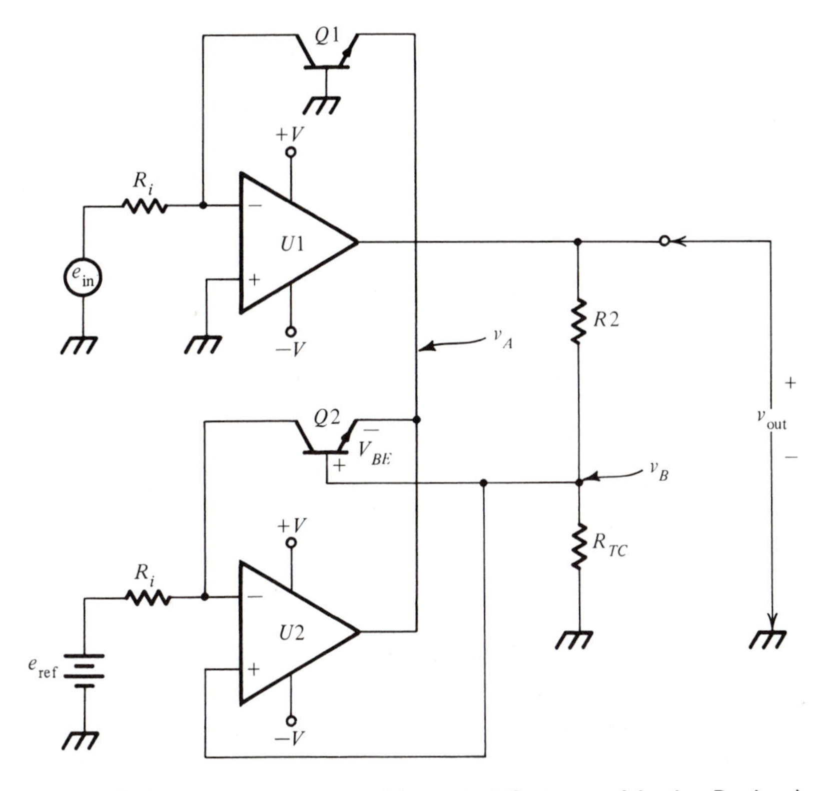

Figure 8-19 Two op amp compensated log amp (*Courtesy of Analog Devices*)

Frequency stability considerations may demand external compensation. Normally this is obtained by placing a capacitor across the negative feedback element. This lowers gain as the frequency rises. However, the effective resistance in the negative feedback loop is

$$r_e \cong \frac{26 \text{ mV}}{I_c}$$

At high levels of input (collector) current, r_e becomes very small, shorting out the parallel compensation capacitor. The capacitor loses its effect. The solution is to add a resistor in series with the emitter, placing a lower limit on the negative feedback resistance through the transistor. A typical frequency compensation network is shown in Fig. 8-20. Values of C_c and R_c are best determined experimentally, since parasitic effects play a large role. Begin with C_c in the 100 pF range and R_c in the 1 kΩ range.

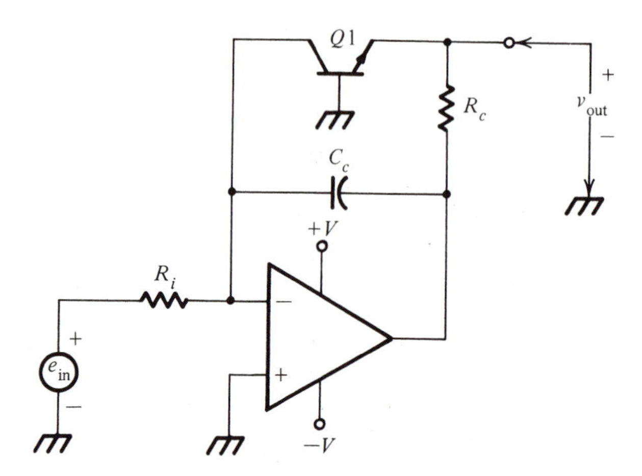

Figure 8-20 Log amp frequency compensation (*Courtesy of Analog Devices*)

8-2.2 Antilog Amps

Two major areas of application for log amps are in complex analog calculations and in signal processing. To raise a signal to a power, or to multiply two signals or to divide two signals, you could first take the log of the signal. Then, exponentiation to any power comes by amplification (multiplication by a constant). Multiplication of two log "encoded" signals is performed by a summer, while division of two log "encoded" signals is performed by a difference amp.

To process signals whose range varies from 10s of μV to 10s of volts requires very high quality, expensive equipment, to have both the resolution (μV) and range (10s of volts). By first taking the log of the signal, each decade of input signal change requires only one volt of output signal change. This log compression of the signal significantly lowers the demands on the processing circuits for both high resolution and high range.

In either analog calculations or signal processing, however, once the signal has been properly conditioned, it must be "decoded." To convert these log "encoded" signals back into real-world terms, you must take the antilog. This is done with the antilog amp. For the log amp, a decade change of input signal causes one volt change on the output. For the antilog amp, one volt change of input signals causes a decade's change at the output.

You can convert the log amp of Fig. 8-19 to an antilog amp by exchanging input and output connections (more or less). This is shown in Fig. 8-21. Compare this with the log amp of Fig. 8-19. The input for the antilog amp is fed into the temperature-compensating voltage divider R2 and R_{TC}, and then to the base of Q2. The output of the antilog amp is fed back to R_i and the inverting input of U1. Other than this input/output reversal, the only difference between the antilog

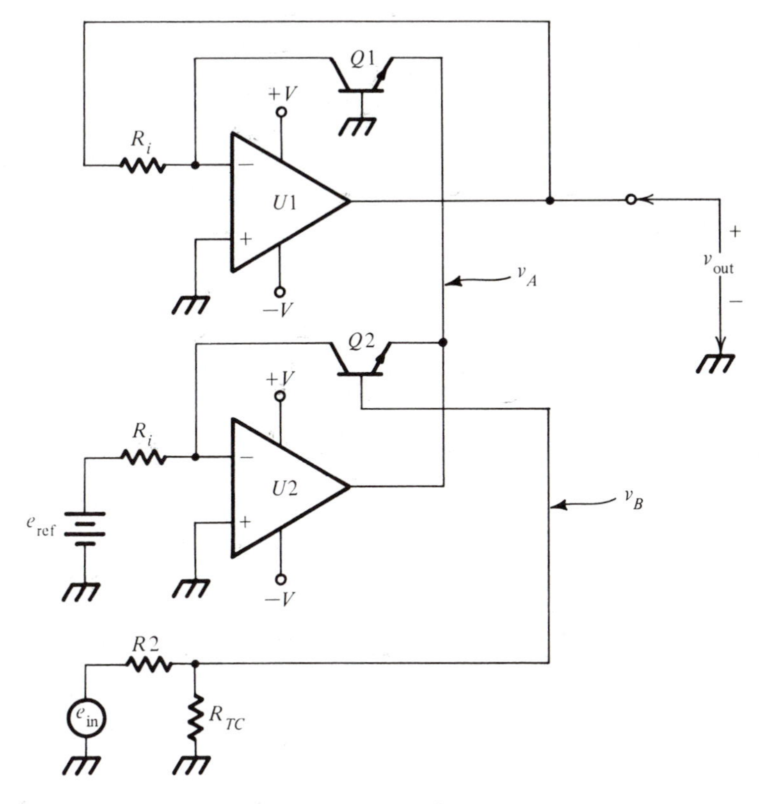

Figure 8-21 Antilog amp (*Courtesy of Analog Devices*)

and the log amp is the noninverting input of U2. For the log amp it is tied to the base of Q2. For the antilog amp, U2 noninverting input is tied to ground.

You can determine the output voltage by first recalling that

$$v_{Q1\,B-E} = \frac{kT}{q} \ln \frac{v_{out}}{R_i I_s} \tag{8-19}$$

$$v_{Q2\,B-E} = \frac{kT}{q} \ln \frac{e_{ref}}{R_i I_s} \tag{8-20}$$

Since the base of Q1 is tied to ground,

$$v_A = -v_{Q1\,B-E} = -\frac{kT}{q} \ln \frac{v_{out}}{R_i I_s} \tag{8-21}$$

v_B is the base voltage of Q2, and is the output from the R2, R_{TC} voltage divider.

$$v_B = v_{Q2\ Base} = \frac{R_{TC}}{R2 + R_{TC}} e_{in} \tag{8-22}$$

The voltage at the emitter of Q2 is

$$v_{Q2\ E} = v_{Q2\ Base} + v_{Q2\ B-E}$$

Substituting from equations (8-22) and (8-20), you obtain

$$v_{Q2\ E} = \frac{R_{TC}}{R2 + R_{TC}} e_{in} - \frac{kT}{q} \ln \frac{e_{ref}}{R_i I_s} \tag{8-23}$$

But the emitter voltage of Q2 is v_A. Equating (8-21) and (8-23) gives

$$v_A = -\frac{kT}{q} \ln \frac{v_{out}}{R_i I_s} = v_{QE2} = \frac{R_{TC}}{R2 + R_{TC}} e_{in} - \frac{kT}{q} \ln \frac{e_{ref}}{R_i I_s}$$

$$-\frac{kT}{q} \ln \frac{v_{out}}{R_i I_s} = \frac{R_{TC}}{R2 + R_{TC}} e_{in} - \frac{kT}{q} \ln \frac{e_{ref}}{R_i I_s}$$

Grouping the ln terms on the right, you have

$$\frac{R_{TC}}{R2 + R_{TC}} e_{in} = -\frac{kT}{q} \left(\ln \frac{v_{out}}{R_i I_s} - \ln \frac{e_{ref}}{R_i I_s} \right) \tag{8-24}$$

Subtracting logarithms is the same as dividing the augments. So equation (8-24) becomes

$$\frac{R_{TC}}{R2 + R_{TC}} e_{in} = -\frac{kT}{q} \ln \frac{\dfrac{v_{out}}{R_i I_s}}{\dfrac{e_{ref}}{R_i I_s}}$$

$$\frac{R_{TC}}{R2 + R_{TC}} e_{in} = -\frac{kT}{q} \ln \frac{v_{out}}{e_{ref}}$$

$$-\frac{q}{kT} \frac{R_{TC}}{R2 + R_{TC}} e_{in} = \ln \frac{v_{out}}{e_{ref}}$$

From equation (8-11)

$$-0.4343 \frac{q}{kT} \frac{\mathrm{R_{TC}}}{\mathrm{R2} + \mathrm{R_{TC}}} e_{\text{in}} = 0.4343 \ln \frac{v_{\text{out}}}{e_{\text{ref}}}$$

$$-0.4343 \frac{q}{kT} \frac{\mathrm{R_{TC}}}{\mathrm{R2} + \mathrm{R_{TC}}} e_{\text{in}} = \log \frac{v_{\text{out}}}{e_{\text{ref}}}$$

$$-Ke_{\text{in}} = \log \frac{v_{\text{out}}}{e_{\text{ref}}}$$

To "undo" the log (take the antilog) use the left side of the equation as the power of 10.

$$\frac{v_{\text{out}}}{e_{\text{ref}}} = 10^{-Ke_{\text{in}}}$$

$$v_{\text{out}} = e_{\text{ref}}\, 10^{-Ke_{\text{in}}} \tag{8-25}$$

For the circuit in Fig. 8-21, increasing the input one volt causes the output to decrease by a decade.

The antilog amp is subject to the same types of errors you saw with the log amp. Bias current, drift, and frequency stability problems can be solved for the antilog amp as they were for the log amp.

8-2.3 Log and Antilog Functional Module

To build a *precision* log amp, you must have an integrated circuit which contains two carefully matched transistors and a voltage divider. The lower resistor in the voltage divider must have a specific positive temperature coefficient. The Analog Devices 755 Log/Antilog amplifier provides these, plus all other components necessary to build either a log or an antilog log amplifier with very little external wiring.

The schematic of the 755N is given in Fig. 8-22. Compare it closely to the log amp of Fig. 8-19 and the antilog amp of Fig. 8-21. A 10 kΩ input resistor has been included. Also, a current input (sum point, pin 5) is available. The reference voltage and resistor have been replaced by a 10 μA reference current. The voltage divider has two upper resistors, giving you a choice of gains ($K = 1$ or $K = 2$).

To simplify drawing and understanding this circuit, often Q1, Q2, U2, I_{ref}, $\mathrm{R_{TC}}$, R2, and R3 are enclosed in a box labeled *antilog element*. This is shown in Fig. 8-23(a). One side of the antilog element is tied to the inverting input of U1.

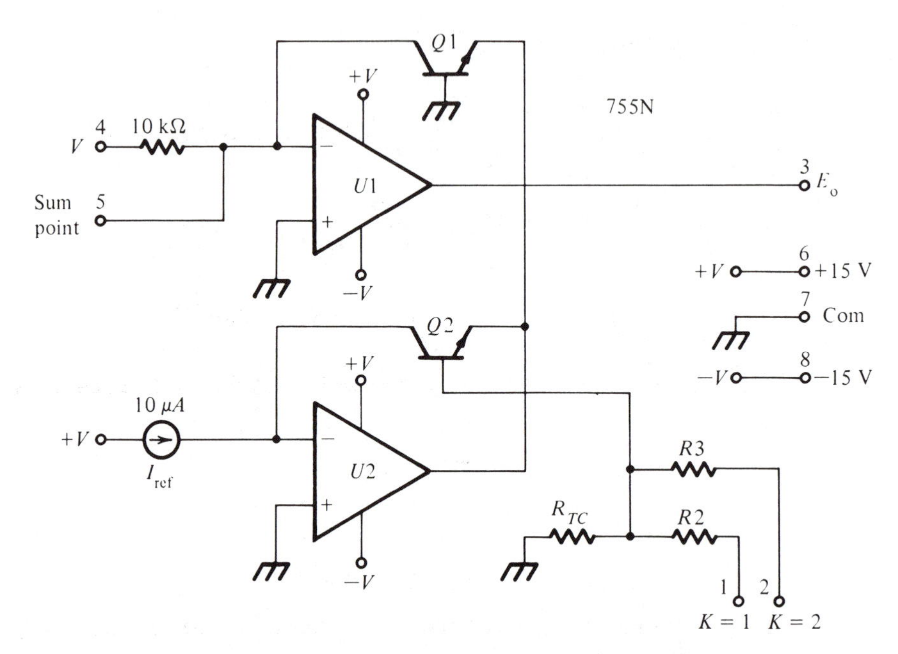

Figure 8-22 755N log/antilog amplifier schematic (*Courtesy of Analog Devices*)

The other side of the element is brought out of the IC. This allows you to connect the antilog element in the feedback loop of U1 to produce a log amp, or in the input loop of U1 to produce an antilog amp.

Connection of the 755 as a log amp is shown in Fig. 8-23(b). This connection provides four decades of input voltage change (1 mV to 10 volts). Over this range the conformity to a log function is 0.5 percent or better. Op amp U1 (also included in the IC) is a high-quality FET input op amp with $I_{bias} \leq 10$ pA and drift $\leq 15\ \mu$V/°C. This reduces DC errors.

Logarithms must be taken of positive arguments. Since the circuits discussed take the logarithm of the ratio (input/reference), the reference and the input must be of the same polarity. For that reason, there are two versions of the 755 module. The 755N has a positive reference current and must have a positive input. (The N is for the NPN transistors). The 755P has a negative reference current and, therefore, must have a negative input.

The log amp of Fig. 8-23(b) provides an output of

$$v_{out} = -K \log_{10} \frac{I_{in}}{I_{ref}} \tag{8-26a}$$

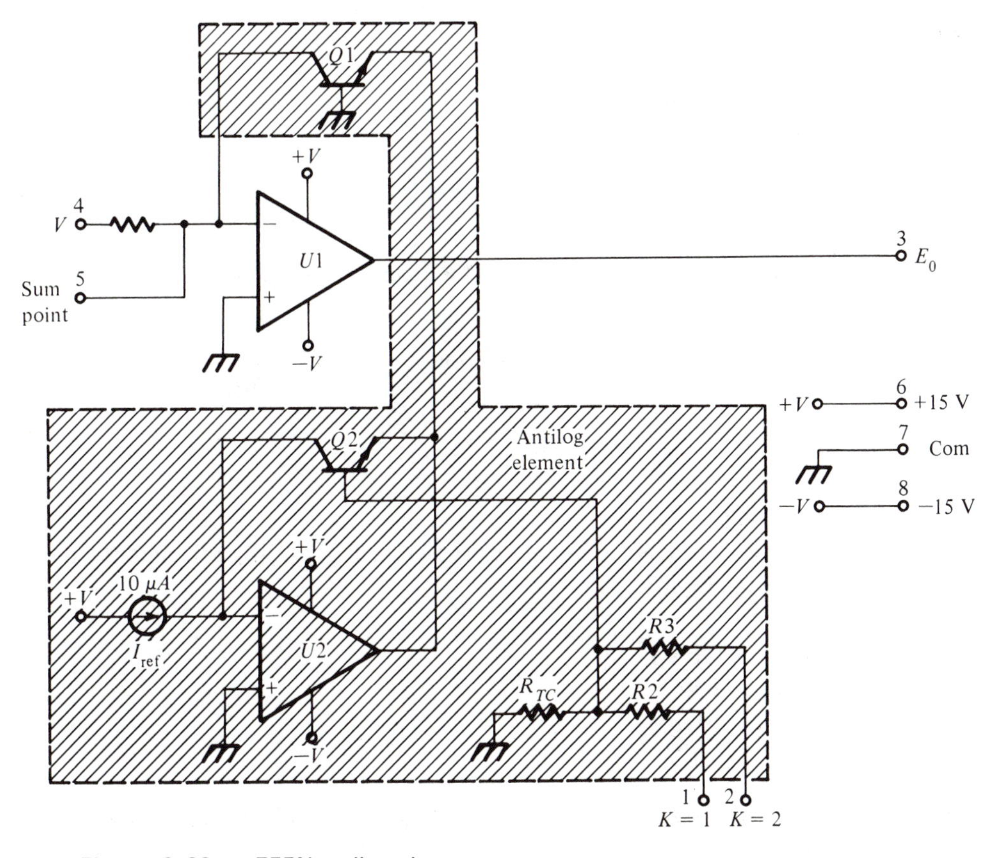

Figure 8-23a 755N antilog element

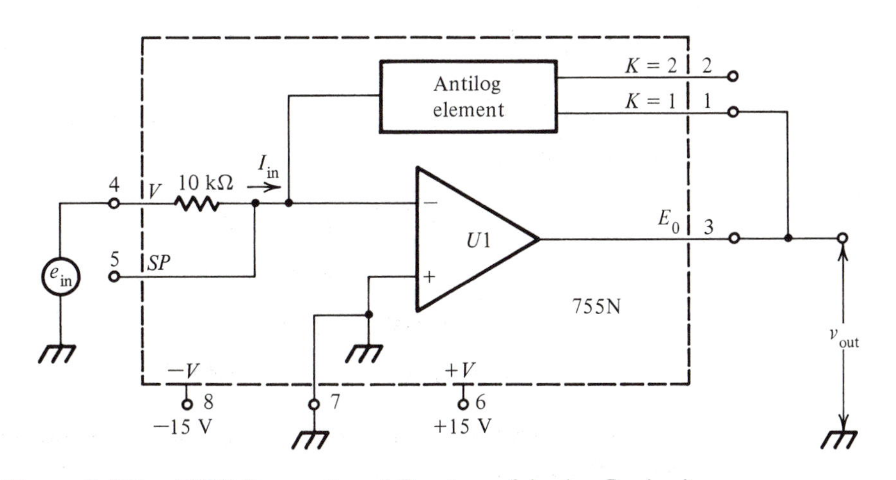

Figure 8-23b 755N Connections (*Courtesy of Analog Devices*)

For a voltage input

$$v_{out} = -K \log_{10} \frac{e_{in}}{R_{in} I_{ref}} \quad (8\text{-}26b)$$

Using the internally provided 10 kΩ R_{in} and 10 μA I_{ref}, you obtain

$$v_{out} = -K \log_{10} \frac{e_{in}}{10\ k\Omega \times 10\ \mu A}$$

$$v_{out} = -K \log_{10} \frac{e_{in}}{0.1\ V} \quad (8\text{-}26c)$$

Example 8-3

a. For the 755N configured as shown in Fig. 8-23(b), calculate the output for the following inputs:

1. $e_{in} = 0.1$ V
2. $e_{in} = 1$ V
3. $e_{in} = 10$ mV
4. $e_{in} = 3.2$ V

b. Repeat the above calculations if the output (pin 3) were connected to pin 2 rather than pin 1.

Solution

a. $v_{out} = -1 \log \dfrac{e_{in}}{0.1\ V}$

$K = 1$ because feedback is tied to $K = 1$, pin 1. Substituting the values above into the equation, we obtain

$e_{in} = 0.1$ V	$v_{out} = 0$ V
$e_{in} = 1$ V	$v_{out} = -1$ V
$e_{in} = 10$ mV	$v_{out} = +1$ V
$e_{in} = 3.2$ V	$v_{out} = -1.505$ V

b. Changing the feedback connection changes $K = 2$. Consequently, the output voltages all double.

To adjust the reference voltage, you can replace R_{in} with an external resistor. Since

$$e_{ref} = R_{in} I_{ref} \quad \text{[see equation (8-26b)]}$$

changing R_{in} effectively changes e_{ref}. This external input resistor should be connected between the input source and pin 5, replacing the internal 10 kΩ resistor at pin 4. To minimize noise pickup, then tie pin 4 to ground. If you use an external input resistor, be sure to limit the maximum input current to less than 1 mA.

$$I_{in} = \frac{e_{in\,max}}{R_{in}} \leq 1 \text{ mA}$$

To alter the log amp scale factor (K), you can place additional resistance in series with the appropriate K terminal. This increases feedback resistances, increasing gain. The recommended resistor values and connections are given in Table 8-1.

TABLE 8-1. 755 scale factor adjustments

Range of K	*Connect in Series to Pin*	*R*
1–2	1	$(K - 1) \cdot 15$ kΩ
> 2	2	$(K - 2) \cdot 15$ kΩ

(Courtesy of Analog Devices)

Connecting the antilog element in U1's feedback loop causes a log output. Connecting the antilog element in U1's input loop produces an antilog output. This is shown in Fig. 8-24. Compare it carefully with Figs. 8-23(a) and 8-21.

The output for the antilog amp in Fig. 8-24 is

$$v_{out} = e_{ref} \times 10^{-e_{in}/K} \tag{8-27}$$

Unlike the log amp connection, which requires only one polarity input, the antilog amp will accept both positive and negative input voltages. If $K = 1$, *adding* +1 V to the input *cuts* the output by a decade.

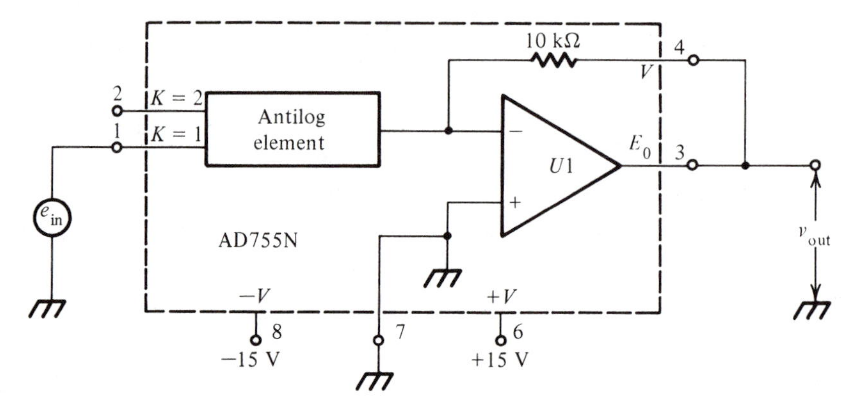

Figure 8-24 755N antilog amp connection (*Courtesy of Analog Devices*)

Example 8-4

Use the outputs from Example 8-3 as inputs to the antilog amp of Fig. 8-24. Calculate the outputs from the antilog amp.

Solution

$$v_{out} = 0.1 \text{ V } 10^{-e_{in}/1}$$

Log amp in	*Log amp out & Antilog amp in*	*Antilog amp out*
0.1 V	0 V	0.1 V
1 V	−1 V	1 V
10 mV	+1 V	10 mV
3.2 V	−1.505 V	3.2 V

8-3 MULTIPLIER INTEGRATED CIRCUIT

The multiplier IC is too often overlooked as a direct, simple solution to complex signal processing problems. Monolithic integration has lowered the cost of multiplier ICs considerably. Applications include signal multiplication for process instrumentation, guidance, chemical analyzers, and servo mechanism control. Frequency doubling and phase angle detection as well as real power computation can be done with multipliers. Multipliers can be configured to output a DC whose value equals the true rms of a signal. Voltage control of an amplifier's gain, an oscillator's frequency, and a filter's cutoff can all be obtained with multipliers. Multipliers can also be used to improve data acquisition through ratioing two signals (to divide out mutual error) and through transducer linearization.

8-3.1 Multiplier Characteristics

A basic multiplier schematic symbol is shown in Fig. 8-25. Two signal inputs (e_x and e_y) are provided. The output is the product of the two inputs divided by a reference voltage e_{ref}.

$$v_{out} = \frac{e_x e_y}{e_{ref}} \tag{8-28a}$$

Normally, e_{ref} is internally set to 10 volts. So equation (8-28a) becomes

$$v_{out} = \frac{e_x e_y}{10\text{ V}} \tag{8-28b}$$

As long as you ensure that both input voltages are below the reference voltage,

$$e_x < e_{ref}$$

and

$$e_y < e_{ref}$$

the output of the multiplier will not saturate.

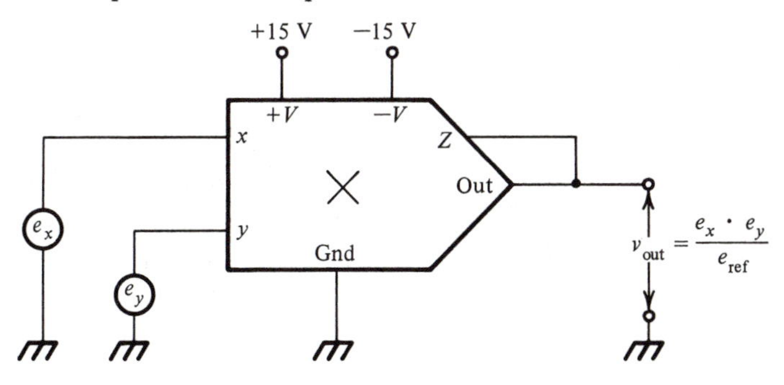

Figure 8-25 Basic multiplier schematic symbol

Depending on the technique that the multiplier IC uses to perform the multiplication, you may have to restrict the polarity of one or both inputs. Refer to Figs. 8-26(a), (b), or (c). If both inputs must be positive, the IC is said to be a one quadrant multiplier [Fig. 8-26(a)]. A two quadrant multiplier will function properly if one input is held positive and the other is allowed to swing both positive and negative [Fig. 8-26(b)]. If both inputs may be either positive or negative, the

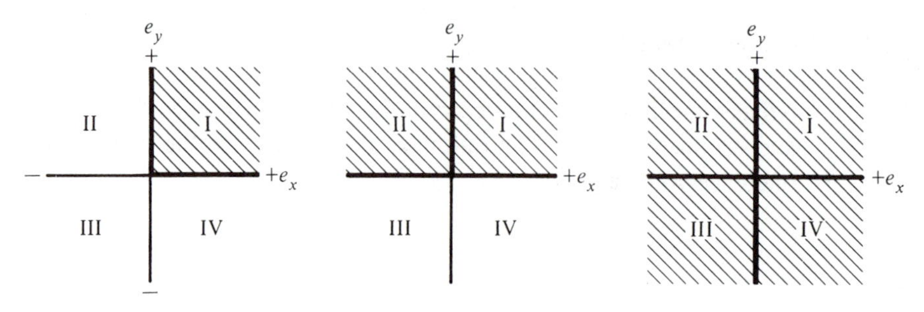

Figure 8-26 Multiplier quadrants of operation

IC is called a four quadrant multiplier. Be sure in selecting a multiplier that you specify one which will properly respond to the polarity inputs that you are using.

There are several ways to build a circuit which will multiply according to equation (8-28). The two dominant techniques are the log-antilog method and the transconductance or Gilbert method. The log-antilog method relies on the mathematical relationship that the sum of the logarithm of two numbers equals the logarithm of the product of those numbers.

$$\ln e_x + \ln e_y = \ln (e_x \cdot e_y) \tag{8-29}$$

Figure 8-27 is a block diagram of a log-antilog multiplier IC. Each input is fed into a log amp (which you studied in the last section). The logarithms of these inputs are then summed. But the sum of the logarithms is the logarithm of the product [equation (8-29)]. The antilog amp, then, "undoes" the logarithm and scales the voltage. In fact, you could build this circuit using three 755 log amps (two for the input log amps and one for the antilog) and an op amp (for the noninverting summer and scaling).

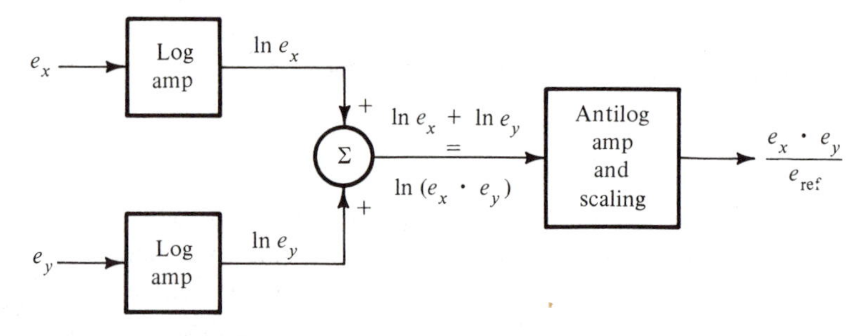

Figure 8-27 Log-antilog multiplier block diagram

The transconductance multiplication technique uses the e_x input as the input of an amplifier. Variations in e_x directly alter the collector currents. Through proper scaling and biasing a direct relationship (triple e_x, triple I_c) can be established. The e_y input is used to control the emitter's constant current generator (see Fig. 8-28). Variations in e_y will directly affect I_y and therefore also I_c (triple e_y, triple I_y, triple I_c). Since tripling e_x triples I_c, and tripling e_y triples I_c again, overall I_c goes up by a factor of nine. Multiplication has been accomplished.

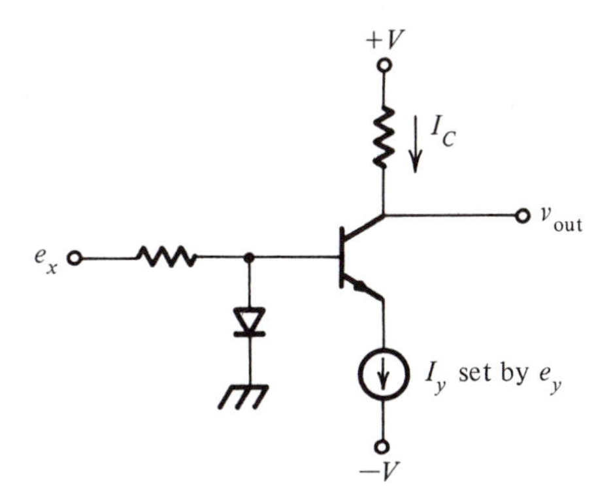

Figure 8-28 Simple transconductance multiplication

Of course, a practical four quadrant transconductance multiplier uses *differential* amplifiers (4) and specially matched base-emitter diodes to compensate for temperature effects, offsets, nonlinearity, and noise. Also, an output differential op amp circuit is provided to give a ground referenced output voltage.

The transconductance multiplier is simpler to integrate into an IC than is the log-antilog multiplier, so you can expect the transconductance multiplier to be cheaper and more readily available.

Log amps require the input and reference voltages to be of the same polarity. This restricts log-antilog multipliers to one quadrant operation. Transconductance multipliers are available in four quadrant models.

One final transconductance multiplier advantage is its speed. Because it is inherently a current mode device, 10 MHz and higher gain bandwidth are available for output signal currents. The log-antilog multipliers are two or three orders of magnitude slower.

Although slower, more expensive, and restricted to one quadrant operation, the log-antilog multipliers do have their strong points. Log-antilog multipliers often have one-tenth of the total error associated with a transconductance multiplier. Also drift (output change with temperature) in log-antilog multipliers is considerably lower than in transconductance multipliers.

8-3.2 Trimming

There are four nonideal DC effects which you can adjust out in most multiplier ICs. The *output offset* is the output voltage present when both inputs are at zero volts. The *X offset* (also referred to as *Y feedthrough*) allows a signal on e_y to appear at the output even where the e_x input is at zero volts. The *Y offset* (also referred to as *X feedthrough*) allows a signal on e_x to appear at the output even when the e_y input is at zero volts. The fourth trimmable effect is the scale factor. It compensates for apparent variations in e_{ref}.

Precise trimming procedures and circuits vary from IC to IC and even from one circuit configuration to another. Be sure to check the manufacturer's specifications. However, as an example, the AD 533 transconductance multiplier trim procedures are given in Fig. 8-29. The Z_0 pin eliminates output offset. The X_0 pin

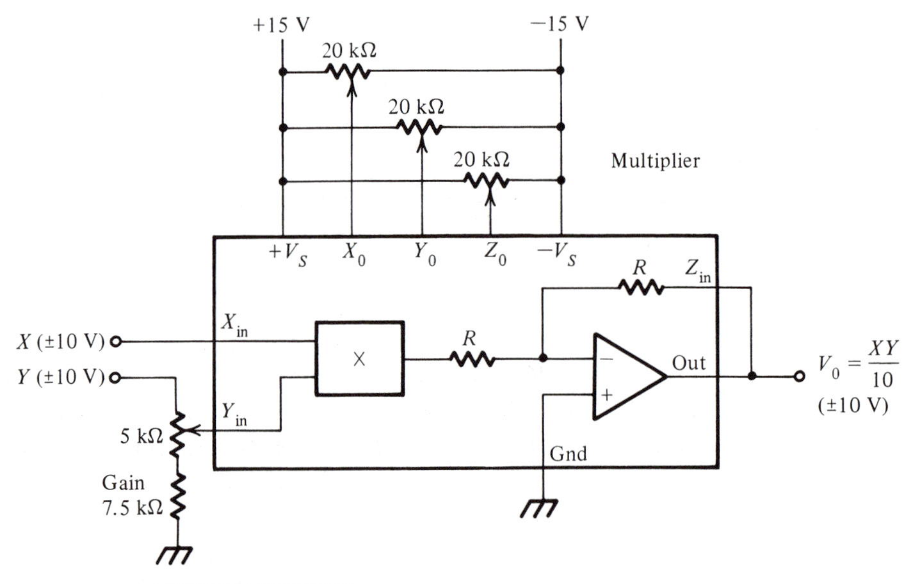

TRIM PROCEDURES

1. With $X = Y = 0$ volts, adjust Z_0 for 0 V dc output.
2. With $Y = 20$ volts p-p (at $f = 50$ Hz) and $X = 0$ V, adjust X_0 for minimum ac output.
3. With $X = 20$ volts p-p (at $f = 50$ Hz) and $Y = 0$ V, adjust Y_0 for minimum ac output.
4. Readjust Z_0 for 0 V dc output.
5. With $X = +10$ V dc and $Y = 20$ volts p-p (at $f = 50$ Hz), adjust gain for output $= Y_{in}$.

NOTE: For best accuracy over limited voltage ranges (e.g., ±5 V), gain and feedthrough adjustments should be optimized with the inputs in the desired range, as linearity is considerably better over smaller ranges of input.

Figure 8-29 Example of multiplier trimming (*Courtesy of Analog Devices*)

trims out X input offset to minimize Y input feedthrough. Similarly, the Y_0 pin trims out Y input offset to minimize X input feedthrough. Full scale gain is trimmed with the 5 kΩ gain potentiometer.

The dynamic, or AC, characteristics of a multiplier are described by using the same terms as an op amp. Gain bandwidth, full power response, slew rate, $f_{1\%}$, and $f_{10\%}$ are all applicable to multipliers as they were to op amps, as described in section 4-2.

However, multipliers do have a unique frequency response problem. When you apply two sine waves of the same frequency to the inputs, the output is a sine wave at twice the input frequency riding a DC level. (This application will be discussed in detail later.) The DC component of the output is completely unaffected by gain bandwidth, or slew rate limiting. However, the AC output is at *twice* the frequency of the input. Overlooking this may lead you into significant frequency limitation problems.

The errors discussed so far are proportional to the first power of the X input or the Y input. With all these fully trimmed out, and with $e_y = 0$, the output may either increase or decrease as the *square* of the signal applied at e_x (ideally, the output would remain at zero). This error is called *nonlinearity*. It can be minimized as shown in Fig. 8-30. If the output goes positive with changes in e_x, use only R_A. For a negative output with changes in e_x, use only R_B. Do not expect to be able to fully cancel nonlinearity for all values of e_x and e_y. Even when fully cancelled at one value of e_y nonlinearity may well appear at other values of e_y.

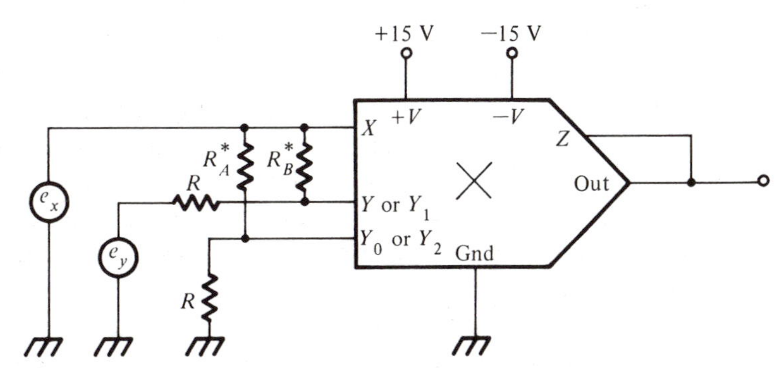

*Use R_A *only* to correct for positive output error. Use R_B *only* to correct for negative output error.

Figure 8-30 Cross connection to minimize second order nonlinearity (*Courtesy of Analog Devices*)

If you need the accuracy of a log-antilog multiplier, but also need four quadrant operation (both inputs going positive and negative), input and output level shifting is a solution. You first add a DC offset to both the X input signal and the Y input signal. These offsets should be enough to ensure that the signals never change polarity. The two inputs have now become

$$X \text{ input} = e_x + e_{ox}$$

$$Y \text{ input} = e_y + e_{oy}$$

$$v_{\text{out}} = \frac{X_{\text{input}} \cdot Y_{\text{input}}}{e_{\text{ref}}}$$

$$v_{\text{out}} = \frac{(e_x + e_{ox})(e_y + e_{oy})}{e_{\text{ref}}}$$

$$v_{\text{out}} = \frac{e_x e_y + e_x e_{oy} + e_{ox} e_y + e_{ox} e_{oy}}{e_{\text{ref}}} \tag{8-30}$$

$$v_{\text{out}} = \frac{e_x e_y}{e_{\text{ref}}} + \frac{e_x e_{oy} + e_{ox} e_y + e_{ox} e_{oy}}{e_{\text{ref}}} \tag{8-31}$$

The output for the multiplier contains two major terms. The first is the desired product $e_x e_y/e_{\text{ref}}$. The second is the cross products of the offset terms. These must be subtracted out.

The schematic is given in Fig. 8-31. Op amp U1 adds an offset e_{ox} to the e_x input. Op amp U2 adds the offset e_{oy} to the e_y input. These offset inputs (which now never change polarity) are applied to the one quadrant multiplier U6. It outputs the full product as given in equations (8-30) and (8-31). One quadrant multiplier U3 produces $e_{ox} e_{oy}/e_{\text{ref}}$. The other two cross products are produced by multipliers U4 and U5. However, since e_x and e_y change polarity, U4 and U5 must be two quadrant multipliers. These three cross products are summed and inverted by U7. This, in turn, is added to the full product of U6. Since the cross products have been inverted by U7, noninverting summation by U8 will subtract out the cross product term. Careful component matching and the use of *precision* op amps will yield a four quadrant multiplier whose errors and drift are 10 to 100 times lower than a single four quadrant transconductance multiplier. Do not forget, however, that the circuit of Fig. 8-31 is still considerably slower than the transconductance multiplier.

In all the multiplier schematics you have seen, there is a Z connection tied to the output. This is the negative feedback loop for the output op amp. Look carefully at Fig. 8-29. This Z to output connection must be made to provide linear operation of the output amplifier. This negative feedback pin allows you to place current or power amplifiers inside the negative feedback loop of the multiplier. This technique, used in several other places in this text, then provides a boost in output power. By sampling and feeding back the signal from the power amp's output, the multiplier's op amp compensates for offsets and nonlinearities in the power amp.

The basic connection and three possible power amps are shown in Fig. 8-32.

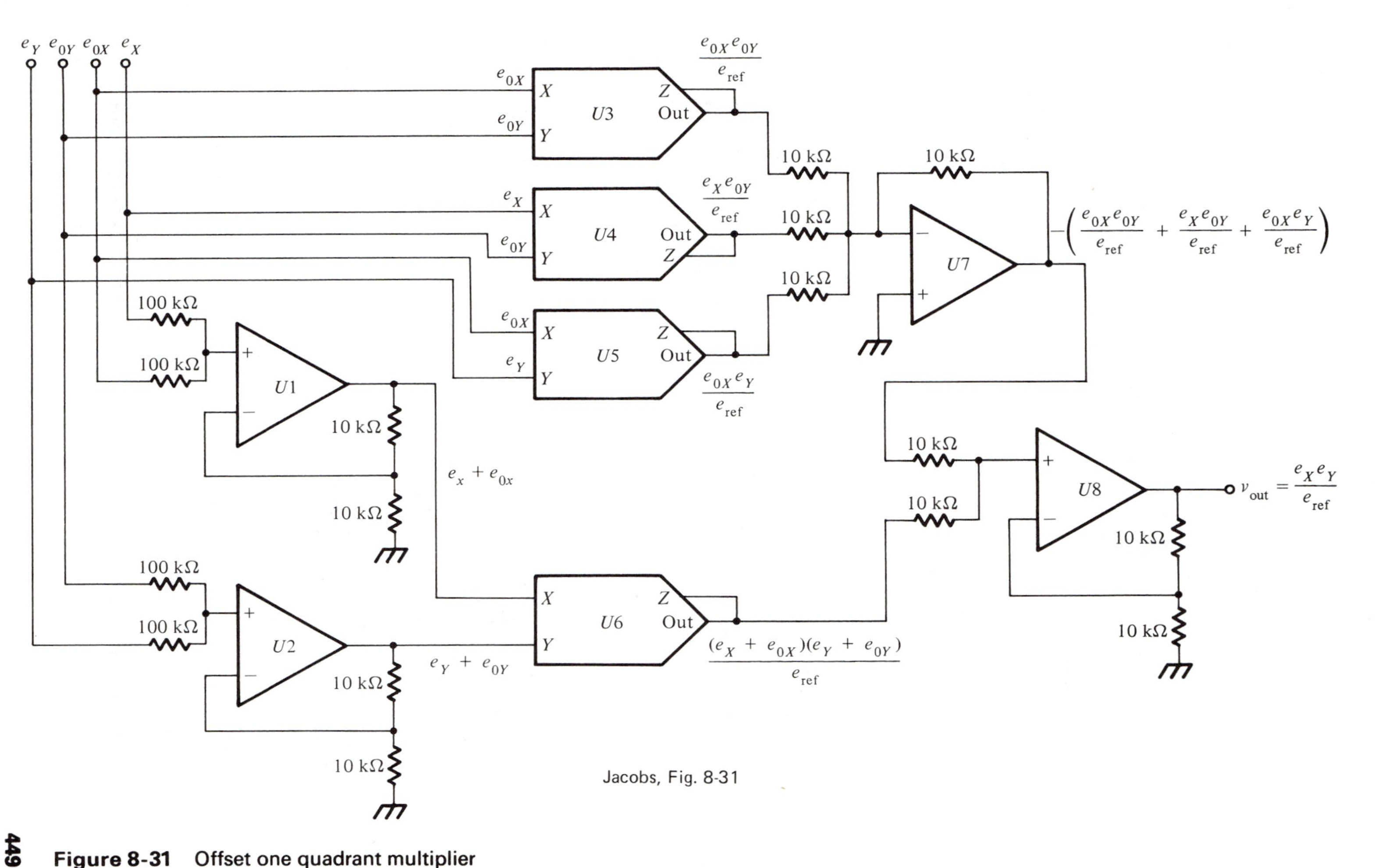

Figure 8-31 Offset one quadrant multiplier

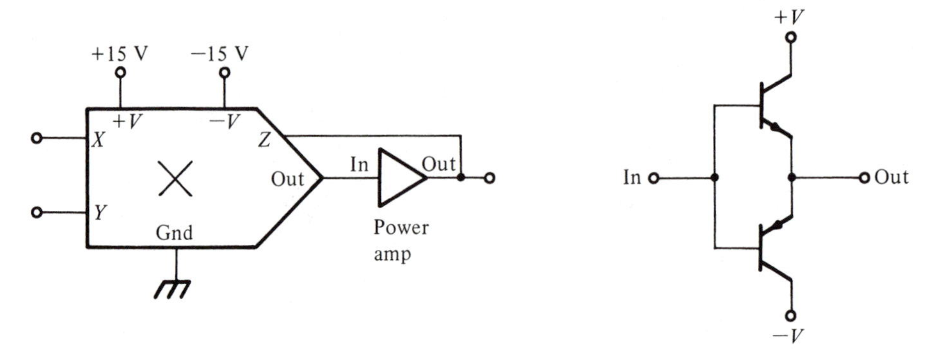

(a) Basic connection

(b) Simplest push-pull power booster

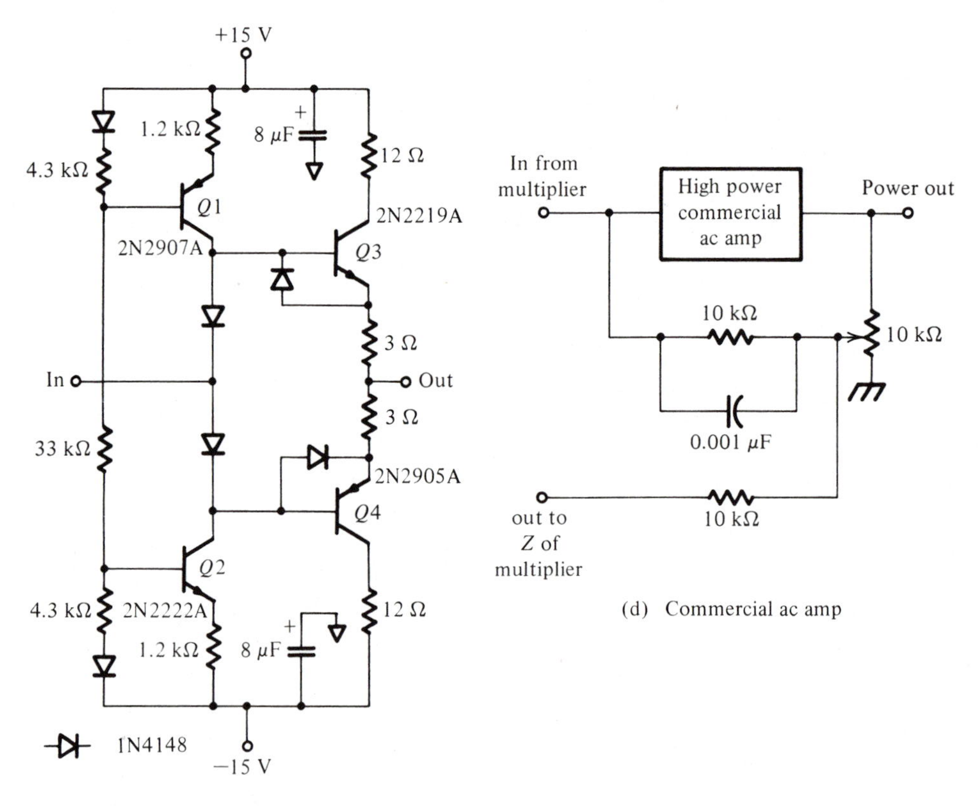

(c) Full push-pull complementary symmetry power amp

(d) Commercial ac amp

Figure 8-32 Power boosting of multiplier (*Courtesy of Analog Devices*)

A single NPN boost transistor (not shown) may be used for one quadrant multipliers, since the output will always be positive. The simple two transistor boost circuit of Fig. 8-32(b) requires that the multiplier provide both signal and bias voltages. As the output crosses zero, the multiplier must jump from +0.6 V to −0.6 V. A high slew rate specification is necessary to minimize zero crossing distortion. The full power amp of Fig. 8-32(c) eliminates this problem by providing its own bias. It can provide ±10 V into 50 ohms. For higher power still, enclose a commercial power amp in the multiplier's feedback loop. As shown in Fig. 8-32(d), if the amp is AC coupled, you must provide DC feedback (two 10 kΩ resistor and 0.001 μF capacitor) around the amp. A feedback reduction potentiometer may be necessary to ensure that the feedback signal does not exceed the multiplier's supply voltage.

One final word of caution: Treat the power supplies and distribution network with great care and respect. Even the best multipliers (or any analog IC) with the full range of trimming and compensation circuits will not work correctly if furnished with poorly regulated power from a high impedance supply. The output impedance of the supply must be low, not only at DC, but more critically at the highest frequency the multiplier is to output. Also, beware of switching transients, especially if you are using a switching power supply. Grounding should be single point, not bussed, especially for the ground return from the output power amp. Decoupling power supply lines with ceramic disc capacitors and keeping associated components physically close to the multiplier IC are both necessary to decrease the chances of noise or oscillations.

8-3.3 Applications of Multipliers

The multiplication of two sine waves of the same frequency, but of possibly different amplitudes and phases allows you to double a frequency and to directly measure *real* power. Let

$$e_x = E_x \sin \omega t$$

$$e_y = E_y \sin (\omega t + \theta)$$

where θ is the phase difference between the two signals. Applying these two signals to the inputs of a four quadrant multiplier will yield an output of

$$v_{\text{out}} = \frac{E_x \sin \omega t \cdot E_y \sin (\omega t + \theta)}{E_{\text{ref}}}$$

But

$$\sin(a+b) = \sin a \cos b + \sin b \cos a$$

$$v_{\text{out}} = \frac{E_x \sin \omega t \cdot E_y(\sin \omega t \cos \theta + \sin \theta \cos \omega t)}{E_{\text{ref}}}$$

Regrouping, you obtain

$$v_{\text{out}} = \frac{E_x E_y}{E_{\text{ref}}} \sin \omega t(\sin \omega t \cos \theta + \sin \theta \cos \omega t)$$

$$v_{\text{out}} = \frac{E_x E_y}{E_{\text{ref}}} (\sin^2 \omega t \cos \theta + \sin \theta \sin \omega t \cos \omega t)$$

But

$$\sin^2 a = 1 - \cos^2 a$$

and

$$\cos 2a = 2\cos^2 a - 1$$

so

$$\cos^2 a = \tfrac{1}{2} + \tfrac{1}{2}\cos 2a$$

$$\sin^2 a = 1 - \tfrac{1}{2} - \tfrac{1}{2}\cos 2a$$

$$\sin^2 a = \tfrac{1}{2} - \tfrac{1}{2}\cos 2a$$

$$v_{\text{out}} = \frac{E_x E_y}{E_{\text{ref}}} [\cos \theta(\tfrac{1}{2} - \tfrac{1}{2}\cos 2\omega t) + \sin \theta \sin \omega t \cos \omega t]$$

$$v_{\text{out}} = \frac{E_x E_y}{E_{\text{ref}}} [\tfrac{1}{2}\cos \theta(1 - \cos 2\omega t) + \sin \theta \sin \omega t \cos \omega t]$$

But

$$\sin 2a = 2 \sin a \cos a$$

or

$$\sin a \cos a = \tfrac{1}{2}\sin 2a$$

$$v_{out} = \frac{E_x E_y}{E_{ref}} [\tfrac{1}{2} \cos \theta(1 - \cos 2\omega t) + \sin \theta \cdot \tfrac{1}{2} \sin 2\omega t]$$

$$v_{out} = \frac{E_x E_y}{2E_{ref}} (\cos \theta - \cos \theta \cos 2\omega t + \sin \theta \sin 2\omega t)$$

Regrouping, you obtain

$$v_{out} = \frac{E_x E_y}{2E_{ref}} \cos \theta + \frac{E_x E_y}{2E_{ref}} (\sin \theta \sin 2\omega t - \cos \theta \cos 2\omega t) \qquad (8\text{-}32)$$

Equation (8-32) shows that the output from a multiplier with two sine waves of equal frequency applied will have two terms. The first is a DC and is set by the magnitudes of the signals and their phase difference. The second term varies with time, but at *twice* the frequency of the inputs (2ω).

Example 8-5

Design a circuit which will output a sine wave at twice the frequency of the input sine wave.

Solution

You can build a frequency doubler by connecting the input signal to both X and Y inputs of a multiplier. For equation (8-32), this means that

$$E_x = E_y = E_{pk}$$

$$\theta = 0 \quad \text{(no phase difference)}$$

$$v_{out} = \frac{E_{pk}^2}{2E_{ref}} \cos 0 + \frac{E_{pk}^2}{2E_{ref}} (\sin 0 \times \sin 2\omega t - \cos 0 \times \cos 2\omega t)$$

$$\cos 0 = 1$$

$$\sin 0 = 0$$

$$v_{out} = \frac{E_{pk}^2}{2E_{ref}} - \frac{E_{pk}^2}{2E_{ref}} \cos 2\omega t$$

The first term is DC and can be blocked by an RC coupler. The second term is a sinusoid shifted in phase from the input ($-\cos$ rather than sin) and at twice the frequency of the input (output frequency of 2ω, input frequency of ω).

This is just a restatement of the fact that multiplication of two sine waves gives two output components. One is at the difference frequency ($\omega t - \omega t = 0$ = DC) and the other is at the sum of the frequencies ($\omega t + \omega t = 2\omega t$).

The schematic for the frequency doubler circuit is given in Fig. 8-33.

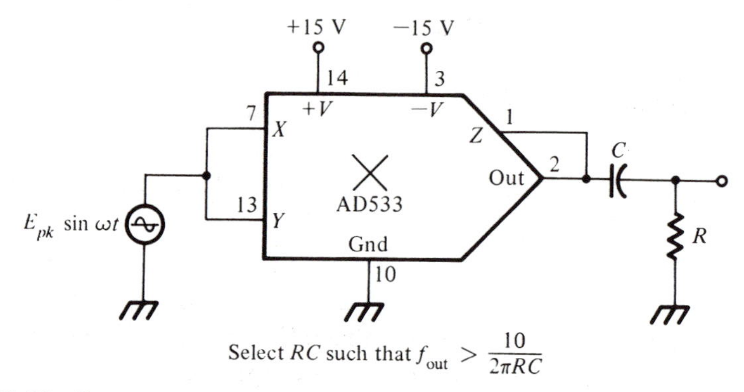

Figure 8-33 Frequency doubler (*Courtesy of Analog Devices*)

Look carefully at equation (8-32). The first term is

$$\frac{E_x E_y}{2E_{ref}} \cos \theta$$

The second term is a high frequency and can be filtered out. The circuit in Fig. 8-34 takes advantage of that relationship to compute the true power that a sine wave delivers to a load. The AD 534 was chosen because it provides differential inputs. This will eliminate the error caused by the 0.1 Ω current sense resistor. No external ground or offset compensation is required for the AD 534. The voltage is measured directly across the load. Current is sensed by the 0.1 Ω resistor and converted to a voltage. This is given a gain of 10 by op amp U1.

The inputs to the multiplier are

$$e_x = E_{peak} \sin \omega t$$

$$e_y = 10 \times [0.1\ \Omega \times I_{pk} \sin (\omega t + \theta)]$$

or

$$e_y = I_{pk} \times 1\ \Omega \sin (\omega t + \theta)$$

The output of the multiplier has a DC level of

$$V_{out\,(DC)} = \frac{E_{pk} \times I_{pk}\,(1\ \Omega)}{2E_{ref}} \cos \theta$$

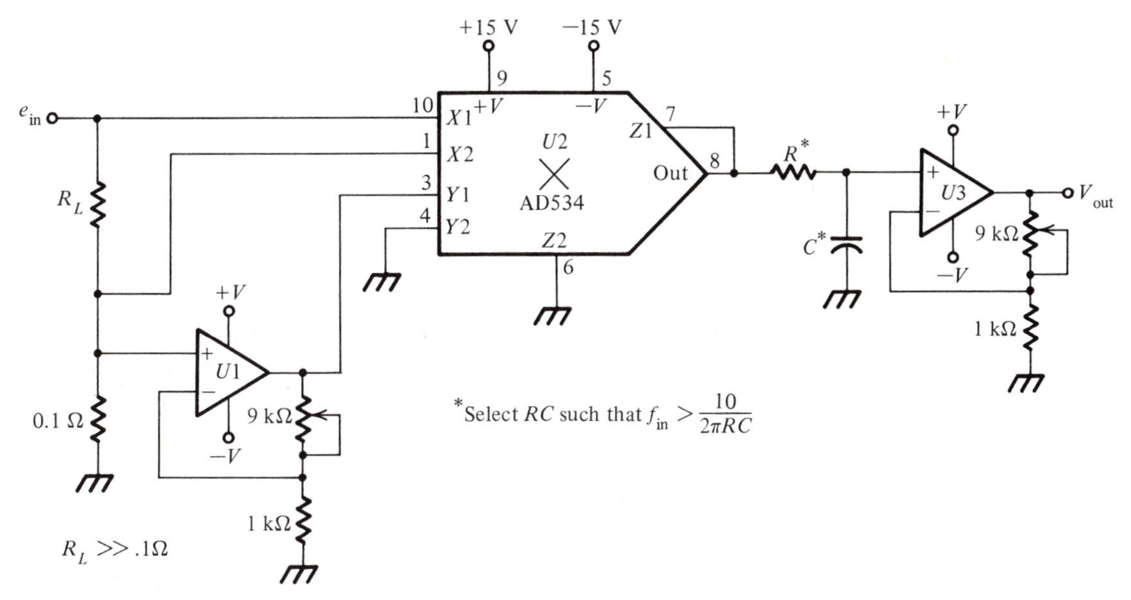

Figure 8-34 True power converter (*Courtesy of Analog Devices*)

The resistor and capacitor at the output of U2 form a low pass filter which shorts the high frequency (2ω) term to ground, passing only the DC component to the output amp. This amp provides buffering for the filter and gives an additional gain of 10 to compensate for the $1/E_{ref}$ scaling down of the multiplier.

$$V_{out} = \frac{E_{pk} \times I_{pk}}{2} \cos \theta$$

So the output voltage is numerically equal to the true power dissipated by e_{in} into R_L.

The AD 534 multiplier can also be configured on a voltage-controlled amplifier. That is, the gain of the signal applied to the Y input is controlled by the voltage applied to the X input. This in turn allows remote, automatic, or computer control of an analog system's performance. Because of its differential inputs and differential feedback, the AD 534 performance is defined by

$$\frac{(e_{x1} - e_{x2})(e_{y1} - e_{y2})}{E_{ref}} = (e_{z1} - e_{z2})$$

where X and Y are the inputs and Z is the feedback. Grounding e_{x2}, e_{y2}, and e_{z2}, and connecting e_x to e_x, e_y to e_y, and output feedback to e_z gives

$$\frac{e_x e_y}{E_{ref}} = e_z$$

See Fig. 8-35(a). The value of feedback voltage is set by the output and the voltage divider R2 and R1.

$$\frac{e_x e_y}{E_{ref}} = \frac{R1}{R1 + R2} e_{out}$$

One additional feature of the AD 534 is its external E_{ref} (or scale factor) adjustment. By altering the resistance between the SF pin and the -15 V supply, E_{ref} can be set between 10 volts and 3 volts.

$$R_{SF} = 5.4\ k\Omega \times \frac{SF}{10 - SF}$$

The circuit in Fig. 8-35(b) is the voltage-controlled amplifier. Resistors R1 and R2 are set to provide

$$\frac{e_x e_y}{E_{ref}} = \frac{1\ k\Omega}{40\ k\Omega} e_{out}$$

or

$$e_{out} = \frac{40 e_{control} \times e_{in}}{E_{ref}}$$

The scale factor is then adjusted to give $E_{ref} = 4$ V or

$$e_{out} = 10 e_{control} \times e_{in}$$

Multiplier ICs can also be used to perform amplitude modulation (AM). Qualitatively this should make sense if you think of the multiplier as a variable gain amplifier [Fig. 8-35(b)]. The carrier is applied to the $Y1$ input as e_{in}. The audio or data to modulate the carrier is applied to $X1$ as $e_{control}$. Constant carrier ($Y1$) and audio ($X1$) inputs cause a constant output. As the audio input varies ($X1$ or $e_{control}$), the gain changes proportionally. This, in turn, varies the amplitude of the output. This is illustrated in Fig. 8-36.

For sine wave inputs, the effect of multiplication can be described mathematically.

$$\sin\theta \sin\phi = \tfrac{1}{2}[\cos(\phi - \theta) - \cos(\phi + \theta)]$$

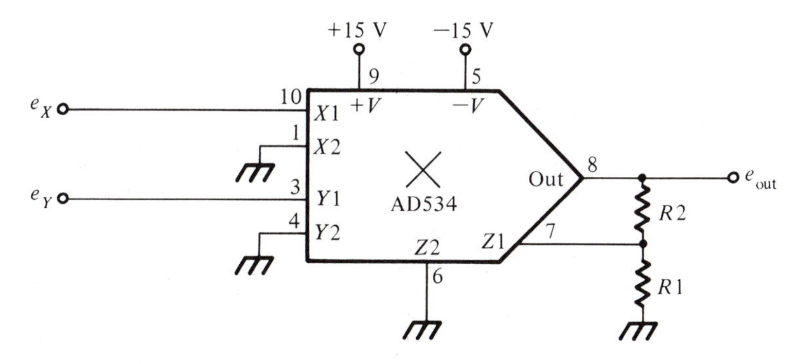

(a) AD534 with feedback voltage division

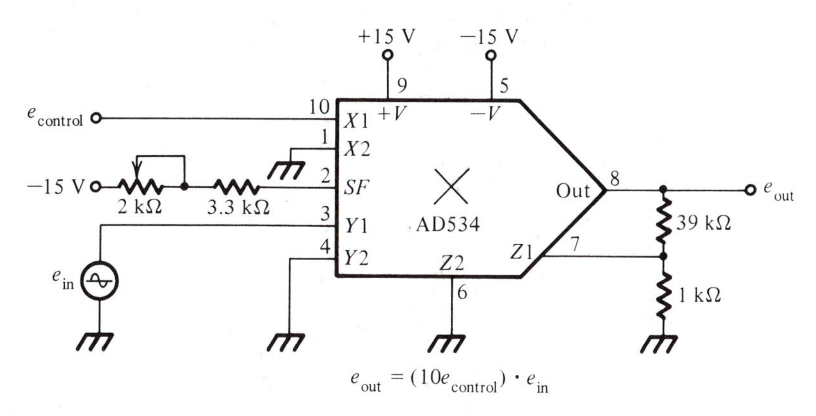

(b) Voltage controlled amplifiers

Figure 8-35 AD534 applications (*Courtesy of Analog Devices*)

So for

$$e_{\mathbf{control}} = X_{\mathbf{input}} = E_x \sin 2\pi f_x t$$

and

$$e_{\mathbf{in}} = Y_{\mathbf{input}} = E_y \sin 2\pi f_y t$$

$$\theta = 2\pi f_x t$$

$$\phi = 2\pi f_y t$$

$$v_{\mathbf{out}} = \frac{E_x E_y}{2E_{\mathbf{ref}}} [\cos 2\pi(f_y - f_x) - \cos 2\pi(f_y + f_x)] \tag{8-33a}$$

$$v_{\mathbf{out}} = \frac{E_x E_y}{2E_{\mathbf{ref}}} \cos 2\pi(f_y - f_x) - \frac{E_x E_y}{2E_{\mathbf{ref}}} \cos 2\pi(f_y + f_x) \tag{8-33b}$$

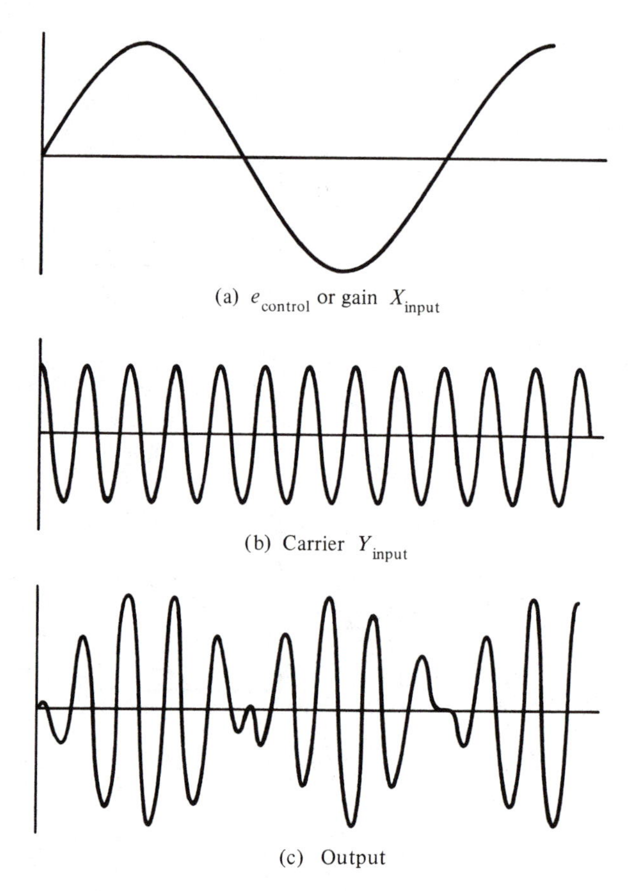

Figure 8-36 Multiplier IC amplitude modulation

The output is two sinusoids, one at a frequency of the *difference* between the two input frequencies and the other at a frequency of the *sum* of the two input frequencies. This is called *balanced* modulation.

Look closely at the output, as shown in Fig. 8-36(c). Although the X_{input} information is indeed encoded on the output, there appear to be two output cycles of the envelope for each input cycle. Also, the second cycle is out of phase with the carrier, while the first cycle is in phase. These observations are characteristics of a balanced amplitude modulator. Demodulation, to recover the control (audio) signal, is difficult.

Standard amplitude modulation adds the carrier to the sum and difference frequency signals.

$$V = E \sin 2\pi f_y t + \frac{E_x E_y}{2E_{\text{ref}}} \cos 2\pi(f_y - f_x) - \frac{E_x E_y}{2E_{\text{ref}}} \cos 2\pi(f_y + f_x) \qquad (8\text{-}34)$$

This produces an output as shown in Fig. 8-37. Adding the carrier to the signal allows you to demodulate easily. Rectification eliminates the negative half. Then filtering removes the carrier frequency variations, leaving only the envelope audio.

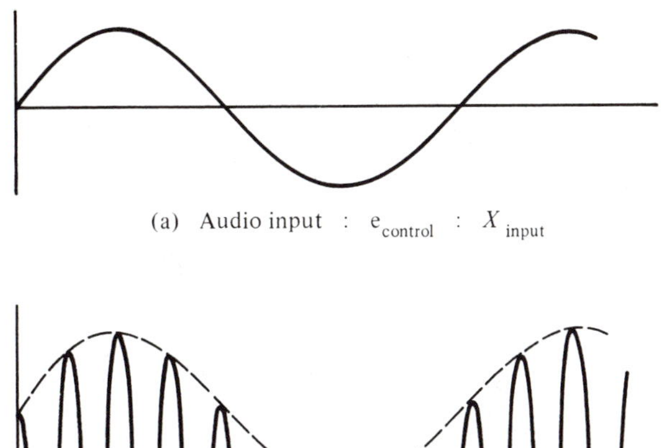

(a) Audio input : e_{control} : X_{input}

Output

(b) Output with carrier added

Figure 8-37 AM with carrier added

The carrier [$Y1$ input for Fig. 8-35(b)] could be added to the multiplier output with a summer. The carrier would be connected to one input of the summer and the balance modulated signal (output of the multiplier) to the other. However, building carrier frequency circuits is more expensive and difficult than building circuits for DC and audio frequency. An alternate approach for adding the carrier to the balanced modulated signal is illustrated in Fig. 8-38. The audio (modulating) signal has been offset from zero. This can be easily and cheaply done with one of the summers discussed in Chapter 2. Because this gain control signal never goes negative, the output amplitude follows the amplitude of e_{control} and the phase never reverses.

Mathematically, this can be proven to produce equation (8-34).

$$e_x = E_{\text{DC}} + E_x \sin 2\pi f_x t$$

$$e_y = E_y \sin 2\pi f_y t$$

$$v_{\text{out}} = \frac{e_x e_y}{E_{\text{ref}}}$$

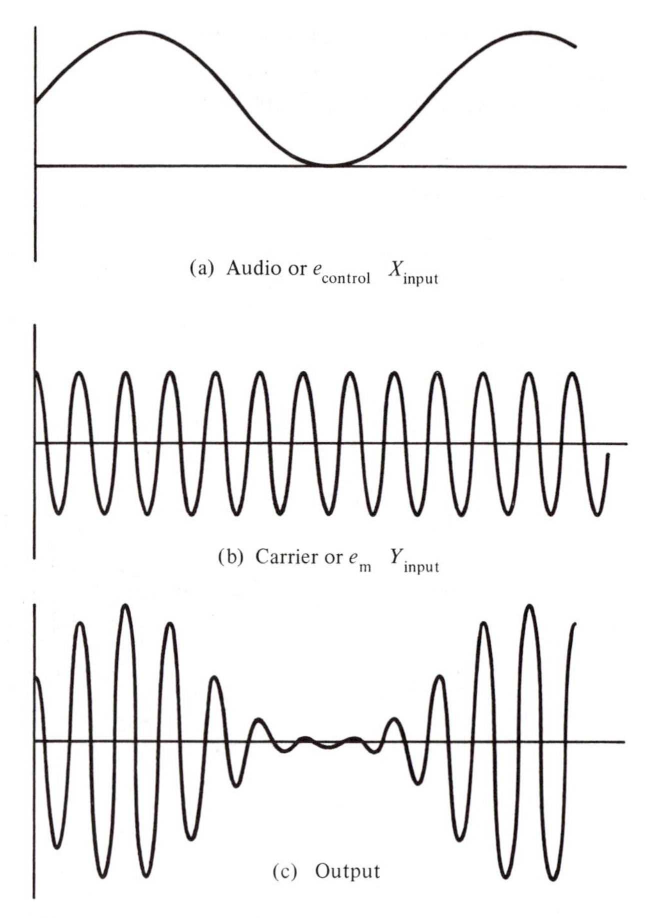

Figure 8-38 Multiplier with offset audio input

$$v_{\text{out}} = \frac{1}{E_{\text{ref}}} [(E_{\text{DC}} + E_x \sin 2\pi f_x t)(E_y \sin 2\pi f_y t)]$$

$$v_{\text{out}} = \frac{1}{E_{\text{ref}}} (E_{\text{DC}} E_y \sin 2\pi f_y t + E_x \sin 2\pi f_x t \cdot E_y \sin 2\pi f_y t)$$

$$v_{\text{out}} = \frac{E_{\text{DC}} E_y}{E_{\text{ref}}} \sin 2\pi f_y t + \frac{E_x E_y}{E_{\text{ref}}} (\sin 2\pi f_x t)(\sin 2\pi f_y t)$$

The second term of the right side is just the product of the sine waves. This was solved in equations (8-33a) and (8-33b).

$$v_{\text{out}} = \frac{E_{\text{DC}} E_y}{E_{\text{ref}}} \sin 2\pi f_y t + \frac{E_x E_y}{2E_{\text{ref}}} \cos 2\pi(f_y - f_x) - \frac{E_x E_y}{2E_{\text{ref}}} \cos (f_y + f_x)$$

Compare this to equation (8-34).

8-3.4 Dividers

To build an antilog amp, you saw in Fig. 8-24 that an antilog circuit element was placed in the input loop of an op amp. The complementary function, the log amp, was built by placing the antilog circuit element in the *feedback* loop of the op amp (Fig. 8-23).

The same is true for multipliers and dividers. Multiplication is achieved by placing the multiplier circuit element in the input loop of an op amp. This is shown in Fig. 8-29. So division, the complement of multiplication, can be accomplished by placing the multiplier circuit element in the op amp's *feedback* loop. Refer to Fig. 8-39.

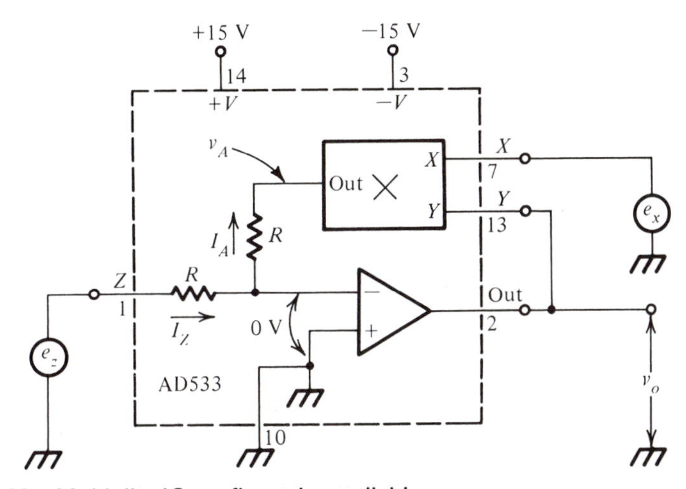

Figure 8-39 Multiplier IC configured as a divider

The output voltage from the divider in Fig. 8-39 is

$$v_O = -E_{\text{ref}} \frac{e_z}{e_x} \tag{8-35a}$$

With a 10 V reference

$$v_O = -10 \text{ V} \frac{e_z}{e_x} \tag{8-35b}$$

You can derive equation (8-35) as follows. First, since the Y input is the output voltage,

$$e_y = v_O$$

Also, note that the op amp's inverting terminal is at virtual ground. Assuming no significant current flows into the op amp itself, we have

$$I_Z = I_A \tag{8-36}$$

Because of the virtual ground at the inverting terminal of the op amp, all of e_Z must be dropped across R.

$$e_z = I_Z \mathrm{R}$$

or

$$I_Z = \frac{e_z}{\mathrm{R}} \tag{8-37}$$

The voltage v_A is determined by the multiplication of e_x and e_y.

$$v_A = -\frac{e_x e_y}{E_{\mathrm{ref}}}$$

or

$$v_A = -\frac{e_x v_O}{E_{\mathrm{ref}}}$$

Because of the virtual ground at the inverting input of the op amp, v_A must be entirely dropped across R.

$$v_A = I_A \mathrm{R}$$

or

$$I_A = \frac{v_A}{\mathrm{R}}$$

$$I_A = \frac{\dfrac{-e_x v_O}{E_{\mathrm{ref}}}}{\mathrm{R}}$$

$$I_A = -\frac{e_x v_O}{E_{\mathrm{ref}} \mathrm{R}}$$

But by equation (8-36), $I_Z = I_A$.

$$I_Z = -\frac{e_x v_O}{E_{\text{ref}} \text{R}}$$

$$e_Z = -\frac{e_x v_O}{E_{\text{ref}}}$$

or

$$v_O = -E_{\text{ref}} \frac{e_z}{e_x} \tag{8-35}$$

Division by zero is, of course, prohibited. Electronically, setting e_x to zero causes v_A to go to zero. This removes the negative feedback, turning the op amp into an inverting comparator. The output, v_O, goes to saturation. Since e_x cannot go to zero, it obviously cannot be allowed to pass through zero. That means that e_x must be restricted to one polarity. Therefore, dividers are at best two quadrant devices.

How small the denominator can become is a critical specification for dividers. Multipliers with 1 percent error usually are restricted to denominator signals of 1 volt or greater. Multipliers with 0.1 percent error will accept denominator signals of 0.1 volt or greater. Even so, offsets and drift are typically two to ten times worse for a divider with a small denominator signal than for a multiplier made with the same IC. Bandwidth also falls off as the denominator signal becomes small.*

The major advantage of using the inverse multiplication technique for division is low cost. If offset, drift errors, noise, bandwidth specifications, or minimum denominator signal size are unacceptable, you can purchase dedicated dividers. These ICs may be built by using either the variable transconductance technique or by using a log-antilog implementation. As with multipliers, log-antilog dividers are restricted to one quadrant operation.

Just as you could square a signal by using a multiplier IC, you can use a divider circuit to take the square root of a signal. This is illustrated in Fig. 8-40. The key that this circuit is really being used as a divider rather than a multiplier is that the input is fed to Z, and the output is connected back to Y. The diode is included to ensure that only positive outputs are produced. (The square root of a number is always positive.) The resistor is necessary to keep the diode biased for small signals. Because of the inverting nature of the IC and the required positive output, you must restrict e_{in} to negative values.

---

* Daniel H. Sheingold, editor, *Nonlinear Circuits Handbook* (Norwood, Mass.: Analog Devices, Inc., 1976), pp. 281–282.

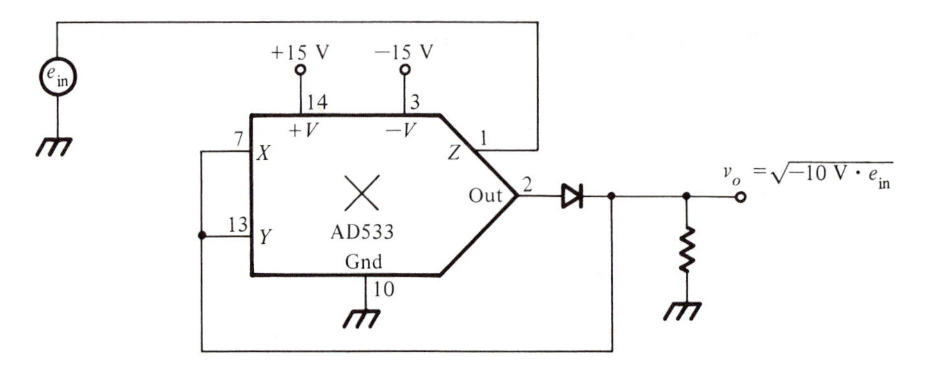

Figure 8-40 Divider used to produce the square root of e_{in} (*Courtesy of Analog Devices*)

To see that Fig. 8-40 does, indeed, provide the square root, recall that

$$v_O = -E_{ref} \frac{e_z}{e_x}$$

But

$$e_x = v_O$$

and

$$e_z = e_{in}$$

$$v_O = -E_{ref} \frac{e_{in}}{v_O}$$

$$v_O^2 = -E_{ref} e_{in}$$

$$v_O = \sqrt{-E_{ref} e_{in}}$$

The e_x input (denominator) can be considered as a voltage-controlled gain.

$$v_O = A_v e_z$$

$$v_O = -\frac{E_{ref}}{e_x} e_z$$

$$A_v = -\frac{E_{ref}}{e_x}$$

Increasing e_x reduces the gain and, therefore, the output. Decreasing e_x increases the gain and the output. This inverse relationship between e_x and gain can be used to build automatic gain controls.

One of the simplest AGC circuits is used in the Wien-bridge oscillator of Fig. 8-41. The AD 534 is used because of its differential inputs. Feeding back the output to $Y2$ (an inverting input) removes the − from the gain equation. Feeding the $Z2$ rather than the $Z1$ input from the RC (Wien) bridge ensures the proper phase relationship to sustain oscillations.

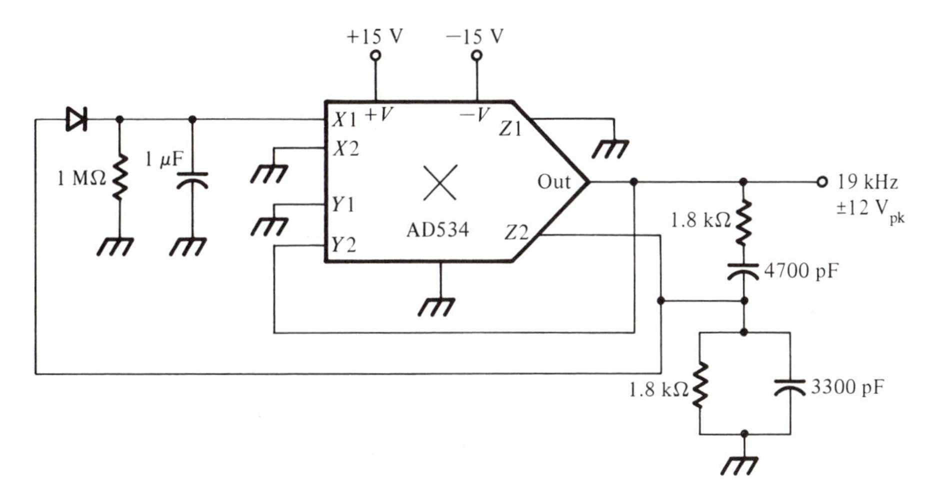

Figure 8-41 Wien-bridge oscillator using a divider (*Courtesy of Analog Devices*)

This input voltage (at $Z2$) from the bridge is rectified and filtered and then applied to $X1$. When the output is small, e_x is small, producing a large gain and, therefore, a large output signal. As the output increases, so does e_x. Increasing e_x lowers the gain, dropping the output amplitude. So the output amplitude is sampled, rectified, and filtered and then used to stabilize the circuit's gain. The larger the output, the larger e_x, the lower the gain becomes. The smaller the output, the smaller e_x, the larger the gain becomes.

This same principle is used in the circuit of Fig. 8-42. Again, you can identify this as a divider because the input is connected to Z and the output is fed back to Y. During the positive half-cycle of the output, the op amp's (integrator) capacitor charges. During the output's negative half-cycle, the op amp's capacitor discharges. The voltage stored on the capacitor is the average value of the output. (Remember, integrating a signal over one cycle gives you its average value.) This is filtered and sent to the X input to control gain. The 100 kΩ resistor controls the integrator's charge rate, which controls the "average voltage" stored on the capacitor and therefore the output from the divider. The 33 kΩ resistor and filter capacitor can be adjusted to remove the low frequency ripple (between cycles), but to allow adequately fast gain adjustment as the amplitude varies over many cycles.

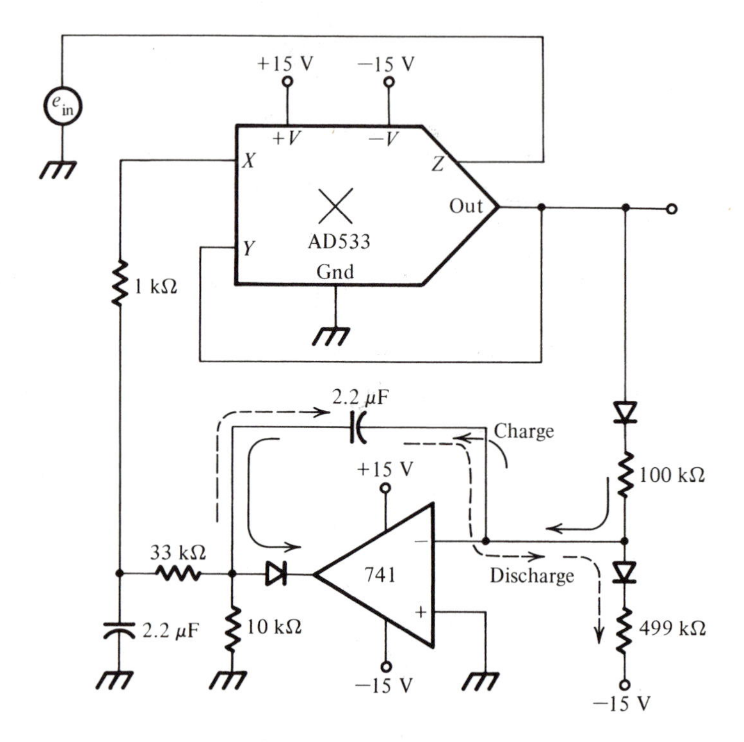

Figure 8-42 Automatic gain control (*Courtesy of Analog Devices*)

The calculation of the true rms value of a signal, regardless of its wave shape, is a desirable feature for many measurement circuits and instruments.

$$\text{rms} = \sqrt{\overline{(e_{in})^2}} \tag{8-36}$$

where $\overline{(e_{in})^2}$ is the average value of the square of the signal. You have seen how to use multiplier ICs to square and take the square root of a signal.

You could directly implement equation (8-36) with three ICs. The first, a multiplier, would square the input. The second block would be a buffered RC network. The capacitor charges and discharges over each cycle of the input. The net charge, and therefore the voltage across the capacitor, is proportional to the average of the signal from the squarer. The final block would be a multiplier IC set up as a divider to perform the square root (see Fig. 8-40).

Although this technique is direct and apparently simple, it is not very practical. The two multipliers required make the technique expensive. Division (square root) severely limits the amplitude range of the input signal. Errors are difficult or expensive to reduce.

As with many other circuits presented, feedback will help solve the problem.

From equation (8-36)

$$e_{\text{rms}} = \sqrt{\overline{(e_{\text{in}})^2}} \tag{8-36}$$

Squaring both sides gives

$$e^2_{\text{rms}} = \overline{(e_{\text{in}})^2}$$

Dividing both sides by e_{rms} (the desired output), you have

$$e_{\text{rms}} = \frac{\overline{(e_{\text{in}})^2}}{e_{\text{rms}}} \tag{8-37}$$

To implement equation (8-37) a three variable IC called a multiplier-divider is used. The 433 programmable multifunction module is shown in Fig. 8-43.

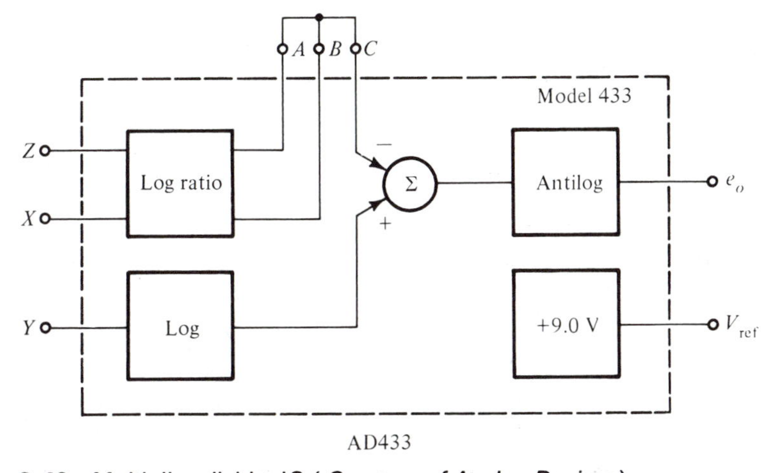

Figure 8-43 Multiplier-divider IC (*Courtesy of Analog Devices*)

In the log ratio block log amps take the log of e_x and e_z. These are then subtracted at the ABC external mode to produce ln e_x − ln e_z. The e_y signal is also conditioned by a log amp. Ln e_x − ln e_z is then subtracted from ln e_y.

$$\ln e_y - (\ln e_x - \ln e_z)$$

$$\ln e_y + \ln e_z - \ln e_x$$

$$\ln \frac{e_y e_z}{e_x}$$

The antilog of this difference (or quotient) is finally taken and proper scaling factors are introduced.

$$e_0 = \frac{10\text{ V}}{E_{\text{ref}}} \frac{e_y e_z}{e_x} \tag{8-38}$$

when A, B, and C are shorted. Because the 433 uses log amps, you must make sure that the input signals never cross zero volts (you are restricted to one quadrant operation). This can be done by preceding the circuit with an ideal rectifier such as those in Figs. 8-12 or 8-13.

The circuit in Fig. 8-44 calculates true rms, using the technique of equation (8-37).

$$e_{\text{rms}} = \frac{\overline{(e_{\text{in}})^2}}{e_{\text{rms}}} \tag{8-37}$$

The 433 outputs $v_0 = e_z e_y / e_x$ when the gain resistor is properly set to compensate for the IC's scaling term. Since

$$e_z = e_y = e_{\text{in}}$$

$$e_x = e_{\text{out}}$$

the output of the 433 is

$$e_{\text{out}} = \frac{e_{\text{in}}^2}{e_{\text{out}}}$$

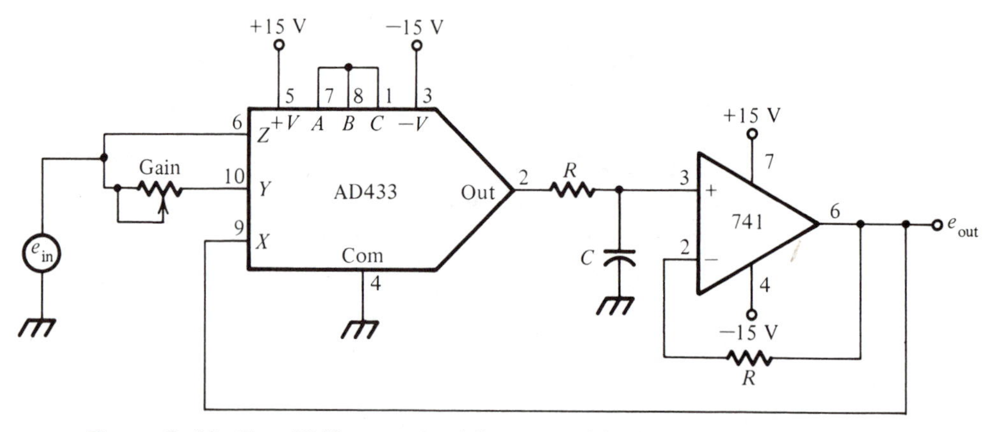

Figure 8-44 True RMS converter (*Courtesy of Analog Devices*)

Properly choosing the values of R and C (the same criteria being used as with the AGC circuit previously discussed) will allow the capacitor to charge to the average value of the output. The output varies very much more slowly than the input cycle, which means that

$$\overline{\left[\frac{(e_{in})^2}{e_{out}}\right]} = \frac{\overline{(e_{in})^2}}{e_{out}}$$

over a small number of input cycles. Consequently, the output does not upset the averaging of e_{in} done by the RC network. The 741 op amp is connected as a buffer (voltage follower). R in its feedback line compensates for bias currents.

Consequently, the output from the buffer is e_{out}.

$$e_{out} = \frac{\overline{(e_{in})^2}}{e_{out}}$$

Compare this to equation (8-37).

$$e_{rms} = \frac{\overline{(e_{in})^2}}{e_{rms}}$$

$$e_{out} = e_{rms}$$

8-4 VARIABLE GAIN AMPLIFIERS

There is a group of amplifiers whose specifications may be programmed externally. In fact, the external programming of parameters may be performed electronically. The ability to electronically alter the characteristics of the amplifier offers some very useful applications, enhancing the "automatic" nature of the circuit built with these amplifiers.

8-4.1 Programmable Op Amps

Programmable op amps such as the LM4250 allow you to alter the bias current, offset, noise, quiescent power, slew rate, and gain bandwidth specifications. All of these parameters depend on the current flowing out of the quiescent current set terminal.

The relationship between each of these parameters and the set current (I_{set}) are illustrated in Figs. 8-45(a) through 8-45(e). Carefully examine each curve. With the exception of unnulled offset voltage, reducing I_{set} reduces each parameter. This means that lowering the input bias current by lowering I_{set} also

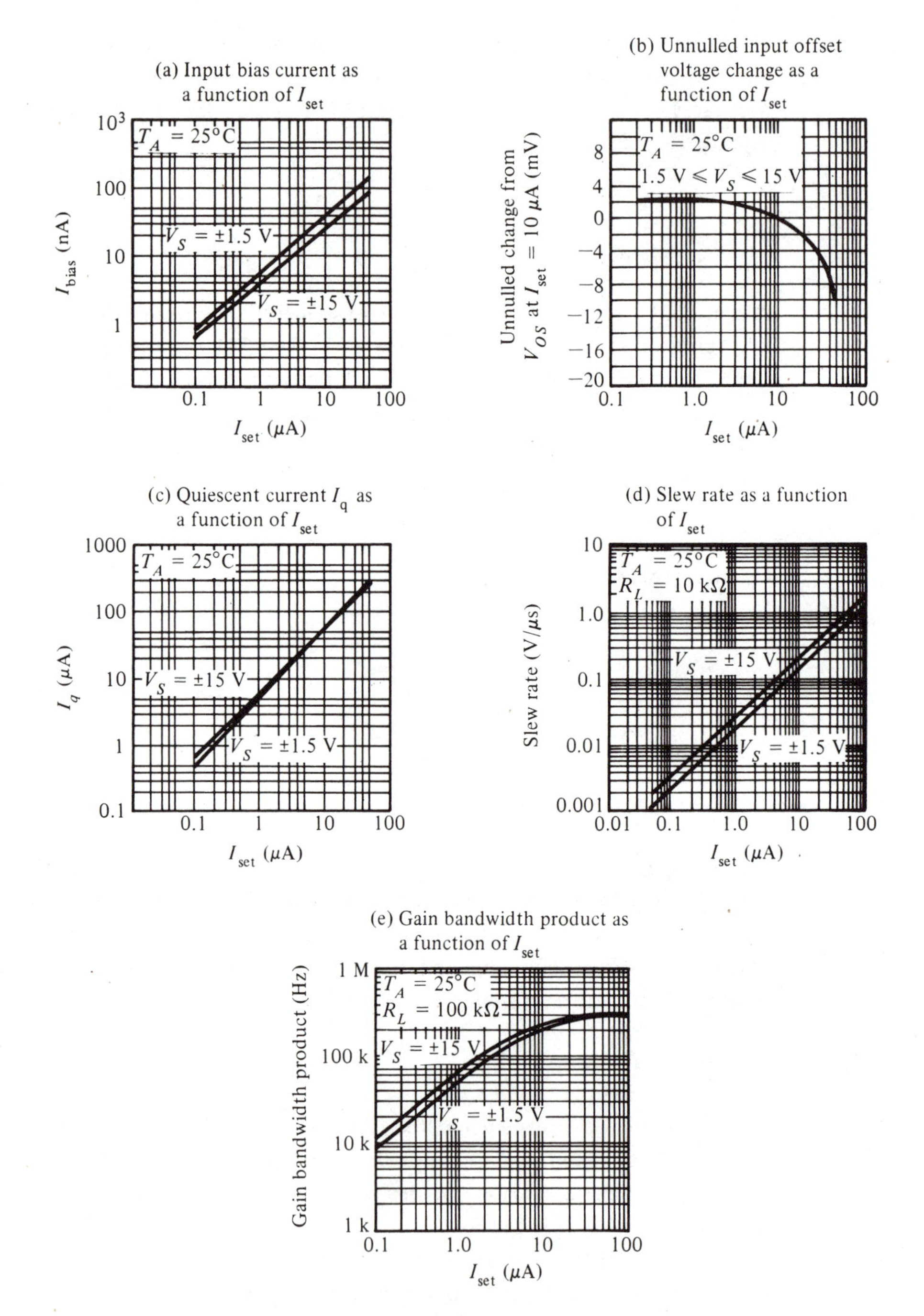

Figure 8-45 LM 4250 amp characteristics affected by I_{SET} (*Courtesy of National Semiconductors*)

allows the op amp to use less supply (quiescent) current. However, lowering I_{set} increases offset voltage, and lowers both slew rate and gain bandwidth. Increasing I_{set} to provide faster response will cause a higher bias current and demand more current from the power supply. Clearly, tradeoffs are required.

I_{set} is determined by placing a resistor between the current set pin (pin 8 in the mini-dip) and the negative supply. The precise value of R_{set} needed depends both on the level of I_{set} desired and the magnitude of the negative supply. Popular combinations are given in Table 8-2.

TABLE 8-2. R_{SET} values for the LM4250 programmable amp

	I_{SET}				
V_S	0.1 μA	0.5 μA	1.0 μA	5 μA	10 μA
±1.5 V	25.6 MΩ	5.04 MΩ	2.5 MΩ	492 kΩ	244 kΩ
±3.0 V	55.6 MΩ	11.0 MΩ	5.5 MΩ	1.09 MΩ	544 kΩ
±6.0 V	116 MΩ	23.0 MΩ	11.5 MΩ	2.29 MΩ	1.14 MΩ
±9.0 V	176 MΩ	35.0 MΩ	17.5 MΩ	3.49 MΩ	1.74 MΩ
±12.0 V	236 MΩ	47.0 MΩ	23.5 MΩ	4.69 MΩ	2.34 MΩ
±15.0 V	296 MΩ	59.0 MΩ	29.5 MΩ	5.89 MΩ	2.94 MΩ

(Courtesy of National Semiconductor)

Electronic control of I_{set} can be accomplished as shown in Fig. 8-46. Variation in $V_{control}$ changes I_{base}, which in turn alters I_{set}. The value of R should be selected to limit the maximum I_{set} of the transistor driven into saturation.

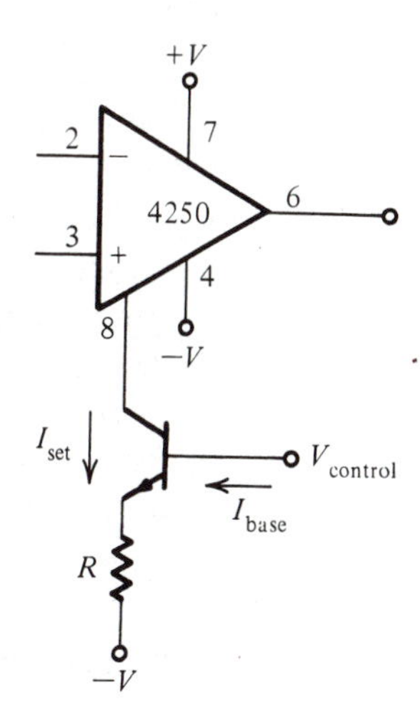

Figure 8-46 Electronic parameter control for the LM 4250 (*Courtesy of National Semiconductors*)

This configuration can be used to allow the op amp to be idling, demanding very little from the power supply (perhaps a battery). An input signal, detected by other circuitry, could then saturate the transistor, bringing the op amp to full performance, but only for the duration of the signal.

When I_{set} is driven below 0.1 μA by turning the transistor off (transistor leakage I must be very low), the op amp also goes off. Its output is no longer actively driven to some voltage. The output floats. The op amp has been put into a high impedance state (just as tristate logic gates which are disabled). This allows you to multiplex analog signals into a common output node. As long as only one op amp at a time is "enabled," the node will contain the signal from the enabled amp. Digital switching and control of analog signals without the necessity (or many of the problems) of analog switches can be accomplished (see Fig. 8-47).

The LM13080 is a medium power programmable op amp. It can deliver up to 250 mA to the load. However, it does not have internal thermal shutdown or current limiting. You must anticipate junction temperature and provide the proper heat sink.

$$T_J = T_A + P\theta_{JA}$$

where

T_J is the junction temperature, which must be less than 150°C.

T_A is the ambient temperature.

P is the package power dissipation.

θ_{JA} is the junction to ambient thermal resistance. Bare case in still air $\theta_{JA} = 120°\text{C/W}$.

Programming is accomplished by selecting the size (or existence) of R_{set}, connected between the input bias pin (pin 7) and the negative supply. Offset voltage and current, input bias current, and slew rate are directly determined by R_{set}. These relationships are shown in Fig. 8-48.

Under normal operation pin 2 (output bias) is tied directly to the negative supply and pin 7 (input bias) is tied through R_{set} to the negative supply. To electronically shut the amplifier off, this junction of pin 2 and R_{set} must be removed from the negative supply. It should then be driven to $+V$. This is illustrated in Fig. 8-49. Resistors R1, R2, R4, and R5 allow single supply operation (see Chapter 5). When the control logic input is high, Q1 saturates and the op amp functions normally as a noninverting amplifier. A low control logic signal turns Q1 off, and R3 pulls pin 2 and pin 7 to $+V$. This turns the op amp off, allowing its output to float.

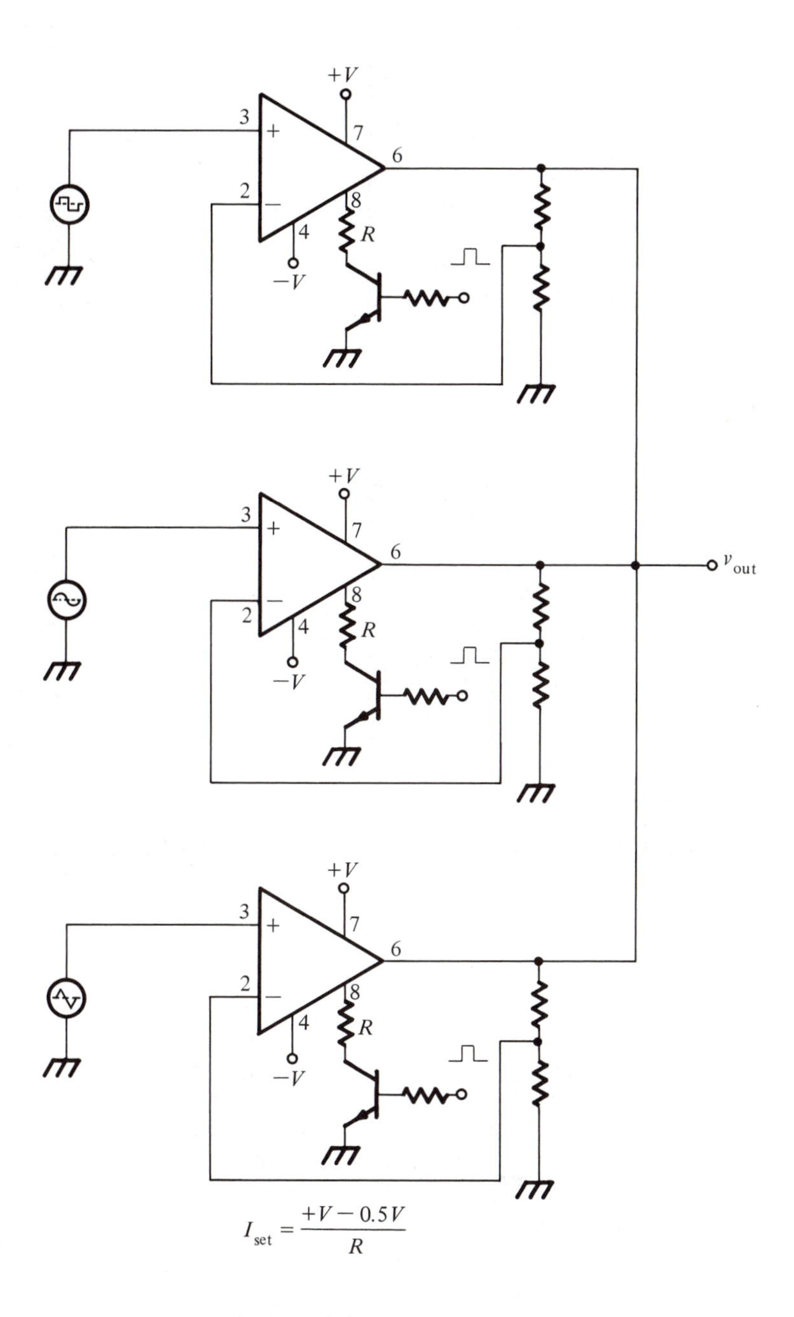

Figure 8-47 Analog multiplexing with LM 4250 programmable op amps

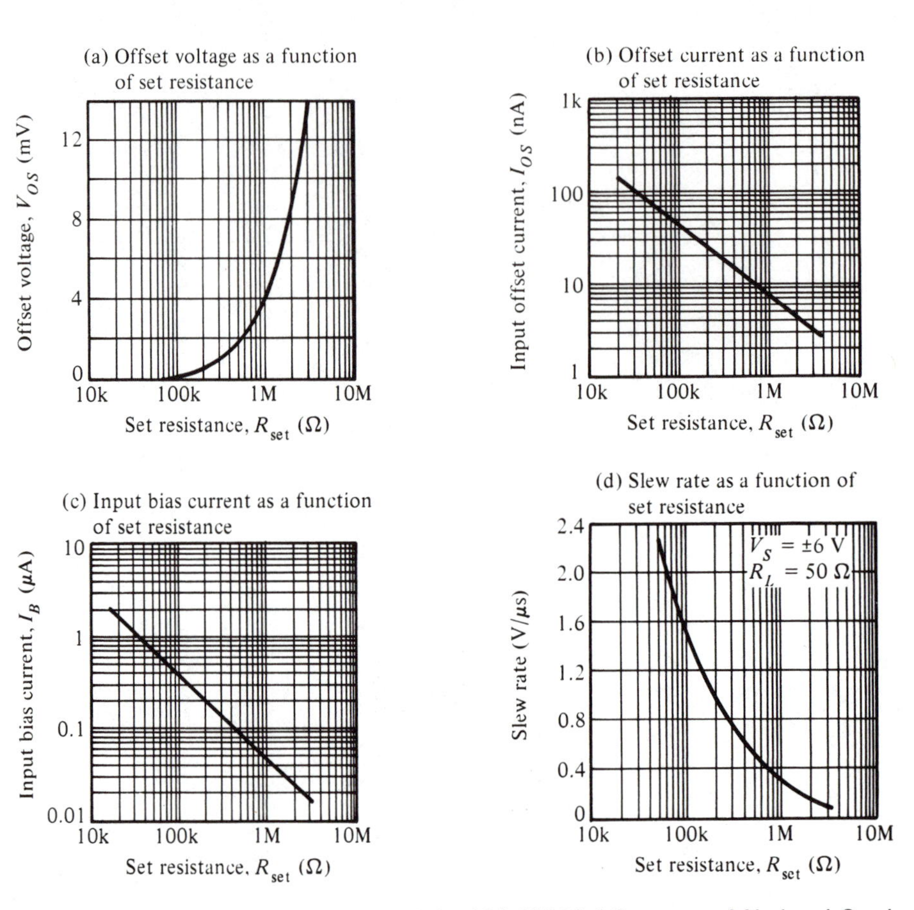

Figure 8-48 Programmable features of the LM 13080 (*Courtesy of National Semiconductors*)

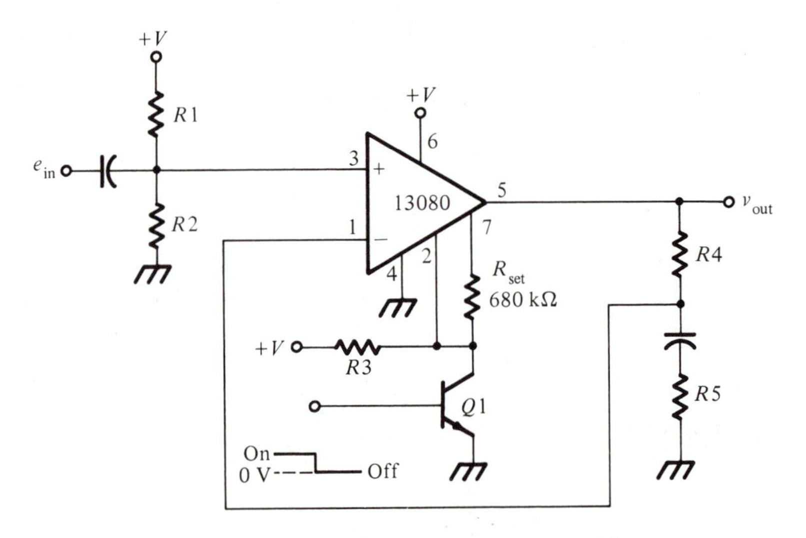

Figure 8-49 Logic control of a LM l3080 power programmable op amp

A second application of the LM13080 is shown in Fig. 8-50. Two square wave generators have been built (see Chapter 6). U2 oscillates at a low frequency. Its output turns U1 on and off. When on, U1 oscillates at an audio frequency and drives a speaker directly. The result is beep—beep—beep . . . .

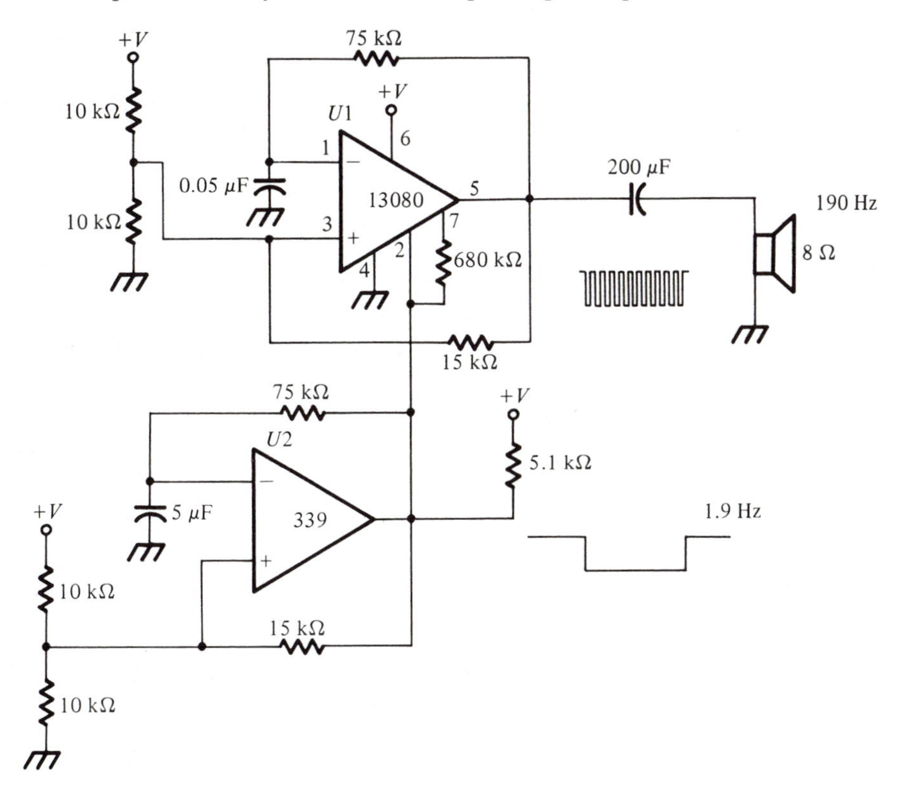

Figure 8-50 Two state alarm (*Courtesy of National Semiconductors*)

8-4.2 Programmable Transconductance Amplifier

The operational transconductance amplifier (OTA) is an IC which outputs a *current* proportional to the difference in potential between its input pins.

$$i_{out} = g_m(e_2 - e_1) \tag{8-38}$$

The "gain" or output divided by input is

$$g_m = \frac{i_{out}}{e_2 - e_1}$$

and must be in units of Siemans (conductance).

The CA3080 is a programmable OTA. Its transconductance or gain (g_m) is directly dependent on the current driven into a current set pin. At room temperature

$$g_m = \frac{19.2}{V} I_{ABC} \tag{8-39}$$

See Fig. 8-51.

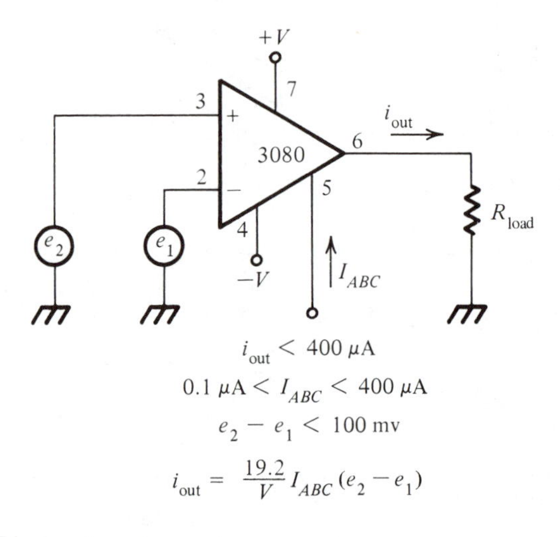

Figure 8-51 OTA circuit and operating restrictions

This relationship holds linearly for

$$0.1\ \mu A < I_{ABC} < 400\ \mu A$$

Substituting equation (8-39) into (8-38), you obtain

$$i_{out} = \left(\frac{19.2}{V} I_{ABC}\right)(e_2 - e_1) \tag{8-40}$$

For the CA3080, equation (8-40) is linear for

$$i_{out} < 400\ \mu A$$

and

$$e_2 - e_1 < 20\ \text{mV}$$

The inputs of the CA3080 are directly connected to the bases of the input differential amplifier. There are no series dropping resistors. Variations in input voltage of greater than 20 mV will begin to alter the bias on the input transistors. This, in turn, affects the linearity of the output signal. You should always restrict input voltages to less than 20 mV.

Also, because the input signals are fed directly into the bases of the first stage with no series resistors, the input impedances are on the order of 10 kΩ to 30 kΩ. This is quite different from the impedance of standard op amps. Anticipate any loading and use op amp voltage followers as necessary to buffer input signals.

A basic voltage amplifier is shown in Fig. 8-52. Again, the input voltage must be restricted to 20 mV or less. At room temperature the output current is

$$i_{out} = \left(\frac{19.2}{V} I_{ABC}\right) e_{in} \tag{8-41}$$

Assuming that all of i_{out} flows into the load resistor R_L (rather than going to bias following stages), you have

$$v_{out} = i_{out} R_L$$

$$v_{out} = \left(\frac{19.2}{V} I_{ABC}\right) e_{in} R_L \tag{8-42}$$

The output voltage is determined by the input voltage and the size R_L chosen. Adjusting voltage gain is simply a matter of adjusting R_L. This assumes that all of i_{out} ($<400\ \mu A$) flows into R_L. Loading errors can be reduced by adding the voltage follower op amp. However, be sure to select an op amp whose bias current (I_B^+) is much smaller than i_{out}.

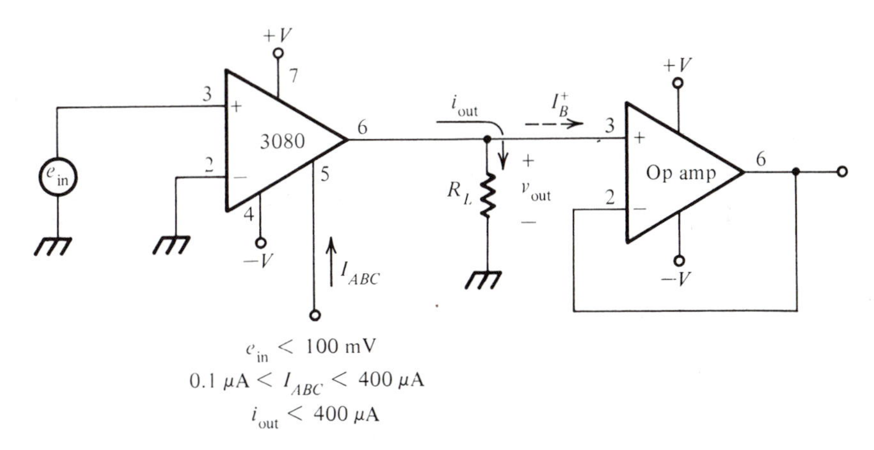

Figure 8-52 Basic OTA voltage amplifier circuit

The amplifier bias current (I_{ABC}) also directly affects the output. This feature provides several very handy applications. There are several ways to set I_{ABC} shown in Fig. 8-53. If you do not plan to alter I_{ABC}, connect the resistor R_{ABC} between pin 5 and ground. I_{ABC} is set by the negative supply and the resistor.

$$I_{ABC} = \frac{|-V| - 0.6\text{ V}}{R_{ABC}} \tag{8-43}$$

You can control I_{ABC} with an externally generated control voltage, $e_{control}$. This is shown in Fig. 8-53(b). Changing $e_{control}$ changes I_{ABC}, which in turn changes g_m, the amplifier's "gain."

$$I_{ABC} = \frac{e_{control} + |-V| - 0.6\text{ V}}{R_{ABC}} \tag{8-44}$$

Substituting equation (8-44) into equation (8-41), at room temperature

$$i_{out} = \left(\frac{19.2}{V}\right)\left[\frac{e_{control} + |-V| - 0.6\text{ V}}{R_{ABC}}\right]e_{in} \tag{8-45}$$

Equation (8-45) shows that the output current is directly proportional to both the input voltage and the control voltage. Said another way, the gain of the amplifier can be remotely (electronically) altered by $e_{control}$.

The third option of setting I_{ABC} is to set it directly with a current source. This current source could be made with a field effect transistor, a bipolar transistor, an op amp, an IC current source chip, or another 3080.

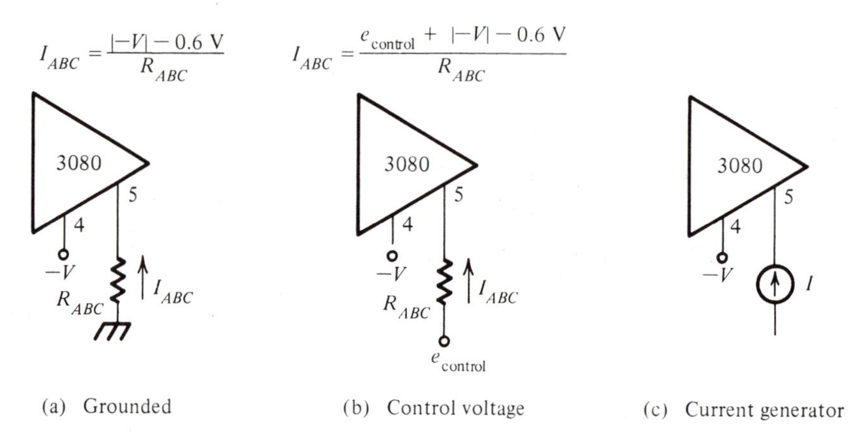

Figure 8-53 I_{ABC} setting circuits

Example 8-6

Calculate the output current for the circuit in Fig. 8-54 for $e_{in} = 10\,\text{mV}$

a. $e_C = +10$ V
b. $e_C = 0$ V
c. $e_C = -10$ V

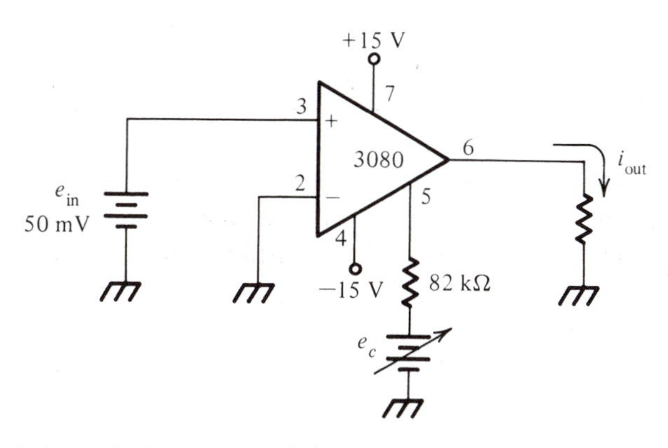

Figure 8-54 Schematic for Example 8-6

Solution

From equation (8-45)

$$i_{out} = \left(\frac{19.2}{\text{V}}\right)\left(\frac{e_{control} + |-V| - 0.6\text{ V}}{\text{R}_{ABC}}\right)e_{in}$$

$$i_{out} = \left(\frac{19.2}{V}\right)\left(\frac{10\text{ V} + 15\text{ V} - 0.6\text{ V}}{82\text{ k}\Omega}\right)(10\text{ mV})$$

a. $i_{out} = 57.2\ \mu\text{A}$

Substituting $e_C = 0$ V and $e_C = -10$ V into equation (8-45) gives

b. $i_{out} = 33.8\ \mu\text{A}$

c. $i_{out} = 10.3\ \mu\text{A}$

In several circuits, the purpose of a resistor is to convert an input voltage into a current (that is the purpose of R_i in an inverting amp). The relationship of input voltage to current out of the resistor is set by the value of the resistor. That resistance, in turn, controls the gain, critical frequency, etc. of the circuit in which it is used. The voltage-to-current conversion performed by the resistor is just like the input-to-output relationship of the OTA. It has an effective "resistance."

$$i_{\text{out}} = \frac{19.2}{V} I_{\text{ABC}} e_{\text{in}}$$

$$\text{R} = \frac{e_{\text{in}}}{i_{\text{out}}} = \frac{1}{19.2/V\ I_{\text{ABC}}} = \frac{10.3\text{ mV}}{I_{\text{ABC}}}$$

So the OTA can be viewed as a programmable resistor whose resistance is set by I_{ABC}.

Controlling I_{ABC} simply with R_{ABC} and e_{control} causes a highly nonlinear relationship between e_{control} and the apparent resistance caused by the OTA. This is because of the $|-V| - 0.6$ V portion of the equation for I_{ABC}.

Example 8-7

Derive the "resistance" equation for the circuit in Fig. 8-55.

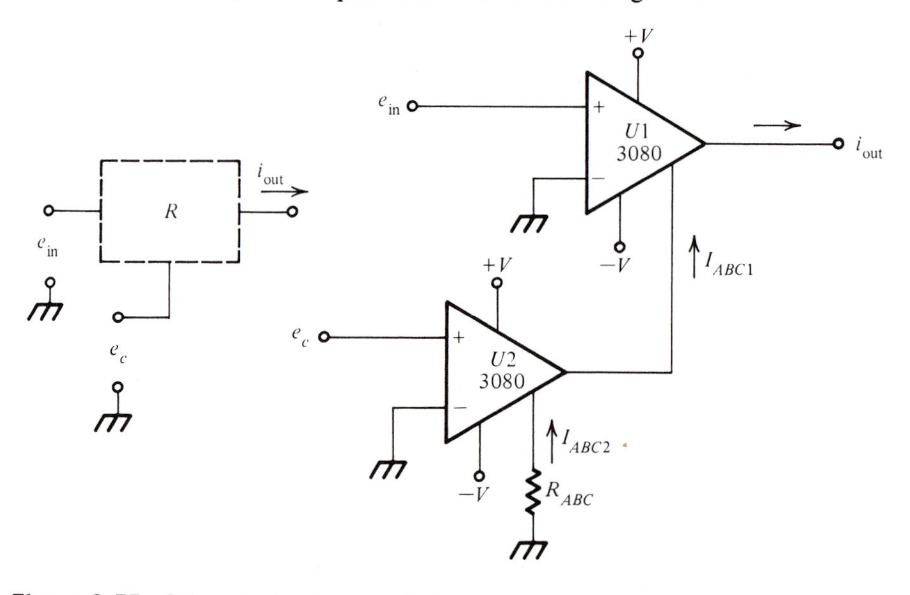

Figure 8-55 Schematic for Example 8-7

Solution

For U2,

$$I_{\text{ABC2}} = \frac{|-V| - 0.6\text{ V}}{\text{R}_{\text{ABC}}} \cong \frac{|-V|}{\text{R}_{\text{ABC}}}$$

$$i_{\text{out}2} = \left(\frac{19.2}{V}\right) I_{\text{ABC}2}\, e_{\text{C}}$$

$$i_{\text{out}2} = \left(\frac{19.2}{V}\right) \frac{|-V|}{\text{R}_{\text{ABC}}}\, e_{\text{C}}$$

For U1,

$$I_{\text{ABC}1} = i_{\text{out}2}$$

$$i_{\text{out}} = \frac{19.2}{V} I_{\text{ABC}1}\, e_{\text{in}}$$

$$i_{\text{out}} = \left(\frac{19.2}{V}\right)\left[\left(\frac{19.2}{V}\right) \frac{|-V|}{\text{R}_{\text{ABC}}}\, e_{\text{C}}\right] e_{\text{in}}$$

$$i_{\text{out}} = \left[\left(\frac{19.2}{V}\right)^2 \frac{|-V|}{\text{R}_{\text{ABC}}}\right] e_{\text{C}} \cdot e_{\text{in}} \qquad \text{(8-46a)}$$

$$\text{R} = \frac{e_{\text{in}}}{i_{\text{out}}} = \frac{\text{R}_{\text{ABC}}}{(369/V^2)|-V| \times e_{\text{C}}} \qquad \text{(8-46b)}$$

OTA U2 forms a voltage to current converter to set I_{ABC} for the main OTA, U1. In this way the offset associated with equation (8-45) is eliminated.

The circuit in Fig. 8-55 can be considered a voltage-controlled resistance whose value is set according to equation (8-46). Remember, however, that e_{in} must be restricted to less than 20 mV and i_{out}, even into a dead short, should be less than 400 μA.

The electronically tunable "resistor" of Fig. 8-55 will allow you to add electronic tuning to many circuits with which you are already familiar. Just be sure to keep in mind the input voltage and output current restrictions.

Example 8-8

For the circuit in Fig. 8-56 calculate the amplifier's gain for e_{gain} voltage of 10 V, 5 V, 1 V.

Solution

OTA U1 and U2 form an electrically tunable resistor, providing R_i for the inverting amp U3.

$$\text{R}_i = \frac{\text{R}_{\text{ABC}}}{(369/V^2)|-V|\, e_{\text{C}}} \qquad \text{(8-46)}$$

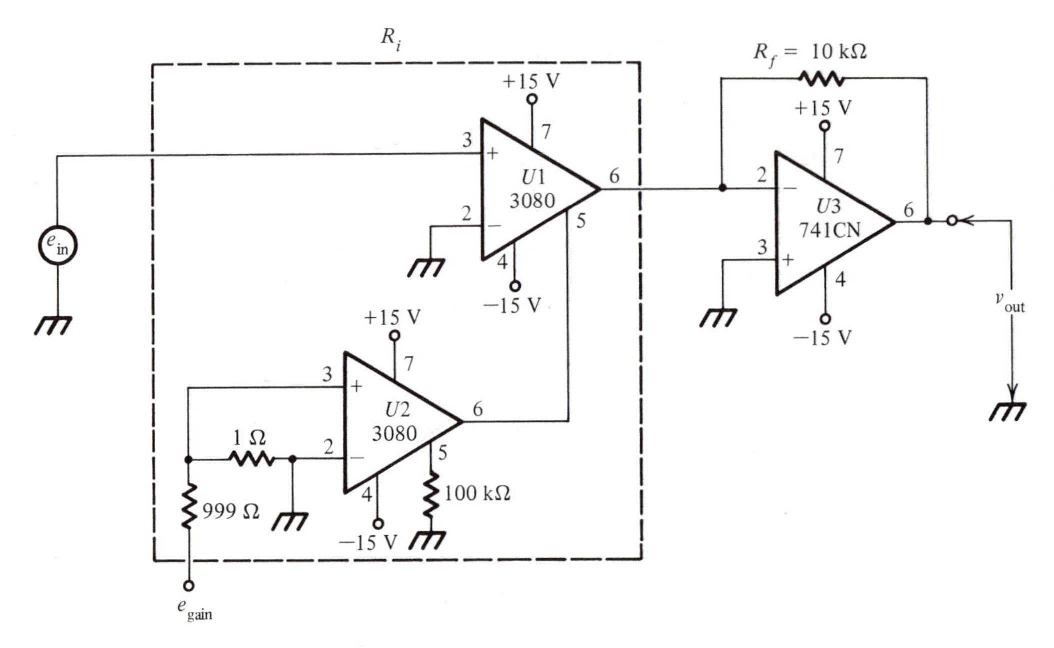

Figure 8-56 Schematic for Example 8-8

The gain setting voltage, e_{gain}, is reduced by 1/1000 by the 999 Ω and 1 Ω voltage divider. Substituting that into equation (8-46), you obtain

$$e_{\text{gain}} = 10\ \text{V} \qquad e_C = 10\ \text{mV}$$

$$R_i = \frac{100\ \text{k}\Omega}{(369/V^2)(15\ \text{V} \times 10\ \text{mV})}$$

$$e_{\text{gain}} = 10\ \text{V}, \qquad e_C = 10\ \text{mV}, \qquad R_i = 1800\ \Omega$$

$$e_{\text{gain}} = 5\ \text{V}, \qquad e_C = 5\ \text{mV}, \qquad R_i = 3600\ \Omega$$

$$e_{\text{gain}} = 1\ \text{V}, \qquad e_C = 1\ \text{mV}, \qquad R_i = 18\ \text{k}\Omega$$

$$A = \frac{-R_f}{R_i} = \frac{10\ \text{k}\Omega}{1800\ \Omega} = -5.5 \quad \text{for } e_{\text{gain}} = 10\ \text{V}$$

$$A = -2.8 \quad \text{for } e_{\text{gain}} = 5\ \text{V}$$

$$A = -.6 \quad \text{for } e_{\text{gain}} = 1\ \text{V}$$

The result is a direct, linear relationship between e_{gain} and A.

Another use of the electronically tunable "resistor" is in an active filter. Variation of the control voltage would shift the critical frequency of the filter. This is illustrated in block form in Fig. 8-57.

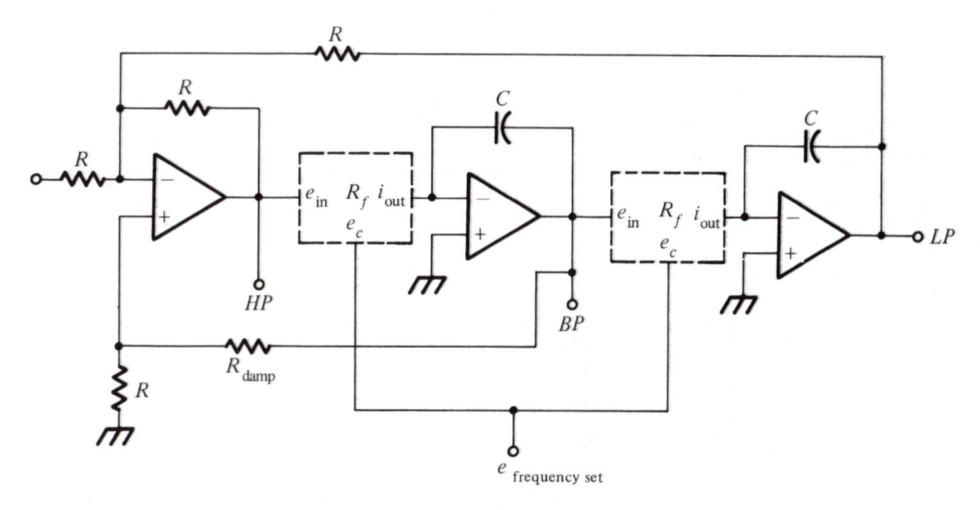

Figure 8-57 Electronically tunable active filter

The circuit in Fig. 8-58 is a general two quadrant multiplier, built with inexpensive general purpose op amps and OTAs. The OTA configuration is the same as in Fig. 8-55. So

$$i_{out} = \left[\frac{369}{V^2}\frac{|-V|}{R_{ABC}}\right]e_y e_x \tag{8-46a}$$

Assuming that the output buffer op amp does not draw any significant current from i_{out}, you have

$$v_{out} = i_{out} R_L$$

$$v_{out} = \left[\frac{369}{V^2}\frac{|-V|R_L}{R_{ABC}}\right]e_y e_x$$

Resistor R_L can be adjusted to provide some convenient reference.

$$\frac{1}{e_{ref}} = \frac{369}{V^2}\frac{|-V|}{R_{ABC}}R_L$$

$$v_{out} = \frac{e_y e_x}{e_{ref}}$$

Compare this with equation (8-28a).

Should the input impedance be too small, or if an input voltage range of greater than 20 mV is desired, input buffers followed by 100 Ω potentiometers can be added.

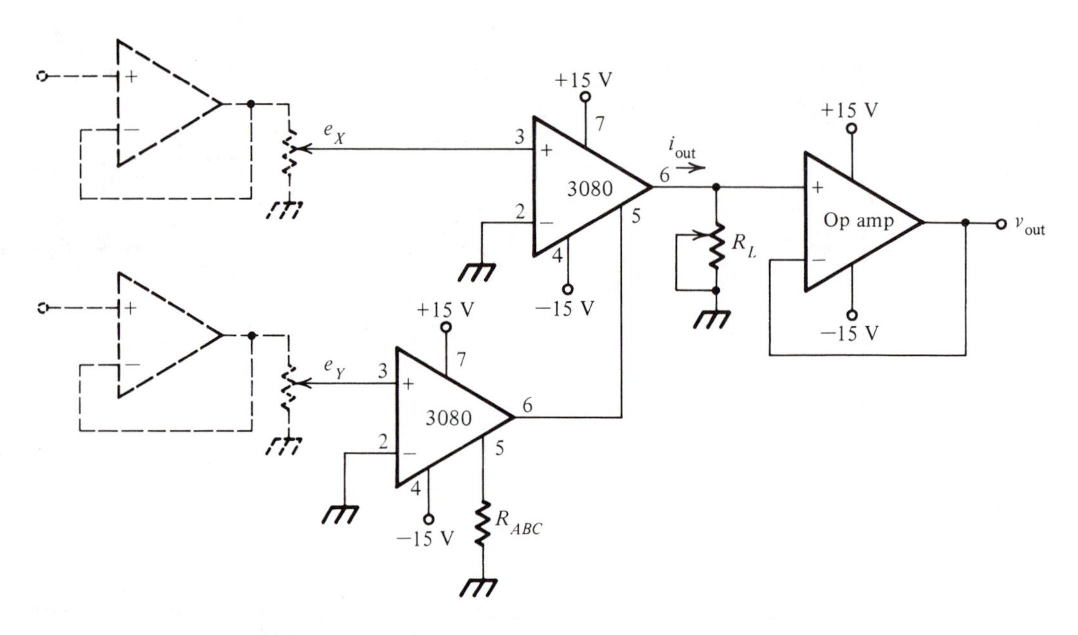

Figure 8-58 Two quadrant multiplier

As with the programmable op amps, variations in I_{ABC} alter several characteristics. These include offset voltage, input bias current, power supply current demanded, input resistance, capacitance, output resistance, and slew rate. These are illustrated in Fig. 8-59. This means that changing I_{ABC} to produce gain adjustment (Figs. 8-54 and 8-56), resistance adjustment (Fig. 8-55), frequency adjustment (Fig. 8-57) or multiplication (Fig. 8-58) will also significantly alter many of the other parameters of the amplifier. Input and output buffering with voltage followers will help. But you must also carefully analyze the effects that varying I_{ABC} will have on slew rate. Frequency compensation can be achieved by placing a series resistor and capacitor between the output and ground.

Temperature will also alter the performance of the CA3080. The proportionality constant $19.2/V$ is valid only at room temperature. More properly, it should be

$$g_m = \frac{q\alpha I_{ABC}}{2kT}$$

where the terms are defined as they were for the log amp. Notice that g_m and, therefore, the entire OTA performance is inversely proportional to temperature (T).

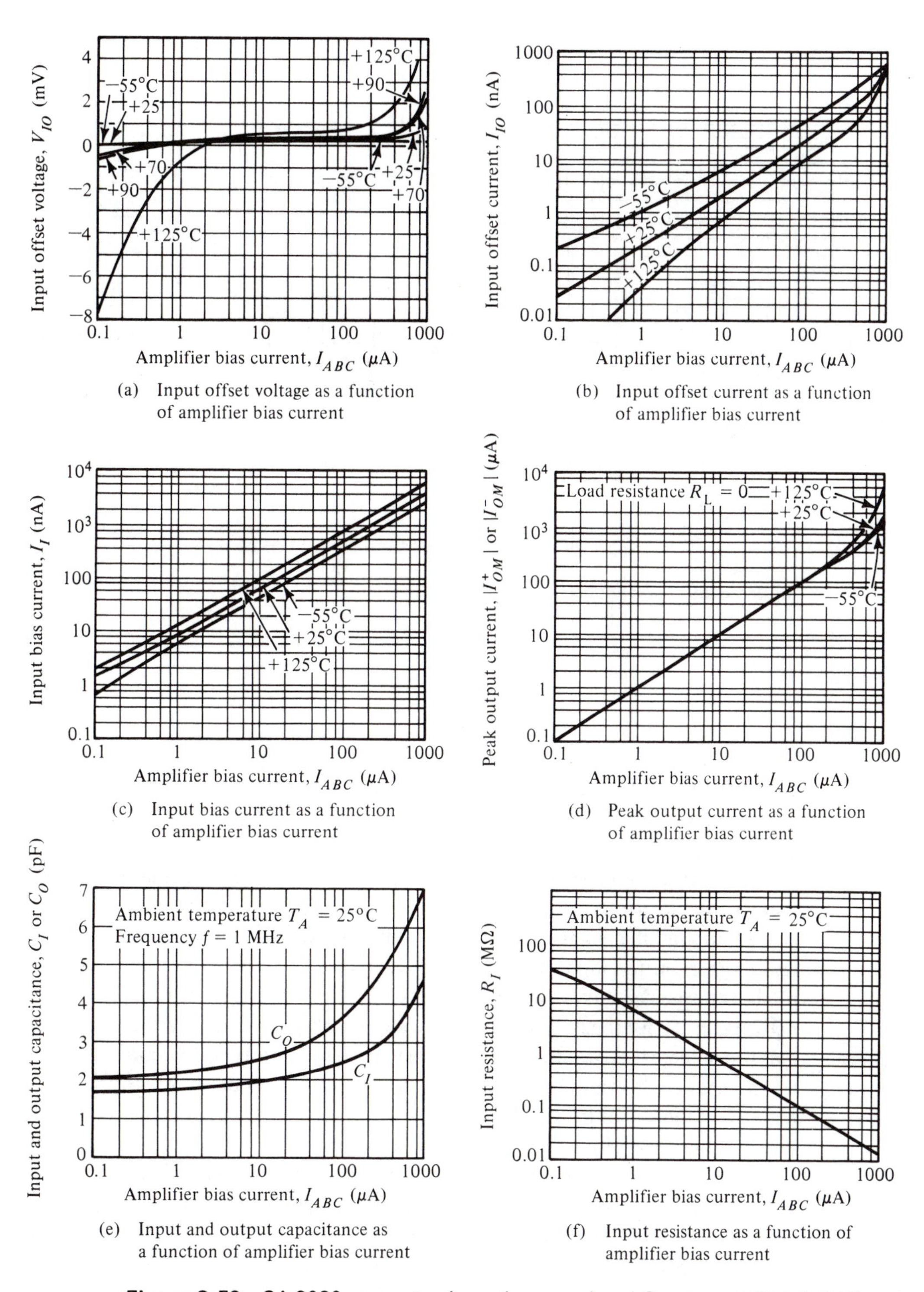

(a) Input offset voltage as a function of amplifier bias current

(b) Input offset current as a function of amplifier bias current

(c) Input bias current as a function of amplifier bias current

(d) Peak output current as a function of amplifier bias current

(e) Input and output capacitance as a function of amplifier bias current

(f) Input resistance as a function of amplifier bias current

Figure 8-59 CA 3080 parameter dependence on I_{ABC} (*Courtesy of RCA Solid State*)

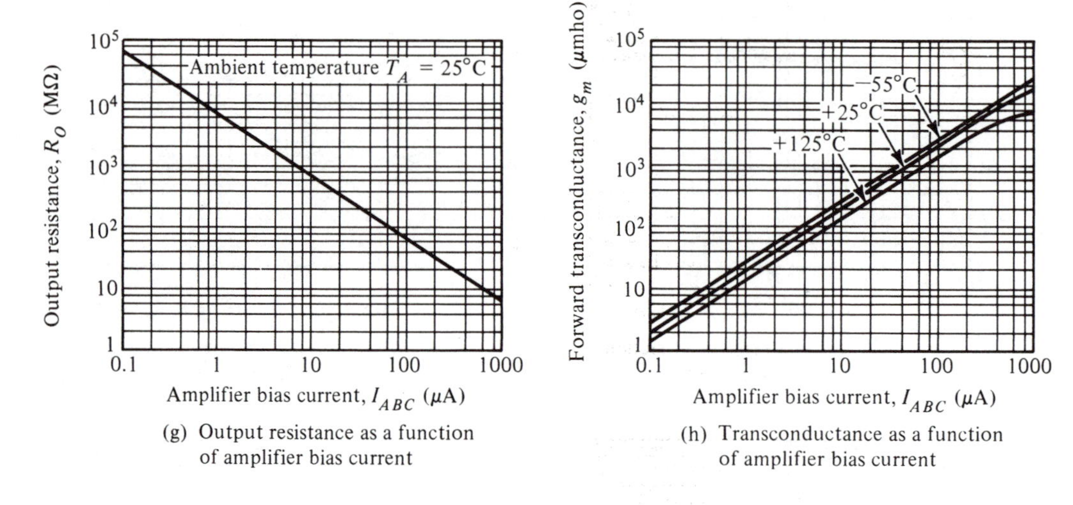

(g) Output resistance as a function of amplifier bias current

(h) Transconductance as a function of amplifier bias current

Note: Supply voltage is same for (a)–(h): $V^+ = +15$, $V^- = -15$

Figure 8-59 *continued*

SUMMARY

Normally, you work hard to produce a linear relationship between the input and output of any circuit or system you build. However, under certain circumstances, a specific, well-defined, nonlinear input-output (transfer) curve is needed.

Given this nonlinear transfer curve, you can build a general transfer function synthesizer. By using a resistor, diode, and bias network in the input loop, the gain of an amplifier increases with input amplitude. Although this is a piecewise approximation, for input signals larger than several volts, a close fit to a given nonlinear transfer curve can be obtained. Placing the resistor, diode, and bias network in the feedback loop causes the amplifier's gain to decrease when output amplitude increases, producing the oppositely shaped transfer function.

Signal rectification and peak detection require a diode. Normally, the first 600 mV of the signal is lost across the diode, since that is the voltage necessary to forward bias the diode. This is totally unacceptable for low-level signals. The problem is solved by placing the diode within the negative feedback loop of an op amp. Assuming an open loop gain of 200K, the first 3 μV of the input signal is multiplied by 200K (since the diode is off and the op amp is operating open loop). This produces the 600 mV offset necessary to bias the diode on. Only 3 μV, rather than 600 mV, of the signal has been lost to get the diode on. Care must be taken to ensure that the op amp does not go into saturation during the off cycle. Component selection is important in minimizing errors in the peak detector.

Log amps find wide applications in measurement and computational circuits. The basic log amp relies on the log relationship between a transistor's collector current and its base-to-emitter voltage. To compensate for other transistor parameters and for variations in temperature, two matched transistors, several op amps, and a temperature-sensitive, matched voltage divider are needed.

Antilog amps (exponentiation), can be built by placing the transistor "antilog" element in the input loop of an op amp. Log amps were built by placing the "antilog element" in the op amp's feedback loops. Precision log/antilog amps are easily configured with a functional IC module. It contains all transistors, op amps, and temperature compensation. You simply connect the antilog element in either the input or feedback loop of the internal op amp by a few external jumpers. Care must be taken to ensure the proper input polarity, since the logarithm of a negative number is undefined.

Multiplication of two signals is also a nonlinear operation. If both signals are restricted to one polarity, the output will fall into one quadrant. If one of the inputs can change sign, the output can be in either of two quadrants. If both inputs may change sign, the output may appear in any of the four quadrants. There are several ways to produce multiplier circuits. You could take the log of each input, add the logs together, then take the antilog. This is highly precise, but allows only one quadrant operation and is expensive. Transconductance multiplication (though less precise) gives four quadrant operation at a lower cost. With careful trimming many of the multiplier's errors can be minimized. With a little imagination, you can use multipliers to simply solve problems which would otherwise require very complex circuitry. Among these are frequency doubling, true power and power factor calculation, voltage-controlled amplification, AM modulation, division, squaring and square root calculations, automatic gain control, and true rms calculation.

Programmable amplifiers allow you to set many DC and AC characteristics by controlling a set current. Though you can not *individually* optimize each parameter, this programmability allows you to electronically alter the amp's performance. In fact, you can even turn the amp off. This provides significant power savings in medium power op amps, and allows analog signal multiplexing without having to use analog switches.

The operational transconductances amplifier (OTA) outputs a current proportional to its input voltage. The relationship between i_{out} and e_{in} (the transconductance) is programmable. Although there are rather severe restrictions on voltage and current magnitudes, OTAs allow you to build programmable gain voltage amplifiers, voltage-controlled resistances, voltage-controlled active filters, and two quadrant multipliers.

PROBLEMS

8-1 For the circuit in Fig. 8-5, draw the transfer curve, given that

$R1 = R2 = R3 = R4 = R5 = R6 = R7 = 1\ k\Omega$
$R_f = 1\ k\Omega$
$\pm V_{ref} = \pm 8\ V$
$R_{B2} = R_{B5} = 4.7\ k\Omega$
$R_{B3} = R_{B6} = 2.2\ k\Omega$
$R_{B4} = R_{B7} = 1\ k\Omega$

8-2 Repeat Problem 8-1 for the circuit in Fig. 8-10. $R_i = 1\ k\Omega$.

8-3 Design a transfer function synthesizer whose output is

$$v_{out} = (e^{v_{in}} - 1)$$

where $0.6 \le v_{in} \le 2.5$ V and $e = 2.71828$

8-4 Design a transfer function synthesizer which outputs a voltage equal to the dBm value of the input. That is,

$$v_{out} = 20 \log_{10}\left(\frac{v_{in}}{0.775\ V}\right)$$

where $0.2\ V < v_{in} < 3.1\ V$

8-5 For the circuit in Fig. 8-12, if $R_i = 10\ k\Omega$ and $R_{f1} = R_{f2} = 22\ k\Omega$, calculate $v_{in} = .1\ V_{pk}$

a. Z_{in}
b. v_{out} peak
c. v_{out} ave (DC value)

(HINT: For a half-valve rectified signal, the DC value is 45 percent of the applied rms, if a gain of 1 is assumed.)

8-6 Design a full wave ideal rectifier which will output an average value numerically equal to the rms of the input. (HINT: For a full wave rectified signal, the DC value is 90 percent of the applied rms, if a gain of 1 is assumed.)

8-7 For the circuit in Fig. 8-12, if $v_{in} = .1\ V_{pk}$
$R_i = 10\ k\Omega$
$R_{f1} = 22\ k\Omega$
$R_{LOAD} = 100\ k\Omega$
what would the output of CR2 be if R_{f2} opened? Careful! You must do a calculation for *each* half-cycle.

8-8 The circuit in Fig. 8-14 must drive a 100 kΩ load. Calculate the value of C

necessary to hold droops of a 10 V peak to less than 1 percent during the first second after the peak is detected.

8-9 If the maximum current out of the op amp in Fig. 8-14 is 10 mA, how rapidly can a 0.1 μF C charge? (answer in V/μs)

8-10 **a.** Equation (8-25) indicates that an antilog amp outputs a voltage proportional to 10 raised to the input voltage. Trace through the derivation to determine what must be done to produce

$$v_{out} = Ae^{(-Be_{in})}$$

when A and B are some constants and $e = 2.71828$

b. Determine the value (or equation) for A and B.

8-11 Repeat Problem 8-4, using a 755N log amp. Draw the schematic and specify all component values.

8-12 Modify the schematic for Problem 8-11 so that by throwing a switch the output voltage will either be

$$v_{out} = \text{dBm} = 20 \log_{10}\left(\frac{v_{in}}{0.775}\right)$$

or

$$v_{out} = \text{dBV} = 20 \log_{10}\left(\frac{v_{in}}{1 \text{ V}}\right)$$

8-13 Using 755N log amp module ICs, design a first quadrant multiplier where

$$v_{out} = \frac{e_x e_y}{10 \text{ V}}$$

Refer to Fig. 8-27. Draw the full schematic and specify all component values.

8-14 Explain why the circuit designed in Problem 8-13 must be one quadrant.

8-15 Design a circuit which outputs a DC voltage equal to the power factor [cos θ in equation (8-32)]. Refer to Fig. 8-60 for a block diagram. Provide all component values, IC numbers, pin numbers, and other details sufficient to build the circuit.

8-16 Explain, including a schematic, how a multiplier can be used to *de*modulate *balanced* AM. Look carefully at equation (8-33).

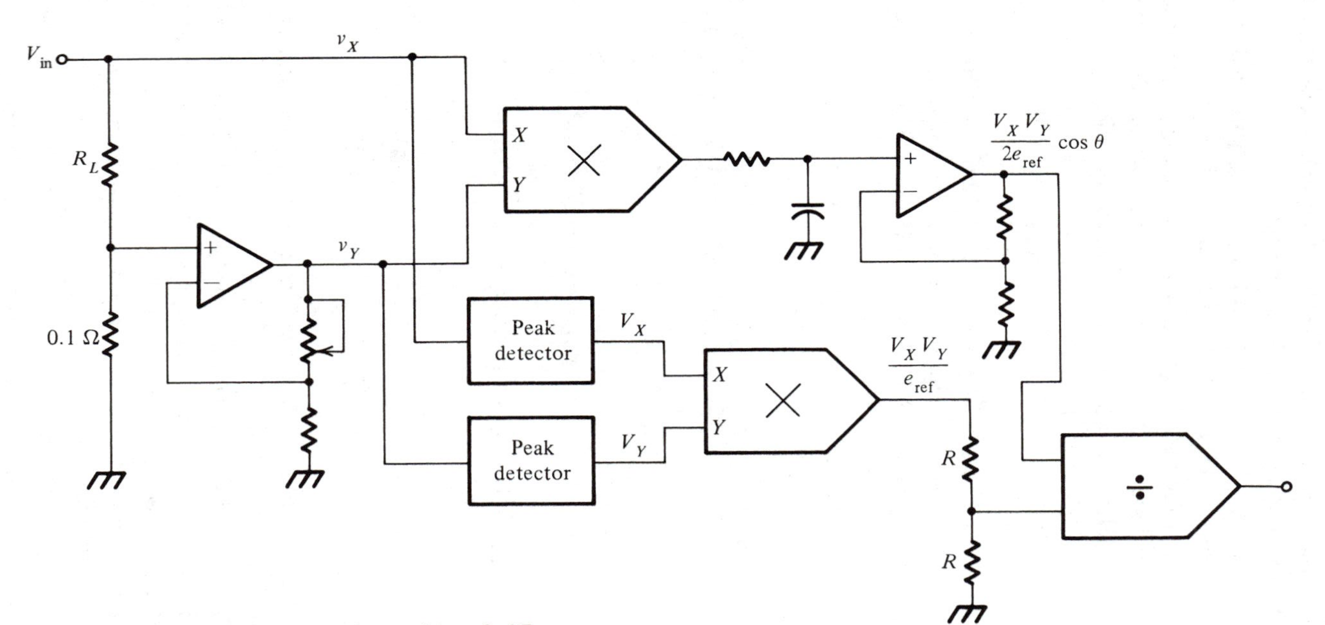

Figure 8-60 Block diagram for problem 8-15

8-17 Using 755N log amp module ICs, design a first quadrant divider where

$$v_{out} = 10 \text{ V} \frac{e_x}{e_y}$$

Draw the *full* schematic and specify all component values.

8-18 For the AGC circuit in Fig. 8-42, how would you increase the amplitude of the multiplier output (that level at which the AGC stabilize the output)? Explain your answer.

8-19 For each resistor and capacitor in Fig. 8-44 (the true rms converter) describe the effect if you were to individually double the component's value.

8-20 For the circuit in Fig. 8-46, calculate I_{set}, given that $V_{control} = -3$ V, $-V = -12$ V and $R = 820$ kΩ.

8-21 For the circuit in Fig. 8-46, what would happen to the following op amp parameters if $V_{control}$ went from -1 V to -10 V? I_{BIAS}, V_{OS}, I_q, slew rate, gain bandwidth. ($R = 150$ kΩ, $-V = -15$ V)

8-22 For the circuit in Fig. 8-47,

a. What must the logic high voltage be to produce $I_{set} = 10\ \mu$A, with $R = 470$ kΩ and $+V = 15$ V?

b. What must the logic low voltage be to turn the amp off?

8-23 Design a circuit using power programmable op amps which will drive a speaker at 200 Hz when the input is below 2 V, turn the speaker off when the input is between 2 V and 5 V, and will drive the speaker at 600 Hz when the input is above 5 V.

8-24 Design a circuit using an OTA (as in Figs. 8-52 and 8-53) which has a gain of 10 when $e_{control}$ is 1 volt and a gain of 20 when $e_{control}$ is 2 volts. Specify all component values.

8-25 Repeat Problem 8-24, using the voltage-controlled amplifier in the schematic in Fig. 8-56.

8-26 Compare the results obtained from Problems 8-24 and 8-25.

8-27 Design an electronically tunable active filter whose f_c can be adjusted across the audio band. Use Figs. 8-55 and 8-57. Specify all component values. Also indicate the relationship between $e_{frequency\ set}$ and f_c.

8-28 Draw the schematic of a true power converter using the two quadrant multiplier of Fig. 8-58. (HINT: You will have to offset e_y to keep it from going negative. See Fig. 8-31.)

$$p_{true} = v.i. \cos\theta$$

Index